A Potpourri of Physics Teaching Ideas

Selected reprints from
THE PHYSICS TEACHER
April 1963-December 1986

Edited by
Donna Berry Conner

Published by
American Association of Physics Teachers

A Potpourri of Physics Teaching Ideas

© American Association of Physics Teachers
  One Physics Ellipse
  College Park, MD 20740-3845

Cover Design by Rebecca Heller Rados

ISBN 0-917853-27-X

Dedicated to Cliff Swartz and Thomas Miner in recognition of their many years of conscientious, meticulous and enthusiastic devotion to The Physics Teacher giving direction, motivation and inspiration to physics teachers of yesterday, today and tomorrow.

TABLE OF CONTENTS

MECHANICS   1—128

Walk-a-graph, run-a-graph	1
A quantitative demonstration of relative velocities	2
Kinematics of a student	3
A demonstration of Newtonian and Archimedian forces	5
A nonlinear spring	7
Cookie sheet friction	8
Functional dependence of elongation of a coil spring on the load applied	8
Motivation for the force vector	9
Further motivation for the force vector	9
Saw-horse on teeter-totter for one person	10
A "Universal" mechanics demonstration	11
Loops from cove molding	12
Which way is up?	13
Balancing	15
Hot wheels in physics	16
Balanced dynamic carts	18
A demonstration of constant acceleration	19
Inelastic collisions using velcro	19
Velocity measurements outdoors with a tape recorder	19
Kinematics and the driver	20
Drag strip timer, an extended use	21
Friction in a moving car	22
Driving safety and straight-line kinematics	24
To stop or not to stop—kinematics and the yellow light	26
Automobile stopping distances	28
On the strength of butterfly wings	30
The answer is obvious. Isn't It?	
(Letters: Newton's third survives, Fan blades, roadrunners, and coyotes?,	
Tongue in cheek?, Obvious??, The obvious answer is correct,	
Old solution still correct, The H.M.S. Newton II:	
An onboard-fan-powered sail craft, Response)	34
Another Newton's sail-boat	35
Sail-boat demonstration	35
The simple pendulum experiment	36
A whipped-cream pendulum	37
Superball problem	38
A bouncing superball—the poor man's projectile	39
Spectacular rocket experiment	39
Rocket engine analog	41
Snowball fighting: A study in projectile motion	42
Predicting the range of a spring cannon	43
Range of a dart gun	45
The hunter and the monkey	46
Monkey and hunter in slow motion	47
Capturing the projectile in the monkey and gun demonstration	47
The water drop parabola	48
(Letter: Wonder anesthetized)	49
The spin on baseballs or golfballs	50
Catapulting the interest and success of physics students	52
The cylindrical wing, why does it fly?	54
(Letter: The flying cylinder)	55
The tippy-top	56
Spinning footballs and class rings	59
The Coriolis effect and other spin-off demonstrations	62
Whirlpool in a swimming pool	63
An experience in observation	64
Large scale use of a liquid-surface accelerometer	66
Teachers' Pets II – Circling carts	67
A personal application of physics	69
Spin art: A roational effects demonstration	71
Centripetal force using a hand rotator	72
A penny for your thoughts	73

The great American vector game	74
The wicked king and the beautiful princess	76
(Letter: Not such a hard bump)	77
A tensile strength lab-contest	78
Using blocks to demonstrate inertia, center of gravity, and friction	79
Dropping a string of marbles	80
The falling meter stick	81
Catch a dollar bill	82
Improved suspension for acceleration of g apparatus	83
Free fall using audio recording tape	83
Weightlessness and other ideas	84
Weightlessness and free bodies	85
The measurement of "g" in an elevator	86
Apparent weight changes in an elevator	88
Using a laser to investigate free fall	91
Physical effects of apparent "weightlessness"	92
A center of gravity demonstration	94
Ways to demonstrate center of gravity	95
With a grain of salt	96
Center of mass revisited	97
Center of mass of a rotating object	100
More center of gravity	100
Center of mass	103
Center of gravity of a student	106

FLUIDS AND HEAT   107–129

The water can paradox	107
The water can explored again	108
Floaters and sinkers	110
How to cool a book by its cover (Vermillion)	111
(Letter: How to cool a book by its cover)	111
"How to cool a book by its cover" revisited	112
Newton's law of cooling or is ten minutes enough time for a coffee break?	113
Tensiometer	115
Brownian motion	116
A question of air pressure	117
Boyle's law using a vacuum gauge	118
Archimedes' principle meets Charles' law	119
Elasticity shown with mirror blank and marble	119
Bernoulli demonstration	120
Repairing thermometers with split Mercury columns	120
The cartesian diver	121
(Letter: The rising cartesian diver)	121
(Letter: More on the cartesian diver)	121
The cartesian diver with pressure head	122
Elasticity demonstration	122
Elasticity of glass	123
(Letter: Another use for a whiskey bottle	123
Teachers' pet III: How thick is a soap bubble	124
Barroom physics, Part I	125
Barroom physics, Part II	126
Divergent barroom physics	127
An application of Archimedes' principle: Eureka! I'm 28% fat	128

ELECTRICITY AND MAGNETISM   130–170

A large-scale electroscope	130
Negative charges from an electrophorous	130
Electroscope shadowgraph	131
Pith ball substitute	131
Electrostatic pong	132
Electroscope discharge rate	132
Electrostatic lobby display	133
Kelvin water dropper revisited	134
Is a swimmer safe in a lightning storm?	136

Dissectible Leyden jar 137
    (Letter: "Dissectible Leyden jar" A comment) 138
    (Letter: More on "Dissectible Leyden jar") 138
Coulomb's law on the overhead projector 139
The Oerstead effect on the overhead 140
Lenz's law 141
Electrostatic charges and copying machines 141
Electrical figures 142
Demonstration of Gauss' law for a metal surface 143
A motor is a generator and vice versa 144
Force between parallel currents on the overhead projector 144
Sealed batteries 145
The smoke detector 146
Ohm's law mnemonic 146
The Omega competition 147
The volt competition 148
Why is the ac power line grounded 149
Electric field using an overhead projector 150
Turn-by-turn transformer demonstration 151
Static electricity demonstration 152
Lenz's law demonstration 152
Voltage surge protection 152
A circuit demonstration 153
Neon lamps and static electricity 153
A simple transistor demonstration 154
ac made visible 154
Mysterious lights in series and parallel 155
Shape of an electric field 156
Force on current carrying aluminum foil 156
Standing waves by a current carrying conductor 156
Touch-panels in elevators, and idiosyncrasies of gas tubes 157
A magnetic tripole 158
A magnetic tripole — what caused it? 159
A three-pole bar magnet? 160
There's still a lot we don't know 161
Recycling a magnet, or The little magnet that could 162
Construction of a simple compass 162
Ceramic magnets 163
Three dimensional views of magnetic effects 163
Why are so few substances ferromagnetic? 164
A question of magnets and keepers 165
Experiments with nickels and magnets 166
The field strength of a permanent magnet 168
Parallel circuit 169
Power lines 170

OPTICS AND WAVES   171—239

A choice observation 171
Blue sky and red sunsets 172
A simplified sunset demonstration 173
Dispersion and inversion 173
Atmospheric refraction 174
Colored lights and shadows 175
Making rainbows in the classroom 176
Solar spectrum projection 177
Making rainbows with a garden hose 177
A pinch of coffee-mate 177
Using a video projector for color-mixing demonstrations 178
A simple reflection experiment 179
Recombination of spectral colors 180
Color mixing for a large audience 181
A different way to use Newton's color wheel 181
Light box, inexpensive but versatile 182
Why is the string colored? 183
Reflection on the study of flat mirrors: Two demonstrations 184

A favorite experiment	186
An introduction to pinhole optics	188
That can't be: a virtual comment	189
Why is your image in a plane mirror inverted left-to-right but not top-to-bottom?	190
Physics at home and in the back yard (Part III)	192
"His specs—Use them for burning glasses"	193
Pinhole glasses?	193
To make a camera obscura	194
Inversion of an image on the retina	195
Inversion of shadows on the retina	195
Behind the eye	196
Image from a pinhole	196
Optics of the rear-view mirror: a laboratory experiment	197
Mirrors in air and water	199
An optical puzzel that will make your head spin	200
Markers for Young's experiment (PSSC version)	201
Cylindrical mirrors	201
Recording timer tape for interference demonstrations	202
Drawing wave diagrams for the interference of light	203
Long lasting soap-gelatin films	205
Soap film interference projection	206
Water lens	206
Standing waves on the overhead projector	207
Sine wave analog	208
Images from a piece of lens	208
The psychedelic student-getter	209
Optics in a fish tank	209
Optical effects in a neutral buoyancy simulator	211
A problem of image formation	212
Ray models of concave mirrors and convex lenses	213
Binocular vision - A simple demonstration	214
Physics for automobile passengers	214
The Cheshire cat	214
The disappearing dropper	215
Ripple tank projection with improved contrast	215
How the world looks underwater-A demonstration for nonswimmers	216
The physics of visual acuity	217
(Letter: Visual acuity revisited)	218
Pyrex "Vanishing Solution"	219
A safe Pyrex "Vanishing Solution"	220
Optical asitgmatism model	221
Wave motion demonstrator	222
Standing wave analogy using pocket combs	224
Beat production analogy using pocket combs	225
Hot standing waves	225
Versatile mount for slinky wave demonstrator	226
Cigar box spectroscope	226
Finding the principal focus with laser and thread	227
Paper waves	228
Traveling waves on a rope	229
Laser gimmick	230
Laser beam splitting for the student lab	230
Laser safety goggles: An unnecessary expense	230
Long-playing diffraction grating	231
Calibrating an inexpensive strobe light	232
Mercury vapor street lamps	232
Diffraction of light apparatus	233
Classical demonstration of polarization	234
Demonstrating interference	235
Can a spot of light move faster than c?	236
An op art wave demonstration	236

SOUND 240—276

A sound level meter: A sound investment	240
Demonstrating resonance by shattering glass with sound	242

Seeing the science of sound 244
Transmitting sound by light 245
Velocity of sound using an oscilloscope 245
Resonance tube for measurement of speed of sound 246
Sound on a light beam 247
Colorful Chladni 247
Biaural hearing 248
Sound waves 248
Resonance and foghorns 249
Sound in a vacuum demonstration 249
Demonstrations of reflection and diffraction of sound 250
Physics on a dinner plate 251
How to hear a mouse roar 252
A classroom apparatus for demonstrating the doppler effect 252
Vibration in pipes 253
Demonstration experiment to show the effect of pressure variation on the velocity of sound in a gas 255
An acoustics demonstration for students interested in music 256
Sound intensity and good health 257
Magnetically driven sonometer 259
Oscilloscope measurement of the velocity of sound 260
The harmonica, an audiofrequency generator 261
A "Sound" crossword 261

TOYS 263–276

Children's toys 263
What makes it turn? 265
A mechanical toy: The gee-haw whammy-diddle 266
A "perpetual motion" toy 267
More toy store physics 268
The physics of toys 269

ODDS AND ENDS 277–337

Ideas for the amateur scientist 277
Humor in the physics classroom 283
  (Letter: A trick demonstration) 287
Proclamation 287
Pocketbook science 288
The physics "flub-stub 290
Hot dog conduction, an edible experiment 291
Hot dog physics 291
Lab cover art 292
Scaling and paper airplanes 294
Enchanting things to think about 294
Further enchanting things to think about 295
A game of gravity 297
An experiment on population growth and pollution 298
The trigonometry "laboratory" 299
Lecture demonstrations for the high school science teacher 301
Backwards clock 304
Drinking duck shutter 304
An introvert rocket or mechanical jumping bean 305
A possible solution to the energy crisis? I-The Archimedes wheel 308
A possible solution to the energy crisis? II-The capillary pump 308
Do-at-home energy exercise 309
A stirring experiment 310
  (Letter: Scooped) 310
Demonstration: A nail driven into wood 311
Graphing Henry Aaron's home-run output 312
Removing the buoyant force 313
L'Eggs Demonstrations 313
"Reciprocating" Engine 314
A device for demonstrating half-thickness of shielding materials 316
Using the overhead projector in simulation of the Rutherford scattering experiment 317

A game for facilitating the learning of units in physics 318
Radiation safety in the lab 319
(Letter: Those Betas can make a difference) 320
Absorption of radiation—A lecture sized demonstration 320
Simulating radioactive decay with dice 321
Two new experiments 322
Throwing dice in the classroom 323
Throwing dice in the classroom II 324
"Mission improbable" problems 327
The history of physics in a high school physics course 328
Legibility in the lecture hall 330
Blitz quiz 331
Problems: Galileo revisited 332
"Physics is phun" quiz 334
Fizzicks Quizz 336

TRICKS OF THE TRADE 338—352

HOW TO STOW IT:
Hang those meter sticks 338
Assemblies of single concept demonstrations 338
Storage and display box for a diffusion cloud chamber 338
Protecting Electronic tubes 338
Small part storage 338
A rack for test tubes 339
Storage of support cable from high ceilings 339
Tool board outlines 339

HOW TO DO IT:
Emptying Ripple tanks 339
Heat Shrinkable tubing 339
Color anodizing 339
A perfect circle without a compass 340
Resuable chalk board graphs 340
"Glue" for plexiglass 340
Fused ammeters 340
Removing scratches from clear plastic 340
A red hot demonstration 340
The speed of apendulum bob 340
Photocurrent detection with an oscilloscope 340
Latex clusters in Millikan's experiment 341
Electrostatic Charging of fur and silk 341

HOW TO MAKE IT:
Ripple tank wave absorbers or shoals 341
Projectile launchers for electric trains 341
Plaster of Paris mold making 341
Scope Shade 342
A quickly made galvanometer shunt 342
Center of mass 342
Fast square wave generator 342
Simple accelerometers 342
Series and parallel electric circuits 342
A resistor board 342
Sound-proofed box for air track blower 343
Simple lab power distribution system 343
Inexpensive immersion heater 343
Simple resistance thermometer 343

HOW TO SHOW IT:
Brownian Movement 344
Projection of diffraction plates with a microscope 344
Bichromatic light source for diffraction demonstrations 344
Vectors with a chalk line 344
$F = q v \times B$; a simple demonstration 344
Simple AC demonstrator 344
Convex lens as a diverging lens 344
Three dimensional magneticfield 345
The vertical view 345

Translucent screens for optical projections 345
HOW TO ADAPT IT:
Regelation made easier 345
A simple timing device 345
Inexpensive tuning forks 345
Using polaroid negatives 346
Better spreader for Polaroid Colorpack II cameras 346
Cartesian diver 346
Poor man's mercury source 346
Minature diffraction gratings 346
A precise time base for the Physics lab 346
Constructing a hyperbola 347
The handy freon can 347
Minimizing glass breakage 348
A cure for the fickle ripple-tank motor 348
Storing slinkies 348
Ripple tank period measurements 349
Rejuvenating bar magnets 349
Tips for laser holography 349
Heavy weights at zero cost 349
Glass tubing microscope 350
Using duco cement as replacement crosshairs 351
Replacing crosshairs with dental floss fibers 351
Work-energy apparatus 351
Lens mount for laser 352
Repairing gold leaf electroscopes 352
Electron beam deflection by a rotating magnet 352
Rust proofing 352
Reinforcing spade lugs 352

INDEX   353–363

Walk-a-graph, run-a-graph

Robert Gardner
Salisbury School, Salisbury, Connecticut 06068

Many students find it difficult to relate graphs of distance or velocity as a function of time to the actual motions from which the graphs are obtained. A qualitative approach to such graphs prior to the study of kinematics or dynamics can often reduce such difficulty.

One way to do this is to ask students to *sketch* graphs of your motion as you take a series of walks and runs in the laboratory or classroom. Your first walk can be one in which you move at a steady pace. After they sketch distance as a function of time for this simple motion, ask them to use their graphs to sketch velocity as a function of time for the same motion.

Then try more complicated motions. Walk at a steady speed, stop, resume walking at a slower or faster pace. Next, walk at a steady speed, stop, walk backwards at a constant speed. Finally, take some walks in which you accelerate and decelerate. Ask students to sketch distance, velocity, and acceleration graphs.

After they have a feeling for these graphs and can interpret the slopes qualitatively in terms of velocities and accelerations, you can sketch some graphs of distance vs time, velocity vs time, and acceleration vs time on the board and challenge them to act out the graphs you have drawn. (See Fig. 1.)

The activities described above do not require much time and students enjoy doing them. After a few minutes most students can act out the graphs successfully. The success they find seems to give them the confidence and understanding they need to draw and interpret the quantitative graphs they encounter later in kinematics and dynamics.

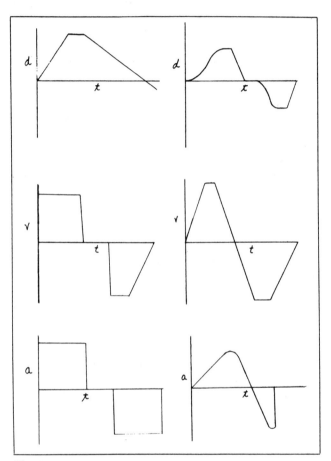

Fig. 1.

A quantitative demonstration of relative velocities

Dean Zollman
Kansas State University, Manhattan, Kansas 66506

A moving reference frame, stopwatches, and a few students can be used to demonstrate, quantitatively, the calculation of velocities in moving reference frames. The demonstration is large enough that it can be seen easily in a lecture hall.

A movable reference frame (Fig. 1) was constructed by mounting a sheet of plywood on an old, portable coat rack. Typewriter correction tape was placed at 25-cm intervals to produce a grid for measurements.

The classroom demonstration involves five students. Three act as timers, one is the moving object, and one pushes the moving reference frame. The experimental arrangement is shown in Fig. 2. While the object and the reference frame both move between the markers, the three timers record the times shown in Table I.

Upon completion of the experiment the speeds of moving person relative to the earth, moving person relative to the reference frame, and the moving reference frame relative to the earth are calculated. The demonstration is repeated with the reference frame and person moving in the same rather than opposite directions. The appropriate speeds are calculated again.

All data, which are remarkably consistent, are placed on the chalkboard before any equations for relative speed are introduced. Many students in a class of nonscience majors are able to deduce the equations by looking at the numbers. Thus, the demonstration acts as a good "exploration" into relative velocities.

This material is based upon work supported by the National Science Foundation under Grant Number SER-7900507. The author thanks Dean Stamel for constructing the moving reference frame.

Fig. 1. A movable reference frame which has been constructed from an old coat rack.

Table I

Timer	Task
1	Measure the time for the person to travel the one meter between the fixed markers
2	Measure the time for the reference frame to travel the one meter between the fixed markers
3	Measure the time for the person to travel one meter as marked by the reference frame

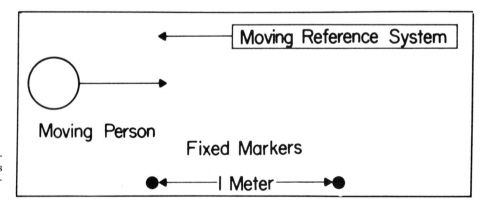

Fig. 2. The experimental arrangement for calculating relative speeds of objects moving in opposite directions.

Kinematics of a student

James H. Nelson
Harriton High School, Rosemont, Pennsylvania 19010

This article describes experiments that involve the student as the object of the experiment. Other such experiments include measuring:
1. Student reaction time.[1]
2. Horsepower generated by a student.[2]

Such experiments generate student interest and thus increase understanding of the physics principles involved. In the first experiment to be discussed the students are challenged to walk or run a straight course at a constant speed.

To do these experiments you will need a 30- to 40-m straight course marked at regular intervals. Since kinematics is often taught in the fall, a convenient course could be laid out on the school football field. Alternatively, a long rope knotted at regular intervals can be used to set a course almost anywhere. A diagram of the course is shown in Fig. 1. The circles with x represent the top view of students who stand in pairs facing each other. Each of these students has a stopwatch, so two time readings are taken at each timing station. The number of timing stations is not important and I use 6 as a matter of convenience.

The student marked R is about to try to move through the course at a constant speed. So that the students have a chance to get up to speed before entering the course they should start a few meters before the first timing station. This will result in a position versus time graph that does not pass through the origin (Fig. 2). The teacher T stands off to one side.

To take a set of data the teacher blows a whistle to start all stopwatches and the student begins moving down

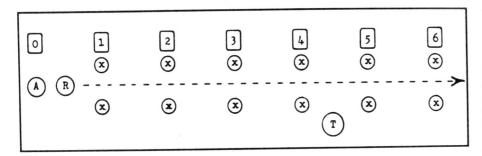

Fig. 1. Diagram of "Kinematics of a student" course showing timers (x), teacher (T), starting point of constant speed runner (R) and constant acceleration starting point (A).

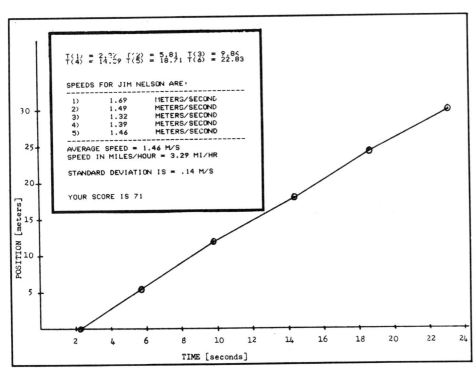

Fig. 2. Typical position-time graph of student trying to maintain a constant speed.

the course. Each pair of timing students look at each other, and when the moving student breaks their line of sight the timers stop their watches. Each timing station has a separate data sheet to record names and times. It is not necessary to run through the course. Students may walk.

At times you may have a student who is reluctant or unable to participate. I encourage everyone to try, but never embarrass or try to force a student to do this activity. Discuss the experiment before you intend to do it. If a student asks to be excused, try to arrange for a substitute. Students will probably ask the teacher to run the course. The analysis of the data is similar to other kinematics experiments.

To determine which student has the most constant speed, I calculate the standard deviation of the five average speeds between the timing stations. The student with the lowest standard deviation is declared the class winner and given a prize (e.g., physics bumper sticker). Depending on your students, you may explain standard deviation in detail, or simply describe it as a measure of consistency.

Students who understand statistics may realize they can get a lower standard deviation by negotiating the course very slowly; therefore, the maximum time allowed for a student to complete the course is 30 s.

So that I can check the calculations easily and accurately, I wrote a BASIC program to find the average speed and the standard deviation for each student. The program also computes a "Speed Score." I wanted a higher score for a lower standard deviation, thus the score is calculated by:

SPEED SCORE = INT(10/SD + 0.5)

where INT is the greatest integer function and SD is the standard deviation.

The second experiment is similar to the first, except this time the students are challenged to move at a constant acceleration. In this experiment the score is based on the average acceleration between intervals divided by the standard deviation.

ACCELERATION SCORE = INT(100*ABAR/SD + 0.5)

where ABAR is the average overall acceleration.

I originally used this formula to prevent students from moving too slowly down the course. Students may realize that with this formula a large acceleration will give them a higher score. But many students do not realize the difference between speed and acceleration. These students start out with a large speed and find they can not maintain their acceleration. When students realize you can have a small speed and a large acceleration at the same time, they start out slowly and increase their speed as they pass each timing station. Figure 3 shows a typical result.

To calculate the average acceleration, I use:
$$d = 1/2(at^2)$$

This requires that the initial speed, position, and time be zero. With six timing stations plus a starting position, you can find five accelerations between intervals. Since Δd rather than Δt is constant for this arrangement, the Δt used to find speed is not equal to the Δt for acceleration. This means that when calculating the standard deviation for acceleration each acceleration must be weighted with a time factor. This is beyond what I expect my students to learn. Thus I have students calculate the five accelerations using the slope of the speed versus time graph, and I use a computer program to determine the acceleration score.

Classroom plaques showing names and top scores for each year help develop a sense of tradition and competition. These experiments also make excellent physics olympics events.

Teachers who are interested in the *Kinematics of a student* experiments can obtain the following from the author:

1. Master copy of a universal kinematics chart.
2. Computer listing in Applesoft for finding speed and acceleration scores.
3. Timing station data sheets.

Please send self-addressed, stamped envelope. I wish to thank Carl Duzen of Lower Merion High School for the original idea for these experiments.

References
1. William Schnippert, "Catch a dollar bill," Phys. Teach. 14, 177 (1976).
2. James Nelson, "Student power," Phys. Teach. 10, 529 (1972).

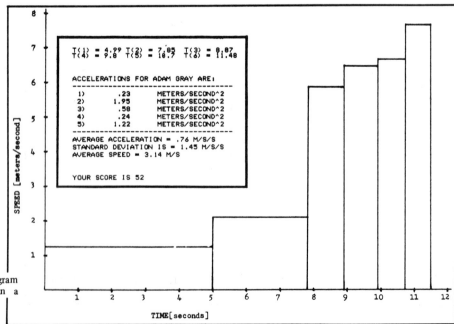

Fig. 3. Typical speed-time histogram of student trying to maintain a constant acceleration.

A Demonstration of Newtonian and Archimedean Forces

Harley J. Haden
Glendale College
Glendale, California

Twenty-two hundred and some odd years ago a Greek by the name of Archimedes was relaxing in the public baths of Syracuse when he suddenly arose and ran through the streets exclaiming, "Eureka! I have found it!" Some say he had the discretion to first drape a towel around himself; others claim otherwise, but in any event he is reported to have been apprehended by the gendarmes and thrown into the local bastille from which his employer, King Hiero II, secured his release. Archimedes' alibi was his jubilation at having discovered a new physical principle which would enable him to determine accurately the percentages of gold and silver in a certain crown belonging to said king. The local papyrus columns probably published his story along with the principle which he discovered. In any event we find it in his writings and it has been passed down through the ages in various forms, one of which is: "Whenever a body is immersed, either wholly or partially, in a fluid there is a buoyant force exerted on it by the fluid which is equal to the weight of the fluid displaced by the body." Since Archimedes' time this has been accepted as a fundamental principle of classical physics. It has been applied to everything from the determination of the densities of ore samples to the weighing of ocean vessels.

Nineteen hundred and some years later a frail young Englishman by the name of Isaac Newton conceived that there was a proportionality between force and acceleration in what is now known as Newton's second law of motion. This law, too, is well-known among physics teachers and has been stated in many different ways ranging from "Force equals mass times acceleration" to "The acceleration of a body is directly proportional to the resultant force acting on it and inversely proportional to its mass."

Since Archimedes' and Newton's times many devices, inventions, and demonstrations have been contrived which employ these laws. Now Archimedes' principle is a principle of statics whereas Newton's second law is a law of dynamics. Nevertheless we can combine them into a general principle that *whenever a body is immersed, either wholly or partially, in an accelerating fluid, there is a force exerted on it by the fluid which is proportional to the mass of the fluid displaced by the body times the acceleration of the fluid.*" The force is in the same direction as the acceleration of the fluid and acts at the center of mass of the displaced fluid.[1] The force might be called the Archimedes inertial force,[2] but, since Archimedes knew nothing about Newton's second law, it might also be called the Newtonian buoyant force, although there is no evidence that Newton himself ever recognized it.

A somewhat interesting application of this force is shown in the car and track apparatus of Fig. 1(a). Here we have a bottle attached to the cart, the bottle containing water and, attached to the bottom center of the bottle, an inverted pendulum consisting of a fishing float anchored by means of a flexible string. When the car moves along the track at constant velocity the float and string remain in a vertical position. But when the car, bottle, water within the bottle, and hence the inverted pendulum are *accelerated* the float moves *forward* within the bottle so that when the float and fluid are in relative equilibrium the string makes an angle θ with the vertical as given by the equation $a = g \tan \theta$ where a is the acceleration of the system and g the acceleration of gravity.

As an anlysis of this demonstration, consider the three situations shown in Fig. 2. In Fig. 2(a) is shown a simple pendulum accelerating to the right. There are two forces on it, the pull of gravity W and the tension T in the weightless string. Simple analysis shows that the resultant force R accelerating the system to the right is $R = W \tan \theta$, and since $R = ma$ and $W = mg$, m being the mass of the pendulum, a simple substitution and cancellation

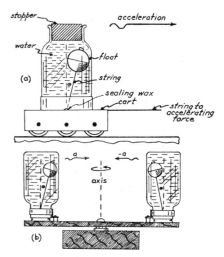

Figure 1. (a) Linear track apparatus (b) Circular motion apparatus

shows that $a = g \tan \theta$. Hence the angle is independent of the mass m of the pendulum, a somewhat interesting conclusion from the standpoint of the student of physics. Next consider the pendulum hanging in a fluid exerting an Archimedes buoyant force B, less than the weight of the pendulum, as shown in Fig. 2(b). There are now four forces acting on the pendulum, the pull of gravity W, the tension T in the string, the Archimedes buoyant force B, and the force F exerted on the body by the fluid, in accordance with the previously stated principle, as it accelerates to the right. Simple analysis shows that in this case the

---
1. Harley J. Haden, "A Demonstration of Centripetal and Centrifugal Forces", Am. J. Phys. **31**, 635 (Aug.) 1963.
2. J. A. Van Den Akker, "Extension of Archimedes' Reasoning", Am. J. Phys. **31**, 943 (Dec.) 1963.

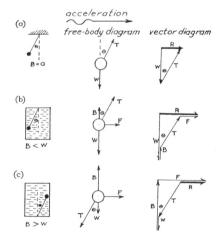

Figure 2. (a) Simple pendulum (b) Simple pendulum in a buoyant fluid where B<W (c) Simple pendulum in a buoyant fluid where B>W

resultant force is
$$R = F + (W - B)\tan\theta$$
Since by Newton's Second Law $R = ma$ and $W = mg$, by Archimedes' principle $B = m'g$ where m' is the mass of the fluid displaced by the body, and by the Newtonian buoyant force $F = m'a$, we have by substitution, transpostion, and factoring
$$(m - m')a = (m - m')g\tan\theta$$
or $a = g\tan\theta$
The same principles apply to Fig. 2(c) for which we may write
$$R = F - (B - W)\tan\theta$$
and simple analysis again shows that $a = g\tan\theta$.
Notice that these results are independent of the values of the densities of the fluids or the masses or volumes of the pendulums. All that is required in Fig. 2(b) is that $W>B$ and in Fig. 2(c) that $W<B$. These conditions naturally occur respectively if in the first case the density of the object is greater than that of the surrounding fluid and in the second if the density of the object is less than that of the surrounding fluid.

A second demonstration which can be very interesting to the student is shown in Fig. 1(b). This is a means of showing that the resultant force on an object in circular motion is centripetal and not centrifugal. The bottles are caused to rotate at some angular velocity ω which need not be very high if the radius r is of the order of one foot. There will then be an acceleration $a = \omega^2 r$ of the system toward the center, and when the pendulum and water are in relative equilibrium the angle θ assumed by the string is given again by $a = g\tan\theta$. Instead of "flying outward" as many students will predict the inverted pendulum will swing inward in a somewhat surprising manner as shown in the figure. This is a convincing way of demonstrating that the resultant force on a body in circular motion is centripetal rather than centrifugal. The free-body and vector diagrams of Fig. 2(c) apply to this case as well as to the case of linear motion, although it may be noted that there is also a centrifugal component of force $T\sin\theta$, an example of where the forces on a body in circular motion are not entirely centripetal.

The Newtonian buoyant force F is as general as the Archimedean buoyant force B and may be applied to bodies partially immersed, immersed in layered fluids of different densities, or in gases as well as liquids. Graetzer and Williams[3] have noticed that a helium balloon in a car will move in the direction of the acceleration of the car. The "Aberrant Candle Flame" is also governed by this same principle.[4] There is a pressure gradient set up in a fluid whenever it accelerates, the existence of which may be proved in a manner mathematically analogous to that of proving the existence of a vertical pressure gradient.[5] It is equal to $-\rho'a$ where ρ' is the density of the fluid. Also the existence of the force F may be proved by mathematical means analogous to the proof[6] of the Archimedean buoyant force B.

Application of this force to other problems is not restricted to horizontal acceleration.

Any acceleration that produces a pressure gradient may be treated in the same manner as that used for horizontal acceleration.

In a freely falling system, the fluid and immersed body fall with the same acceleration. There is no pressure gradient and the net effect of F and B is zero. The only net forces on the fluid and the submerged body are their weights as defined by Sears.[7]

The same situation would exist in an orbiting satellite.

3. Hans G. Graetzer and P. W. Williams, "Behavior of a Helium Balloon in a Car", Am. J. Phys. **31**, 302 (Ap.) 1963.
4. Modern and Classical Instruments Co., Livermore, California.
5. J. L. Meriam, *Mechanics—Part I—Statics*, Second Ed., 226 (John Wiley and Sons. 1959).
6. Max Planck, *General Mechanics*, Intro. to Theor. Physics, Vol. I, 201 (The Macmillan Co., 1949).
7. F. W. Sears, "Weight and Weightlessness," The Physics Teacher **1**, 20 (Ap.) 1963.

It must be remembered that Archimedes' principle of flotation is an "earth-bound" principle, applying only to fluids that have a pressure gradient as a result of their weight, whereas the Newtonian buoyant force (or Archimedean inertial force if you like) would work anywhere in the universe. For example, one could suspend a "float" in a fluid by resilient means as shown in Fig. 3 and

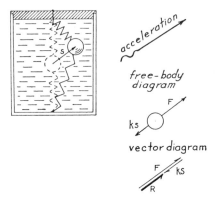

Figure 3. Float suspended in a fluid by resilient means

if the force constant k were adjusted to be the same in all directions (or "independent of direction" if you like) the displacement s of the "float" relative to its equilibrium position in the container would be in the same direction as the acceleration a of the container, fluid, and float (if the density of the "float" were less than that of the fluid). The free-body and vector diagrams are shown respectively at the right (Fig. 3), and from them it can easily be shown that the displacement is
$$s = (m' - m)a/k$$
where m' is the mass of the displaced fluid, m is the mass of the submerged body
or $s = V(\rho' - \rho)a/k$
where ρ' is the density of the fluid, ρ the density of the body, and V its volume. From this we see that the displacement can be increased by having a greater difference in densities or a lower force constant. The acceleration would be relative to a "Newtonian frame of reference" (i.e. the so-called "fixed" stars). But, unlike velocity, the acceleration of a particle is the same in *all* reference frames moving relative to one another with constant velocity.[8]

8. Robert Resnick and David Halliday, *Physics*, Part I, 62 (John Wiley and Sons, 1960).

A nonlinear spring

R. M. Prior
Department of Physics and Astronomy, University of Arkansas, Little Rock, Arkansas 72204

A common example of simple harmonic motion is the motion of a mass attached to a spring. This system is the central part of numerous laboratory experiments and classroom demonstrations. Its simplicity and reliability make it popular with physics teachers and students.

Most springs obey Hooke's law over a wide range of extensions. The failure of springs to obey Hooke's law usually occurs when the elastic limit is exceeded; i.e., when the spring is stretched too far. The region beyond the elastic limit not only produces restoring forces not directly proportional to the spring's extension but also produces permanent deformation and damage to the spring.

Another type of spring commonly used in physics laboratories violates Hooke's law at the other extreme of small extensions. Light-weight stainless steel springs are commonly used on air tracks and air tables for simple harmonic motion experiments.[1] These springs are made from 0.25-mm to 0.5-mm diameter wire in a closely wound helix; in the relaxed position most of the adjacent coils of the helix are in contact.

The nonlinear behavior of this type of spring is easily demonstrated by a simple experiment. A spring is allowed to hang vertically from a support, small masses are successively attached to the bottom of the spring, and the positions of the bottom of the spring are recorded. The results of this procedure for a 1.6-cm diameter, 5-cm long spring (Ealing catalog number 34-1453) are shown in Fig. 1 where the hanging mass is plotted versus the elongation of the spring. The nonlinear behavior for the first few centimeters of elongation is obvious. The curve is straight beyond 3-cm elongation with a slope of 1.61 g/cm which corresponds to a spring constant of 1.58 N/m.

Examination of the spring provides an explanation of the nonlinear part of the curve. In the spring's unstretched condition some of the adjacent coils of the spring are separated and some are still in contact. The observed elongation of the spring is due to only part of the spring actually stretching; the number of participating coils is less than the total number. This behavior, in effect, produces a stiffer spring as evidenced by the greater slope at the left side of the graph. A certain minimum amount of force is necessary to stretch the spring so that all of the

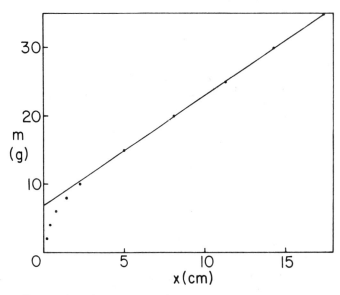

Fig. 1. Plot of mass m hung from spring versus the elongation x produced. The straight line is drawn by eye to demonstrate the linearity of the results for x greater than a few centimeters.

coils are separated from their neighbors; for the spring in Fig. 1 this force is about 0.1 N. Once the individual coils are separated the expected linear relationship between applied force and elongation is observed.

In normal use on air tracks or air tables this type of spring is always stretched several centimeters so that it is used only in the linear response region. In this application the result is simple harmonic motion. In other situations care should be taken to confine the region of operation to the linear region if the normally desired Hooke's law behavior is to be obtained. On the other hand, the behavior of the spring for small elongations provides a demonstration of a simple system with a reproducible nonlinear response.

Reference

1. The Ealing Corporation, 22 Pleasant Street, South Natick, MA 01760, "soft springs" and "weak springs."

COOKIE SHEET FRICTION

Determination of frictional forces and demonstration of the laws of friction is difficult due to small surface irregularities encountered with equipment normally used for friction experiments. As has been suggested by Mainardi,[1] consistent friction force measurements can be obtained by use of a felt surface in contact with a Teflon surface.

An inexpensive Teflon surface may be prepared by cutting 5 in. x 16 in. strips from a Teflon coated cookie sheet and then securing the strip to a board with small wood screws. A 6 in. x 3 in. x 1 in. wooden block was covered with felt on the bottom and side faces. By using these two areas it is possible to demonstrate that the frictional force was approximately independent of the area of contact.

The relationship between frictional force and the normal force for both the static and kinetic case can be studied by loading the block with different masses (500-3000 g) and measuring the force required to start the block and the force needed to move the block with constant speed. From the slope of a plot of the frictional force vs the normal force the coefficient of friction can be obtained. An average value of 0.21 was found for the coefficient of static friction (μ_s) and 0.19 for the coefficient of kinetic friction (μ_k). With no load on the block larger values of both μ_s and μ_k were found. This difference in the coefficient of friction is probably due to the poor surface contact with small normal forces when one of the surfaces is formed by a fiber-like material such as felt.

Due to the fact that μ_s is only slightly greater than μ_k some problem was encountered in differentiating between static and kinetic friction. The repeatability of the measurements was quite acceptable. For example, a series of twenty measurements of the frictional force exhibited a fractional standard deviation of ±3%.

PATTON H. McGINLEY
Longwood College
Farmville, Virginia 23901

Reference

1. *Physics Demonstration Experiments*, edited by Harry F. Meiners (Ronald, New York, 1970), Vol. 1, Chap. 8, p. 154.

Functional Dependence of Elongation of a Coil Spring on the Load Applied

Francis W. Sears
Dartmouth College
Hanover, N. H.

MANY of the experiments carried on in a physics research laboratory consist of a study of the way in which the value of some physical quantity changes when the value of another quantity is varied in a systematic way. We say that the first quantity shows a *functional dependence*[1] on the second. The results of such an experiment can be reported in the form of a table of corresponding values, a graph, or a mathematical equation.

An interesting experiment with simple equipment which requires a little analysis on the part of the student involves a coiled spring, a screen door spring that has an initial "set." The spring unloaded is under a "tension" and does not stretch until an appreciable "load" is placed on it. These springs are available from our local hardware store for as little as 25 cents.

---

[1] Alfred Romer, Am. J. Phys. **29**, 630 (1961).

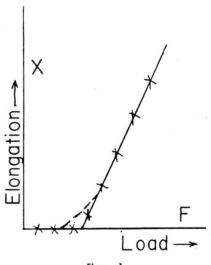

Figure 1

MOTIVATION FOR THE FORCE VECTOR

"Are there any members of the football team in the class?"

The football team members or any two powerful young men are brought to the front of the room and instructed to hold a 10-m rope taut between them at waist level.

"Now we need someone who is very strong."

The smallest girl in the class is selected and instructed to push the center of the rope to the floor. She easily accomplishes this feat. The demonstration is repeated on the pretext that the boys weren't really ready. The beads of sweat appear as the boys try again to win the contest with the young lady.

"This is truly amazing. How can it be that one young lady can overpower two members of our football team?"

The explanation, of course, lies with the fact that a rope being pulled horizontally will have no vertical component to balance the downward push of the girl. Equilibrium will be reached only when the rope has reached the floor. For example, if a 10-m long rope is held 1 m above the floor and the girl weighs 400 N (approximately 100 lb), each boy would have to exert a force of 1000 N (approximately 250 lb) to prevent the rope from touching the floor.

The effect becomes even more dramatic if you take the entire class out into the corridor and put four boys at each end of a 50-m rope. Imagine the

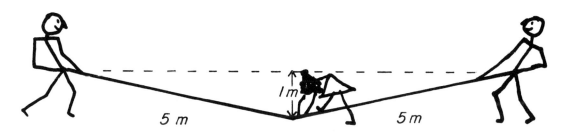

consternation of your students when you prove that one frail female is stronger than eight muscular men.

MICHAEL BERNSTEIN
*Lafayette High School
Brooklyn, New York 11214*

FURTHER MOTIVATION FOR THE FORCE VECTOR

Alfred M. Eich

Charles F. Brush High School, Cleveland, Ohio 44124

Michael Bernstein in the Notes section of *The Physics Teacher* [9, 148 (1971)], presented an interesting and dramatic variation of a demonstration that I have used.

A piece of tough string or nylon fish line (approximately 20 lb test) is strung between two poles fitted with clamps as shown in the diagram. One pole is a table support rod and the other is fastened into a tripod stand. A heavy weight, W, of about 10 lbs (I use a starter casing discarded by our autoshop) is hung on a wire from the string. A rubber pad is placed under the weight to protect the table. If the string is whitened with chalk it will show up better against a dark background. It is observed that the string easily supports the weight when θ is small. Now, slowly pull on the tripod stand to lift the weight. Of course, you must pull harder and harder. At some large value of θ, the string will break with a loud "snap" and the weight will thump to the table.

The physics of the demonstration may be explained as follows:

If θ_1 and θ_2 are equal, it can be shown that the tension, T, in the string is given by

$$T = \frac{W}{2\cos\theta}.$$

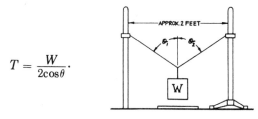

The student's attention is called to the fact that as θ approaches 0°, T approaches W/2 and more pertinent, as θ approaches 90°, T approaches *infinity!* As Dr. William Whewell so aptly expressed it in his *Elementary Treatise on Mechanics* in 1819,[1]

> No power on earth, however great,
> Can pull a string, however fine,
> Into a horizontal line
> That shall be absolutely straight.

Of course, the conclusion drawn from this demonstration can be extended from the piece of string to wires, rods, and the strongest beams one can imagine, so long as they are supported at their ends only.

Reference

1. C. Fadiman, *Fantasia Mathematica* (Simon and Schuster, New York, 1958).

Saw-Horse on Teeter-Totter for One Person

Terry Lee Templeton
Everett, Washington

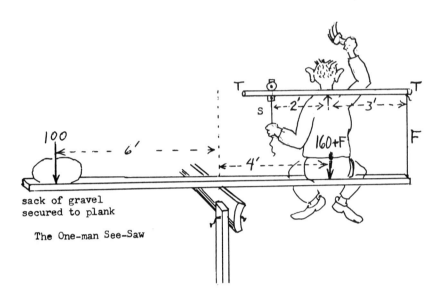

The One-man See-Saw

Although developed originally as a convenient one-man-operated painter's scaffold, the arrangement (see Fig.) may be of interest to parents of only one child. Obviously, a 160-lb man seated 4 ft from the pivot of the 14-ft plank (below) develops too much torque for the 100-lb counterbalance at 6 ft. However, let him place across his shoulders a section of light-weight aluminum tubing (T–T), tie the right end of the tubing to the plank by string F, and pull down on the left end of the tubing. This assembly can readily be tried out using as a pivot a saw-horse set on the ground, so that the operator's feet, or the right-hand end of the plank on the ground, will ensure stability at the start.

Let the operator pull on string at S with a reasonable force, say 24 lb; this produces tension of 16 lb on the string F. This creates additional counterclockwise torque, $16 \times 7 = 112$ lb-ft which adds to the 600 lb-ft due to the sack of gravel. Of course the 16 lb tension (F) adds to the man's weight, causing his seat to press against the plank with $160 + 16 = 176$ lb force, which, still at 4 ft from the pivot, creates $176 \times 4 = 704$ lb-ft torque (clockwise). The two counterclockwise torques add to 712 lb-ft; hence, the whole assembly slowly accelerates counterclockwise, raising plank and man clear of the ground. It can be readily calculated, using dimensions as above, that by pulling with 20 lb force with his left hand the operator can hold the plank in equilibrium.

After a little practice, the operator can readily control his up-and-down movement over a fairly wide range, which is advantageous when painting vertical siding. By attaching string S to a trolley-wheel set on the aluminum tubing rather than tying it, the operator can also adjust his torque corrections by moving his arm, avoiding the tedium of exerting force in one position.

Perhaps the real advantage of torque adjustment has not been stressed above: that the operator is not limited to the 4-ft position. He can move to 3 ft from the central support and still balance himself.

A POTPOURRI OF PHYSICS TEACHING IDEAS—MECHANICS

A "Universal" mechanics demonstration

George W. Ficken, Jr. and Angelo A. Gousios*
Cleveland State University, Cleveland, Ohio 44115

A common puzzler[1] for introductory physics students is to ask them the reading of the spring balance shown in Fig. 1, where the hanging masses are equal. While setting up such a demonstration, we stumbled upon a far more interesting (and complicated) one which is related to Meiner's

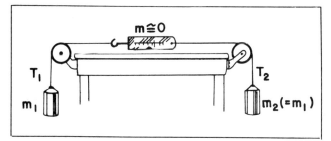

Fig. 1. Physical set up of the demonstration.

description[2] of how a swinging 500 gm mass can be made to pick up a 1000 gm mass.

If mass m_2 is held fixed while mass m_1 is raised to some angle θ_{max} with the vertical and then both are released, the swinging mass m_1 gradually raises mass m_2 in a nonsteady, periodic manner. This occurs because of the variations in the tensions T_1 and T_2 of the strings connected to m_1 and to m_2. Although the apparatus is very elementary, a great deal can be learned from an attempt to analyze this motion using the principles of mechanics covered in an introductory course.

For simplicity the mass of the spring balance, mass of the pulleys, radius of the pulleys, and frictional effects will be neglected (friction will be considered later). As a further initial simplification, let us keep m_2 fixed in place and analyze the interplay between m_1 and its connecting string, wherein lies the key to the system's behavior. When m_1 is near the top of its swing, only a component of its weight equal to $m_1 g \cos \theta$ acts parallel to the attached string; therefore the mass m_2 at the other end, if released, would produce a larger force on the system and accelerate the scale to the right. When m_1 is near the bottom of its swing, T_1 is greater than $m_1 g$ because the string must also supply the centripetal force required to make m_1 follow a curved path; therefore the mass m_2, if released, would produce a smaller force on the system, allowing the scale to accelerate to the left.

With the entire system free to move, the behavior of this system appears to require a great number of mechanics principles and physical insights to describe even qualitatively what is happening. They are as follows:

1. The concept of centripetal force and the associated free body force diagrams for a body undergoing motion in a vertical circle.

2. The fact that at some critical angle θ_c during each quarter swing, $T_1 = m_2 g$, or

   $$m_1 g \cos \theta_c + m_1 v_T^2 /r = m_2 g$$

   and the system is momentarily balanced. Since $m_1 = m_2$

   $$\cos \theta_c = 1 - v_T^2/gr, \quad (1)$$

   where r is the suspension length of the cord attached to m_1 at this instant and v_T is the component of m_1's velocity tangent to the instantaneous radius of curvature of the arc of its swing. (A radial component v_r is also present, being equal to the velocity of m_2). For $\theta < \theta_c$, m_2 accelerates upward and for $\theta > \theta_c$, m_2 accelerates downward.

3. The application of the law of conservation of mechanical energy to the system, with the initial maximum potential energy of m_1 when at θ_{max} [$m_1 gh = m_1 g r_0 (1 - \cos \theta_{max})$, where r_0 is the initial suspension length of the cord attached to m_1] being converted into potential energy of m_1 at θ, kinetic energies of m_1 and m_2, and potential energy of m_2 (which changes because of the vertical motion of m_2).[3]

4. The effect of linear impulses on m_2 caused by net positive (upward) or negative forces. These net forces vary in size while acting during their portion of each quarter cycle. The net linear impulse on m_2 per quarter cycle is equal to the change in linear momentum of m_2 per quarter cycle.

5. The recognition that if frictional effects are appreciable, an angular range around θ_c exists for which no acceleration can take place in either direction.

6. The fact that the period of the oscillation of this simple pendulum changes continuously as the length of the cord to m_1 changes.[4]

Analysis indicates that the positive linear impulses are greater than the negative ones for all θ_{max}. Thus for the

---

* A senior Physics major at Cleveland State University.

idealized case of no friction and a weightless cord of sufficiently great length attached to m_2, the average velocity of m_2 per cycle would keep increasing (with no reversal in direction other than the one occurring shortly after the system is released), but at an ever decreasing rate. This would appear to be a conversion from a situation in which the maximum velocity of m_1 is principally along the x axis to one in which its velocity is principally along the y axis. The extra potential energy of m_1 when displaced to θ_{max} turns into kinetic energy of m_1 and m_2.

In our actual setup, having appreciable friction, if a large θ_{max} was used, m_2 was at first observed to ascend continuously, with varying velocity. But, as the amplitude of the oscillation died down, a point was reached where m_2 could be seen to reverse direction twice each cycle, when m_1 neared the end points of its oscillation.

References
1. D. Halliday and R. Resnick, *Fundamentals of Physics* (John Wiley and Sons, New York, 1970), p. 87.
2. R.M. Sutton, "Particle Dynamics," in *Physics Demonstration Experiments*, edited by H.F. Meiners. (Ronald, New York, 1970), Vol. I, p. 158.
3. One might wish to attempt to solve for v_T of m_1 when at some critical angle θ_c (say the first θ_c reached following release of both m_1 and m_2) by using conservation of energy principles. Because of the (initial) downward motion of m_2, a very complicated expression for v_T results. It contains θ_{max}, θ_c, the vertical distance through which m_2 drops, and the instantaneous velocity of m_2. Therefore, it is not possible to obtain an explicit expression for θ_c by substitution for v_T in eq. (1), something one might also wish to determine. For the much simpler case of m_2 held fixed, the result is $\cos \theta_c = (1 + 2 \cos \theta_{max})/3$.
4. It was tempting to speculate that the angular momentum of m_1 about its suspension point (assuming a negligible pulley radius) might be constant for those moments when $\theta = 0$, twice each cycle. But a theoretical analysis indicates that the alternate positive and negative torques caused by the gravitational pull on m_1 when $\theta \neq 0$ do not cancel out. Rather, the angular momentum of m_1 keeps increasing as the suspension length of m_1 increases without limit.

LOOPS FROM COVE MOLDING

Many advantages can be found in this easily built loop-the-loop. The most important aspect, cost, amounts to less than one dollar which is nice to hear with science budgets dwindling. The flexibility of the track allows easy adjustment for any desired angle of the entrance ramp and another plus is the ability to handle any spherical object up to 2 in. in diameter. The only materials needed are: an 8-ft vinyl strip of cove molding (the type used with paneling for rooms), a board or two, each approximately 4 in. x 12 in. x ½ in. and some epoxy glue.

Spread some glue along 2 in. of the molding on the backside at a place 48 in. from the end. Position this where you want it on the board and clamp it down until the glue sets. Again spread some glue along 2 in. of the back about 4 or 5 in. from one end of the molding. Bend the molding to form a loop and clamp the glued portion to the board.

After the glue has dried, usually overnight, the loop-the-loop is ready for use. Rest the free end of the molding on adjustable support rods to make an easily adjustable entrance ramp for the loop.

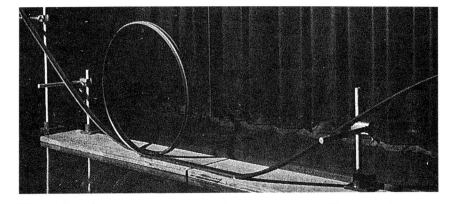

Fig. 5. Loop-the-loop apparatus made by gluing sections of cove molding to wooden bases.

As an extra activity, another piece of molding can be fastened, without forming a loop, to another board in a similar manner (spread glue only on 2 or 3 in. of one end of the molding). Carefully position the molding so it will butt with the end by the loop.

Join the ends of the molding strips together by clamping both boards with C-clamps to a table top. This forms entrance ramps on both ends of the loop. Continuous loop-the-loops can be demonstrated by having the two ends elevated.

The ramp without the loop makes an ideal inclined plane for acceleration, energy, momentum, and many more experiments and demonstrations.

MICHAEL L. BERRY
West Central High School
Francesville, Indiana 47946

A POTPOURRI OF PHYSICS TEACHING IDEAS—MECHANICS

Which way is up?*

Lester Evans
Tates Creek High School, Lexington, Kentucky 40502

J. Truman Stevens
Department of Science Education, University of Kentucky, Lexington, Kentucky 40506

The physics teacher closes the door, pulls the shades, turns off the lights, and prepares his students to travel on a space voyage. After a few moments of preliminary briefing, the class begins the countdown. At "blast-off" the teacher hits a switch turning on a hidden, but noisy, vibrating motor. After a couple of minutes of space flight, the teacher challenges students to "prove" that they're not accelerating upward at a rate of 9.8 m/sec^2. Thus a delightful discussion of "Which way is UP?" has begun.

The inspiration for this activity comes from Bartlett's excellent article in *The Physics Teacher* on the same topic.[1] An attempt has been made to present a more simplified explanation for teachers who do not wish, or do not have sufficient time, to present a more elaborate or mathematical discussion of the topic. This activity also works well with nonphysics students.

After some discussion regarding the direction of acceleration, the instructor introduces a hand-held accelerometer. The construction of this device is simple and the equipment is available to every physics teacher (see Fig. 1). The apparatus is first described to the students, then they are asked to make predictions concerning the motion of the float relative to the flask as the flask is accelerated. The apparatus is then accelerated in a straight line while students observe only the accelerated portion

Fig. 2.

of the path. An obstruction to block the view of the flask from the students during deceleration is helpful to prevent confusion (see Fig. 2). The demonstration can then be reversed to illustrate the effects of negative acceleration.

Finally, it is established that the device can be used to detect any acceleration. When the flask is moved in a circular path, the float leans toward the center of the circle. This, again, illustrates the direction of the acceleration (see Fig. 3).

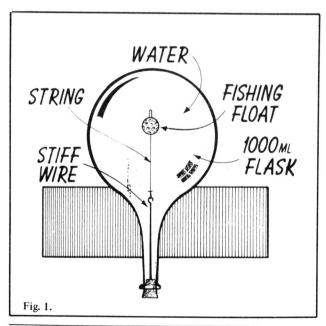
Fig. 1.

Fig. 3.

*This punctuation has been used by one of the authors for many years as a "frustration mark" and appears to have a proper place in the introduction of many physics principles.

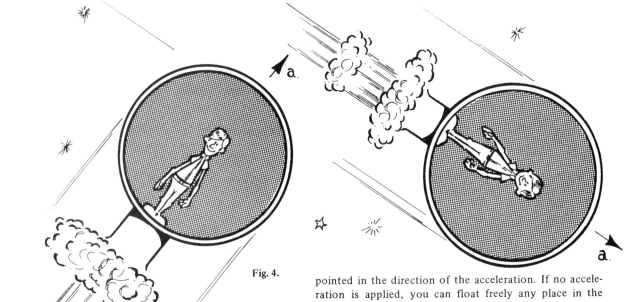

Fig. 4.

pointed in the direction of the acceleration. If no acceleration is applied, you can float freely any place in the room. You conclude that anytime you can stand on a wall, the ship must be accelerating in a direction opposite to the wall on which you are standing. You have been oriented and educated to only inertial forces (see Fig. 4).

Now suppose (unknown to you) the ship descends and lands on the planet Earth. You can then stand on a wall at point X in the diagram (see Fig. 5). Using your prior learning, you logically assume that the ship is accelerating at 9.8 m/sec² in a direction opposite this wall. After a prolonged period of time, you may even be concerned that the ship may exhaust its fuel supply under the extended acceleration.

Let's now assume that the ship has wheels and the ability to accelerate along the earth's surface at 5m/sec². You could then stand at point Y and your head would be parallel to the vector sum of the 9.8 m/sec² and the

Why does the float lean (or point) in the direction of the acceleration? This can be answered with the following futuristic fable.

Let us assume you were born in outer space. Your family was forced to leave their home planet which was doomed by a catastrophic disaster. You then spend your early years in space. We can further assume that your spaceship has rocket engines surrounding its spherical surface such that any thrust can accelerate the ship in a particular direction without a new orientation of its axis. In your spherical room, you learn that when a particular thrust is used you can stand on the side of your room near that thrust. In other words, you can stand with your head

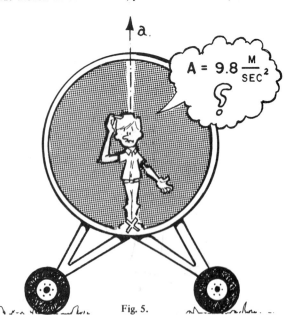

Fig. 5.

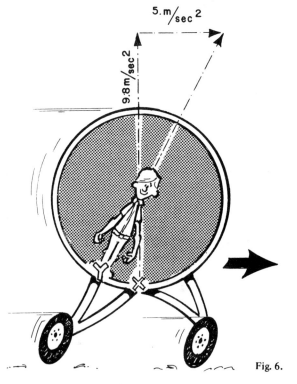

Fig. 6.

5 m/sec². In any case you can stand "up" — up being the direction of the apparent inertial acceleration (end of fable). (Fig. 6)

At this point it is relatively easy for students to see that the conditions in the "classroom spaceship" would be exactly the same in either of the following cases:
1. gravity acting as it normally does, or
2. the room being accelerated up at 9.8 m/sec² in the absence of a gravitational field.

Now, back to the accelerometer. The float behaves as if its only reaction is to inertial forces. You might say that it behaves like the space wanderer: it points up — up being the vector sum of all apparent inertial acceleration.

To open new avenues of discussion, can free fall (be it falling out of bed or falling through space) be explained as a cessation of your upward acceleration while the earth continues to accelerate toward you at 9.8 m/sec², inertially?

Reference
1. A. A. Bartlett, "Which Way is dΩ or the 'Force of Gravity' in Some Simple Accelerated Systems," Phys. Teach. 10, 429 (1972).

Balancing

Robert N. Jones
*Philadelphia College of Pharmacy and Science,
Philadelphia, Pennsylvania 19104*

A simple, interesting, and thought-provoking demonstration is the one shown in the pictures. It is a little more sophisticated than just having a student support a stick on his index fingers and bringing them together without tipping the stick.

As illustrated, [Fig. 1(a) - (c)], the idea is to support a board on two spring scales with several weights placed at random on the board.

Have the class read the total of the two scales and then ask a student to bring the scales together without tipping the board and weights. Usually the student will move the scales toward each other rather carefully and finally, when they meet, will breathe a sigh of relief and will get approval of his classmates.

You might ask him or others to try it again. Finally, you quickly do it and show the scales will automatically meet without tipping anything.

Fig. 1. *A simple demonstration of balancing and center of gravity. A student slides the two scales together without tipping the board and weights.*

Do the demonstration again and have the students observe:

(1) The two different scale readings as well as the total at the start and throughout the demonstration.
(2) Which scale slides first and why (friction, action and reaction, etc.)?
(3) When do both scales move toward each other?
(4) Where is the overall center of gravity when the scales meet?

HOT WHEELS PHYSICS

Stanley J. Briggs
T. Roosevelt High School
Wyandotte, Michigan 48192

Once in a while I take a trip through the toy section of our local department store to see if there are any new items that could be used in teaching some scientific principles. In the past, these trips have yielded such items as tops, superballs, water-propelled rockets, etc. Recently the Mattel Toy Company introduced their low friction, gravity-operated race car sets, "Hot Wheels." The "Stunt Action Set," Stock No. 6279, consists of a miniature car, 16 ft of plastic track, a loop-the-loop frame, two jump ramps, and table clamp. Several interesting experiment demonstrations, using this equipment, have been developed.

Experiment 1: Loss of Energy Due to Friction

Figure 1 shows one end of the track mounted with the table clamp, which is supplied with the set, at a height of about 1 m above the floor or demonstration table. The other end of the track is also elevated to about the same height. The car is released from the starting point 1 m above the floor or table, and the height to which the car rises on the other end of the track is recorded. It is important to secure the track to the floor with double stick tape or masking tape rolled over on itself, so that some of the car's energy is not transformed into motion of the track. The ratio of the final height to the initial height is equal to the ratio of the amount of potential energy transferred back into the car going up the left ramp to the potential energy the car had at the top of the right ramp at point A. This ratio of h_1 to h_2 could be called the "efficiency" of the system for the car going from point A to point B on the track. The distance along the track from point A to point B may be called the "standard length," so that the ratio of h_2 to h_1 (efficiency) will represent the fraction of the initial energy the car will have when it reaches point B on the track.

This ratio must be used in later experiments to determine how high the car must be started, so that it will have a predictable total amount of energy (potential and kinetic), when the car reaches point B.

Sample Data

When the car was released from a vertical height of 100 cm, it went up the other end to a height of 80 cm. When point A was lowered to 50 cm and the car released, the car went up to 40 cm, which shows the efficiency of the system over the "standard length" was about 80%.

Experiment 2: Loop-the-Loop

Figure 2 shows a loop-the-loop section placed in the track at about the "standard length" from the starting end of the track. The problem is to predict the height from which the car must be released, so that it will successfully travel all the way around the loop without falling away from the track. In order for this to take place, the centripetal force (F_c) must just be supplied by the weight of the car (mg) or

$$F_c = (mv^2)/R = mg \text{ or } v^2/R = g. \quad (1)$$

The total energy of the car at point B must be its kinetic energy as it travels at speed v in Eq. (1) plus its gravitational potential energy at point B, $mg2R$. This total energy is supplied by the loss in gravitational potential energy of the car as it travels down the track from the starting height. This means:

$$\tfrac{1}{2}mv^2 + mg2R = mgh \text{ or } \tfrac{1}{2}v^2 + 2Rg = gh. \quad (2)$$

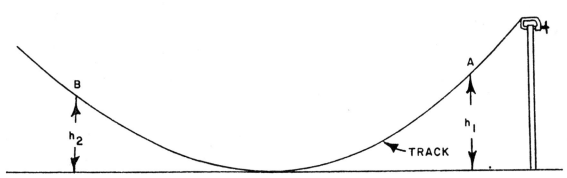

Fig. 1. Schematic showing track setup for Expt. 1.

A POTPOURRI OF PHYSICS TEACHING IDEAS—MECHANICS

Solving Eqs. (1) and (2) for v^2 and setting them equal to each other, we find that

$$Rg = 2gh - 4Rg \text{ or } h = (5/2)R. \quad (3)$$

This means that if there were no friction, the car should be released from a height of $\frac{1}{2}R$ above the top of the loop. Since our system does have friction, we must release the car from a point just enough higher to make up for the frictional loss. If the efficiency of the system is K, then we must release the car from a height of h/K.

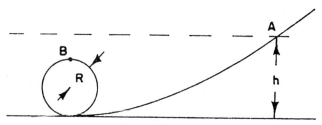

Fig. 2. Schematic showing track setup for Expt. 2.

Sample Data

In our set the diameter of the circular loop is 22 cm. This means that without friction the car must be released from 22 cm + 5.5 cm, or 27.5 cm. When we divide 27.5 by 0.8 (the efficiency of our system), we get about 34.5 cm. Next one releases the car from this height, and the car makes it around the circle in fine shape. The students will probably request that you try a lower height to see what happens. A reduction of about 10% produces a very noticeable drop away from the track. If one releases the car from a height of 20% to 30% above the required height, the loop-the-loop frame will jump off the floor! (Reaction to the centripetal force being supplied by the track in addition to gravity.)

Experiment 3: Dare-Devil Jump Ramp I (Range)

The "Hot-Wheels" set mentioned above also supplies a pair of jump ramps that will launch a moving car into a trajectory predictable by projectile motion mechanics. The problem here is to calculate the height from which the car must be released for the car to follow the trajectory from one ramp to the other. If the car leaves the ramp at an angle of θ, then the horizontal component of the velocity,

$$v_x = v \cos\theta, \quad (4)$$

where v is the velocity of the car up the jump ramp. The vertical component of the velocity:

$$v_y = v \sin\theta. \quad (5)$$

The kinetic energy, $\frac{1}{2}mv^2$, at the bottom of the ramp equals the loss in gravitational potential energy from the starting point on the track:

$$mgh = \frac{1}{2}mv^2 \text{ or } v^2 = 2gh. \quad (6)$$

The range or distance between the ramps is R.

$$R = v_x T, \quad (7)$$

where T is the total time the car is in the air. Using the uniform acceleration of gravity, we find that

$$T = (2 v_y)/g. \quad (8)$$

From Eqs. (7) and (8) we see that

$$R = (2v_x v_y)/g. \quad (9)$$

Substituting Eqs. (4) and (5) for v_x and v_y, we get:

$$R = [2v^2(\sin\theta \cos\theta)]/g, \quad (10)$$

and substituting Eq. (6) for v^2, we find that

$$R = 4 h \sin\theta \cos\theta. \quad (11)$$

Therefore, the height from which the car must be released is

$$h = (R)/(4 \sin\theta \cos\theta). \quad (12)$$

Sample Data

The jump ramps in our set have angles of about 15°. With the ramps set at a range of 40 cm, the release height must be

$$(40 \text{ cm})/(4 \sin 15° \cos 15°) = 40/4(0.26)(0.97) \quad (13)$$

or $h = 40$ cm above the height of the top of the jump-ramp point B. Since the efficiency of the system is about 0.8, the release height must be 40 cm/0.8 = 50 cm. (It does not look as if it would jump that far, but it does!)

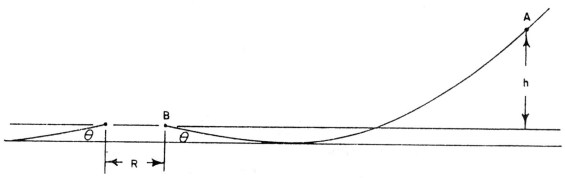

Fig. 3. Schematic showing track setup for Expt. 3.

$$H = h (\sin\theta)^2. \tag{15}$$

Sample Data

From the last example we found that $h = 40$ cm. Solving for H: $H = 40$ cm $(\sin 15°)^2 = 40$ cm $(0.067) = 2.7$ cm. It should be noted that this height should be measured from the level of the top of the jump ramp.

Conclusion

Student interest has been high with these experiments, and perhaps the reader can think of other experiments to perform with the equipment to help justify its purchase from the petty cash fund (less than $4.50). The basic experiments suggested here are, of course, not new ideas to the physics class room, but the use of the track and race cars may help spark some students' interest in the study of mechanics.

It should be noted that the use of the so-called "efficiency factor" depends upon the assumption that the energy loss due to friction depends only on the length of travel on the track. If the length of the track is held constant in the experiments, the efficiency should remain the same. This, of course, is not quite true, and perhaps it should be pointed out to the students that the amount of friction will also depend upon the angle of the track. In our experiments, the track angle does not change very much, so we are justified in our assumptions. Some students could use this as a further investigation to determine the effect of the track angle with efficiency.

Students performing Expt. 4—"Dare-Devil Jump Ramp II (Fence Jump)."

Experiment 4: Dare-Devil Jump Ramp II (Fence Jump)

A small fence of paper or cardboard may be constructed and placed between the jump ramps. The problem here is to calculate the maximum height of the fence, which the car in Expt. 3 will just clear in moving through the trajectory between the two ramps. If we call the maximum vertical component of the trajectory H, we know that this will be equal to

$$\tfrac{1}{2}v_y\, t, \text{ where } t = v_y/g \text{ or } H = \tfrac{1}{2}v_y^2/g. \tag{14}$$

From Eq. (5) we substitute for v_y and Eq. (6) for v^2 for

Balanced Dynamic Carts

The equivalence of inertial and gravitational mass may be shown by tying two dynamics carts together and placing them in the center of a 10-ft board, balanced on a 2 in. by 2 in. fulcrum, as indicated in Fig. 2. Burn the string and have the class observe that the system remains in balance as the force of the compressed spring bumpers propels the carts in opposite directions along the board. Adding mass to one of the carts will result in a reduced velocity because of the inertial mass, but the system will remain in balance because the torques developed by the gravitational forces are equal until one of the carts reaches the end of the board.

Fig. 2. Apparatus for demonstrating the equivalence of gravitational mass and inertial mass. (Not to scale.)

A Demonstration of Constant Acceleration

Ernest Hammond
Morgan State College
Baltimore, Maryland

An experiment to demonstrate the laws of gravity without using waxed paper tapes, golf balls, or expensive photographic equipment (Polaroid) is performed with an electronic stroboscope light source and a common laboratory water faucet used as a source of water droplets.

The water droplets are allowed to drip from the water faucet at a constant frequency. The electronic stroboscope being a source of short, accurately timed light flashes equal to the above frequency instantly stops the water droplets in flight. Since the time intervals of the stroboscope are equal, the immediate and live classroom observation reveals an increasing separation of the droplets during their fall. Thus the conclusion that the droplet velocity is continually increasing or that the droplets are accelerated.

The above experiment proves the rule of constant acceleration to even the most skeptical student. Also, it permits accurate measurements for the calculation of the acceleration due to gravity.

INELASTIC COLLISIONS USING VELCRO

A standard apparatus to study inelastic collisions utilizes a soft wax to bind the two objects when they hit.

Students introduce errors when their masses fail to stick together and find that using wax is a messy operation: the apparatus must be scraped clean after use and spattered wax must be removed from floors and lab tables.

As an alternative approach, I began experimenting with Velcro* last year. Velcro, a fabric fastener commonly used in the apparel industry, consists of two strips of cloth, one having a "pile" side and the other having a "hook" side. They grip tightly when pressed together and may be separated repeatedly with no loss of adhesion.

To use Velcro on most any inelastic collision apparatus, first remove all wax from the impact objects and then glue the Velcro strips across the front of each mass using an adhesive such as Scotch-Grip 1300 Rubber Adhesive.† It appears to make no appreciable difference which object receives a given side of the Velcro. As long as the centers of the masses are aligned, the Velcro will stick on almost every trial.

* John Dritz and Sons, New York, N.Y.
† 3M Company, Adhesives, Coatings and Sealers Division, St. Paul, Minn.

G. SCOTT HUBBARD
Vanderbilt University
Nashville, Tennessee 37235

VELOCITY MEASUREMENTS OUTDOORS WITH A TAPE RECORDER

Measurements with *The Tape Recorder as a Small Interval Timer* by Steward Schultz in the October 1971 *Physics Teacher* can be made with some tape recorders by developing the tape with iron filings which stick to the magnetized portion of the tape. However, this method does not work with all tape recorders even with the recording gain set at maximum. For tape recorders where this method will not work, the tape can be stopped as the recorded portion reaches the playback head and marked at this point with a ball point pen or a small sliver of splicing tape.

This method has been used to directly measure the speed of a hunting arrow. The reaction time of the students usually exceeds the flight time of the arrow. The microphone picks up the snap of the bow string and the impact of the arrow. With a small correction for the speed of sound we can calculate the average velocity with an uncertainty of 1 or 2%. The higher the recording speed, the smaller the uncertainty. Masking tape or a piece of rubber tubing may be used to enlarge the capstan to produce higher recording speeds. The recording speed can then be calibrated and checked for variability by taping a time signal such as CHU, Canada or WWV, Fort Collins, Colorado.

The variability of a PSSC ticker tape timer or the frequency of an audio generator may also be determined with a tape recorder. Direct comparison of the developed tape and a millimeter scale under a dissecting microscope will facilitate measurement.

We have developed our own version of the Project Physics' "Leslie's 50" with a 160-m bicycle sprint. Strips of ¼ in. plywood 1 in. wide and 3 ft long with one surface covered with aluminum foil are fastened to the side wall at 10-m intervals with masking tape. A piece of wire 0.5 cm above the foil insulated by the air gap completes the switch. The foil and wire are connected such that when the bicycle wheel presses the wire to the aluminum foil the circuit between a battery and an earphone is closed. The earphone is taped to a microphone and as the cycle wheel crosses the switches, clicks are recorded. The developed tape yields a distance–time graph for the sprint. At each 10-m interval a click is produced by the front and rear wheel. Thus we have the time it takes for the cycle to travel its own length at each switch as well as the time to go between switches. These two times provide an introduction to the concept of instantaneous velocity versus average velocity.

A stereo tape recorder can be used to make a direct measurement of the speed of sound. A loud click or bang near one microphone is recorded at a high tape speed at the maximum separation achievable with the microphones. The difference in position of the recorded tracks, tape speed, and microphone separation distance will determine variables for the calculation of the speed of sound.

DONALD J. PRUDEN
State University of New York at Albany
Albany, New York 12003

Kinematics and the driver

Barton Palatnick
California State Polytechnic University, Pomona, California 91768

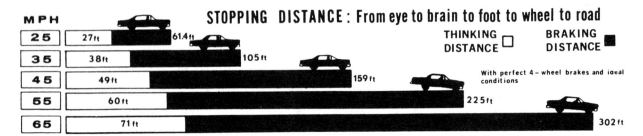

The formulas of kinematics, $s = v_o t + \frac{1}{2} at^2$ and all the rest, can be readily developed by the teacher, memorized (and hopefully understood) by the student, and then applied to the usual variety of problems dealing with falling stones, baseballs, railroad trains, etc. I have come across one particular application which is of exceptional interest and which I would like to present here. It could be used in class as an example, assigned as a homework problem, or used in some variation as a test question.

The example I have in mind has to do with driving, something that interests nearly everyone in our society. The beginning driver is likely to come across, in either a book on driving or a state driver's manual, a list of numbers and perhaps an illustration of stopping distances at different speeds. The stopping distance is the distance the auto travels from the time the driver decides he wants to stop until the stop is actually accomplished. The numbers give some indication that the stopping distance increases faster than the speed, but the entire table can in fact be very well understood in terms of simple kinematics. Let me give the details in the form of the answer to the following question which I have given on an exam:

"The California Driver's Handbook has a chart giving the stopping distances of cars with different initial velocities. The stopping distance is broken down into 'thinking distance' and 'braking distance.' You are to examine the numbers from the chart and make whatever comments about them you can in view of what you know about kinematics."

mph	Thinking Distance (ft)	Braking Distance (ft)	Total (ft)
25	27	34	61
35	38	67	105
45	49	110	159
55	60	165	225
65	71	231	302

This problem was not the only one dealing with kinematics on the test—there were some others more of the "plug-in" type. This problem was more challenging in its approach than the others, and was more difficult in that there is no single formula, from those standard ones the student has learned, that applies. The situation must be sized up in terms of two formulas, one which applies for the "thinking distance," where the speed is constant, and one for the "braking distance," where the acceleration (or deceleration) is constant. The sort of reasoning I was looking for is as follows. (Incidentally, the numbers in the table were for perfect 4-wheel brakes and ideal conditions.)

First to deal with the "thinking distance." It takes a certain amount of time, starting from the driver's decision to stop, until his foot hits the brake pedal; this is the reaction time. The car is moving at constant speed during this interval, so the formula that applies is $s = vt$. Since the reaction time t is a constant, the ratio s/v (the "thinking distance" divided by the velocity) must also be constant. Let's check this:

$$27/25 = 1.1,$$
$$38/35 = 1.1,$$
$$49/45 = 1.1,$$
$$60/55 = 1.1,$$
$$71/65 = 1.1.$$

So we understand this aspect of the table. If the unit conversion is carried out, something I didn't really expect on the test because of time limitations, one gets 0.75 sec for the reaction time of the driver — a quite reasonable number.

Now for the "braking distance." Making the assumption that the deceleration is constant and wanting to involve the distance traveled, one is led to the formula $v^2 - v_o^2 = 2as$. Here $v = 0$ (a is negative) and we conclude the ratio $v_o^2/2s$, or just v_o^2/s for our purposes, should be constant. Again we check:

$$(25)^2/34 = 18$$
$$(35)^2/67 = 18$$
$$(45)^2/110 = 18$$
$$(55)^2/165 = 18$$
$$(65)^2/231 = 18$$

So the second part of the table is now comprehensible, too, and hence the entire stopping distance. Converting units and calculating $v_o^2/2s$ gives 20 ft/sec² for the deceleration. This is a bit greater than ½g, again a very reasonable result.

This, then, is the example, certainly not run-of-the-mill and capitalizing on the student's very keen interest in driving. It shows how a data table of numbers which at first seem only loosely connected with each other can be quickly and logically understood through some of the most basic formulas of physics. There are other approaches to the data that could be taken; for instance, one could calculate the reaction time and deceleration from the distances given for 25 mph, then proceed to calculate and fill in all the other distances in the table. This problem was given in an introductory (noncalculus) college physics class; the answers showed the entire range of understanding, from total awareness of what was going on to total confusion, such as trying to apply $s = v_o t + \frac{1}{2}at^2$ for the entire stopping distance.

DRAG STRIP TIMER, AN EXTENDED USE

A drag strip timer and a tape recorder may be connected together to produce a permanent record of small time intervals. One end of a soda straw is connected to the tape recorder speaker and the other end of the straw is connected to a felt tip pen or other writing instrument to form a trace on the drag strip timer (Fig. 1). The trace may then be duplicated by Xerox or by Thermofax master and ditto for class distribution.

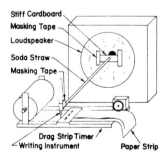

Fig. 1. Traces are produced on a drag strip timer by a large extension speaker operated by a tape recorder.

A fine line pen can be made from glass tubing drawn to a fine capillary with a fine wire in the bore to aid the flow of ink. To produce traces for projection, an acetate sheet may be substituted for the paper in the drag-

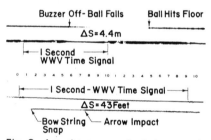

Fig. 2. Actual traces made during experiments to measure free fall of a ball and time of flight of an arrow.

strip timer. Smoked glass and a fine needle will also work.

Illustrated in Fig. 2 are traces made by a dragstrip timer while doing an experiment to measure free fall acceleration and to measure the speed of an arrow in flight according to the technique described by Schultz a few months ago.[1] The scale in the illustration shows the length of the trace in centimeters. The best traces are made from recordings in an acoustically dead environment. I also found it necessary to damp the buzzer with masking tape to prevent the appearance of residual vibrations and reverberations during periods when the buzzer is off.

DONALD J. PRUDEN
The Milne School
Albany, N.Y. 12203

[1] S. Schultz, Phys. Teach. **9**, 413 (1971).

Friction in a moving car

Fred M. Goldberg
West Virginia University, Morgantown, West Virginia 26506

This paper describes an experiment performed this past year by 80 students in an introductory noncalculus physics course, and in modified form by 120 students in a physical science course. The experiment was to measure a coefficient of friction for a moving car. The students in the classes were divided into groups of three or four and measurements were made in the students' own cars. The basic equipment consisted of a stopwatch, level and meter stick. Although explicit directions were provided for parts of the experiment, the students had to determine their own methods for measuring one of the parameters. One of the aspects of the experiment that the students commented very favorably on in their evaluation was that they really did not know what results to expect. This was in sharp contrast to the majority of the other in-class physics experiments in which the students knew exactly what results they were supposed to obtain. Those students who looked up in their textbook the value for the coefficient of static friction between rubber and asphalt (which varied between 0.5 and 0.9) were quite surprised at the results they got.

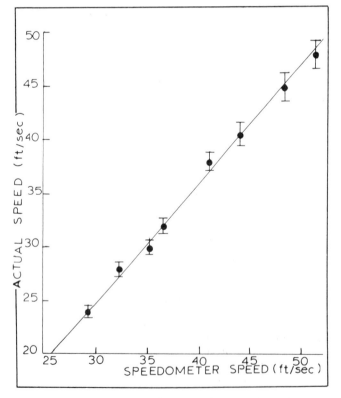

Fig. 1. Calibration of speedometer. The actual speed of the car, computed from measurements of distance and time, is plotted as a function of the speedometer reading, converted to units of ft/sec.

The experiment was performed on a 360 ft section of a parking lot that had a slight incline. Natural markings on the lot indicated the initial and final points of the section to be used. The students accelerated the car to a certain initial speed, put it in neutral, and started the stop watch when they passed the initial point. They then measured the time to coast the 360 ft. The procedure was repeated for about ten different initial speeds varying between about 20 mph and 40 mph.

The equation which relates the coefficient of friction, μ, to the measured parameters (initial velocity, distance, and time) can be derived from the following equations:

$$N - mg \cos \theta = 0,$$
$$-f - mg \sin \theta = ma,$$
$$f = \mu N,$$
$$S = v_i t + \tfrac{1}{2} a t^2.$$

The first two equations are Newton's second law written for components perpendicular and parallel to the incline. N represents the normal force and θ the angle of the incline. The third equation represents the assumption that the frictional force f is proportional to the normal force, from which μ is defined. The last equation is the kinematic equation relating the distance s, initial velocity v_i, acceleration a, and time t. Using these four equations we can solve for μ, the coefficient of friction, in terms of v_i, t, s and θ, all of which had to be measured by the students:

$$\mu = \frac{2(v_i t - S)}{gt^2 \cos \theta} - \tan \theta \qquad (1)$$

The initial speed was read from the speedometer, and it had to be calibrated. This was done in the following way. The students drove over the 360 ft section maintaining a constant speed on their speedometer and measured the time it took. Dividing the distance by the time gave the actual speed. This value was compared to the speedometer reading converted to similar units. The experiment was performed for several different speeds and the results were plotted. A typical graph is shown in Fig. 1. The results for almost all the groups fell on fairly good straight lines, but the slopes were rarely unity. The range went from the speedometer reading being about 5% too low to about 10% too high.

How the students were to measure the angle of the incline was left up to their own ingenuity. Since it turned out to be about 1° it was no easy chore. Some of the groups used an elaborate transit borrowed from friends in engineering, but the majority of the students used a meter stick and level.

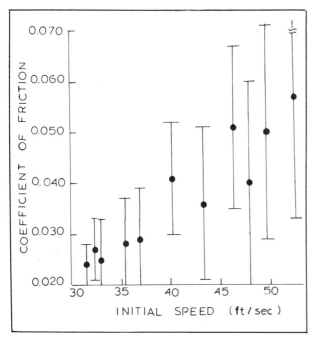

Fig. 2. Computed value of the coefficient of friction [from equation (1)] as a function of the initial speed of the car.

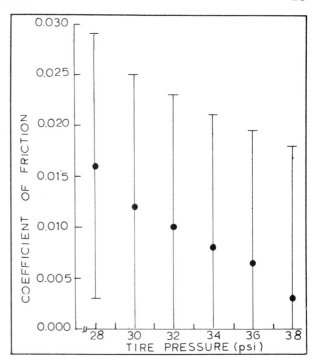

Fig. 3. Computed value of the coefficient of friction as a function of tire pressure for a constant initial speed of 26 mph.

The better groups measured the angle several places along the section and used the average.

The result of the data for one particular group is shown in Fig. 2 where the computed coefficient of friction is plotted as a function of initial speed. The error bars represent the computed uncertainty in the value of the coefficient of friction [from equation (1)] corresponding to the uncertainties in the measured quantities: initial speed (± 1 mph), time (± 0.1 sec), and distance travelled (± 5 ft). There is an additional uncertainty in the coefficient of ± 0.008 for each computed point. This corresponds to the standard deviation from the mean of the values of tan θ as measured by all the groups in the physics class. The values for the coefficient, displayed in Fig. 2, were fairly typical of the results obtained by other groups although the total range of values among the 54 cars participating in the experiment was from 0.004 to 0.066. All these values, however, were considerably less than the value of 0.5-0.9 listed in textbooks for the coefficient of static friction between rubber and concrete. Most students realized at this point that something other than static friction was causing the car to slow down.

From Fig. 2 it appears that there is an increase in μ with increasing initial speed. This would be expected due to the effect of increasing wind resistance. As the initial speed increases the wind resistance becomes greater and the car would slow down faster than it would if wind resistance were not present. This would result in a longer time to travel the set distance of 360 ft and hence the effective value of μ computed from equation (1) would be larger.

As part of their write-up, the students were asked to list the most important factors causing the friction. These were: the friction between the tires and the road, the internal friction between moving parts, the friction between the brake drums and linings, and the effects of the wind. The use of a coefficient of friction as the ratio of the friction force to the normal force is probably not applicable for many if not all of these effects. Further experimentation might establish the relative importance of the internal friction versus the external friction. Perhaps some groups might raise up the rear wheels of their cars with a garage jack and measure the deceleration of the spinning wheels. This would give an indication of the internal frictional effects only and could be compared to the results on the road.

The external friction between the tires and the road is an example of rolling friction.[1] Instead of a sliding or rubbing effect there is an energy loss resulting from the deformation of the tire at the area of contact with the road surface. Assuming rolling friction contributes significantly to the total friction coefficient measured by the students, another interesting experiment would be to measure the coefficient as a function of the tire pressure. Recently some students in my physical science class performed a similar experiment as an extra credit project. Starting from an initial speed of 26 mph they timed how long it took to coast the 360 ft. They repeated this experiment for six different values of the tire pressure, varying between 38 and 28 psi, and computed the corresponding value of μ. Their results are shown in Fig. 3. The uncertainties were computed as previously described. Although the uncertainties are large, the graph seems to indicate that the coefficient of friction increases with decreasing tire pressure. This would be expected since the less the air pressure in the tire the more it would be deformed and therefore the greater the energy loss.

Finally, I would like to indicate some student reaction to

this experiment. Because they worked in groups, outside of class, the students sometimes had difficulty finding a common time to get together. In some cases one or more members of a group did a disproportionate amount of the work for the experiment, while other members contributed very little. In general, though, the students reacted very favorably to this out-of-doors, partially unstructured, and interesting laboratory experience.

Reference

1. See for example Harvey E. White, *Modern College Physics* (Van Nostrand, Princeton, 105, New York, Reinhold, 1972) Sixth Edition, p. 120.

Driving safety and straight-line kinematics

Van E. Neie
Purdue University, West Lafayette, Indiana 47907

The recent note by Palatnick[1] dealt with an interesting problem in kinematics as it relates to driving safety. I have for several years suggested that students work out the problem of minimum stopping distances, car separation as it is related to reaction-time, etc. One assumption that Mr. Palatnick and others make is that all cars have the same negative acceleration when stopping. This is a fairly reasonable assumption since the values would not be likely to differ more than a few percent under normal circumstances. However, there are situations that one could imagine for which these decelerations are different and this difference can have a bearing on the minimum separation distance between cars required to prevent rear-end collisions.

This minimum distance when both cars have the same negative acceleration is simply

$$S \geq V_o t_R \quad (1)$$

where V_o is the common speed of both cars and t_R is the reaction time. This expression merely points out that if both cars can stop in the same distance *after* the brakes are applied, then the trailing car will travel an additional distance equal to its speed multiplied by the reaction time. If one assumes a reaction-time of 1 sec and an average car length of 15 ft, we get the often quoted "rule" that one car length separation should be maintained for every 10 mph of speed (14.7 ft/sec).

If the accelerations are *not* the same the situation becomes more complicated. The expression for the minimum distance of separation is now

$$S \geq V t_R - \frac{1}{2} \left[\frac{a_1 - a_2}{a_1 a_2} \right] V^2 \quad (2)$$

where a_1 and a_2 are the values of the accelerations for the leading and trailing cars, respectively. Examination of expression (2) reveals some important facts: The minimum distance of separation depends on the *difference* in accelerations as well as their *numerical values*; and this distance is no longer linear, but quadratic, in V. Whether the distance of separation is greater or lesser than $V t_R$ depends on the accelerations of the cars. If the front car has a greater negative acceleration (better brakes), then $a_1 - a_2$ is negative and $S > V t_R$, i.e., the trailing car would have to increase its separation distance from the front car to avoid a collision.

The following table illustrates the seriousness of the situation when accelerations are different. Two speeds, one rather slow (20 mph) and the other quite fast (68 mph), have been selected to emphasize the effect of V. In addition, two acceleration pairs, each pair differing by the same amount, are utilized to illustrate the acceleration effects.

A convenient way of presenting these ideas graphically is to employ overhead projection. A separate graph of S vs. t for

TABLE 1[2]

Minimum Distance of Separation to Avoid Rear-End Collision for Various Values of Speed and Acceleration

Speed	Acceleration of Leading Car	Acceleration of Trailing Car	Minimum Separation Distance for Equal Accelerations	Minimum Separation Distance for Given Accelerations
30 ft/sec	-20 ft/sec^2	-15 ft/sec^2	30 ft	38 ft
100 ft/sec	-20 ft/sec^2	-15 ft/sec^2	100 ft	185 ft
30 ft/sec	-15 ft/sec^2	-10 ft/sec^2	30 ft	45 ft
100 ft/sec	-15 ft/sec^2	-10 ft/sec^2	100 ft	267 ft

each car is plotted (using the same scale) and reproduced on transparency sheets. If the two coordinate axes are then superimposed, the two curves will intersect at the point (S, t), these values being associated with a collision. An additional advantage to using transparencies is the ability to adjust the initial separation distances so that the effect on collision distance and time can be shown.

References

1. Palatnick, Barton, "Kinematics and the Driver," The Phys. Teach. 12, 229 (1974).
2. The author prefers to use SI units exclusively, but the use of English units is employed here to facilitate comparison with existing driver's manuals and other published data on highway safety.

To stop or not to stop — kinematics and the yellow light

J. Fred Watts

Department of Physics, The College of Charleston, Charleston, South Carolina 29401

Teachers and students enter the physics laboratory with different expectations. While the teacher is often looking for pedagogical value, the students may be looking for relevance and excitement. It is not easy to find laboratory exercises which satisfy both teachers and students. This note describes one such exercise involving kinematics.

Automobiles are both familiar and interesting to the average student. This makes automobiles ideal objects for the study of kinematics.[1] Our exercise involves the student with the automobile in his most common role — that of driver. Almost every time we make even a short automobile trip we are faced with the problem of the yellow light. That is, when the traffic light turns yellow, should we stop or try to get through the intersection before the light turns red? Most of us at one time or another have made the wrong decision. Unfortunately, this has sometimes resulted in a citation to appear in traffic court.

It is well known, at least to traffic engineers,[2] that some yellow lights are so short in duration that there exists a dilemma zone; i.e., a region where the driver is too close to stop and too far away to clear the intersection before the light turns red. State laws vary on this subject but in South Carolina the law states that "yellow alone or 'caution' when shown following the green or 'go' signal or in combination with the green signal means that vehicular traffic facing the signal is thereby warned that the red or 'stop' signal will be exhibited immediately thereafter and such vehicular traffic shall not enter or be crossing the intersection when the red or 'stop' signal is exhibited."[3]

We explain the concept of the dilemma zone to our students and quote the above law. We then arm them with

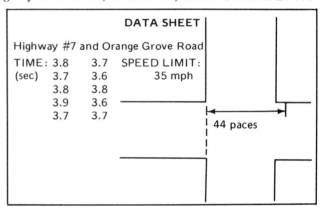

Fig. 1.

stopwatches, meter sticks, and a table of stopping distances from the South Carolina Driver's Handbook (Table I). Their assignment is to evaluate several intersections to determine if any dilemma zones exist. A typical data sheet for an intersection is shown in Fig. 1. When the required data have been obtained the students return to the lab and complete their calculations in order to determine the presence, or the absence, of dilemma zones. Table II is a sample of the results from several intersections.

Since a car must not be crossing the intersection when the light turns red, a dilemma zone will exist if a car traveling at the speed limit cannot, during the time the light is yellow, travel a distance equal to the width of the intersection plus the length of the car (we use 17 ft as an aver-

	Table I		
	Stopping ability of standard passenger cars on dry, clean level pavement		
Speed (mph)	Driver reaction distance (ft)	Braking distance Range (ft)	Total stopping distance range (ft)
20	22	18-22	40-44
25	28	25-31	53-59
30	33	36-45	69-78
35	39	47-58	86-97
40	44	64-80	108-124
45	50	82-103	132-153

			Table II			
Intersection	Yellow light (sec)	Speed limit (ft/sec)	Intersection width (ft)	Stopping distance (ft)	Dilemma zone (ft)	Minimum yellow light (sec)
A	3.5	66	139	153	78	4.7
B	3.7	51	190	97	115	6.0
C	4.4	66	153	153	33	4.9
D	3.5	66	51	153	---	3.4
E	3.0	59	123	124	87	4.5
F	3.4	51	142	97	83	5.0

age) plus the stopping distance. When a dilemma zone is found the students are asked to calculate the minimum time that the yellow light should be on. If no dilemma zones are found we often hear groans of dismay, and we are frequently asked to allow the students to go out again to check other intersections. There is great delight among the students when they find that the "establishment" has passed a law which cannot be obeyed. If there is no dilemma zone for a car approaching the intersection going the speed limit the students are asked to find a lower speed for which a dilemma zone exists. Since there are always such speeds some of the disappointment from not originally finding a dilemma zone is erased. This experiment may be used early in the semester as described or modified for later use by adding questions about average acceleration while stopping and the coefficient of friction between tires and road.

In conclusion, we feel that this experiment is one which gives students experience in taking data and doing relatively simple calculations and at the same time connects what they are studying in the classroom with the "real" world outside. Interest in and excitement about physics are increased and seem to carry over during the remainder of the year. It is an enjoyable experiment from the viewpoint of both the student and teacher.

References
1. Barton Palatnick, Phys. Teach. **12**, 229 (1974).
2. Paul L. Olson and Richard W. Rothery, *Operations Research 9*, 60 (1961).
3. South Carolina Code of Laws, 56-970 (2).

Automobile stopping distances

L. J. Logue
Department of Chemistry and Physics, Southern Technical Institute, Marietta, Georgia 30060

A recent article in this journal[1] points out that the usual model of frictional forces predicts that the stopping distance for an object sliding on a plane surface is independent of the mass of the object. This is an interesting result and should certainly be discussed in any treatment of frictional forces in an elementary physics course. But we should be aware that the result depends on certain assumptions about the properties of frictional forces. We must be very careful about extrapolating these results to apparently similar situations in which these assumptions may not be valid. For example, the article mentioned above implies that two automobiles of greatly different masses, starting from the same speed, will have the same stopping distances with the wheels locked. In fact the characteristics of rubber tires sliding on pavement violate many of the usual assumptions about friction and make equal stopping distances unlikely.

Experimental data on the braking acceleration of automobiles is widely available in the popular automotive press.[2] These data show that wide variations do exist in the performance of actual cars. An explanation of some of the causes of these variations can be obtained using principles and techniques which are available to most college physics students. Since the force responsible for stopping a car comes from the interaction between the road and the tires, a study of the behavior of tires should shed some light on the problem.

There are several factors which affect the stopping performance of automobile tires. These include (but are not limited to) contact area between tire and road, the temperature of the tire, amount of slip between tire and road,[3] and the vertical load carried by the tire. These factors are not entirely independent of each other. For example, contact area depends not only on the size of the tire but also on the camber angle (angle between the plane of the tire and the road surface) which in turn depends on the design of the suspension system. The camber angle will change with changing load for most independent suspension systems in current use. The temperature of the tire will also depend on the load since increasing the mass of the vehicle increases the kinetic energy which must be dissipated in stopping. Some of this energy will show up as internal energy of the tire. Some typical tire characteristics are shown in Fig. 1.[4]

These curves show the coefficient of friction for a typical automobile tire as functions of temperature, percent slip, and load when all other factors are held constant. Note

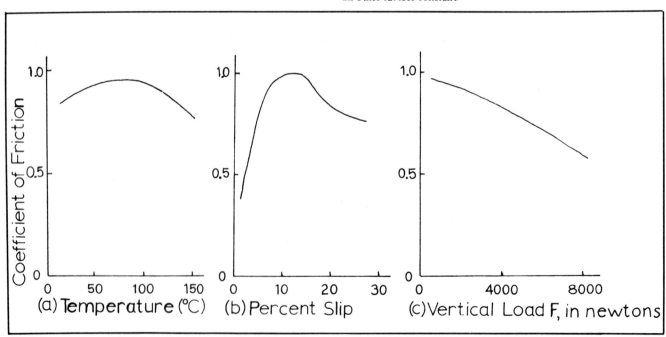

Fig. 1. Plots of coefficient of friction for a typical automobile tire and road surface versus (a) temperature of the tire, (b) percent slip and (c) vertical load. Each curve was obtained by attempting to hold all other factors constant.

A POTPOURRI OF PHYSICS TEACHING IDEAS — MECHANICS

that the temperature and slip curves have peaks while the load curve is monotonically decreasing. The slip curve is especially interesting since the maximum value of the coefficient of kinetic friction is obtained when substantial slip occurs. This maximum coefficient of kinetic friction is in fact greater than the coefficient of static friction, in contrast to the examples used in most physics texts. The fact that the load curve is not a horizontal line shows us that the mass of the vehicle may have a significant effect on stopping performance.

To estimate the effect of vehicle mass on stopping distances we can look at an example which can be analyzed using elementary techniques. We will choose a sample vehicle and tire and calculate the braking acceleration. A free-body diagram of the vehicle is shown in Fig. 2. Note that for a four-wheeled vehicle the forces B_i and F_i ($i = 1,2$) applied to the tires are each distributed over two wheels. Then $B_i = 2B_i'$ and $F_i = 2F_i'$ where B_i' and F_i' are the forces on one tire.

The acceleration will be horizontal and is given by

$$B_1 + B_2 = ma \qquad (1)$$

The net vertical force will be zero, so that

$$F_1 + F_2 = mg \qquad (2)$$

Once a steady state condition is reached (how long this takes depends on the detailed design of the vehicle) there will be no net torque about an axis through the center of mass,

$$F_1 d = (B_1 + B_2)h + F_2(L - d) \qquad (3)$$

To complete the description we need the relationship between the frictional forces and the vertical loads. If we made the simplest assumption, that $B = \mu F$ and that μ is constant, we would obtain the usual result that $a = \mu g$, independent of load and dimensions.

A more realistic approximate solution can be obtained by using a linear approximation to the load curve in Fig. 1(c). We can write the equation of such a line in slope-intercept form as

$$\mu = \mu_o - AF$$

where μ_o is the coefficient of friction extrapolated to zero load and A is a positive constant. The frictional forces are then given by

$$B_1 = (\mu_o - AF_1)F_1 \qquad (4)$$

and

$$B_2 = (\mu_o - AF_2)F_2 \qquad (5)$$

In this approximation we can solve the numbered equations simultaneously for the acceleration. Combining Eqs. (1), (2) and (3) yields

$$F_2 L = mgd - mah \qquad (6)$$

while Eqs. (1), (2), (4), and (5) give

$$(mg - F_2)^2 + F_2^2 = (\mu_o mg - ma)/A \qquad (7)$$

Substituting F_2 from Eq. (6) into Eq. (7) finally yields (after a great deal of algebra)

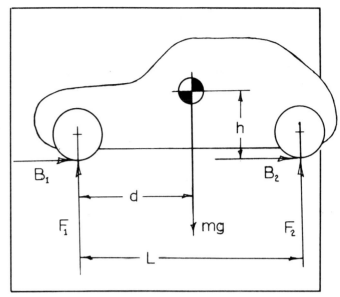

Fig. 2. Free-body diagram for a vehicle under braking showing relevant dimensions.

This form of the solution emphasizes the way in which the braking acceleration depends on the dimensions of the vehicle since the individual wheel loads change as shown by Eq. (6).

The data shown in Fig. 1(c) were obtained for a constant slip near the peak of the slip curve, Fig. 16. Using a different value of slip would shift the load curve vertically without appreciable change of shape. The values shown are reasonably approximated by using $\mu_o = 1$ and $A' = 0.4/8000$ N $= 5 \times 10^{-5}$ N^{-1} as the intercept and slope. These values give

$$B_i' = (1 - 5 \times 10^{-5} F_i') F_i'$$

as the one-wheel frictional forces. The two-wheel frictional forces are then given by

$$B_i = 2B_i' = 2[1 - 5 \times 10^{-5}(F_i/2)] F_i/2$$

or

$$B_i = (1 - 2.5 \times 10^{-5} F_i) F_i$$

so that we will use $A = 2.5 \times 10^{-5}$ N^{-1} in Eq. (8) for four-wheeled vehicles.

To demonstrate the effect of this tire characteristic on braking acceleration consider a car with a wheelbase $L = 3$ m, even static weight distribution, $d = 1.5$ m, with its center of mass at height $h = 1$ m above ground level, and a total weight of 8000 N. Equation (8) predicts an acceleration of 0.87 g for this car. If we simply add 4000 N to the weight of this car without changing the dimensions or the tires the predicted acceleration drops to 0.81 g which would give a 7% greater stopping distance for the heavily loaded car. This is certainly a significant difference to

$$\frac{a}{g} = \frac{-1}{2mgA}\left(\frac{L}{h}\right)\left\{\frac{1}{2}\left(\frac{L}{h}\right) + \frac{(L-2d)}{L} mgA \pm \sqrt{-(mgA)^2 + mgA\left[2\mu_o + \frac{(L-2d)}{h}\right] + \frac{1}{4}\left(\frac{L}{h}\right)^2}\right\} \qquad (8)$$

anyone seriously concerned with vehicle performance. Note that for the lighter car the static wheel load is 2000 N on each wheel. This load would give a coefficient of friction of 0.90. The acceleration of the car is less than 0.90 g because of the load transfer during acceleration. The rise in rear wheel coefficient is more than compensated for by the drop in front wheel coefficient. This effect could be reduced by increasing the wheelbase or by lowering the center of mass.

It is also interesting to examine the effect of changing the dimensions while holding the weight fixed. Consider an 8000 N car with $L = 4$ m, $d = 2$ m and $h = 1$ m. The predicted acceleration of this car would be 0.88 g, slightly greater than our first example. The location of the center of mass will also have an effect. Consider moving the center of mass forward to $d = 1.5$ m. This drops the predicted acceleration to 0.85 g. This loss could be recovered by lowering the center of mass to $h = 0.5$ m which would again give an acceleration of 0.88 g. Beware of placing too much weight on these results since they are obtained on an "all other things being equal" basis. In particular, changing load distributions will change the tire temperatures. This effect may or may not be beneficial, according to where we are on the μ versus T curve.

What does this tell us about the teaching of friction in elementary physics courses? The usual simplified analysis of motion involving frictional forces does provide a useful introduction to the principal effects of such forces. Indeed, there is a fairly wide range of applications in which it is adequate for quantitative analysis. But teachers should be aware of the limitations of the model and avoid extrapolation to areas in which deviations from the simple model are significant. This does not mean that we cannot call the objects in our example problems cars, but simply that we should specify the assumptions that are involved in any model used to describe the behavior of real physical systems.

References

1. Robert D. Grimm, Phys. Teach. 16, 559 (1978).
2. State driver's manuals also often claim to give such data, but the stopping distances given are normally not experimental data. Instead they are estimates based on assumed values of acceleration which are generally rather poor compared with current experimental values.
3. Slip is defined as the difference between the angular velocity of the wheel and the angular velocity for pure rolling at that speed, normally expressed as a fraction of the rolling angular velocity.
4. Paul Van Valkenburgh, *Race Car Engineering and Mechanics* (Dodd, Mead and Co., New York, 1976) p. 10.

On the strength of butterfly wings

George W. Ficken, Jr.
Cleveland State University, Cleveland, Ohio 44115

This summer while cruising along the turnpike near Toledo, I noticed that a monarch butterfly had become jammed into the crack between the hood and fender of our car. Assuming it had been killed by the 55 mph impact, I continued onward. The air blowing past the wings, most of which protruded upward, caused them to flap so wildly that they became blurred. Occasionally in their erratic motion they beat upon the metal surfaces. I fully expected these "fragile" appendages to disintegrate rapidly from such violent activity. But three miles farther on they were still intact! Later they appeared to have diminished in size and lost much of their color, partially conforming to my dire prediction. But then they popped upward again into full view, almost as colorful as before!

We stopped in Toledo, I opened the hood, and there was the poor creature, battered but alive! It even attempted to fly on its almost-complete wings, adding to my guilt feelings for not having stopped.

I submit this report because of a suspicion that there is a paucity of wind tunnel data on butterflies in the literature.[1]

Reference

1. In a private communication with J. Goddard (a neighbor), I learned that a specimen of a different species once became lodged under his windshield wiper. He stopped, placed it carefully on the rear seat, observed a tear in its body, and continued onward. Eventually the wings were ever so carefully brought downward from their upright (stunned?) position and then slowly tested up and down. Upon receiving a "ready" signal from his recovered passenger, my friend stopped again and allowed the doughty creature to resume its flight.

A POTPOURRI OF PHYSICS TEACHING IDEAS — MECHANICS

The Answer is Obvious. Isn't It?

By Robert Beck Clark
Department of Physics, Texas A&M University, College Station, TX 77843

In the early 1960's I had the opportunity to spend a delightful year in Seattle at the University of Washington. By the luck of the draw, I was given the responsibility of supervising a small army of fellow Teaching Assistant's (TA) who were assigned as graders for the introductory physics course for engineers which had hundreds of students enrolled.

For the second examination, the professor of one of the large sections included a short question at the beginning of the test which said something like:

> One often sees, in the comic strips, a character holding a fan up to the sail to propel a boat forward.
> This is forbidden by _____.

The students had recently covered Newton's laws of mechanics and almost all of them responded with the stock (and expected) answer of Newton's third law or action equals reaction or conservation of momentum.

During the grading, I was approached by the thoughtful TA who was responsible for grading this particular question. He explained that he had graded over a hundred examinations and that almost all of the students had given the standard answer. However, one student had claimed that the boat could indeed be propelled forward in this fashion and had given a simple explanation in which the TA could find no errors. To make matters worse, the student had apparently spent most of the hour on this question and neglected the remainder of the examination, even though this particular question was worth only five points.

Soon the room was full of TA's assembled around the blackboard trying to find a flaw in the student's reasoning. Finally in desperation we decided we would have to approach the professor with the problem. Certainly, he would be able to find the flaw we were missing in what was obviously an incorrect answer. I was picked as the reluctant spokesman. His first response was similar to ours, but after 20 minutes at the blackboard he concluded that the student was actually correct. The happy end to the story was that the student was rewarded with an excellent grade on the examination for his effort and a group of TA's learned several important lessons.

Oh, you would like to see the student's simple explanation? It went something like this. Consider an air molecule at rest in front of the fan blade. When the fan blade strikes the molecule, the molecule goes forward with linear momentum $\vec{p}$. By conservation of linear momentum, the momentum imparted to the fan blade (and boat) is $-\vec{p}$. Now we assume the sail is set at a right angle to the centerline of the boat. If the molecule experiences an elastic collision with the sail, then since the mass of the sail (and boat) is much-much greater than the mass of the molecule, the molecule will bounce off the sail with momentum $-\vec{p}$. The change in the momentum of the molecule is then $\vec{p}f - \vec{p}i = -\vec{p} - \vec{p} = -2\vec{p}$, and by the conservation of linear momentum the sail and boat gain a momentum $+2\vec{p}$. Now if we combine the momentum acquired by the boat during the collisions of the molecule with the fan and the sail, we find $\Delta \vec{p} = -\vec{p} + 2\vec{p} = +\vec{p}$ giving a net increase in the forward momentum of the boat.

If you are unhappy with the insistence of an elastic collision between the molecule and the sail, you can consider any case between that and a totally inelastic collision, in which the molecule stops and the sail acquires momentum $+\vec{p}$ giving $\Delta \vec{p} = 0$. Therefore, any case intermediate between elastic and total inelastic would result in a reduced, but finite forward motion.

How would you grade this student? How would Sir Issac Newton have graded this student? How would Daniel Bernoulli have graded this student? How would Al Bartlett have graded this student? What would happen if an experiment were actually done?

In retrospect, I think my own hang-up was with my intuition insisting that the momentum of the fan, sail, and boat must be conserved, while failing to consider the important role of the air in the process. This means that the fan, sail, and boat system is not isolated. I also remember being somewhat relieved when I realized that the student's restrictive assumption that the molecule originally be at rest, could be generalized to the more realistic case in which a finite volume of air which has random molecular motions, but zero total linear momentum, could be considered in place of the molecule at rest, without altering his basic argument. It is also interesting to consider the related case in which the sail is set at 45° and the fan is placed so that it blows along a direction perpendicular to the centerline of the boat. One's intuition does not seem to object nearly as strenuously to the prospect of the boat being propelled forward under these conditions.

It was also recently noted by the author that the "clam-shell" thrust reversers, which are employed on the Boeing 737, to break the plane on landing, exhibit an anlaysis behavior in which the engine plays the role of the fan and the "clam-shell" the sail. □

In closing, I wish to thank Professor James H. Earle of the Texas A & M University, Department of Engineering, Design Graphics, for sharing his time and talents in creating the amusing illustration which accompanies this article.

A POTPOURRI OF PHYSICS TEACHING IDEAS—MECHANICS

LETTERS

Editor's Note: All our letters this month are in response to a note by Robert Beck Clark. "The Answer is Obvious. Isn't It?" appeared in the January 1986 issue of TPT, **24***, 38–39.*

Newton's Third Survives

Robert Clark, in his article, "The Answer is Obvious. Isn't It?", solves his own problem in the penultimate paragraph of his otherwise convincing argument. The assumption is that the impinging air molecule is initially at rest (or that a finite volume of air acting on the sail has zero total linear momentum).

The assumption of zero initial momentum remains true, but once the boat gains a velocity, v, the relative velocity of the air molecule to the boat is −v. Thus, the change in momentum of the air molecule on impinging on the fan blade is 2p. This gives a momentum of −2p to the boat. The boat still gains a momentum of 2p, when the air molecule hits the sail. So, the total change in momentum of the boat (and air molecule) is zero. Thus, any motion initiated cannot be sustained and Newton's third survives.□

Pamela Lambert, *Concordia College, 7128 Ada Boulevard, Edmonton, Alberta, Canada T5B 4E4*

Fan Blades, Roadrunners, and Coyotes?

Concerning "The Answer is Obvious. Isn't It?" in the January 1986 issue, where a Texas group concluded that a fan could propel a boat forward, the issue is not settled that easily. Fan blades have two sides: For a gain of $+\vec{p}$ of momentum on the one side, there is a loss of $-\vec{p}$ on the other side; total = + $(+\vec{p}) - (-\vec{p}) = +2\vec{p}$ at the fan. Since there must be an air space between fan and sail, not all the momentum gained by the air affects the sail: Some diffuses around it, not to mention what passes through the sail, which is not usually airtight. So, the forward effect is *less* than $2\vec{p}$ in magnitude. The net force on the boat is *backward*.

Perhaps the team at Texas A & M was influenced by living in Roadrunner/Coyote country.□

Stuart L. Mills, *Head, Physics Dept., Fredericton High School, 365 Prospect St., Fredericton, N.B., Canada E3B 3B9*

Tongue in Cheek?

I don't know if the author of the article "The Answer is Obvious. Isn't It?" had his tongue in his cheek, but the answer he gives is wrong. *Some* of the air leaving the fan will strike the sail, but not all of it, and certainly not at the speed with which it leaves the fan. But there will be a force on the sail, to the left. However, a larger force, to the right, is exerted on the fan by the air which the fan is pushing to the left. So, the boat accelerates to the right until resistance forces create equilibrium with the boat moving at a constant velocity to the right.□

C.J. Millar, *Master of Studies and Senior Physics Master, The King's School, 25 The Precincts, Canterbury, Kent, England*

Obvious??

The recent note by R.B. Clark cautions us well to look *carefully* before we leap. He points out that if the first molecular collision with a sail is anything other than totally inelastic, there would be some forward momentum given to the boat (by that molecule). He seems concerned about the grade a student would receive (from us or the "masters") by proposing the possibility of propelling a boat that way, and asks "What would happen if an experiment were actually done?"

I perform such an "experiment" virtually every year (often several times) as a classroom demonstration of the "stock (and expected) Newton's third law or...conservation of momentum". The demonstration is easily done with an apparatus called a "fan cart" (or fancart) available from such supply houses as Cenco, Sargent-Welch, Frey, and most likely others. The results are always the expected (before the Clark note) namely, that the fan on the cart does NOT drive the cart (unless the sail is removed). It seems possible that the boat (cart) could be driven forward by suitably "dishing" the sail to redirect the air rearward (as Clark accomplishes with his completely elastic collision), but it is far simpler to pull the sail down and turn the fan around.□

J. Wallingford, *Professor, Pembroke State University, Pembroke, NC 28372*

The Obvious Answer Is Correct!

The answer to "The Answer is Obvious. Isn't It?" is "No, but the obvious answer is correct!" This thought-provoking article clearly shows that careful analysis is required to determine whether a fan in the back of a sailboat can propel the boat by blowing air against the sail.

Sargent-Welch's FAN CART, No. 0894K is a wheeled version of an apparatus to demonstrate this situation. It turns out that the "obvious" answer that the cart doesn't move is correct if the flat, rigid sail is oriented perpendicular to the airstream from the electric fan. (What if it were at a 45° angle?) Although other factors are certainly involved, I believe the principal reason is that the air collides essentially inelastically with the sail and moves off at right angles to it. Then if the air gave the fan a momentum −p, it gives the sail +p and the net result is zero, so the cart doesn't move. Of course if the sail is removed, the cart accelerates rapidly.

Jet engine "clam shell" thrust reversers work differently. I suspect they change the direction of the airflow by closer to 180° than 90°. This is similar to a person in the back of the boat throwing hard rubber balls forward which then bounce elastically straight backward off a hard, flat plywood sail. Since the final momentum of a ball would be −p, the boat must receive a momentum of +p, just as if the ball had been thrown to the rear in the first place. Thus, the boat would move forward.□

Carl T. Rutledge, *Assoc. Prof. of Physics, East Central University, Ada, OK 74820*

Old Solution Still Correct

The article by Robert Clark, "The Answer is Obvious. Isn't It?" was an interesting look at an old problem, but I believe there is an error in his solution.

First, let me briefly repeat Mr. Clark's analysis. A sailboat with a fan on board is at rest on a lake in still air. The fan is turned on and pointed at the sail. An air molecule gains momentum $\vec{p}$ from the fan, and the boat gains momentum $-\vec{p}$. When the air molecule hits the sail, it bounces off and imparts a momentum of 2p to the sail. The result is the sailboat gains a forward momentum of $+\vec{p}$.

So far the analysis is correct, but the problem is not yet complete. As the sailboat moves forward, it collides with a "stationary" air molecule in front of the sail. Since the boat is much more massive than the air molecule, the molecule gains momentum $\vec{p}$, and the sailboat loses momentum $\vec{p}$. The *net* result is that the sailboat has zero momentum and will not move forward.

If the fan is faced towards the stern of the boat, the air molecule gains a momentum of $-\vec{p}$, and the boat gains a momentum of $\vec{p}$. As the boat moves forward, the sail falls slack and does not transfer momentum to the still air in front of the boat. Thus, if the effect of the air is taken fully into account, the "old" solution to the problem is still correct.

James Guillory, *Physics Instructor, Henderson County Junior College, Athens, TX 75751*

The H.M.S. Newton III:
An Onboard-Fan-Powered Sail Craft

Robert Beck Clark's article, "The Answer is Obvious. Isn't It?" discusses the question of whether a sailboat could be propelled forward by a fan attached to the deck blowing air into a sail set perpendicular to the centerline of the boat. This is a popular question in elementary textbooks, such as the widely used *Fundamentals of Physics*, by Halliday and Resnick.[1] The standard answer is "no," with an explanation. In the instructor's manual for *Conceptual Physics* by Paul Hewitt,[2,3] the explanation is: "No net momentum is achieved because of impulses that tend to be equal and opposite. The impulse of the fan on the air tends to push the fan to the left. The impulse of the air against the sail tends to push the boat to the right. No net impulse means no change in momentum." Mr. Clark reported a different answer given by a student. He stated that the combined change in momentum acquired during the collisions with the air molecule resulted in a net forward momentum. In all cases we found, no experimental evidence is presented. In order to resolve these theoretical conflicts once and for all, a model was built approximating Clark's description, and the experiment was conducted.

The first attempt at testing Mr. Clark's theory consisted of mounting a 10-in. circular spinnaker sail to the chassis of a cart outfitted with a fan and rollerskate wheels, which moved briskly (in the expected direction) without the sail. (Parts are similar to the Fan Cart, Central Scientific Company catalogue part #72717-025.) The cart was carrying a Rowe 550 series DC motor powered by six C-cell batteries, spinning an 8-in. diam fan. The sail was made from a stainless steel welding rod and a sheet of plastic. It was made larger than the total span of the blades to prevent the air blown by the fan from moving forward past the sail. The cart failed to move forward on a flat surface; however, it did seem to accelerate forward when on a ramp which compensated for friction, so we were encouraged to continue.

To eliminate friction, the motor, batteries, fan, and sail were mounted on a 9.5-in. air track slider. Still, this resulted in no forward movement. It was noted that with the sail mounted at the bow, the nose of the slider was driven into the track. Moving the sail toward the middle balanced the forces acting on the slider when the fan was running. This required placing the battery pack perpendicular to the centerline of the boat immediately in front of the fan blade. This configuration was unstable and resulted in a sideways rocking motion. The battery pack was then removed completely. This resulted in reduced wobbling and a much lighter "boat." Still no success was achieved. (About this time, the professor who shared our lab began to avoid us.) We noticed that the top of the sail was vibrating rapidly. This was wasting energy, so a mast (actually a pencil) was mounted to the back of the fan motor and a line tied from it to the top of the sail. Success at last was ours! The boat moved steadily forward even when the track was tilted slightly uphill.

To unveil and christen the craft, the H.M.S. Newton III, our (reluctant) faculty sponsor, Dr. Allen Tucker, was convinced to ask the department chair for a special colloquium. Bets were placed, including a bottle of champagne vs a can of beer. The Newton III sailed forward much, to the surprise of the audience.

In conclusion, we would like to thank the Physics Department who (unknowingly) funded the project (six C-cell batteries), our faculty sponsor who supplied the champagne for the christening, and our best critic, Dr. Hamill, for the motivation to test Mr. Clark's theory.

References

1. David Halliday and Robert Resnick, *Fundamentals of Physics*, 2nd ed. (New York, NY: John Wiley & Sons, Inc., 1981), p. 146.
2. Paul G. Hewitt, *Conceptual Physics*, 5th ed. (Boston, MA: Little, Brown & Co., Ltd., 1985), p. 78.
3. Paul G. Hewitt, *Instruction Manual to Accompany Conceptual Physics*, 5th ed. (Boston, MA: Little, Brown & Co. Ltd., 1985), p. 39.

Keith Martinez and **Michael Schulkins**, *Department of Physics, San Jose State University, San Jose, CA 95192-0119*

Response

It is pleasing to learn that there were a number of individuals who took the time to consider the student's perplexing solution to the fan-driven sailboat problem reported in our original note. I was particularly delighted that there were three individuals or groups (J. Wallingford, C. Rutledge, and K. Martinez and M. Schulkins) who either had the previous experience or took the time necessary to subject this problem to an experimental investigation. Obviously, this clumsy form of propulsion is hampered by substantial effects of friction, not the least of which result in an inevitable loss of velocity for the molecules as they travel from fan to sail. It is not surprising, therefore, that the traditional fan cart setup described by Wallingford, Rutledge, and Martinez and Schulkins fails to result in observable forward propulsion.

It is most interesting that with great efforts at friction reduction Martinez and Schulkins were able to observe forward propulsion with their H.M.S. Newton III.

Finally, it may be in order to note that the original question did not focus primarily on the efficiency of this admittedly anemic form of propulsion, but rather on the question of whether or not it is *strictly forbidden* by Newton's Third Law (or momentum conservation), as may appear obvious at first glance.

P.S. A check with the local chapter of the Audubon Society revealed that a roadrunner had not been observed in Brazos County in the past 40 years. I wish we could say the same about coyotes.

Robert Beck Clark, *Department of Physics, Texas A & M University, College Station, TX 77843-4242*

A POTPOURRI OF PHYSICS TEACHING IDEAS — MECHANICS

ANOTHER NEWTON'S SAIL-BOAT*

My father, Bob Brown, in his book *200 Experiments for Boys and Girls* (to be published, Spring 1973, by World Publishing, New York) includes another demonstration of Newton's sailboat. I believe you would like to tell your readers how it is made since it is very simple:

A balloon is the air source. It and the sail are mounted separately on blocks of wood, and these blocks rest on a piece of balsa which in turn rests on dowels. The dowels must have a smooth surface on which to roll.

Pick up the balloon as air comes from it, and the air blowing against the sail moves the device one direction as wind moves a ship. Put it down, pick up the sail, and the reaction as air comes from the balloon moves the device the other way.

My father first tried using soda straws instead of dowels, but the greater friction caused by the slight flattening of the straws prevented easy operation of the device. It works beautifully on dowels.

Fig. 1. *Another Newton's Sailboat*

ROBERT J. BROWN, Jr.
Asheville, North Carolina 28806

*Steven R. Smith and Jerry D. Wilson, Phys. Teacher 10, 208 (1972): Jerry D. Wilson. Phys. Teacher 40, 300 (1972).

Sailboat Demonstration

Paul G. Hewitt,
City College of San Francisco, San Francisco, California

The sailboat provides one of the most interesting illustrations of vector resolution. A simple demonstration with which I have had considerable success in introductory physics classes for both nonscience and science majors utilizes an easy-to-build sailcart and an electric fan.

The sailcart is a small block of 2 in. × 4 in. wood mounted on four wheels. Slots are cut in the block which allow a sheet metal sail (about 1 ft sq) to be placed at various angular positions. Three slots are sufficient: one parallel to the wheel axles, one perpendicular to the axles along the keel, and the third about 20° from the keel slot. When the sail is placed in the slot parallel to the wheel axles, the cart simulates a Columbus type ship, and for forward propulsion the fan must be held at the back of the craft.

The interesting case is when the sail is positioned in the angular slot. Students are impressed when the cart is propelled forward by the fan when held in front of the cart as shown in Fig. 4. I do this on a long lecture table walking with the fan and holding it at a more-or-less constant angle to the accelerating cart. A discussion of vector resolution at this point is accompanied by increased student interest.

Discussion is stimulated when the sail is placed in the keel slot and the class is asked where the fan should be placed for forward propulsion. It is then demonstrated that no fan position will propel the sailcart, because the perpendicular force of air impact on the sail has no component along the keel.

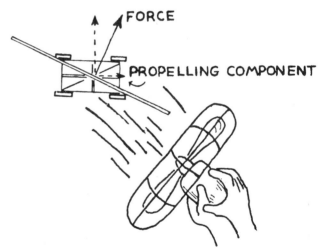

Fig. 4. Sailcart on wheels moves forward into the wind.

The demonstration is repeated later in the course when Bernoulli's principle is studied. At this time I compare cart speeds using the flat sail (Fig. 4) and a curved sail. The curved sail is an identical sheet of aluminum bent into a uniform and shallow arc. The curved sail provides a noticeable increase in cart speed when sailing into the wind. Fitting the curved sail into the straight slots is no problem if the slots are wider than the metal sheet. Taping the lower part of the flat sail insures a snug enough fit.

One of several alternatives to the slot method of sail attachment would be to have the sails mounted on pegs to fit in a hole in the cart. Such an arrangement provides any angular position desired.

The simple pendulum experiment

Robert Kern Curtis
Hackensack High School, Hackensack, New Jersey 07601

In treating the simple pendulum this year, I assigned the experiment to be done at home, before any classroom discussion of it. I thought it a good project for the Christmas holidays. In addition to the usual measurements with bobs of varying masses and pendulums of different lengths, I had instructed my students to measure and plot the period of a pendulum as a function of its amplitude. I told them to test a maximum displacement of from ten to ninety degrees, at least every ten degrees.

When classes resumed in January, I was amazed to find that *all* my physics students had written that the period was independent of the amplitude and had plotted graphs appropriate for this conclusion. I was so surprised by this that I had my ninth-grade science class do the same experiment as a group while I observed them. They seemed to have no preconceived notions about what ought to happen. At my direction, they measured the length of a pendulum from the center of the bob to the edge of a pendulum clamp and adjusted it to be as close to 0.985 m as they could get using a meter stick. Thus, I had them make a "seconds pendulum," that is one whose period for "small angles" is two seconds. They timed ten complete periods for each maximum angle with an ordinary 1/5th-second stopwatch. Their data are shown in Table I.

The expected values, also listed in Table I, were computed using the relation:

$$T = 2\pi \sqrt{\frac{l}{g}} \left(1 + \frac{1^2}{2^2} \sin^2 \frac{\theta}{2} + \frac{1^2 \cdot 3^2}{2^2 \cdot 4^2} \sin^4 \frac{\theta}{2} + \ldots \right)$$

Considerable damping occurred for very large angles, and this accounts for the difference between the experimental and expected values for the periods of eighty and ninety degrees.

I presented the contradictory data from the ninth graders' experiment to my physics students. *They* were surprised! The ninth graders' results were consistent with what I had expected, but they were inconsistent with what the juniors and seniors taking physics had expected; the physics students had let their expectations overwhelm the experimental evidence. They explained that both their physics text and their mathematics books gave the formula:

$$T = 2\pi \sqrt{\frac{l}{g}}$$

so they *knew* that, like the mass of the bob, the amplitude made no difference.

In this experiment, the students and I grew in our appreciation of the influence prejudice has on an experiment and of how vulnerable we are to this influence. Further, I saw how easily students can be misled. I strongly recommend that a measurement of period as a function of amplitude be included in the traditional simple pendulum experiment. Those interested in more sophisticated experiments should see "Experiment 1" in *Laboratory Experiments in High School Physics* by Miner and Kelly (Ginn & Co., Boston, 1967) as well as S. D. Schery in *The American Journal of Physics*, **44**, 666(1976).

TABLE I
The period of a simple pendulum as a function of its maximum angle of displacement

Maximum angle	Experimental value of the period	Expected value of the period
10 degrees	2.00 seconds	2.00 seconds
20	2.00	2.01
30	2.02	2.03
40	2.06	2.05
50	2.10	2.09
60	2.14	2.14
70	2.18	2.19
80	2.22	2.27
90	2.28	2.35

A Whipped-Cream Pendulum

Edward H. Leonard
Westfield State College
Westfield, Mass.

The impression created by many a classic demonstration can frequently be enhanced by the addition of some imaginative or dramatic embellishment. This seems to be even more the case when the student can recognize as a part of the lesson some household or commonplace device with which he is familiar. Such student comments as, "You should have seen what happened in the physics class today," not only make the instructor's efforts seem more justified but also may generate a more lasting appreciation of the fascination of the subject.

In this demonstration the Lissajous figures produced by the sand pendulum can be graphically traced out by a stream of whipped cream from any one of the many available topping dispensers. The release mechanism consists of a washer in which have been drilled two holes to accommodate two springs (available in a set assortment from either of the catalog stores at less than 50¢ a set) to the free ends of which are connected fish hooks from which the barbs have been removed. The trigger device is simply a U-shaped strip of metal which can be removed to release the springs immediately prior to setting the pendulum into motion.

This demonstration can be a spectacular lecture-terminating vehicle. Assuming that care has been used to prepare a hygienic surface, students can be asked, as they take a closer look, to help clean up the pattern for which purpose disposable, barbeque-type spoons can be supplied as students are leaving the lecture room.

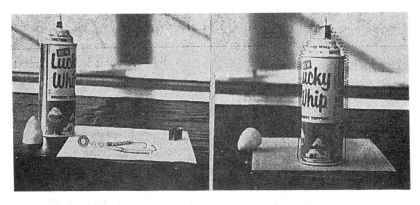

Superball problem

G. Stroink
Department of Physics, Dalhousie University, Halifax, Nova Scotia, Canada B3H 3J5

If one puts a small superball on top of a big superball[1] and then releases this combination from a height h, dramatic things happen as soon as this combination strikes the floor.

The small ball will take off and can reach a height nine times as high as h. Why?

Just after the combination hits the floor, the big ball moves up with a velocity v whereas the small ball is still moving down with a velocity v. This velocity v is given by

$$\frac{1}{2}mv^2 = mgh$$

So we have this situation:

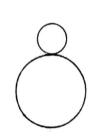

Now place yourself on the big ball and try to anticipate what happens.

For an "observer" on the big ball, the small ball is approaching you with a velocity of $2v$. To you it looks like this:

Assuming the collision is elastic, then the velocity of the small ball after the collision between big and small superballs is again $2v$, but now in the opposite direction, like this:

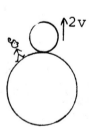

The small ball moves away from the big ball with a velocity $2v$, but don't forget the big ball moved up with a velocity v relative to the floor. So the small ball moves up with a velocity $3v$ relative to the floor:

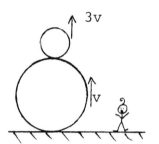

The small ball would have gone up with a velocity v if it had just hit the floor, but now it takes off with a velocity $3v$.

Because how high it goes depends on v^2 [$mgh = (\frac{1}{2}mv^2)$], a 3 times larger take-off velocity means that it goes 9 times as high.

Application

If a satellite with velocity v manages to have an "elastic collision" with a big planet with velocity V, by swinging around it,

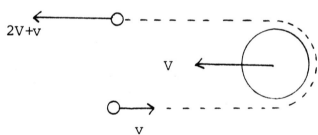

then its velocity after the collision, at the same distance away from the planet, will be $v + V$ (the relative velocity of the two objects) *plus* V, the velocity of the big object.

This illustrates the "sling shot" effect which is being employed for space probes on a "Grand Tour" of a number of the planets of the solar system.[2]

References

1. Superballs can be purchased from Wham-O Manufacturing Company, San Gabriel, CA 91700.
2. L. Epstein, "Pioneer Jupiter backlash," Phys. Teach. **11**, 299 (1973).

A POTPOURRI OF PHYSICS TEACHING IDEAS — MECHANICS

A bouncing superball — the poor man's projectile

Paul Latimer
Physics Department, Auburn University, Auburn, Alabama 36830

In the elementary treatment of projectile motion, the velocity is resolved into vertical and horizontal components which are separate examples of uniformly accelerated motion. The vertical acceleration is g, the horizontal acceleration zero. Many students find it difficult to accept the idea that the two motions are actually independent. Then the instructor may present a demonstration to make the point. While playing with a "superball," we developed a procedure which demonstrates the independence of these two motions simply and easily. ("Superballs" are made by Whamo Corporation, and are usually available at toy stores.)

The demonstrator places the ball on the palm of his hand at about head level. He allows it to roll off the hand in the forward direction, and walks along beside the ball at a constant rate. The student can see that the instructor walks with a uniform velocity; comparison with the instructor reveals that the ball must also be moving with a uniform horizontal velocity. As this constant horizontal motion continues, the ball bounces up and down many times with slightly decreasing amplitude. The vertical motion is anything but constant. The student sees that the changes in vertical motion do not influence the uniform horizontal motion; hence, the two must be independent. We have tested students by asking them what was proven or demmonstrated and how. The results have been encouraging.

We use the superball demonstration as a partial alternative to the traditional demonstration where a blow gun is shot at a target that begins to fall when the gun is fired. There, impact of the falling projectile and target is used to demonstrate that the difference in their horizontal motions does not influence the vertical motions. Available film loops and other visual aids also provide spectacular illustrations of projectile motion.

One problem in teaching physics is that the student can become so fascinated with the apparatus, methodology, or showmanship of the demonstrator that he looses sight of the physical principles being demonstrated. We do not encounter that problem with our little bouncing ball. However, it does implant the idea that the vertical and horizontal motions of the projectile are independent.

Spectacular rocket experiment

Evan Jones
Sierra College, Rocklin, California 95677
P. Peter Urone
*California State University
Sacramento, California 95819*

In the book, *Daybreak*,[1] Joan Baez describes what her famous physicist father called the "demonstration of the century." During a lecture at a summer course at Harvard University, Dr. Baez sat astride a carbon dioxide fire extinguisher mounted on a little red wagon and proceeded to propel himself around the room. Our students react to the demonstration (Fig. 1) as did those of Dr. Baez: with awe followed by applause and shouts of approval.

This memorable demonstration can be developed into an exciting laboratory experience to dramatically illustrate a wide spectrum of principles. It can be used to exemplify Newton's second and third laws, impulse, variable mass systems, the kinetic theory of gases, and many other concepts. For example, the impulse and momentum considerations, applied to a variable mass system, can be used to obtain the exhaust gas velocity and the average rocket thrust.[2] If v is the gas velocity relative to the rocket, m is initial rocket mass, Δm is the mass of released gas, ΔV is the increase in rocket velocity, F is the backward friction

Fig. 1. Rocket and driver ready for "demonstration of the century."

force of floor against rocket, and Δt is the time of run, then:

$$v \cong - \left[\frac{m(\Delta V) + F(\Delta t)}{\Delta m} \right]$$

and thrust $\cong - \left[v \frac{(\Delta m)}{\Delta t} \right]$.

Calculation of the kinetic energy imparted to the rocket and to the gas reveals the fact that the gas receives by far the most energy, despite both gas and rocket having received an impulse of the same magnitude.

The obvious cooling of the exit gas can lead to a discussion of adiabatic processes, and even to the mechanisms by which clouds are formed in the atmosphere. The kinetic theory of gases can be introduced to determine the root-mean-square molecular speed of the exit gas.[3] The acoustic speed may be calculated for comparison with the experimentally-obtained exit gas velocity.[4] The exit velocity (210 m/sec) compares well with the calculated acoustic velocity of CO_2 (217 m/sec) at 195 K (sublimation temperature at atmospheric pressure).

To perform the experiment we first remove the exhaust horn from a ten-pound-charge fire extinguisher, rest the cylinder in the wagon, and bolt the exit nozzle firmly(!) to the wagon. As the driver opens the valve, the rocket accelerates through a measured distance (usually 10 m). Final rocket velocity may be measured with a photocell gate or with a stopwatch. Mass of released gas is determined by weighing the gas cylinder before and after the run. Frictional force is obtained by pulling the loaded rocket with a spring scale, at constant velocity.

For comparison, the rocket thrust on the gas can be measured directly with a force transducer (Fig. 2). The rocket nozzle is clamped to an aluminum arm to which a pair of tension springs is attached, as shown. One end of the arm pivots on a hinge, and the other end activates a linear potentiometer (e.g., Pacific Scientific Co., RP04-0101-1) as the force at the nozzle changes the equilibrium position of the arm. The linear potentiometer is wired as a voltage

Fig. 2. Transducer for measuring nozzle force, showing pivoted arm (1), with springs attached, linear potentiometer (2), and recorder (3).

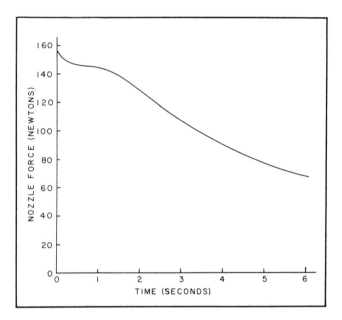

Fig. 3. Exit nozzle force vs time for carbon-dioxide rocket. Data obtained from static firing, using force transducer.

divider, and operates a recording potentiometer. This transducer is easy to build, safe to operate, and has fast, accurate response.

Figure 3 shows the force-time relationship obtained with the transducer. The area under this curve can be compared to the total impulse on the gas, as calculated for the 10-m run.

Typical data and results are shown below:

Distance:	10 m
Time:	3.5 sec
Final rocket velocity:	3.8 m/sec
Initial rocket/driver mass:	81 kg
Gas released:	1.9 kg
Rolling friction:	22 N
Average exit velocity:	210 m/sec
Total impulse on gas (experimental):	390 N-sec
Total impulse on gas (from Fig. 1):	450 N-sec
Average Rocket power:	127. W
Average Gas power:	9700. W

Any gas under high pressure demands careful handling. The valve and hose should be protected from damage, and the hose should be secured to the wagon at several points. Wagon speed should never exceed 4 m/sec. Blast off!

References

1. J. Baez, *Daybreak* (Avon, New York, 1966), p. 63.
2. R. Resnick and D. Halliday, *Physics Part I* (Wiley, New York, 1962), pp. 198-203.
3. *Ibid*, pp. 577-578.
4. *Ibid*, p. 587

Rocket Engine Analog

Robert F. Coutts
Van Nuys High School
Van Nuys, California 91409

Slightly modified PSSC lab carts can be used to illustrate Newton's third law in the operation of a rocket. The newer carts with metal loops at both ends should be used. (See Fig. 6.)

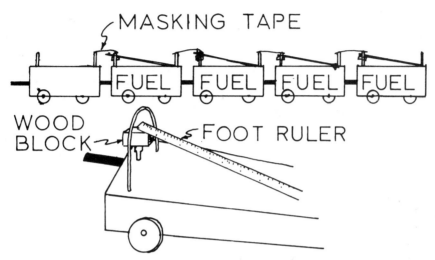

Fig. 6. Starting at the right, striking the rulers in succession simulates operation of a rocket engine.

About six carts should be prepared by taping a foot ruler with a small block of wood along the top of the cart with the block resting on top of the release button. The plunger is locked in position so that when the ruler is struck, the plunger is released. The carts are all carefully aligned and then connected with masking tape, which is pressed onto the tops of the metal loops. This keeps them together during the "launch" while not robbing the springs of their explosive force.

To fire the rocket, the instructor uses a meter stick to strike each ruler in succession till the last (front) cart is released. It is advisable to stand to the front of the "rocket" at the beginning of the "launch" so that you will not have to chase it. Also, it pays to practice a few times before demonstrating.

The "exhaust velocity" of the rear cart can be determined using a bell clapper timer and tape. The stronger spring provides an exhaust velocity of about 100 cm/sec and the weaker 35-40 cm/sec. If the assumptions are made that all the masses are equal and all the springs have the same force constant, the velocity of the last cart (payload) can be calculated using the conservation of momentum. This may be measured by flash photography for a check.

The demonstration generates considerable interest while illustrating good physics.

Snowball Fighting: A Study in Projectile Motion

Peter N. Henriksen
The University of Akron, Akron, Ohio 44325

The objectives of most introductory physics courses are to present and formulate the fundamental principles and ideas of physics, and relate these principles and ideas to the student's environment and academic discipline. The latter objective is usually approached via examples and problems. No matter how trivial these examples, if they stimulate the student to participate in and do experiments in physics, then they have served the purpose of giving the student better insight into the subject at hand.

As an example, in projectile motion many of the problems and examples in text books have to do with gunnery which is presently a very unpopular topic and one in which the student has had little experience. An alert teacher can find supplementary problems and examples which are appropriate to his particular group of students and their environment. An example of projectile motion and gunnery is found in snowball fighting which is applicable in regions where there is occasional snowfall. Such a problem involves one person trying to hit another with a snowball, and maximizing the probability of a hit if one has a basic understanding of projectile motion.

Assuming that one throws with a constant initial velocity v_0, for any range R less than the maximum range, the angle of projection is double valued, i.e.,

$$\theta_0 = \tfrac{1}{2} \sin^{-1}(Rg_2/v_0)$$

which has solutions

$$\theta_1 = \theta_0$$
$$\theta_2 = \pi/2 - \theta_0,$$

and is the common result[1] that there are always two trajectories which have the same range but different times of flight. The difference in times of flight being

$$\Delta t = t_2 - t_1 = 2v_0/g\,[\sin \theta_2 - \sin \theta_1]$$

The strategy in snowball fighting is based on this result and proceeds as follows: first one throws a high ball which has the greatest time of flight, (the main purpose of this throw is to attract the attention of the target person), then after the time interval Δt, and while the target person is distracted by the first ball, one throws the second ball along the lower trajectory. Hence, the thrower has two projectiles hitting the target area simultaneously while the target person is only aware of one of these, thus increasing the probability of a hit. After the target person becomes aware of the strategy, he still has to cope with the problem of two snowballs arriving at the same time.

Assuming that the range is 100 ft. and $v_0 = 72$ ft/sec, then $\theta_1 = 20°, \theta_2 = 70°$ and $\Delta t \simeq 3$ sec, which is ample time to create a diversion.

Obviously this example has seasonal and geographical applications, however, it serves the purpose of pointing out that if one provides examples applicable to the student environment and which produce student involvement, then a learning stimulus is provided for the student to proceed to the more abstract examples frequently used in introductory text books.

Reference

1. Sears, F.W. and F.W. Zemansky; *University Physics*, Fourth Edition, Addison-Wesley Publishing Co., Inc., Reading, Mass.

A POTPOURRI OF PHYSICS TEACHING IDEAS—MECHANICS

PREDICTING THE RANGE OF A SPRING CANNON
William E. Cooper

Geneva Community High School, Geneva, Illinois 60134

This experiment is a variation of the sling shot demonstration found in the *Teacher's Guide,* Part 3, Demonstration 34, of Harvard Project Physics. The Project Physics uses two stretched rubber bands to hurl a folded lead strip through the air. This apparatus consists of a ramp inclined at a 45° angle and a 5–6 in. spring which is launched from the ramp. As shown below, the spring is stretched 20–30 cm and then released, so that the potential energy stored in the spring is converted into kinetic energy of the spring itself, as it becomes the projectile.

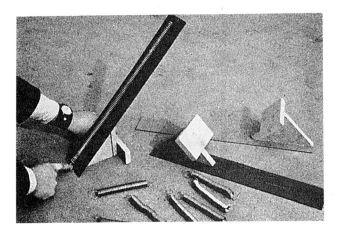

I constructed a dozen launching ramps from odd scraps of masonite and ¾ in. pine, glueing and tacking all the assemblies together in about an hour. Make sure the ramps are at least 22 in. long so there is plenty of room to stretch the springs enough to give them a maximum range of 7 m. Most of the springs we use are segments of the spring used with the slinky in PSSC experiment II-6.

My physics classes perform this experiment during the study of different forms of mechanical energy (Chap. 24, PSSC). As an introduction to the experiment, I give each student a copy of the derivation of the range equation, $R = kx^2/gm$ (for 45°) towards the end of the period on the day before the lab and then immediately ask the class this question: "Without a single practice shot, do you suppose you could fire a small cannon and hit the target at which you were aiming?" While asking this question, I am at the board sketching a cannon with its barrel inclined at a 45° angle, showing the trajectory of the cannon ball. A discussion of the question primes the class for the lab session the next day. (See below for the derivation of R that I give to the students.)

In the lab, the students use a pan balance to obtain their spring's mass and plot a graph of F vs x for the spring to obtain the spring constant from the slope of the graph. Next, they crank these values into the range equation along with the distance, x, that they decide to stretch their spring to fire it. Finally, they measure off the predicted range and place a cardboard box, with one side cut out, at the predicted impact point (Fig. 2).

Most of the students get "on target hits" the first time, and are really pleased with their results.

Fig. 2. Students measure off the predicted range.

Range Equation

R --- range --- the horizontal distance traveled by the spring

t_n --- the time the spring takes to go up and to come back down (total flight time)

$R = v_\rightarrow t_n$

$2t_\uparrow = t_n$

$R = 2 v_\rightarrow t_\uparrow$

$t_\uparrow$ --- the time the spring takes to travel to the highest point in its trajectory

$t_\downarrow$ --- the time it takes the spring to travel from the highest point in its trajectory back to the ground ($t_\uparrow = t_\downarrow$)

g --- the acceleration of gravity (9.8 m/sec^2)

m --- the mass of the spring (kilograms)

v --- the speed of the spring at the instant it leaves the launching ramp (meters/sec)

$v_\uparrow$ -- the speed of the spring in the vertical direction at the instant it leaves the ramp (vertical component of v)

$v_\rightarrow$ -- the speed of the spring in the horizontal direction (horizontal component of v)
$v_\rightarrow$ is constant throughout the spring's flight

thus $v^2 = v_\rightarrow^2 + v_\uparrow^2$ but $v_\uparrow = v_\rightarrow$
$v^2 = 2v_\uparrow^2 = 2v_\rightarrow^2$

(*) or $\dfrac{v}{\sqrt{2}} = v_\uparrow = v_\rightarrow$

k --- the spring constant of the spring

x --- the distance the spring is stretched in order to fire it

F_k -- the kinetic energy of the spring at the instant it leaves the ramp (cannon)

U_s -- the potential energy of the spring just before it is fired

$v_\uparrow = gt_\uparrow$ $\dfrac{v_\uparrow}{g} = t_\uparrow$

$R = \dfrac{2v_\rightarrow v_\uparrow}{g}$

using (*)

$R = \dfrac{2\, v\, v}{g\sqrt{2}\sqrt{2}} = \dfrac{v^2}{g}$

$F_k = \tfrac{1}{2}mv^2$

$\dfrac{2F_k}{m} = v^2$

$R = \dfrac{2F_k}{gm}$

$F_k = U_s = \tfrac{1}{2}kx^2$

$R = \dfrac{kx^2}{gm}$

RANGE OF A DART GUN

Analyzing the independent components of distance and velocity in projectile motion is a concept that frequently presents difficulty to the student. With some modifications to a demonstration described by Eich in this column,[1] one can develop safe, inexpensive experiments for laboratory use with a spring action dart pistol.

A satisfactory dart pistol may be purchased at most toy stores for little more than one dollar. It should be securely mounted with a buret or condenser clamp to a heavy duty ringstand and leveled (Fig. 8). Several toy guns may be mounted on a laboratory table creating a "shooting gallery" effect. A metal ball bearing should be taped to the suction cup of the plastic dart to reduce the muzzle velocity. Excellent results may be obtained using a 1.9-cm metal sphere having a mass of approximately 28 g. It is important to maintain the same weight distribution each time the dart is fired, so a pencil mark should be made on the taped end. This mark should be in the same position each time the dart is inserted into the gun.

Four sheets of plain paper should be taped together to make a single large sheet. Do the same with four sheets of carbon paper. Mount the dart pistol parallel to the floor and fire the weighted dart while one observer notes the approximate point where the dart strikes the floor. Tape the plain paper on the floor at this point and cover it with carbon paper, carbon side down. Shoot the projectile at least six times. Measure the vertical distance (S_y) from the floor to the muzzle of the gun. Since the only significant force acting on the projectile after firing is its weight, the time of free fall from the end of the muzzle to the floor can be computed from:

$$S_y = \tfrac{1}{2}gt^2. \qquad (1)$$

Solving for t in Eq. (1) yields:

$$t = \sqrt{2S_y/g}. \qquad (2)$$

For the horizontal displacement (S_x) measure the horizontal distance from the end of the pistol to the mean position of all the points made on the plain paper by the weighted dart. The lack of any horizontal force (neglecting air resistance) after firing makes it easy to calculate the muzzle velocity (V_x) using:

$$V_x = S_x/t, \qquad (3)$$

and the value of t obtained from Eq. (2).

The results of the muzzle velocity (V_x) computed with the dart pistol mounted in the horizontal position may be compared to the muzzle velocity (V_{vert}) of the dart pistol mounted in a vertical position according to the following procedure:

Mount the dart gun so that it points straight up. Place the apparatus on the floor and pull the trigger. Another observer should determine the maximum height (S_{vert}) reached by the projectile from the muzzle of the gun. This may be accomplished with several meter sticks taped together or with a marked tape on a wall or on a piece of wood molding. Use the maximum height measurement to compute the time it takes for the projectile to rise vertically.

$$t = \sqrt{2S_{\text{vert}}/g}. \qquad (4)$$

Using this value of time, calculate the muzzle velocity (V_{vert}) from:

$$V_{\text{vert}} = gt. \qquad (5)$$

A stop watch might be used to determine the vertical component of time by halving the projectile's round trip time to the muzzle of the gun; it is difficult, however, for a student holding a stop watch to make this measurement.

Students will find that the final results of muzzle velocity determined with the gun in the horizontal and vertical positions show close agreement.

Once the muzzle velocity of the dart gun has been determined, students may follow up the experiment by predicting the range for a given elevation angle. Place the apparatus on the floor. A protractor may be used to measure the elevation angle; however, better results are obtained with a square of cardboard ruled with the desired angle and held behind the apparatus. Compute the range from:

$$R = 2V^2 \sin\theta \cos\theta/g, \qquad (6)$$

Where V is the average of the two determinations of muzzle velocity with the gun in the horizontal and vertical positions and θ is the angle of elevation.

On the floor measure and mark the approximate range from the muzzle of the pistol. Pull the trigger. Keep in mind that the actual range is measured from the position of the muzzle to a point at the same height at the other end of the trajectory. In spite of the built-in errors, students find the results amazingly close to the theoretical predictions.

The total involvement and enthusiasm of the students makes this variation of the determination of muzzle velocity a most rewarding laboratory exercise.

EDWIN PAUL HEIDEMAN
Pleasantville High School
Pleasantville, New York 10570

Reference

1. Alfred M. Eich, Jr., Phys. Teach. 7, 116 (1969).

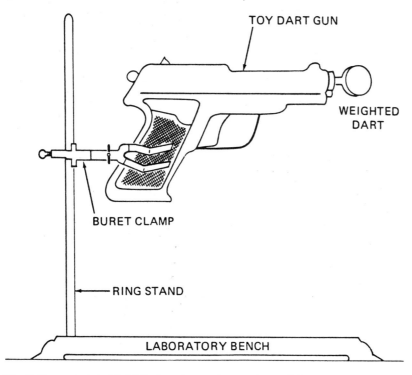

Fig. 8. *Dart gun is held rigidly by a buret clamp prior to firing.*

THE HUNTER AND THE MONKEY

It is quite easy to set up this popular demonstration with the Project Physics Universal Power Supply (Holt No. 03-074900-x). The projectile leaving the gun interrupts a light beam on a photo resistor. The darkened resistor blocks the current to an electromagnet and the monkey falls.

For the gun I used a 2-ft piece of ½-in. copper tubing bore-sighted on the suspended monkey. Attached to the breech end was a length of rubber tubing. A sharp puff by the hunter gives the projectiles a satisfactory range.

The projectile was a 5-in. length of wooden dowel which fitted quite loosely in the gun. This long projectile keeps the resistor dark until the magnet releases the monkey.

The photo resistor (Canada #NSL364C), which has a dark resistance of 50MΩ, drops the current sufficiently to activate the transistor switch in the power supply.

The photo resistor is connected between the positive terminal of the fixed 6-V dc output and the positive terminal of the transistor switch input of the power supply. The magnet is then connected across both terminals of the transistor switch output.

The magnet I used was a doorbell coil with the vibrator temporarily jammed to make continuous contact.

The photo resistor was illuminated with a 6-V lamp from a Millikan Apparatus.

Fig. 3. Monkey, held by an electromagnet, is released when the projectile leaves the muzzle of the gun and breaks a light beam.

The current to the magnet can be controlled by adjusting the distance between the light source and the photo resistor or by aiming it slightly to one side.

The magnet should be just strong enough to support the monkey. If the magnet is too strong it will recapture the monkey before it falls very far.

The photo resistor can be made more sensitive by slipping a piece of rubber tubing over it to shield it from the room light.

Physics can be fun!

JAMES MOORE
Canton High School
Canton, Massachusetts 02021

MONKEY AND HUNTER IN SLOW MOTION

The "Monkey and Hunter" demonstration is a favorite for most physics teachers. Here's one you can set up on the spur of the moment.

Place physics books under two legs of a table to produce a small incline (Fig. 6). Use a 1-m length of corner molding with a V groove as your gun. Hold the molding up a slight angle from the table and aim at the stationary "monkey" (a steel ball held by a student). Allow your steel ball bullet to roll down the ramp barrel toward the monkey. When the bullet reaches the table (bang!), the monkey sees the bullet and falls from his tree (the student lets go of his steel ball). The monkey will fall straight down as the bullet follows a parabolic path for a hit. The total time for the event to occur is relatively long, 1 to 2 sec, because of "diluted" gravitational acceleration along the slight incline. Therefore, all students can *easily observe* the entire event.

You can even "fire" upwards as

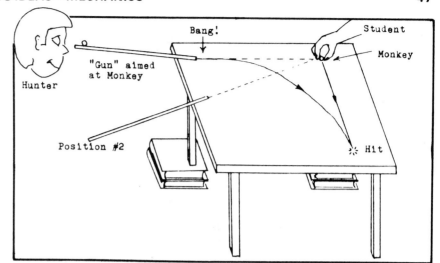

Fig. 6. Tilted table permits a slow motion observation of the monkey and hunter demonstration.

shown in position 2, for a score. By the way, you might miss. Don't worry. This will initiate a tremendous amount of discussion from your class. But don't let them make a monkey out of you.

WILLIAM P. BROWN
Irondequoit High School
Rochester, New York 14617

CAPTURING THE PROJECTILE IN THE MONKEY AND GUN DEMONSTRATION

Kenneth Wright

Central Michigan University, Mt. Pleasant, Michigan 48858

In "Three Demonstrations of Projectile Motion" in the September 1967 issue of *The Physics Teacher*, William F. Poole, Jr. suggests a variation of the Monkey and the Gun experiment. We, too, have often performed this demonstration using the traditional apparatus with the gun about 2 ft long with a metal contact at the end connected in series with batteries and an electromagnet. As the little ball leaves the gun, the circuit is broken and the electromagnet releases the can.

It is, however, often difficult to know whether the ball actually hits the can; and if it does, considerable time may be required to find the ball. This difficulty is easily overcome. Use a can about the size of a beer can. Cut out the front of the can; wrap the can rather tightly in paper. (The paper may be taped around the

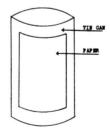

Fig. 1. The target.

can.) If the projectile hits the paper, the paper will be torn and the ball will go into the can. You have captured the ball. There will be no question whether the hit was made.

The water drop parabola

Billy Tolar
Jefferson High School, Port Arthur, Texas 77640

The water drop parabola is an old demonstration experiment[1] which never fails to excite student interest.

Setting up the experiment

In order to study the parabola unhindered by variations in water pressure, a pneumatic trough is mounted about 1 m above a 10-ft section of gutter. Three hoses are hooked to the trough: one from the faucet to fill it; one from the bottom to the glass part of an ordinary eyedropper; and one from the side of the pan to the sink, to act as an overflow hose. The faucet is adjusted so that water is flowing through the overflow whether or not water is flowing through the eyedropper. The end of the gutter under the dropper is on the edge of the sink and the other end is capped. The eyedropper is wired to an endless screw[2] of the type known as a wormgear, which is turned to vary the initial angle of the water. Two clamps are attached to the hose leading to the eyedropper. One is a pinch clamp used to cut off the flow of water. The other is a small screw clamp to control the flow, and therefore the initial speed of the water. The hose leading to the eyedropper is placed between the reed and electromagnet of a 60-Hz vibrator[3] securely clamped to the table. A strobe light is available nearby.

Doing the experiment

With the water coming out of the eyedropper but without using the vibrator, the stream of water moves in an almost parabolic arc, since air resistance is small. By varying the angle of the eyedropper and keeping the initial speed of the stream constant, it is easy to show that maximum horizontal distance is obtained for an angle of about 45° from the eyedropper to the horizontal.

Many calculations involving the stream of water are possible. By measuring the total horizontal distance and maximum vertical distance traveled by the water, it is possible to compute the initial velocity of the stream, which can be checked by a calculation using the measured angle of elevation of the eyedropper. The initial speed of the stream can also be determined by measuring the time for the stream to fill a 100-ml graduated cylinder, and the diameter of the stream, using a 10-X graduated loupe, and then making the necessary calculations. These different methods of determining the initial velocity usually agree to within ±10%.

The vibrator is then turned on causing the stream of water to break up into regularly-timed drops which follow almost the same trajectory. This has the dramatic effect of reducing the many streams formed when the stream breaks up randomly into drops, to a single, apparently solid stream. Since the water pressure is almost perfectly constant, the parabola formed by the vibrator looks from a distance like a glass arc.

Then the strobe light is turned on and adjusted to 60 Hz. The following effects are best seen from a distance

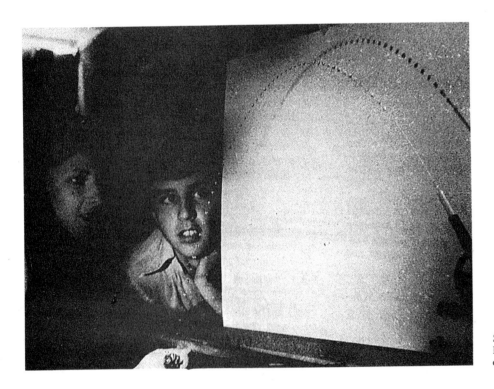

Students David Fouts and Tessa Romero observe a water drop parabola. Photo was taken by Tony Pitts.

of not more than two or three meters.

The drops appear to hang almost motionless in the air and can be studied in detail. It is immediately noticeable that the space between the drops is greater at the ends of the parabola than at its high point, indicating a greater velocity at the ends. The number of drops between two points may be counted and multiplied by 1/120 sec, giving the time interval between the two points. In short, almost any measurement that can be made on a multiflash photograph, can be made with the stationary drops.

Equally interesting are the effects produced by making the frequency of the strobe light slightly faster or slower than 60 Hz, thus making the drops appear to move backward or forward. Since the drops can be made to move forward as slowly as one wishes, the formation of the drops can be studied in great detail. Sometimes as the vibrator is turned on, but before the strobe light is turned on, two streams are formed rather than one. When the strobe is turned on it can be seen that the two streams are formed as a large drop is bumped from behind by a slightly faster moving small drop. The small drop veers off to one side and the large drop veers off slightly to the other side — conservation of momentum in mid-air! As the drops are run backward, students are delighted to see drops being reformed out of the water in the gutter and moving back into the eyedropper, apparently violating the second law of thermodynamics. It is also interesting to note that the drops formed are not round and that they tumble as they fall.

How the picture was taken

The photograph (Fig. 1 and cover) was taken, developed, and printed by Tony Pitts, a physics student, who is also one of our school photographers. Tony dodged certain parts of the picture to improve the contrast.

The picture is a one-second exposure at f/5.6 on Tri-X film taken with a Minolta SRT-102 without flash but with the strobe light blinking. The students observing the parabola are David Fouts, a physics student, and Tessa Romero who stopped by to see the experiment.

We plan to blow the picture up to poster size and place it in our main foyer with a caption advertising our physics classes.

References
1. Harry F. Meiners, *Physics Demonstration Experiments* (Ronald Press Co., New York, 1970), Vol. I, pp. 123-125.
2. Sargent-Welch Scientific Co. 0795.
3. Central Scientific Co. 85102.

WONDER ANESTHETIZED

I enjoyed the cover story, "The Water Drop Parabola" in the May issue of *The Physics Teacher*. Once students have learned about the parabolic trajectories that are so beautifully demonstrated in that article they should be encouraged to learn some of the marvelous physics that is involved when a cylindrical stream of water breaks up into a regular series of droplets. Every school library should have a copy of *Soap Bubbles; their Colours and the Forces that Mould Them* by C. V. Boys.[1] This is an old (almost ancient) standard work but it is beautiful for the elegant simplicity with which it deals with problems relating to surface tension and with many phenomena that are shaped by the forces of surface tension. The book has sections on "Soap Films, their tension and curvature" and "Liquid Cylinders and Jets" which explain how a stream of liquid breaks into a chain of droplets.

Even though interesting contemporary problems arise because of surface tension[2] the physics of surface tension phenomena does not appear as frequently in introductory physics texts as once was the case. This physics has been replaced by elegant and abstract analysis of "more fundamental" topics. I fear that in this trend, wonder and excitement are often anesthetized and that curiosity and understanding may thus be lost.

Albert A. Bartlett, *University of Colorado at Boulder, Boulder, Colorado 80309*

References

1. Reprinted by Dover Publications, Inc., New York City, 1959.
2. H. P. Burstyn, A. A. Bartlett, Am. J. Phys. 43, 1975, 414.

String and sticky tape experiments

Section Editor: **R. D. Edge**
University of South Carolina, Columbia, South Carolina 29208

The spin on baseballs or golfballs

The experiments we remember best tend to be those which are the most fun, and this month is devoted to one or two, induced by watching an evening of demonstration experiments organized by Rae Carpenter at the last Southeastern Section A.P.S. meeting. Dick Minnix of Virginia Military Institute gave some delightful demonstrations of Bernoulli's principle: People always argue over the effects of putting spin on a baseball or golf ball.

Dick had a cardboard tube cut, as shown in Fig 1, to fit a light Styrofoam ball about two inches in diameter, of the sort available in the dime stores around Christmas time. The inside of the tube was lined with sandpaper. To use this device the ball is thrown at the audience so that it rolls along the tube as it is projected outward (Fig. 2). If

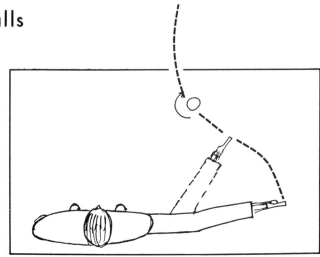

Fig. 2. Seen from above.

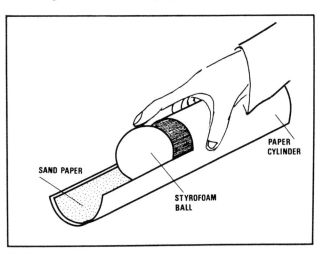

Fig. 1.

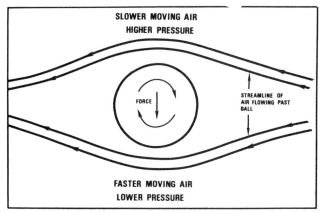

Fig. 3. Viewed from above.

it spins about a vertical axis, it will curve as shown in the figure. The physical principle involved is shown in Fig. 3 — the flow of air past the ball is encouraged by the rotation of the rough ball on one side, and discouraged on the other. Bernoulli's principle tells us that the pressure is lowest where the air speed is highest, and the ball will tend to move in this direction. The ball may be made to curve right or left, up or down. Other objects which may be made to spin nicely to demonstrate the same effects are a toilet roll tube, or two Styrofoam cups taped with their bases together. To launch these objects, three or four rubber bands are joined together to form one long piece and then wrapped several times around the object while stretched tightly. The tube may then be fired in a sling-shot manner by grasping it in one hand, the end of the rubber band in the other hand, moving your hands apart to stretch the rubber band, and then releasing the tube. The spin rate of the tube may be altered by changing the amount of stretch placed in the rubber band as it is wrapped around the tube. With ample stretch, it is possible to make a light toilet roll tube do a

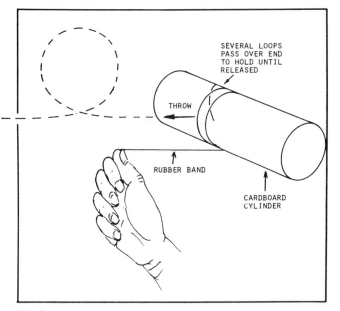

Fig. 4.

complete loop the loop as shown in Fig. 4.

Another simple experiment was sent me by Iain MacInnes of the Jordanhill College of Education, Glasgow. Three coins are held as shown in Fig. 5, a nickel between two quarters (originally a half-penny was held between two pennies). The bottom coin is then released, holding on to the top coin. It is nearly impossible to release both sides of the quarter at the same instant, and the side first released starts at once to turn about the opposite side. This turning motion continues as the coin falls, and the nickel falls below the quarter which is very odd to the uninformed observer. MacInnes points out that if it takes a fall of about ten inches for the coins to rotate 180°, many students believe it will take twenty inches for the coins to land again with the nickel above the quarter. In fact, of course, it is 40 inches. Once released, the coins rotate at a constant speed and so form a clock, since there is no torque, and the distance fallen will be proportional to the square of the angle of rotation.

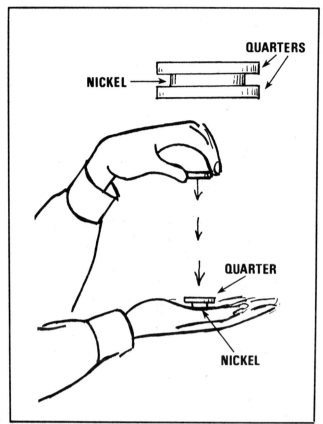

Fig. 5.

Catapulting the Interest and Success of Physics Students

Charles Hartman,
Covington Catholic High School, Park Hills, KY 41011

I have been teaching physics at a Catholic high school for boys in Northern Kentucky (Covington Catholic) for the past five years. During this time we have maintained an enrollment of 40 students in Physics (approximately 30% of the senior class). This number is divided between two sections, the honors class and the academic class. The basic difference between these two classes is one of mathematical sophistication. The academic students are taking pre-calculus along with physics, while those in the honors section are either taking a faster-paced pre-calculus or a class in calculus.

Our student body basically comes from families in which one or both parents are well educated. Therefore, they encourage their sons to enroll in as many science and math classes as possible. This, along with the chance to compete against each other and those at other schools, may be the reason that our enrollment in physics is relatively high. For the last few years our physics classes have participated in local science contests. It was very evident at these times that much learning took place, not only on the students' parts but also on the teacher's. While preparing for a contest, students must focus their attention on one particular aspect of science, and in doing so they get a clearer picture of the connection between pure science and that of technology. Preparing for a bridge-building contest allows students to see the practical application of such topics as concurrent forces, parallel forces, the first law of equilibrium, the second law of equilibrium, and other related topics. This, coupled with the fact that students are allowed in the contest environment to use not only their knowledge of physics but also their imagination and past experiences, makes this type of experience a valuable tool for reinforcing the laws of physics and for nurturing those seemingly rare qualities known as creativity and insight.

With this in mind, I decided to increase this type of experience by having a number of in-house contests during the year. The first of these activities, a catapult contest, was held after covering the topics of linear motion, projectional motion, force, and torque. Other such contests are a bridge-building contest and an electromagnet contest.

The object of this contest was to propel a large egg over a one-meter-tall wall by a catapult that was placed one and a half meters from the wall. The only restrictions on the catapult were that it must have a mass no greater than one kilogram and a height no more than one-half meter. The method of scoring was as follows: 1) To score, the student must propel the egg over the wall safely; 2) for each centimeter that the egg went past the wall, the student would score one point; and 3) if the

Fig. 1. The academic class.

egg returned to the Earth without breaking, a bonus of 100 points would be earned.

Students were advised to research the design of their catapult through discussion with their history teachers and/or research at the school or local libraries. Within a few days, each student had decided upon one of two basic designs—sling shot or lever mechanism. By the sling-shot design, I mean a design in which the egg is placed directly on an elastic object (spring, rubber tube, or rubber band) and propelled by the restoring force of that object once it is released from its distorted position. With the lever-arm design, the egg is placed on a lever which is accelerated forward by the restoring force of objects like those mentioned above. Those students who took the sling-shot approach were interested in Hook's law, elastic limit, and other related ideas, while the lever idea gave rise to concerns about torque and lever arms. Most students decided against trying to protect their egg. As they saw it, the extra mass needed to protect the egg would reduce their total distance and not be worth the extra 100 points. (If you could get an extra one meter, this would be equivalent to an unbroken egg.)

As the day of reckoning got closer, students' interest waxed and debates raged as to which of the two basic designs would take all the marbles. When all was said and done, the sling shot won out with a winning distance of 36.83 m. Second place went to a lever-arm device from the academic class with a distance of 27.32 m. The average distance for the competition was 8.16 m. The honors class had an average distance of 7.76 m while the academic class came in with an average of 8.57 m! Six of the top ten entries came from the academic class! Although the winning design came from the honors class, I was pleasantly surprised at the strong showing of the academic class.

This experience has shown me that the academic students can more than hold their own with the honors students when the challenge takes place outside of the textbook and written test. I believe that this approach of application rather than memorization holds the key to motivate those students who have had marginal success in science. For too long, educators have been overly concerned with the future Nobel prize winner and less concerned with "ordinary Joes" that make up 99% of the population of the United States. If science is to make a lasting imprint on mankind, it is this vast majority on which we must make an impression.

Oh, by the way, you may be thinking about the mess produced by all those unsuccessful landings. Well, "we" decided early on that the more successful class would watch the less successful class clean up the "physical" mess.

Fig. 2. A device from the honors class.

The cylindrical wing, why does it fly?

This question was asked by **Gary Ronald Login,** *a student in Rutgers University Physics Department, New Brunswick, New Jersey 08903 RPO 5544.*

I've been told by professors of aerodynamics that almost all things fly. The questions arise when an object doesn't fly. However, as a beginning student in the study of physics and a neophyte in the field of fluid dynamics and Bernoulli effects, I am compelled to ask the question: *Why* does it fly? Furthermore, I ask the question about an object that a few of my college friends and I devised one cold winter day in the hallway of our physics laboratory.

This object is a cylinder with a weighted perimeter at one end (Fig. 1 shows how to construct it). The launch of such a device is almost as unique as its design. The cylinder is grasped with the thumb and middle finger on opposite sides of the weighted perimeter. Hold this heavy end down and parallel to the ground. Throw the cylinder forward and upward letting it roll off your finger. As you will see, the cylinder immediately flips into its flying position, with its axis horizontal and weighted end forward.

A clockwise rotation and forward throw will give the cylindrical wing a horizontal motion and straight path for most of its flight (25 to 30 ft is easy). Toward the end of its path, as it begins to slow down and fall, it will move to the left.

I talked about this to at least a dozen physicists and professors in aerodynamics. After many hours of trying to understand their explanations, one professor's statement, "If it flies don't question it," became quite clear.

Why does the cylindrical wing fly?

Send answers to the editor. The best correct answer will be published.

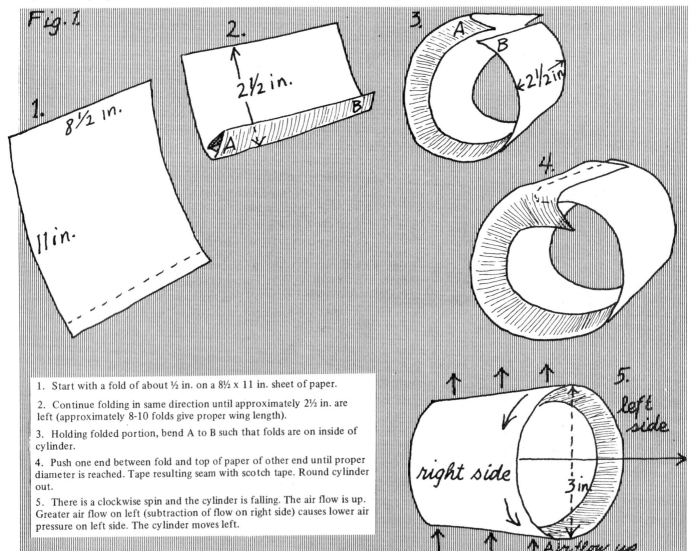

1. Start with a fold of about ½ in. on a 8½ x 11 in. sheet of paper.

2. Continue folding in same direction until approximately 2½ in. are left (approximately 8-10 folds give proper wing length).

3. Holding folded portion, bend A to B such that folds are on inside of cylinder.

4. Push one end between fold and top of paper of other end until proper diameter is reached. Tape resulting seam with scotch tape. Round cylinder out.

5. There is a clockwise spin and the cylinder is falling. The air flow is up. Greater air flow on left (subtraction of flow on right side) causes lower air pressure on left side. The cylinder moves left.

THE FLYING CYLINDER

Here is an attempt to explain the flying cylinder [Phys. Teach. **16**, 662 (1978)].

When launched with *no spin*, and with considerable speed, and with its axis nearly parallel to the launch velocity, the edge-weighted symmetric cylinder will "fly" as a glider flies. It has stability under these conditions because its center of mass is ahead of its center of pressure and it sort of "weather vanes" under these conditions. For the same reason (c. of m. ahead of c. of p.), it has a forward (down) pitching torque which results in a stable glide path. This path is steep because the lift-to-drag ratio is quite low.

When the cylinder is thrown with axial spin angular momentum, which is colinear with the launch velocity, the cylinder will follow a nearly straight line path (conservation of angular momentum) until the forward speed has decreased to where the large weather vaning torques available are comparable with the forward (down) pitching torque. Then this latter (gravitational) torque acts to rotate the (spin) angular momentum according to Newton's second law (rotary form),

$$\frac{d}{dt}\vec{L} = \vec{\tau}$$

In other words it is then a flying gyroscope which exhibits precession due to gravitational torque after the weather vaning torques have decreased sufficiently. Bernoulli and baseball-type path curvature is not involved in the turning which takes place as the speed decreases. Figure 1 shows a similar commercial toy which is called a "SKYRO." It is generally thrown overhand with as high a speed and spin as the launcher can muster. Flights of 100-150 ft are possible. As with the paper cylinder, when forward speed decreases it precesses according to

$$\frac{d}{dt}L = \tau$$

I must acknowledge the assistance of one of our graduate students, Ed Burlbaw, who owns the "SKYRO," and of another graduate student, Forest Rennick, who has very sore arms from intensive cylinder test flying.

Robert Liefeld
New Mexico State University
Las Cruces, New Mexico 88003

Ed. note: In an earlier letter Professor Liefeld pointed out that in making the cylindrical wing the folds should be parallel to the 11-in. dimension of the paper, and not as indicated in the drawing accompanying the note.

We have also received a very interesting letter from the Advanced Placement Physics class of the Pingry

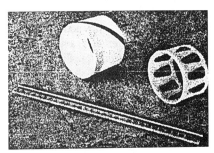

Fig. 1.

School, Elizabeth, N.J. (of which Gordon Rode is the teacher) describing a series of experiments they did with the flying cylinder and offering an ingenious analysis of the forces on the device.

The Tippy-Top

George D. Freier
University of Minnesota
Minneapolis, Minnesota 55455

An examination of the construction of the Tippy-Top shows that it has a section of a spherical surface on one end with a rather large radius of curvature and a stem for spinning it on the opposite end. When placed onto a horizontal surface with no spin, the top will rest on the large spherical end. This implies that the center of gravity of the top is below the center of curvature of the spherical surface and thus makes the system stable in this position (Fig. 1).

If the top is given a spin in either direction about its axis of symmetry, it no longer remains stable in the static position but flips over in a short time so that it stands with the stem end in contact with the supporting surface. The new position is stable, while the top is rotating, and a close inspection shows that the direction of the original angular momentum vector remains approximately unchanged in the flipping process. The amazing thing about this is that parts of the top are turning in an opposite direction about a directed axis of symmetry in the top. How can the addition of spin change the stability and apparent direction of motion of the top?

In order to understand the strange motion of the Tippy-Top it may be best to look first at the motions of a *sleeping* top. The sleeping top is just an ordinary top spinning rapidly about its axis of symmetry on a pointed end with a small radius of curvature. The center of gravity of this top is far above the center of curvature of the pointed end so the top is unstable when there is no spin (Fig. 2).

When spin about the axis of symmetry is added to the ordinary top two motions are observed: (1) the spinning motion, and (2) the precession about a vertical axis. The precession is the response to the gravitational torque and is a much slower motion than the spin. In Fig. 3 the spin is represented by the vector, **S**, and the precession by the vector, **P**. The two vectors can be added to give the resultant angular velocity, $\vec{\Omega}$. At the instant shown in Fig. 3 the gravitational torque, **L** is directed into the paper, and in a time Δt it will add the angular impulse $\mathbf{L}\Delta t$ to the angular momentum, **J**.

J is directed approximately along $\vec{\Omega}$, but they are not necessarily parallel because the moments of inertia about different axes may be different. The fact that **J** and $\vec{\Omega}$ are not parallel leads to nutation of the top which is a fast wobbly motion. Frictional forces usually produce torques in an ordinary top which tend to align **J** and $\vec{\Omega}$ along an axis of symmetry, and after a short time of spinning the nutation vanishes.

Since the gravitational impulse, $\mathbf{L}\Delta t$, is normal to **J**, it moves the end of **J** in a horizontal circle to give the precession, **P**, as shown in Fig. 4.

As the above motion proceeds a third motion is observed, namely, the axis of the top becomes vertical and the top is said to be sleeping. This last motion is caused by friction at the tip. Fig. 5 shows an enlarged view of the tip where the force of friction opposes the motion. At the contact point, K, the elements of the top are moving out from the paper for a downward directed spin so the force of friction, **F**, which opposes the motion, is directed into the paper. The torque, **L'**, due to the force, **F**, about the center of gravity gives an additional angular impulse, $\mathbf{L'}\Delta t$, which changes the direction of **J** towards the vertical and thus makes the top sleep (Fig. 6). We should note carefully that $\mathbf{L}\Delta t$ does not bring **J** closer to a vertical line while $\mathbf{L'}\Delta t$ does.

The sleeping top is much like the case of the Tippy-Top once the Tippy-Top has tipped far enough to spin on the stem. We must still determine why the Tippy-Top leaves its statically stable position once it is given a spin about its axis of symmetry. Before considering the Tippy-Top motion it may be best to study another similar problem which can be analyzed in a relatively simple way.

If a football is given a spin about a vertical axis while in its statically stable position it will rise and spin stably on its pointed end. Fig. 7 shows a football which is rolling without slipping while it is spinning. For the direction of spin shown, the material points of the football are moving out of the paper at the contact point, K, so the force of friction, **F**, on the football is directed into the paper at this point. The force of friction does two things that must be carefully considered. First, it pushes the football into the paper leading to a curved path on the supporting surface. The path is curved because there is a gravitational

Fig. 1. The Tippy-Top is stable while resting on the large spherical end.

A POTPOURRI OF PHYSICS TEACHING IDEAS—MECHANICS

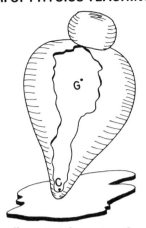

Fig. 2. In an ordinary top the center of gravity is above the center of curvature of the pointed tip so that the system is statically unstable in this position.

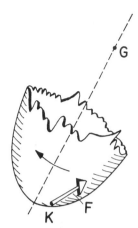

Fig. 5. Enlarged view of tip where the frictional force, **F**, at the point of contact, **K**, produces a torque, **L**′, about the center of gravity, G.

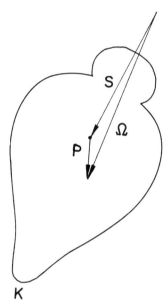

Fig. 3. The spin, **S**, added to the precession, **P**, gives the total angular velocity, $\vec{\Omega}$.

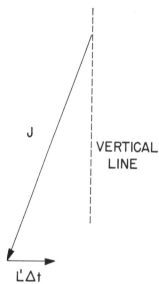

Fig. 6. The torque, **L**′, produced by the frictional force gives an angular impulse, **L**′Δt, which tends to move the angular momentum vector, **J**, parallel to a vertical line and thus made the top sleep.

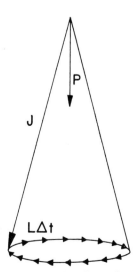

Fig. 4. The angular impulse, **L**Δt, due to the gravitational torque, **L**, is added to the angular momentum, **J**, so that only the direction of **J** changes. This motion of **J** gives the precession, **P**.

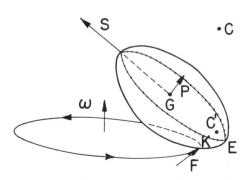

Fig. 7. A football given a spin will rise and spin on its pointed end.

torque which produces the precession, **P**, as shown. As the football precesses its direction of motion is changed. If the rate of precession is just right the path on the supporting surface will be a circle traveled in the sense indicated by the vector, $\vec{\omega}$, which has the same direction as the vertical component of **P**. The second effect of **F** is that it produces a torque about the center of gravity, *G*, which tends to move the angular momentum vector closer to a vertical line as in the case of a sleeping top.

As the angular momentum becomes more vertical the gravitational torque becomes greater, which in turn makes the precession greater. If the football precesses faster than it rotates around the circular path of translational motion, the contact point, *K*, will move towards the end, *E*, on a spiral path along the football. We then have two effects: the response to the force of friction and the increased precession, both tending to make the axis of the football be vertical. When one effect enhances the other we have an unstable situation, so that the football spinning in its statically stable position becomes unstable.

The effect that restores stability is the change in curvature of the football surface. As the football rises on its end due to the frictional force the center of curvature of the region of its surface which includes *K* changes from *C* to *C'*. If the center of curvature is the point, *C'*, the approach of the angular momentum of the football to a vertical position, caused by friction, will make the gravitational torque smaller as the football rises. Under this condition the precession rate is smaller instead of larger, the polhode (locus of contact points on the football) is a smaller circle closer to the end, *E*, and the rate of travel along the supporting surface is at a much smaller rate. The football then essentially spins on its end like an ordinary top, and, just as in the case of an ordinary top, the statically unstable position of the football becomes dynamically stable. The initial motions of the football also show how the statically stable position is dynamically unstable when spin is added.

The Tippy-Top motion is not too different from the motion of the spinning football. The end of the Tippy-Top with the small radius of curvature is at 180° from the contact point in the statically stable position instead of at 90° as in the case of the football. The conditions of instability apply sufficiently long enough to carry the axis of symmetry more than 90° away from the vertical. The stem can then touch the supporting surface so that the point of contact is on the stem which has a small radius of curvature. Once this contact is made, the motion becomes dynamically stable while spinning on the stem, just as in the case of a football spinning on its pointed end or an ordinary top rising to a sleeping position.

The complete details of the motion are complicated. During the unstable parts of the motion the frictional forces separate the axis of symmetry from the angular momentum vector and produce violent nutation. During this nutation the angular momentum remains approximately vertical so the axis of symmetry must change its orientation in space. When the motion again becomes stable the frictional forces align the axis of symmetry with the vertical angular momentum, and this realignment results in the axis of symmetry being oriented in a new direction. In the case of the Tippy-Top the top tips completely over; in the case of a football, it turns through 90°.

The final rate of spin will always be less than the initial rate of spin as some of the kinetic energy of rotation has gone to potential energy in raising the center of gravity of the system. During the tipping process the steps of infinitesimal rotations will have components parallel to the gravitational torque, so that work can be done against this torque in bringing about the new orientation. The detailed study of the motion necessary to show these relations is beyond the scope of this discussion.

Spinning Footballs and Class Rings

Questions:

1. Take a small plastic football of the type sold in toy stores and lay it on a flat surface. While maintaining it in the horizontal position, give it a good, hard spin using both hands. After a few seconds of spinning in this mode it suddenly stands on one end and spins like a top. Eventually it gets tired of this and returns to horizontal spinning. What causes these changes in rotational axis? This question asked by 7th grader **Matthew Treptow**, son of chemistry professor Richard S. Treptow of Chicago State University, Department of Physical Sciences, Chicago, IL 60628.

2. When a class ring is placed jewel-end-down on a table and then spun like a top, why does it immediately invert and spin jewel-end-up? This closely related question was asked by **Mike Flynn** a student of David M. Pope, a physics teacher at Lincoln Academy in New Castle, ME.

Answers:

The answers were prepared by **Ralph Baierlein**, professor of physics at *Wesleyan University, Department of Physics,* Middletown, CT 06457.

The Football

To understand why the football rises to spin on its end, we need to note one fact and to recall another:
a) slipping occurs where the spinning football touches the table, and the ensuing frictional force produces a torque; and
b) angular velocity depends not so much on the instantaneous value of the torque as its past history, i.e., on the cumulative effect of past torques.

Suppose we spin the football about a vertical direction but do not have the long axis perfectly horizontal; this initial situation is sketched in Fig. 1. At the point where the football touches the table there is a vertical contact force $\vec{F}_c$. The football is spinning about its center of mass (by specification), and so the point of contact is moved around on the table. This dragging generates a frictional force parallel to the table; at the instant shown in the sketch, the frictional force is directed toward us (out of the page). When we take torques about the center of mass, that force produces a frictional torque $\vec{T}_f$. The torque vector lies in the plane perpendicular to the table and passing through the center of mass, the contact point, and the football's long axis. The plane plays a central role in our analysis; let's call it the "long-axis plane."

The dynamics of the football must follow from the angular momentum equation: Rate of change of angular

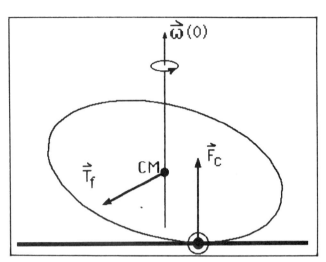

Fig. 1. View from side. The gentle ellipsoidal curvature ensures that the point of contact is not below the center of mass (CM). The initial angular velocity $\vec{\omega}(0)$ is purely vertical and passes through the center of mass.

momentum equals frictional torque plus contact torque. To simplify our analysis, we will keep only the essentials. Then we may drop the torque produced by the contact force. Moreover, although the football has different moments of intertia about its long and short axes, that is not crucial; so we will suppose the moments to be equal. The fundamental equation reduces to:

$$I \frac{d\vec{\omega}}{dt} = \vec{T}_f .$$

Here I is the moment of inertia about any axis; $\vec{\omega}$ is the angular velocity; and $\vec{T}_f$ is the frictional torque. As an acceptable approximation, we may take the torque to be perfectly horizontal.

Because the vector $\vec{T}_f$ points horizontally in the long-axis plane, the angular velocity will acquire a component perpendicular to the vertical direction. Our task is to find out how much.

In a short time interval Δt, the increment in the horizontal part of $\vec{\omega}$ is:

$$\Delta \vec{\omega} = \frac{\vec{T}_f t}{I} .$$

To excellent approximation, the long-axis plane is carried around uniformly with the initial vertical angular velocity. The successive increments $\Delta \vec{\omega}$ will add like small chords of a circle; this is shown in Fig. 2. The cumulative effect is the vector sum of these increments; it also is shown in the sketch. We see that $\vec{\omega}$ acquires a component

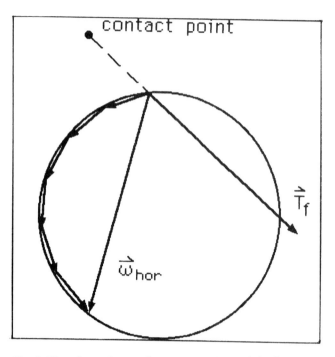

Fig. 2. View from above. The contact point and the frictional torque $\vec{T}_f$ lie in the long-axis plane. (The plane has swung around by 135° relative to Fig. 1.) The chords represent the increments $\Delta\vec{\omega}$ whose vector sum yields the horizontal part of $\vec{\omega}$, denoted $\vec{\omega}_{hor}$. (In the limit of infinitesimal Δt, the "$\Delta\vec{\omega}$"s will be tangent to the circle.)

perpendicular to the long-axis plane. The perpendicular component implies that the long axis itself is being rotated in the long-axis plane. The sense of rotation raises the high point on the long axis even higher—and so the football proceeds to stand on its end.

A Few Comments

Here is another way to see that $\vec{\omega}$ acquires a component perpendicular to the long-axis plane. Draw all the increments $\Delta\vec{\omega}$ with their tails at the same location. The little vectors will bristle out like the quills of a porcupine. The first little vector will point along the initial direction of $\vec{T}_f$; the last, along the current direction of $\vec{T}_f$. The vector sum of all the increments will lie precisely halfway between those two vectors.

While the football makes one revolution about the vertical, the horizontal component of $\vec{\omega}$ grows from zero to a maximum magnitude and then decreases back to zero again. When it is non-zero, it always lags behind $\vec{T}_f$; its part that is perpendicular to the long-axis plane always rotates the long-axis in the same sense (further toward the vertical).

The curvature of the football's ellipsoidal shape keeps the frictional drag in the same sense for all of the 90 swing: from the long-axis being almost horizontal to being vertical. You can see this by holding the football

Fig. 3. Wolfgang Pauli and Niels Bohr with a spinning top, probably a "tippie top," which flips over for reasons similar to those that explain the football's behavior. (Photo courtesy of the Margrethe Bohr Collection, AIP Niels Bohr Library.)

in various orientations on a table and looking at it from the side.

If you rotate the long-axis a little past the vertical, however, the contact point moves to the other side of the center of mass, and the drag reverses in sense. The frictional torque $\vec{T}_f$ changes sense and tends to return the long-axis to the vertical. Thus the vertical orientation, when achieved, is dynamically stable.

The Decline and Fall

When the football is spinning with its long axis up, the rubber flattens a bit at the contact end. Thus there is a frictional drag over a disk-like area. This drag saps kinetic energy. The torques from each little patch on the disk sum up (by symmetry) to a torque that is purely vertical (or anti-vertical); there is no net horizontal component such as what caused the football to rise on end. The net torque does, however, nibble away at the football's angular momentum.

As the angular momentum decreases in magnitude, the vertical orientation becomes unstable. Conservation of energy and angular momentum—over a short period of time—can help us to see this. Imagine a stray breeze that tips the spinning football a bit from the vertical. The center of mass drops a bit, and some gravitational potential energy is made available for motion. The football's spin about its long axis is now inclined to the vertical, and so that spin gives less vertical angular momentum than it previously did. To maintain a constant amount of vertical angular momentum, the football must precess bodily about the vertical direction. The precessional motion represents kinetic energy. Only if the demand for such energy is less than what can be supplied by the drop in the center of mass can the football continue tipping over. What is necessary is a sufficiently small magnitude for the vertical angular momentum.

Once the football becomes unstable to tipping over, it will precess and nutate. More energy will be dissipated by friction, and the center of mass will sink as low as it can go. The long axis will become horizontal, but the residual angular momentum will keep the football rotating—for a while, until that motion, too, is damped out.

The Class Ring

The behavior of a spinning class ring has fundamentally the same explanation. The center of mass is not at the center of the circular finger hole; rather, it is displaced toward the jeweled end. A spin about the center of mass leads to torques like those for the football.

Indeed, let us idealize the ring's outline as a circle but keep the center of mass off-center. Let me ask you to provide the sketch: Draw a circle and, to represent the table, a horizontal tangent line; then place the center of mass anywhere to the left of the vertical line that goes through the contact point (but fairly close to the circle's center). Your sketch will be similar to Fig. 1; our previous analysis indicates that the center of mass will rise. That motion sends the jeweled end upward. (An initial situation with the center of mass low and directly above the contact point is dynamically unstable. It doesn't persist, and the situation in your sketch takes over.)

Spinning objects induce a fascination all their own. Curiosity about them can start early and never cease. For evidence, Fig. 3 shows Wolfgang Pauli and Niels Bohr bent over a spinning top.

The Coriolis effect and other spin-off demonstrations

Lester Evans
Tates Creek Senior High School, Lexington, Kentucky 40502

In attempting to "make clear" to our students some of the not too obvious physical principles, we teachers use elaborate vector diagrams, mathematical wizardry, a considerable amount of hand waving, and many statements such as "it can be shown." After these exasperating experiences, usually brought on by a lack of good demonstration or laboratory equipment, the teacher is left with the feeling that his efforts were in vain with regard to the real concept. It was difficult, for example, for me to feel successful with the Coriolis effect. I am proud to say I have solved this problem completely with the demonstration described below (Fig. 1). To my delight, I was also surprised by several spin-off ideas that occurred to me while performing the exercise.

The demonstration allows for direct student involvement (which is the best kind!) and the equipment has other applications making it even more worthwhile to construct. Some of these "since you have it set up" applications will be described later.

The equipment consists of a heavy rotating platform (Cenco #74790-000 or equivalent), a 2-in. by 12-in. by 12-ft long pine board, two large "C" clamps, two backs from broken chairs, one trash can, several tennis balls, and two brave students of about equal mass. Tables and chairs are moved from the center of the room, a convenient spot in or outside the building is located, so that a circular area about 16 ft in diameter is available. Going outside is not a bad idea since a little exposure to curious sightseers of the fun we have in physics may help future enrollments. Place the rotating platform in the center of the area with the board centered across the platform and clamped firmly. The trash can (or bucket) is positioned at the center and the two students are seated at the ends of the board facing inward. The chair backs were added as a safety measure but the exercise can be performed without them.

Arm each of the students with a few tennis balls and gently set the whole thing rotating at about one revolution in 5 or 6 s. Ask one of your students to gently toss a ball into the trash can — !! You are now a successful teacher. The concept is clear and the vector diagrams are now meaningful. The effect is very dramatic and was a grand surprise to this old physics teacher the first time it was

Fig. 1. Two brave students preparing for a spin.

used in my classes.

Tossing the ball from student to student is also interesting and some may readily achieve the skill to toss it to themselves on the opposite side of the circular path. During this latter "fun and games" segment, have the rest of the class determine if the horizontal flight of the ball is a straight line or curved. You may suggest they hold a book or their hands to block their view of the rotating system.

Now, while you have it set up anyway, try some of the following spin-off ideas. For this part, have your two students seated but facing outward as near the ends of the board as possible, knees up and feet on the board. Give the system a slow rotation and command your students to slowly slide themselves inward at about 10 cm per revolution. Since they cannot observe each other, you may need to give them verbal instructions to keep the system balanced. Conservation of momentum! Now draw your vectors.

In this demonstration, you also may experiment with a "student fit to the equation" $F_c = mv^2/r$. If given a sufficient initial rotational speed, the students find it very difficult, if not impossible, to force themselves inward beyond a certain point until the system slows.

The angular momentum of this low friction system is nearly constant, and each student can be considered a "point" mass with angular momentum mvr. Thus, reducing the distance from the center causes an increase in the tangential velocity and the centripetal force (mv^2/r) [For constant (mvr), $F_c = mv^2/r \to m^2v^2r^2/mr^3 = K/r^3$], the effort required of the students to push themselves inward, increases dramatically.

A test of the centripetal force equation can also be demonstrated by having one student sit in the center holding a spring scale attached to a string and to a dynamics cart. The cart can be held at any radius of rotation and can be loaded with up to 10 Kg mass. The radius and period are easily obtained and the force can be observed by the student on the spring scale.

From here, I will just say that variations of these or other demonstrations are left to the imagination of the innovative physics teacher. The demonstrations suggested are very visible to all students even if they do not participate. The sequence of demonstrations described consumes a full class period, but for me, at least, it is one of the most productive 55 minutes of the year.

With regard to safety, the board is rather heavy and due caution should be taken while it and students are in rotation. Keep the exercise serious and do not let the "fun and games" get out of hand.

Whirlpool in a swimming pool

George W. Ficken, Jr.
Cleveland State University, Cleveland, Ohio 44115

This past summer my daughters accidentally discovered a fascinating way to make better use of the cramped quarters available for swimming in a 10 ft diameter, 2 ft deep swimming pool. Positioning themselves at nearly opposite ends of a diameter, they would begin shuffling forward around the perimeter, partially dragging the water along with them. The water would gradually pick up speed, and in order to maintain sufficient frictional force to continue the acceleration process, a sloshing sort of jogging (very tiring indeed!) would eventually be necessary to achieve the maximum effect. Meanwhile the whole pool had become a slow-moving "whirlpool," and it was time to play their game. Floating on their backs near the edge, and tangent to the pool, they would be carried 'round and 'round, without brushing against the rim. Since I knew how unnaturally one had to swim in order to circumnavigate the pool in still water without touching the rim, the idea immediately appealed to me and I became involved.

The apparent stability of one's body while floating around in its orbit was what intrigued me. If I initially launched myself in the direction of the water's tangential velocity, only minor adjustments were usually required to maintain proper floating orientation, even at the start when the greatest stability problems would be anticipated. It should be pointed out that the physical setup being described is more complicated than one might assume to hold after the irregularities caused by the stirring method have smoothed out. For example, spinning leaves in a teacup are transported in the secondary flows induced by the drag at the boundary of the cup.[1,2] Nonetheless, at the low angular velocities attainable for the "whirlpool," every orientation attempted seemed to produce excellent results:[3] head first or feet first while positioned tangentially along the path; head or feet near the pool's center while stretched out along a radius and behaving like the spoke of a rotating wheel -- all such fun! And the rotation continued far beyond my conservative expectations. Approximately twenty revolutions were possible when I alone produced the "whirlpool," although near the end I traveled at a snail's pace. My maximum speed produced a period of about eight seconds, but I shall train diligently all year for an assault on these records.

References:

1. For a theoretical discussion with associated lecture demonstrations of such phenomena, see D.J. Baker, Jr., Am. J. Phys. 36, 980 (1968).
2. J. Higbee, Phys. Teach. 12, 29 (1974).
3. A scaled down version of the "whirlpool" conducted in a circular dishpan, using a piece of wood loaded so as to be nearly submerged, allowed a better controlled experiment. A tendency toward instability was noted here for the wood if positioned tangentially, with either the upstream or downstream end often slowly swinging away from the wall until the more stable radial position was approximated.

An Experience in Observation

Merle Fisher
Ricks College
Rexburg, Idaho 83440

A delightful and exciting opportunity to participate in a genuine scientific experience can be provided for a class in the following sequence of experiments.

To begin, a lighted candle is dropped from a height of 6 to 10 ft and is observed to continue to burn when it is caught at the bottom of its fall. (Provided a brave student, unafraid of hot wax, is available to make the catch and keep the candle quite upright.)

In contrast with this observation, it is found that when the lighted candle is placed in a closed gallon glass jar and dropped from a similar height the candle flame is extinguished when the jar is caught at the bottom of its fall. A class can then be led to discuss the nature of the candle flame "blow out." The candle flame is "blown out" so quickly that it is quite impossible to discern whether it is "blown out downward" or is "blown out upward". (If at this point the class is asked for a "guess" on the question, the predominant opinion will likely be that the flame is blown out downward.)

In determining the downwardness or upwardness of the candle flame "blow out" an instructor has an ideal scientific situation to be explored. It is a determination which cannot be made with the student's unaided senses and indirect observations must be utilized to resolve the problem. Fortunately the problem can be resolved through a study of acceleration so it can be integrated directly into the regular physics instruction and used to add zest to such instruction.

A "tangential" and a "centripetal" accelerometer can be utilized with a rotating platform to demonstrate that under conditions of uniform circular motion the tangential acceleration is zero while the centripetal acceleration has a measurable value (Fig. 1). Lighted candles, enclosed in gallon glass jars, are then placed on the rotating platform and it is observed that under uniform circular motion the candle flames are directed inward toward the center of the motion (Fig. 1). The observation can be stated: "The candle flames are bent in the direction of the acceleration." This statement forms the clue to a kinematic solution to the problem of the direction of the candle flame "blow out" when the candle is dropped in the gallon glass jar.

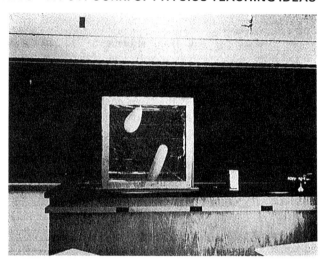

Fig. 2. Cage, accelerating toward right. Air-filled balloon top, falls backward. Hydrogen balloon, bottom, "blown" to the right, in direction of acceleration.

After some discussion of the situation some members of the class might be willing to conclude that at the instant the bottle with the candle is caught it experiences a tremendous upward acceleration and the candle flame is "blown out" in the direction of this acceleration—upward. But every hypothesis must be tested and when a class has reached this point in the experience, they will have an almost limitless number of ideas to be tried. Some suggestions will be: Make the drop longer, from the top of a three-story building, for example; use smoke or dust of some sort in the jar to test convection current effects; change the heights of the candles in the jars; etc. Also, there will always be discussion as to why the candle in the bottle goes out when caught at the bottom of its fall but the one out of the bottle does not.

The discussion will likely lead to the conclusion that the candle flame of hot gases is less dense than its environment. Is it possible to observe other less dense objects moving in their environment? Someone will suggest a hydrogen or helium-filled balloon. Such a balloon, when placed in a closed cage, is found to move forward, in the direction of the acceleration, when the cage is accelerated forward. This is in contrast with an air-filled balloon in the same cage which falls backward in the accustomed manner under forward acceleration (Fig. 2). This is truly an "Ahah" discovery for students who have never seen it before, and they will be riding with helium-filled balloons in their cars whenever such balloons are available to them. Again one has demonstrated that objects which are less dense than their environment, like candle flames, are "blown" in the direction of the acceleration.

Another interesting test of the hypothesis can be made by tying fish floats on a string so that they float near the

Fig. 1. One version of a rotating platform with accelerometers and bottles attached.

Fig. 3. Rotating platform containing floating object.

top of inverted erlenmeyer flasks and then rotate these flasks in uniform circular motion (Fig. 3). One observes that the floats (objects which are less dense than their environment) move inward toward the center of the motion—in the direction of the acceleration. One thus has been able to observe several bits of phenomena which justify the conclusion that the candle flame is "blown out" upward—in the direction of the acceleration—as originally proposed.

It will be noted that the above description is a purely kinematic one. In an actual class situation, it will likely be impossible to confine the discussion to such a point of view; and such dynamical problems as pressure gradients, convection currents, etc., will be injected into the discussion. This, however, enhances the opportunity for the teacher to utilize the phenomena to teach additional principles of physics by helping students distinguish between the notions of kinematics and dynamics—notions which are sometimes confusing to beginners.

Teachers will find these simple experiments and the associated "home-brew" apparatus a worthwhile addition to their physics teaching tools.

Large Scale Use of a Liquid-Surface Accelerometer

Lawrence E. L'Hote
Monroe City R-1 High School, Monroe City, Missouri 63456

The following note describes a quantitative use of a liquid surface accelerometer in both linear and circular motion.

Each pair of physics students was provided with pre-cut pieces of plexiglas and accelerometers were constructed as outlined in *The Project Physics Handbook*.[1]

Linear Acceleration

The day before the demonstration was scheduled, I marked off a 300 foot course on a little-traveled, flat section of an asphalt country road. A windowed van school bus was found to be an ideal vehicle to use because it could hold the

TABLE I Student data for linear acceleration

Distance, ft.	Elapsed time, sec.	Accelerometer Scale Reading	Experimental Acceleration, ft/sec^2	Theoretical Acceleration, ft/sec^2
300	9.9	2.25	6.9	6.1
300	10.5	2.0	6.1	5.4
300	9.0	2.5	7.7	7.4

Fig. 1. Physics students using accelerometers on playground merry-go-round.

A POTPOURRI OF PHYSICS TEACHING IDEAS—MECHANICS

entire class and could be maneuvered easily. The students placed their accelerometers level against the van windows and, using the lower gears of the automatic transmission, I drove through the 300 foot course. Elapsed time and accelerometer scale readings were recorded for five runs in both first and second gears. Several "trial" runs were made to familiarize the students. Also, it takes a little practice for the driver to produce an even acceleration rate throughout the course length. Typical student data is shown in Table I. Although the class didn't take any data for deceleration, the reaction of the liquid surface to slowing was obvious.

Centripetal Acceleration

Four students were seated on a playground merry-go-round as shown in Figure 1. The students were to center their accelerometers at specified distances from the center of the merry-go-round I.e. 30 cm, 50 cm, 80 cm and 100 cm). Other students were assigned as "pushers" and one student was to act as timer. Data from several students are shown in

TABLE II Student data for centripetal acceleration. One revolution of merry-go-round in 4.2 seconds.

Radius of Rotation, cm	Accelerometer Scale Reading	Experimental Acceleration cm/sec^2	Theoretical Acceleration cm/sec^2
30	.8	.78	.68
50	1.2	1.18	1.12
80	1.8	1.76	1.79
100	2.5	2.45	2.24

Table II. It should be noted that as one used the accelerometer toward the center of the rotating platform the accelerometer scale becomes less accurate. At the 30 cm position the accelerometer, which is about 22 cm wide, does not indicate the true acceleration at exactly the 30 cm mark because the edge toward the center of rotation is 10 cm closer to the center while the outer edge is 10 cm further out from the center. The photograph in Figure 2 was taken at a rather high rotation rate in order to illustrate the difference in left and right scale readings at the 30 cm position. Also, note the slightly concave shape of the liquid surface. Data at the 30 cm position reflects the amount of error involved. Although the students were advised to take readings on both left and right scales and average to eliminate some of the error, the students seemed to favor just reading the right-hand scale. It could be that the persons riding the merry-go-round considered the problem of just hanging on far greater in importance than taking precise data.

Reference

1. Gerald Holton, F. James Rutherford, and Fletcher G. Watson (directors), *The Project Physics Handbook*, The Project Physics Course (Holt, Rinehart, and Winston, Inc., New York, 1970) pp. 46-48.

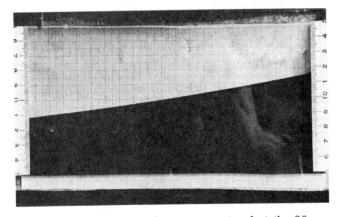

Fig. 2. Liquid surface accelerometer centered at the 30 cm position of a rotating merry-go-round.

Teachers' Pets II — *Circling carts*

Robert Gardner
Salisbury School, Salisbury, Connecticut 06068

Many students find it hard to believe that a body moving with a constant speed can be accelerating. I have found that bubble or cork accelerometers are very useful to students striving to understand centripetal forces and accelerations. Once students accept the fact that the bubble or cork moves in the direction of the acceleration, they can see that there really is an inward acceleration when they place an accelerometer on a rotating turntable. Even more convincing (and certainly more fun) is to have students watch an accelerometer as they ride on small carts. Such carts can be made by nailing a 12 in. x 20 in. piece of 3/4 in. plywood to a pair of 2 x 3's mounted on roller skate wheels.

Two bubble accelerometers are taped to a cart as shown in Fig. 1. One accelerometer should be parallel to the cart's axis; the second should be perpendicular to it.

Fig. 1. Bubble accelerometers taped to top of cart.

Small carpenters' spirit levels or test tubes nearly filled with water can serve as accelerometers.

Students can readily see that there is an acceleration in the direction of the force when their cart is pushed forward or brought to rest.

Now ask them to watch the accelerometer aligned perpendicularly to the cart's axis as the cart is pushed in a circle. To make the cart move in a circle tie a rope to a large staple or a hook that has been screwed into the cart. One student holds the other end of the rope to a fixed point on the floor while another student pushes the cart and rider at a steady speed. It is quite clear that an inward acceleration exists aboard the moving cart.

What happens if the cart's speed is greater or less but still constant? What happens if the cart moves at a steady speed but in a circle with a larger radius? a smaller radius? (To maintain about the same speed for different circles, the same student should push the cart throughout. He can generally maintain a fairly steady pace.)

To make this experiment more quantitative some students checked the calibration of a Project Physics accelerometer by attaching it to a PSSC lab-cart and accelerating the known mass by means of a falling mass hung over a pulley.

They found it took a mass of less than 10 g hanging on the string to accelerate the lab-cart and "cargo"; consequently they ignored the frictional force in their experiment. The data they collected is found in Table I.

Fig. 2. Here a liquid accelerometer of the type used in the Project Physics course is attached to the cart. It can be used in a more quantitative approach to the study of centripetal acceleration.

TABLE I

Total mass of lab cart, accelerometer, and adjusting masses	Mass used to accelerate lab cart & contents	Total mass accelerated	Ideal acceleration (F=ma)	Reading of liquid level on accelerometer
2.0 kg	0.5 kg	2.5 kg	1.96 m/sec^2	2
2.0	1.0	3.0	3.26	3 +
2.5	0.5	3.0	1.63	1.5 +
3.0	1.0	4.0	2.45	2.5
2.0	0.2	2.2	0.89	1 −
2.0	0.1	2.1	0.47	0.5

On the basis of this data the students decided that the manufacturer had calibrated the accelerometer to read the acceleration in units of m/sec^2 and that it was calibrated as accurately as we could hope to read it.

With the calibrated accelerometer shifted to the larger cart they circled the auditorium stage as they measured the acceleration, and also the radius of the circle, and the time required to make one revolution at "constant speed." Then they compared the average value read on the accelerometer with the acceleration predicted mathematically from the data. To read the accelerometer students crouched about the circular path of the cart so they could read it at eye level.

The results (which are summarized in Table II) seemed to give the students added confidence in the mathematical analysis as well as a better gut feeling for centripetal acceleration.

TABLE II

Radius of circle (m)	Time to make one revolution (sec)	Calculated acceleration $4\pi^2 R/T^2$ (m/sec^2)	Acceleration according to accelerometer (m/sec^2)
2.8	10.2	1.06	1
3.3	12	0.90	1 −
3.3	9.5	1.44	1.5
2.0	6.2	2.06	2
2.0	11.1	0.64	0.5 +

A PERSONAL APPLICATION OF PHYSICS

Dorothy Russell
Thomas Jefferson High School, Bloomington, Minnesota 55437

The Practicality of Physics to an Art Student: The enrollment in a science course may be directly related to the practicality of its content. Students these days are becoming more selective and analytical in the choice of their high school courses. Evidence of this trend is given by the following paper turned in to Richard Snydle, physics instructor at Thomas Jefferson Senior High School in Bloomington, Minnesota, by one of his students named Dorothy Russell.

One of the stated goals of the Harvard Project Physics course is "to help students see physics as the wonderfully many-sided human activity that it really is." I remember last fall, Mr. Snydle, when I asked you whether I would be able to make use of concepts I would learn in physics in my ceramics class. You told me then that you personally make use of physics know-how in any number of day-to-day activities, but that you too kind of wondered how physics would be of any use to me in art. Granted, some of the following considerations of applied physics are definitely not mandatory to creating aesthetic pottery, but knowledge of them can serve as either an expedient in actually making the piece, a means of preventing predictable collapses of the clay, or simply provide a physics-and-art student with ideas linking the two subjects. It's kind of fun to study a concept and then see how it can actually be applied:

Physics at the Potter's Wheel

Once the ball of clay is ready for throwing (it has been kneaded, known as wedging to remove any airpockets, scraps of debris, etc.), the student turns on the wheel so

Fig. 1. Julie Streff demonstrates the application of physics in her ceramics class.

that the wheel head is rotating in such a way that pressure will be applied first with the heel of the hand, and then follow the motion out to the fingertips. This means that "normal people" want the wheel rotating counterclockwise, but lefties want it rotating clockwise. At this point, the center of the wheel is very lightly dampened with a sponge to impart in the clay atoms a greater affinity for the metallic atoms of the wheel.

The ball of clay is ready to be thrown. According to the Galilean description of free fall, $d = v_i t + \frac{1}{2} at^2$, the ball of clay, regardless of its mass, falls with a constant acceleration due to gravity of 9.8 m/sec², or 32 ft/sec², toward the center of the wheel (and beyond that, the center of the earth).

At this point, a relationship is established between the clay and the rotating wheel, known as momentum. It depends on the product of the mass of the ball of clay and the velocity of the wheel. Throughout the throwing process, the momentum will change frequently: the speed of rotation will be altered, and/or the mass of clay will decrease as excess clay is removed, increase as water is added to the clay.

Centering the clay ball is essential: this means that the clay is formed into a symmetrical shape at the exact center of the wheel as it rotates. The ball of clay is centered when, according to the law of inertia, the only way that you can detect its motion is because you can see the wheel rotating in your frame of reference. In the event that your reference frame were moving at the same velocity as that of the wheel, you could not be sure whether or not you had centered the clay.

The centered ball of clay is experiencing a centrifugal force as the wheel rotates in uniform circular motion. Even if the speed of a rotating point on the wheel remains constant, the velocity of the wheel is changing constantly, due to changing direction. If at this point you lost interest in your ball of clay, you could instead determine the centripetal acceleration of the wheel $a_c = v^2/R$. By first determining the frequency of rotation in revolutions per second, you could determine the period T as $T = (\text{rev/sec})^{-1}$. Then (circum. of wheel head/period) would give you the velocity term for the original equation.

But getting back to the ball of clay, let's say you decide to make a bowl. You must press your thumbs down into the center of the clay to open it up, and then begin to flare out the hole in order to give the bowl a rounded shape inside. An even pressure (force/unit area) must be exerted. Otherwise the side of the bowl will be subjected to Newton's unbalanced forces. The bowl will consequently look like it challenged the law of inertia (and lost).

As the side of the bowl is raised, physics helps to explain and prevent another difficulty. Since the outer edge of the bowl is obviously at a greater radius from the center than any other area of the bowl, its speed is greater than the inner parts. The forces produced which act on the clay are similar to those felt on a merry-go-round — the greatest force is felt at the outer edge. This force, the centrifugal force, is given by the equation $F_c = mv^2/r$ which indicates that although the force varies inversely with the radius, it also varies directly with the square of the speed. This force and the force of gravity are acting on the bowl against the potter's desires. To compensate for this effect, the hand motion must be quickened as the edge of the bowl is approached.

Gravity imposes the final setback in making a bowl. Unless you allow an adequate base to support the wall of the bowl, it will collapse under the burden of gravitational force. Ceramics (by common sense) and physics (by the second law of thermodynamics) agree that all familiar processes are, under natural conditions, irreversible. So much for making bowls, right?

Spin Art: A Rotational Effects Demonstration

Dean Zollman
Kansas State University, Manhattan, Kansas 66506

An inexpensive device to demonstrate effects of rotational motion can be purchased at many toy stores. "Spin Art"[1] has a small stage at the center of a larger piece of plastic. A piece of paper with paint on it is placed on the stage. Using a battery-operated motor, the paper and stage are spun. The result is a design similar to ones reproduced in Fig. 1-3.

The Spin Art toy is made to spin in only one direction and at only one speed. These limitations were easily overcome by replacing the stock switch with a reversing switch and putting a variable resistor in series with the battery-motor circuit. With this arrangement the student can investigate questions such as: Does the distance the ink moves depend on speed of rotation? (Compare Figs. 1 and 2). Does the curvature of the lines depend on speed of rotation? (Figs. 1 and 3). The answers to these questions can lead to useful

Fig. 1. Result of a high speed rotation in a counter clockwise direction.

Fig. 3. Result of a high speed rotation in a clockwise direction.

discussions of "forces" one experiences in a rotating frame of reference.

Spin Art can also be used in large lecture classes. I have cut pieces of clear acetate to fit on the stage. The paint does not stick to the acetate very well. However, the result can be projected on an overhead projector. The questions listed above can then be discussed.

The toy seems to grab students' attention rather easily. The attractiveness to the students plus the cost (about $5.00) makes this device very useful for qualitative demonstrations of rotational effects.

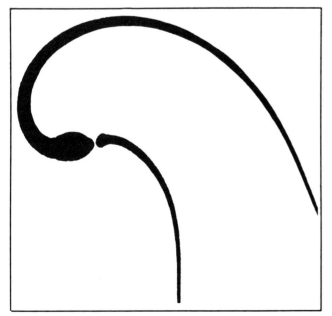

Fig. 2. Result of a low speed rotation in a counter clockwise direction.

Reference

1. Manufactured by Rapco, Inc., Chicago, Illinois 60624.

CENTRIPETAL FORCE USING A HAND ROTATOR

The relationship $F = mv^2/r$ can be verified in absolute units using some rather standard equipment found in most laboratories. Two equally matched 2000-g spring balances are mounted on a block on a hand-rotating device or a variable-speed motor-driven rotator. All mounting may be done with masking tape since rotation speeds will be slow (Fig. 6).

In operation, two number 5-1 rubber stoppers of equal mass are placed on the ends of the hooks of each balance. The apparatus is rotated at a speed which will stretch the springs to a reading between 600 and 1000 g. All readings must be taken with a stroboscope in subdued light. The background reading is obtained by removing the stoppers and rotating the device at a speed equal to the loaded speed. (This reading is based on a slightly different radius than when the springs are loaded, but the difference is small and with stiff springs on the 2000-g balances it can be neglected.)

The radius of the circle is measured from the center of the circle to the center of mass of the rubber stoppers with the springs stretched to the position they had during rotation. The rotation frequency of the device shown was obtained using a stop watch to determine the frequency of the hand crank; the result was then multiplied by 16, the step-up ratio.

Using the mass of one stopper, and the other data, the value mv^2/r can be compared with the value obtained by direct reading of the balances (minus the background). The values on the spring balances must be converted to newtons. Differences in the calculated values of the centripetal force and the values read directly from the spring balances should be within 5%; we have obtained values within 2%.

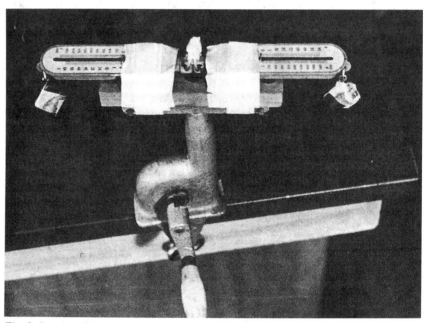

Fig. 6. Centripetal force apparatus uses a rotator and two spring balances.

David Chesnut
Canal Winchester High School
Canal Winchester, Ohio 43110

A Penny For Your Thoughts

John Dixon
The Thacher School
Ojai, California

THE equivalence of an accelerated frame of reference and a gravitational frame is classically shown by swinging a bucket of water in circular motion up over your head, occasionally with disastrous failure. The classical demonstration is certainly adequate in getting the point across, but it employs rather crude methods and is not particularly impressive in the classroom. A more exacting and delicate method can be demonstrated with an ordinary wire hanger and a penny.[1]

Grasping the hanger by the hook and midway along the longest side, stretch it into an elongated, diamond shape. Now bend the hook slightly so that it points back toward the opposite end of the diamond. If the end of the hook is not flat, you should flatten it with a file.

Dangle the hanger from your index finger and carefully balance the penny on the end of the hook as shown in the diagram. The balancing process is usually a bit tricky, but this delicacy makes the remainder of the demonstration all the more impressive. A little practice, a steady hand, and a lot of patience are all desirable at this point!

Having precariously balanced the penny, gently and smoothly begin to swing the hanger to the left and right without losing the coin. When you have a feeling of the rhythm, smoothly swing the hanger all the way around and continue with a circular motion in the vertical plane. If you're careful, the penny will remain "balanced" on the end of the hook even when you slow the motion down until the hanger is about to fall off your finger.

With a little practice you can shift the plane of motion into the horizontal plane up over your head. Once you have done this, it is not difficult to slow the motion sufficiently to return the hanger to its original, stationary hanging position with the penny precariously balanced all the while! It looks almost as though the penny had been slyly glued to the end of the hook; some students will be convinced that you have fooled them.

Perhaps this demonstration requires a bit more finesse than the classical one, but the point will be made in an extremely exacting manner that won't quickly be forgotten. A discussion dealing with imaginary forces and potentials in accelerated frames of reference might easily follow the demonstration.

---

[1] Pennies with the Lincoln Memorial on the back are most easily balanced. The memorial provides a good, relatively flat surface on the penny.

The Great American Vector Game

Thomas Ritter
West Windsor-Plainsboro High School, Princeton Junction, New Jersey 08550

While sitting at my desk planning for my physics class, I was suddenly struck with the idea of "The Great American Vector Game." My junior-senior physics classes at West Windsor-Plainsboro High School were just finishing a unit on Newton's laws of motion and vector addition (*Project Physics*, chapter 3) and I sensed that they were having some difficulty with vector addition by mathematics (as opposed to scaled drawings). We had already done our share of word problems, and I didn't want to bore and frustrate them any longer. I then recalled the force tables (Sargent-Welch, 0740) I had stored away somewhere and wondered how I might use them as a motivational tool for solving vector problems. Thus was created in my mind "The Great American Vector Game." We had so much fun with it, I felt it was worth sharing with my colleagues.

Most physics teachers are familiar with the force table and how it can be used to demonstrate addition and balance of force vectors. I created a TV Game Show atmosphere around that concept, operating on the premise that if Richard Dawson of "Family Feud" can be successful, why can't we physics teachers?

I divided the class into six three-person teams. The number and size of the teams would depend, of course, upon the number of students in any given class, but I found that six teams worked out well. I then split the teams into two divisions, green and gold. These are our school colors and a little school spirit didn't hurt.

Fig. 1. The author (c.) with students Dennis Meseroll (l.) and Kent Shin (r.) about to pull the holding pin and verify their answer. (Photos by students Dennis Meseroll and Brian Yen.)

Table I

Great American Vector Game Rules

1. In rounds one and two each team (A, B, C) plays against each other within their division (Green, Gold).
2. The highest scoring team from each division plays the winner from the other division for the "Great American Vector Game Championships" to determine which division wins the game.
3. Each division is given a different problem in each round.
4. There is a time limit to solving each problem. However, the first team to turn in its answer will verify first, and if correct, scores higher than the second team to submit its answer, and they higher than the third.
5. At the end of the calculation time each team, in order, verifies its answer on the "Great American Vector Game Vector Verification Table." If correct they are awarded points. If incorrect, no points are awarded.

Table I lists the rules which give guidelines on game procedure. Play begins with two preliminary rounds which serve to determine division winners. Each team within a division works to solve the same problem in a given round whereas the teams in the other division are given a different problem to solve. The game then moves into the championship round where the two division winners compete in an attempt to become the "Great American Vector Game Champions."

Table II shows the time limits for solving the problems and the scoring for the preliminary and championship rounds. The times are arbitrary but should be of sufficient length to allow teams to compute solutions. The degree of difficulty of the problems should be adjusted to the level of understanding of the class. However, this level should increase as the game progresses from the preliminary through the championship rounds. This can be accomplished by transforming from vectors at right angles to ones with acute or obtuse angles, and in the championship round to a three-vector problem.

The scoring differential in each round rewards accuracy and speed. It also allows all of the teams a chance to verify their calculation on the force table and hopefully achieve some degree of success. Those with incorrect

Table II
Great American Vector Game
Time Limit and Scoring Sheet

Game phase	Time limit for calculation	Scoring order of finish	score
Round one	5 min.	First	3
		Second	2
		Third	1
Round two	5 min.	First	6
		Second	4
		Third	2
Championship	15 min.	First	10
		Second	8
All wrong answers: no score			

Table III
Great American Vector Game
Team and Score Sheet

	Green		Gold
Team	Score	Team	Score
A	rnd. 1: rnd. 2: chmpshp: Total:	A	rnd. 1: rnd. 2: chmpshp: Total:
B	rnd. 1: rnd. 2: chmpshp: Total:	B	rnd. 1: rnd. 2: chmpshp: Total:
C	rnd. 1: rnd. 2: chmpshp: Total:	C	rnd. 1: rnd. 2: chmpshp: Total:

answers receive no score for that round. This cautions those who tend to be too hasty with their calculations.

In preparation for the game several aids can be made. For instance, a transparency of the rules could be made and used in the beginning of the game. This is just to show that we are a legitimate outfit and not some fly-by-night game show! A transparency blocking out team and division names with space for putting in student's names and team scores should also be available (see Table III).

Additional components of the game include Great American Vector Game Problem and Answer Cards which are given to each team by the M.C. (teacher) at the beginning of each round. Samples can be seen in Fig. 2. Each should be titled with the Round, Division, and Team. The students seemed to enjoy this aspect and it helped to set the game-show atmosphere. This was also enhanced by strategically placing the contestants' tables in a semicircle around the M.C.'s table with its overhead projector and two "Great American Vector Game Verification Tables" (force tables) at either side.

At the end of each round when the time is up and all answer cards have been turned in, the teacher gathers the contestants around the "Great American Vector Game Vector Verification Tables" to verify . . . or discredit each team's answer in the order in which they were submitted. If "hammed-up" enough, the suspense builds as the forces are positioned and the holding pin is removed to test the accuracy of each answer.

After all the excitement and congratulatory remarks are over, time can be spent reviewing the problem solutions and allowing the students to retest their answers on the tables and perhaps determine their amount of error. I would estimate that the entire game can be completed in one and a half to two class periods each 45 minutes in length.

If the idea of physics games is contagious with your students, you might consider creating balanced teams and pursuing the concept throughout the year using other topics in physics and keeping a running score for each team. At the end of the year winners could be determined and perhaps winners from each class could be pitted against those of the other classes in the "Grand Local Great American Physics Championships." Perhaps some problem dealing with the application of several concepts learned throughout the year could be used.

As an example of a contest in another topic, consider the *Project Physics* Experiment 11, "Prediction of Tra-

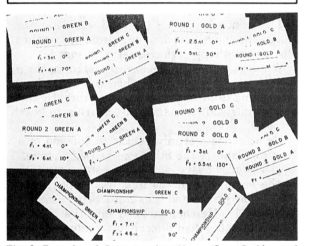

Fig. 2. Examples of Great American Vector Game Problem and Answer cards.

jectories." The idea is to have the student, armed with his knowledge in projectile motion, determine where a sphere which is projected from a table will hit the floor. Using a little "pill holder" paper cup (a tad larger than the metal sphere used in the experiment) the teacher locates the position, calculated by the team, where the sphere is supposed to hit. It is carefully and slowly positioned as the "suspense" builds. The run is made. Bingo, right in the cup . . . or perhaps not so! Points are awarded for accuracy, and the target cup, the nemesis of all projectile motionists, moves on to the next team.

Most topics will lend themselves to such antics if one just sets the imagination free. Accuracy with a beam of light through a series of lenses and/or reflectors, accuracy in determining the release point for a ball in the loop-the-loop apparatus, minimum speed for a ball to just get over the top of a hill, and determination of the amount of current needed in a current balance which causes a certain deflection of the loop are only some of the many activities conducive to the game idea. The point is that the concept of friendly competition in a light-hearted atmosphere can be created in the physics class at the culmination of a unit of study. It serves to motivate students and summarize the information learned in the unit. And, of course, if nothing else it's just plain fun.

THE WICKED KING AND THE BEAUTIFUL PRINCESS

Samuel Derman

New York University and Hunter College
New York, N. Y.

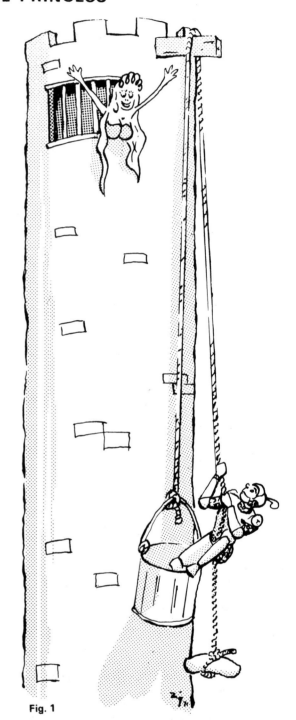

Fig. 1

Once upon a time there was a terribly wicked King who had a daughter who was very beautiful. This daughter was in love with a handsome prince, and before long the prince and the princess had become inseparable. The King, however, was wicked and did not believe in happiness and so he had his daughter locked up in a prison at the top of a tall tower.

The prince learned of this and was determined to rescue the one he loved, so he started out for the tower where the unhappy princess sat imprisoned.

When he arrived at the base of the tower the prince looked up and noticed that there was a wooden beam protruding from the top of the structure. He immediately contrived a method to use this to reach his princess.

He attached a sturdy bucket to one end of a very long rope and to the other end he tied a stone. Then with a mighty heave he threw the stone across the top of the beam so that the rope was looped across the beam. The prince had thus constructed a simple pulley (Fig. 1). He then stepped into the bucket, and since the pulley had a mechanical advantage of *two*, he proceeded to hoist himself up.

In due time the prince reached the top and was rewarded with a long embrace by the King's daughter. The prince, however, could not return the embrace, nor could he begin his work to release the princess, since letting go of the rope would cause the bucket to fall. So he began searching for a way to fix the rope to the tower wall.

Luck seemed to be smiling on the young man because close by, he discovered a metal hook imbedded in the stone wall. The prince tugged on the hook with one hand (the other hand holding the rope tightly), and finding it secure, he proceeded to tie the rope to the hook. But the instant he did that, the supporting

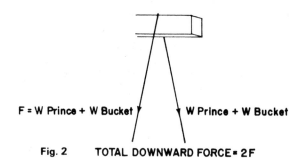

Fig. 2   TOTAL DOWNWARD FORCE = 2F

beam broke and the bucket, together with the poor prince, came crashing to the ground.

What had happened was this. The King, who was very wicked, also happened to know his physics very well (no connection between the two), and he had originally designed the beam to support the weight of the prince and the weight of the bucket, but no more.

During the time the unsuspecting prince was hoisting himself up, the total load on the beam was simply his weight plus the weight of the bucket. But as soon as one end of the rope was hooked onto the tower, the situation changed drastically. Now the weight of the prince plus the weight of the bucket all exerted a force on one end of the rope while the tower, via the hook, pulled down on the other end with an equal and opposite force (Fig. 2). The total force on the beam was now twice the original weight. This exceeded the supporting strength of the beam, causing it to break. That was the end of the prince and this the end of this story.

This story, modified slightly by time, was told originally by my first-year physics teacher. I have used it on my own physics classes and have found that it still never fails to delight the listener as well as the teller.

NOT SUCH A HARD BUMP...?

In a recent note [TPT **9**, 387 (1971)] S. Derman describes how a prince pulls himself up a tower by a rope thrown over a beam; but when the rope is attached to a hook, the beam breaks and the prince crashes to the ground. The physics is excellent, but the answer is wrong. When the prince reached the top and secured the rope to the hook, only a short piece of rope remained between the hook and the bucket. Hence, when the beam breaks, the prince only drops a few feet, and does not crash to the ground.

GARY D. GORDON
Comsat Laboratories
Clarksburg, Md. 20734

A tensile strength lab-contest

Arnold Gorneau
Benjamin Franklin High School, Philadelphia, Pa. 19130

Fig. 1. *Increasing the mass supported by a strip of paper.*

We try to provide a variety of labs to hold the interest of our culturally-deprived students. The Tensile Strength Lab-Contest suits those who want a lab that is easy to understand. It offers an opportunity to manipulate lots of equipment, it encourages creativity, and it is unstructured. Motivation arises from a built-in element of competition, and since some worthwhile concepts of physics are involved, even the instructor is pleased.

Students (preferably in pairs) carry out their own experiments on a trial-and-error basis. They are charged to "Find the tensile strength of a strip of paper." A variety of materials is made available, and the students choose what they want to use in order to investigate the problem. The instructor's explanation is intentionally cursory. The gist of the experiment is in *how* the student uses the materials available, and how he constantly changes (hopefully improves) his design.

Students use a single segment of tape torn from a roll of the type used with acceleration timers. It is white paper, about one-half inch wide. They usually take a piece about a foot long, attach one end in a clamp on a ring support, and form a loop at the bottom. It is this loop which will support weights that are added until the tape rips. The amount of weight supported at this time is recorded.

Students are constrained to cause the tape to break in the stretched portion only. If it breaks in a loop (at the top or bottom) this is probably due to shear forces, and the maximum tensile strength of the paper has not been obtained. Hence, the trial is unsuccessful, and the student must *redesign* and try again.

Equipment available includes rubber bands, short wooden dowels, paper clips, masking tape, "C" clamps (to hold the iron ring stands stationary), and weights. I found that slotted weights and weight hangers are best. Beginning attempts show surprisingly little knowledge of these materials and justify the need for this experience. Some unsuccessful attempts I observed include:

Loops held together using paper clips. (The paper slips through the clips).

Weights are placed within the bottom loop. (The loop cannot hold many weights).

Small weights are attached (one or two hundred grams) and the student waits (for paper fatigue?) for the rip to occur.

The top loop is omitted, and the paper is held in a test tube clamp (at the sharp edge).

A paper clip is twisted open and pierced through the bottom loop. The first weight added tears the clip through the loop.

Learning now begins — and with little teacher intervention. Someone attracts the class's attention. His design is supporting 600, 800, 1000 g! He is using rubber bands in the loops. The loop will crunch together and resist ripping — for a while. Now other groups begin to get ideas. The loops are reinforced with tape. The maximum weight now passes 2000 g. Still, no one gets a good, clean break far from either loop. Good designs now begin to appear. Loops are formed around wooden dowels Both ends are supported by "shock absorbing" rubber bands. Groups now run out of room on the weight hangers, and out of weights.

To see an accumulation of over 3 kg supported by a thin strip of paper is memorable. To see the evolution of a good design — better. To *create* a good design — is rewarding and motivating.

Allow at least an hour for this experiment. If your group should meet with quick success (really very unlikely) suggest doubling the tape, or redesigning the equipment to find the tensile strength of thread, a wood splint, etc.

This experiment is not set up for mechanical engineers. It contains technical flaws. But the lack of structure has great appeal to adolescents. Try it. They'll like it!

Using Blocks to Demonstrate Inertia, Center of Gravity, and Friction

James R. Keady
Sonora Union High School, Sonora, California

Most teachers are familiar with the experiment in which one pulls a cloth from under either a glass of water or a baseball bat standing on its end. For an interesting variation in which one may demonstrate center of gravity and friction as well as inertia, use a 4 in. × 4 in. × 4 in. wood block with some 2 in. × 2 in. × 4 in. wood blocks. Stack about five blocks on end on top of a sheet of paper placed near the edge of a table, then quickly jerk the paper downward from under the blocks. Few in the class are surprised that the blocks do not fall because most have seen this attempted with a ball bat in elementary science.

To make it more interesting, stack the blocks higher to see if they will fall. Before the paper is pulled from under the blocks, the class and teacher can speculate about what will happen and *why*. This brings out a discussion of center of gravity; whether the coefficient of friction has been changed; forces; and inertia. Interest rises with each additional block for the pupils are sure that the stack will fall this time.

Now is the time a pupil will ask, "What happens if the paper is pulled slowly?" By starting the paper very slowly the blocks move with the paper and do not fall. This surprises most of the class. Repeat the experiment pulling the blocks slightly faster. The blocks will now fall as shown in Fig. 1. This raises the question, "Why does the top of the stack stay behind the bottom of the stack?" This can lead to further discussion of inertia.

Ask, "What would happen if the stack were started in motion and then you stopped moving the paper?" The stack will fall forward as in Fig. 2.

Now add the large block to the stack just above the bottom block as shown in Fig. 3. Again questions are raised about the new stability of the stack, the center of gravity, friction, and inertia and which factor will have the greatest effect. After the paper is jerked from underneath and the results observed, the large block is placed at the bottom of the stack. Again discuss the changed situation, including the broader base and its effect on the stability.

After this experiment, the final experiment usually suggested by the pupils is, "What will happen if the large block is placed on top of the stack?" Be sure to stack the blocks as high as the maximum height attained in previous experiments. The result is usually surprising.

This classroom demonstration may be used in general and physical science, but is especially effective in physics after friction, inertia, and center of gravity have been studied. Questions from parts of this demonstration may also be used effectively as examination questions. An enjoyable demonstration such as this invites active participation, thought, and argument by members of the class.

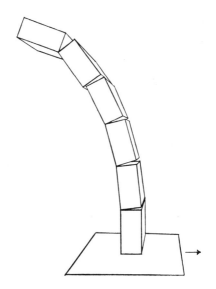

Fig. 1. Blocks falling after paper is pulled forward moderately fast.

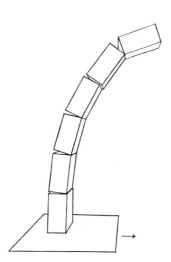

Fig. 2. Blocks falling after paper is slowly pulled forward, then suddenly stopped.

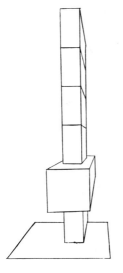

Fig. 3. Stack of blocks with large block before being moved.

String and sticky tape experiments

Section Editor: R.D. Edge, Physics Department, University of South Carolina, Columbia, South Carolina 29208

Dropping a string of marbles

The ear can provide quite an accurate measurement of time in the absence of a clock. We are very sensitive to changes in beat in music, and Galileo is thought to have timed spheres rolling down an inclined plane by singing a song with a very steady rhythm. Even today, darkroom workers use this method of timing where the slightest amount of stray light can fog a film. (For example, "Onward Christian Soldiers" has a beat of about half a second.) An experiment on the acceleration due to gravity, g, can easily be performed using this ability. You require five marbles, a piece of string, and some sticky tape. The string should be as high as the room, but we will suppose this is 8 ft. The marbles are taped to the string at intervals proportional to the squares of the whole numbers, i.e.,

Number	0	1	2	3	4
Square	0	1	4	9	16
Distance	0	6 in.	24 in.	54 in.	96 in.
Difference		6 in.	18 in.	30 in.	42 in.

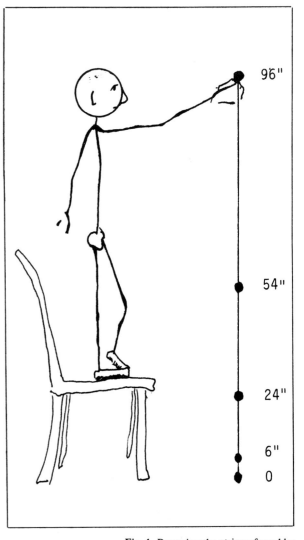

Fig. 1. Dropping the string of marbles

Now, stand on a chair holding the string as shown in the figure. The bottom marble should not quite touch the floor. Drop the string, and listen to the clicks. They are more audible if you drop the string into a trash can, or onto a metal plate.

You can repeat the experiment with a string having marbles spaced at uniform 2-ft. intervals. Do you hear the time between clicks get shorter as the higher marbles from this last string strike the floor? Qualitatively, the higher marbles have been accelerated for a longer time, and are travelling faster, covering the same distance in a shorter time as they approach the floor than do the marbles starting near the floor.

Quantitatively, we have the familiar formula

$$\text{distance} = \frac{1}{2} g \,(\text{time})^2$$

We spaced the marbles on the nonuniform string so the square roots of successive distances are proportional to whole numbers. The time taken between successive clicks should then be constant, about 0.176 sec. Shift one of the marbles up or down the string to test the sensitivity of your ear to the time between clicks. A change of 20% is easily detectable, and some people can do much better than this.

# A POTPOURRI OF PHYSICS TEACHING IDEAS — MECHANICS

THE FALLING METER STICK

J. Thomas Dickinson

Washington State University
Pullman, Washington 99163

Textbook problems involving phenomena which can be duplicated in the physics laboratory often make excellent student experiments. It offers the student the opportunity to test the "system" (i.e., the physics he applies to solve the problem) as well as fulfill the urge to test himself (i.e., the correctness of his reasoning and manipulation of equations). With the apparatus at hand, the student often is motivated to explore interesting modifications of the problem, thus encouraging him to "milk" the problem further. Consider the problem of the falling meter stick[1] as an example. A meter stick is released from a vertical position with one end on the floor. The student is asked to calculate the speed of the moving tip when the stick hits the floor, assuming the end of the stick on the floor does not slip. This standard problem suggests a simple piece of apparatus which can be used for several experiments in the elementary physics lab.

Apparatus

The apparatus (Fig. 1) consists of a meter stick mounted on a ¼-in. steel shaft with a ball bearing. The stick is free to rotate the full 360° about this axis and this rotation is essentially frictionless. A small steel slug attached to the tip of the meter stick allows an electromagnet to be used for holding the stick in any desired initial position and releasing it when desired. The position of the stick as a function of time is recorded on Poloroid film (e.g., Type 47) using either a strobe lamp or strobe disk with photofloods. The camera is usually positioned so that all of the stick is in view over the entire 360°. A black cloth is used for a background and a small piece of reflecting tape on the end of the meter stick is used to improve contrast.

The position data are analyzed by cementing the photo in the center of a page of the student's lab notebook with rubber cement, "extending" the meter stick with lines drawn on the paper, and measuring the angular positions of the stick with a protractor. The time interval between exposures of the moving stick is determined using a photocell and a surplus binary scaler. Figure 2 shows a simple circuit for producing pulses suitable for counting. The pulses are generally counted for 60 or 100 sec. The student is asked to determine the uncertainty in the time interval which brings up a consideration of reaction time (manual stop–start of counter), precision of the clock

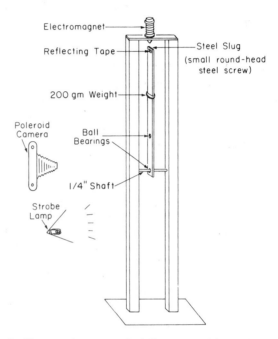

Fig. 1. Diagram of apparatus for falling meter stick experiments.

used to time the count, etc. A triple beam balance is used for determining masses.

Among possible alterations of the apparatus is the addition of a small weight (200 g) that can be clamped to the stick at any position. Also, a second bearing has been mounted at the 30-cm mark to provide an alternative axis of rotation. A friction fit is employed between the shaft and inner race of the bearing, so that changing axes of rotation is a simple matter. Finally, provision has been made for attaching a second stick to the end of the first, again with a shaft and bearing, producing a coupled pendulum.

Predictions

In the write-up for the experiment, the students are given a few basic equations and definitions to help them focus their attention on the problems of interest; e.g., parallel axis theorem, conservation of energy equation ($mgh = \frac{1}{2} I\omega^2$), equation for the center of mass of a system of particles, and the period of a physical pendulum. The student is asked to choose or devise one or more experiments, calculate predictions,

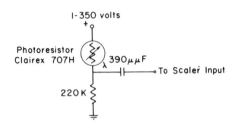

Fig. 2. Circuit for producing pulses from light flashes, either from strobe lamp or through rotating strobe disk.

do the experiments, and compare the experimental and predicted results.

Experiment

The suggested experiments are as follows:

(1) Determine the velocity of the tip (or the angular velocity) as the meter stick passes through the horizontal position. Repeat for the bottom of its swing. (Consider the ratio of these two velocities.)

(2) Determine the moment of inertia of the meter stick about the axis of rotation by measuring the period of the swinging stick.

(3) Repeat (1) and (2) for a second axis of rotation.

(4) Add a mass m_2 at an arbitrary position along the meter stick and repeat (1) and (2). Before making calculations and carrying out measurements, make a qualitative guess as to the effect of the added mass on the way the stick falls and oscillates.

(5) Construct a coupled pendulum by attaching the second stick to the end of the one meter stick. Observe the motion of the system for a variety of initial positions at release. Try to describe qualitatively the motion of each stick. Do you see repetition of the movements observed? Do you see something resembling "beats"?

Conclusion

Students obtain very satisfying results for the quantitative parts of the experiment (typical agreement within 3%) and are quite enthusiastic about trying several parts of the experiment. The effects of adding the 200-g mass often contradicts their guesses. The coupled pendulum is treated as a novelty by most of the students; however a few have attempted to determine experimentally the time dependence of the angular positions (relative to the vertical) of each stick. One pair of students studied the damping of the swinging meter stick, starting from the vertical position. Another pair investigated the maximum angular deflection the stick experienced when it received the same impulse at different positions (points of percussion). The impulse was delivered by a rubber-band launcher indented the same amount for each release. The initial position of the stick was in the vertical down position.

Thus, a standard textbook problem has suggested a relatively simple piece of apparatus which leads to a variety of interesting experiments.

References

1. This problem appears in several texts in various forms; e.g., D. Halliday and R. Resnick, *Physics I* (Wiley, New York, 1966), p. 294, problem 39.

Catch a dollar bill

William Schnippert
Miss Porter's School, Farmington, Connecticut 06032

Here is a motivating activity that may be used before doing the familiar experiment in which the acceleration due to gravity is measured by dropping a meter stick between the fingers.

Bet a member of the class that he cannot catch a dollar bill if you drop it between his fingers, starting at the midpoint on the dollar. You can demonstrate how "easy" this is by releasing the dollar with one hand and catching with the other.

If you have the victim rest his/her hand on the table so the hand can't follow the drop the dollar can't be caught.

Improved Suspension for Acceleration of G Apparatus

Sam Hirschman
Forest Hills High School, Forest Hills, N.Y.

The inexpensive apparatus* which uses a swinging meter stick to obtain the value of g can be improved by using a drilled ball for the falling body and a suspension system which consists of an eye screw and two finishing nails. With this suspension, it is possible to reduce the student error, in determining the acceleration of gravity from 8% to 2% or less.

Hammer two finishing nails partway into a wooden block, as shown in Fig. 1. File a notch in the nail at *A* to serve as a bearing for the screw eye. With the meter stick hanging *vertically,* pass the cord over the nail at *B* and rotate the entire wooden block until the middle of the ball just touches the zero mark at the end of the meter stick. If the ball were to be released in this position, it would just graze the surface of the meter stick all the way down.

When the meter stick is pulled back by the cord, there is no physical contact between the meter stick and the ball. When the string is released, the ball falls instantly and then collides with the meter stick when the stick has reached the vertical position (see Fig. 1).

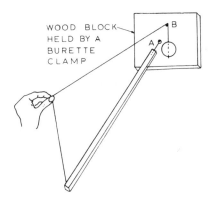

Fig. 1.
Ball and meter stick are released together and collide when the stick becomes vertical.

---
* The Phys. Teacher **5**, 137 (1967).

Free Fall Using Audio Recording Tape

Charles Rudisill,
Price Junior High School, Price, Utah.

The Pohl free fall apparatus* operates by dropping a long paper-covered cylinder past a leaky ink cartridge that is rotated by a 1700 rpm motor. The distances between the ink spots are then measured and analyzed to determine the acceleration of gravity. It is suggested that a strip of audio tape recorder tape be used in place of the paper cylinder and that a small bar magnet be substituted for the ink cartridge (Fig. 3). As the magnet spins, the tape will be magnetized strongly each time one of the poles approaches. If the magnet is placed off center and counterweighted, the period between magnetizing pulses will be twice as long. After each trial, the tape can be examined by sprinkling iron filings over its length. Some filings will stick to the portions that are strongly magnetized. To overcome air resistance, it will be necessary to attach a weight to the bottom of the tape as shown in the illustration.

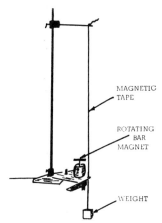

Fig. 3. Tape is magnetized as it falls past a rotating bar magnet.

** *Physics Handbook* (The New York State Education Department, Albany, 1959), p. 101.

String and sticky tape experiments

Section Editor: **R. D. Edge**
University of South Carolina, Columbia, South Carolina 29208

Weightlessness and other ideas

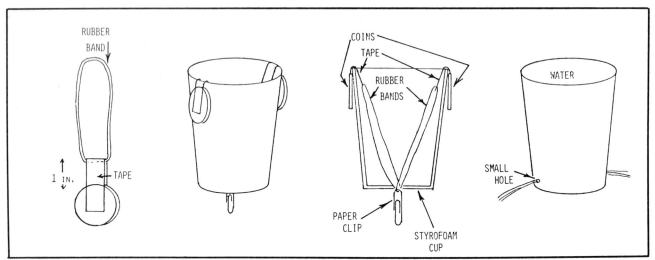

Fig. 1.

Once again Rae Carpenter and Dick Minnix of V.M.I. came up with some very interesting simple experiments at our recent Southeastern Section A. P. S. meeting. Two nice ones relate to the effects of free fall. Two little weights are hung by rubber bands over the edge of a Styrofoam cup. The weights used were small chemical balance weights, but instead, coins such as quarters or half dollars can be taped to rubber bands, leaving about an inch of free tape as shown in Fig. 1. The other ends of two rubber bands go through a hole in the bottom of the cup, and are held there by a paper clip. The bands should be under light tension. Now drop this system, and the weights hop into the cup. This is because the coins become "weightless" when dropped while the tension remains in the rubber bands. In the second experiment, which is even simpler, you poke two holes close to the bottom of a paper cup, as shown. The cup is filled with water, which pours out of the two holes. Now climb a step ladder, hold the cup high at arms length, and drop it, preferably into a trash can. No water runs out as the cup falls — again, the water is "weightless" in the falling system.

Another simple experiment that is fun requires a vacuum cleaner with an outlet that can blow rather than suck. You take a plastic garbage bag and tape the top end with duct tape so that it is air tight, except for the vacuum hose, which is sealed poking into the bag through the opening, as shown. Put a piece of thin plywood on the bag, then sit someone on the bag, and turn on the cleaner (Fig. 2). The bag blows up, lifting quite a heavy individual — and tipping him over unless he is steadied! If the board is, say 50 cm by 50 cm (2500 cm^2, 387 in.^{2}) and if the individual weighs 75

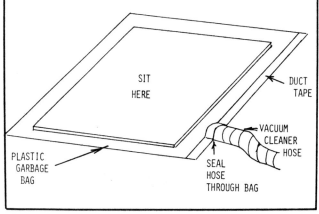

Fig. 2.

kg (165 pounds) the pressure is 2940 pascals (4.26 pounds/sq. in.) which even a relatively inefficient vacuum cleaner can supply.

Peter Rents, editor of W. M. Freeman and Co., has an interesting suggestion for no-cost batteries. A side product of shooting a packet of the new (and expensive) Polaroid film is the "empty" film pack, which contains a "Polapulse" battery. This can easily be removed, and the batteries have plenty of punch left in them. A notice asking students to bring these would probably produce a plentiful supply, at least in those schools where parents could afford the cameras! The batteries give 6 V, and the one I tried gave 6 to 8 A for a few seconds on short circuit, becoming quite warm in the process.

Weightlessness and free bodies

D. Easton
Lacombe Composite High School, Lacombe, Alberta, Canada T0C 1S0

In the March 1981 issue of *The Physics Teacher*, R. D. Edge described a demonstration of the "weightlessness" of free-falling objects.[1] The demonstration made use of a Styrofoam cup with two elastic bands fastened to the bottom on the inside. The elastic bands were stretched over the lip of the cup, one on each side, and weights were suspended from them to keep them stretched (Fig. 1). Weightlessness was demonstrated by dropping the cup. During the free fall the weights became "weightless" and the elastic bands pulled them into the cup.

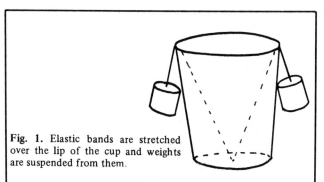

Fig. 1. Elastic bands are stretched over the lip of the cup and weights are suspended from them.

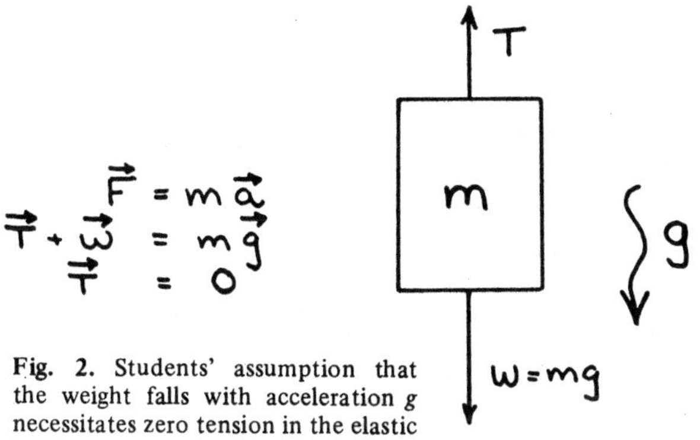

This same demonstration lends itself to another analysis that does not require the introduction of the concept of weightlessness. Using free-body analysis to isolate the forces on the weights and assuming a freefall acceleration of g for the weights (I know this is not correct, please read on), students are able, before seeing the demonstration, to calculate the tension in the bands to be zero (Fig. 2). This, of course, is incompatible with a stretched rubber band. The student finds himself in an either-or situation that necessarily tests his faith in Newton's second law of motion. Either Newton's law is correct, in which case the elastic bands are not stretched over the lip of the cup; or, the bands remain stretched over the lip of the cup and the tension is not zero. A test of faith of this nature does no one harm and for some, the weights "popping" into the cup will come as somewhat of a revelation.

Fig. 2. Students' assumption that the weight falls with acceleration g necessitates zero tension in the elastic bands.

The mechanism by which the weights are pulled into the cup can also be made clear by close scrutiny of the free-body diagrams for the weights and the cup. They reveal that the assumption that the cup and weights each fall with an acceleration of g is incorrect (Fig. 3). Rather, the cup and weights initially have very different and changing accelerations (Fig. 4). They do a short "acceleration dance" that ends with the weights inside the cup.

Reference

1. R. D. Edge, Phys. Teach. **19**, 190 (1981).

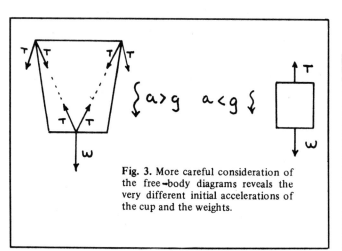

Fig. 3. More careful consideration of the free-body diagrams reveals the very different initial accelerations of the cup and the weights.

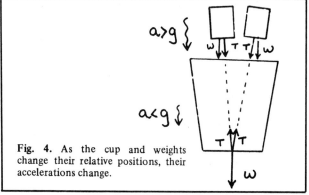

Fig. 4. As the cup and weights change their relative positions, their accelerations change.

The measurement of "g" in an elevator

Haym Kruglak
Western Michigan University, Kalamazoo, Michigan 49001

The Behr free-fall apparatus was used to obtain records of a freely falling body in the cage of an hydraulic passenger elevator. Tapes were made for the acceleration and deceleration during the upward and downward motions. The bob was released as soon as the elevator was started. Control data were taken with the elevator at rest. The average velocity-time graphs for the accelerated motion are shown in Fig. 1. It is evident that the elevator accelerated uniformly during the recorded time interval. The values of "g" for the stationary, ascending, and descending conditions were calculated from the slope to be 9.84, 10.36, and

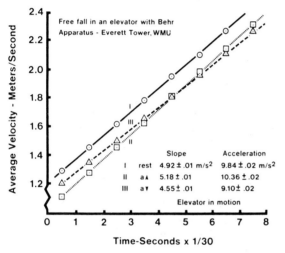

Fig. 1. *Average velocity – time graphs from Behr free-fall data in a stationary and accelerating elevator.*

9.10 m/sec², respectively. The uncertainty from reading the graph was - .02 m/sec². Thus, the acceleration relative to ground of the ascending elevator was 0.52 m/sec²; of the descending, - 0.74 m/sec².

With the elevator in what was judged to be uniform motion, the "stop" button was pressed and a record was made immediately afterward. The average velocity-time graphs for this motion are displayed in Fig. 2. The deceleration of the elevator was also uniform. The values of "g" during this phase were 9.76, 8.70, and 11.20 ± .02 m/sec², respectively, for the stationary, ascending, and descending conditions. Thus, the deceleration of the upward-moving elevator was -1.06 m/sec²; of the downward-moving, 1.44 m/sec². Numerically, the deceleration was greater than the acceleration by a factor of 2.

The student may be asked to collect data in an elevator or be given a prerecorded tape which he uses to make the customary measurements and calculations associated with the Behr free-fall exercise. As part of the laboratory report, the student can be asked to calculate the acceleration of the elevator relative to ground.

In using the Behr free-fall apparatus it is important to have a 120-V ac outlet for the spark timer operation. The new elevator for which the above data were obtained did not have an outlet. To comply with the fire marshal's regulations it was necessary to have an electrician make a temporary connection through a trap in the cage ceiling. The electrician rode on the cage roof while the data were being collected!

Values of "g" in the moving elevator were also calculated from the observed maximum elongations and compressions of a light spring from a Jolly balance. The values of

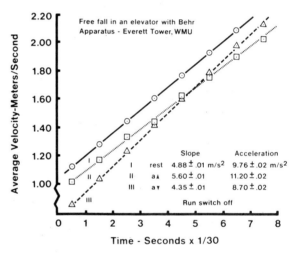

Fig. 2. *Average velocity – time graphs from Behr free-fall data in a stationary and decelerating elevator.*

"g" were in good agreement with those obtained from the Behr apparatus data as shown in Table 1.

It is easy, inexpensive, and instructive to use a personal weight scale to demonstrate the weight change in an accelerated frame. The data for five trials of one student are given in Table 2. The average scale readings and the known value of "*g*" can be used to calculate the acceleration and deceleration. The precision and accuracy is, of course, lower than those obtainable with the two methods described above.

These three exercises provide students with the opportunity to use several laboratory techniques, to relate the acceleration to the motion experienced in an accelerated reference frame, to practice Newton's second law and Hooke's law, and to observe weight changes in an accelerated frame.

The advantages to the instructor are: laboratory exercise variety, data tapes with an unknown for the laboratory practical test, and greater student motivation.

I am indebted to Professors J. Dewitt, J. Herman, and N. Nichols for helpful suggestions, and to Mr. James Avery for able technical assistance.

A POTPOURRI OF PHYSICS TEACHING IDEAS — MECHANICS

Table 1. *Acceleration of a freely falling body in an elevator. (Everett Tower, Western Michigan University, March 1971)*

		Acceleration (m/sec^2)			
		Up		Down	
	Rest	Behr	Spring	Behr	Spring
g	9.84 ± 0.02	10.36 ± 0.02	10.48 ± 0.07	9.10 ± 0.02	9.01 ± 0.08
Δg	0.04	0.56	0.68	-0.70	-0.79
Δg/g		5.4%	6.5%	-7.0%	-8.1%
		Deceleration (m/sec^2)			
		Up		Down	
	Rest	Behr	Spring	Behr	Spring
g	9.76 ± 0.02	8.70 ± 0.02	8.60 ± 0.12	11.20 ± 0.02	10.64 ± 0.08
Δg	0.04	-0.90	-1.20	1.40	0.84
Δg/g	0.4%	-9.2%	-12.3%	14.2%	8.6%

Table 2. *Weight variation in an elevator. (Everett Tower, Western Michigan University, February 1971; Observer: J. Avery) (See text, p. 467.)*

		Acceleration		Deceleration	
Weight	Rest	Up	Down	Up	Down
W	181.5 p	186.5 ± 0.5 p 186.0 187.0 188.0 188.0	170.0 ± 0.5 p 170.0 173.0 170.0 170.0	172.5 ± 0.5 p 170.0 172.0 172.0 172.0	187.0 ± 0.5 p 187.0 188.0 188.0 188.0
$\overline{W}$	181.5 p	187.0 ± 0.5 p	170.5 ± 0.5 p	171.5 ± 0.5 p	187.5 ± 0.5 p
ΔW	0	5.5 ± 0.5	-11.0 ± 0.5	-10.0 ± 0.5	6.0 ± 0.5 p
ΔW/$\overline{W}$	0	3.0%	-6.5%	-6.5%	3.3%

Apparent weight changes in an elevator

Larry Jensen
Baldwin High School, Pittsburgh, Pennsylvania, 15216

The elevator doors slipped silently together. Ron shifted on the bathroom scale under his feet (Fig. 1). Someone tapped the button for the 64th floor and a surge of acceleration jumped through our soles. A smartly dressed woman across from me gave us an intrigued look of unanswered curiosity. The fellow on my right, young and bearded, glowed with satisfaction at being involved in this mystery, certainly not the customary entertainment on the express to the Top of the Triangle, high atop the U. S. Steel Building in Pittsburgh.

The pointer on the scale left 188 pounds, quickly moved up to 214 (Fig. 2) and then plopped back to 188 (Fig. 3).

"We're doing a physics experiment," I said, launching into one of the fastest moving lectures I've given lately. "The upward force as the elevator accelerates seems to increase our weight temporarily. When we stopped accelerating a moment ago, our weight returned to normal. The elevator force and the tug of gravity are equal now. In a few seconds we'll all seem to weigh less as the deceleration takes place."

The woman flashed us a smile at the anticipation of this entertaining new idea. Nature neatly tied off the lecture. The scale indication of Ron's weight did indeed drop as we rapidly slowed — to 164 pounds in about 5.5 seconds (Fig. 4).

Two intrigued people stepped away from us at the 64th floor and we returned groundward. We had piqued their curiosity and given them an entertaining fact to carry away with them. It had never been our intention to amuse a captive audience in an elevator, although I suppose more gutsy people than ourselves might invent a whole repertoire of mini-demonstrations to spring on the populace unawares. Ron and I had simply wanted to slip quietly in and out of the U. S. Steel Building to record on film a common force example, one he and his sister had nicely verified over the vacation break.

I listened with a great deal of joy when Ron described

A POTPOURRI OF PHYSICS TEACHING IDEAS—MECHANICS

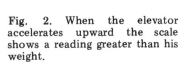

Fig. 1. The experimenter stands on a scale in the elevator. He weighs 188 pounds.

Fig. 2. When the elevator accelerates upward the scale shows a reading greater than his weight.

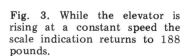

Fig. 3. While the elevator is rising at a constant speed the scale indication returns to 188 pounds.

Fig. 4. Approaching the top floor the elevator loses speed and the scale indicates a force less than his weight.

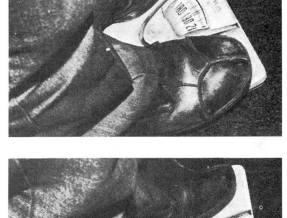

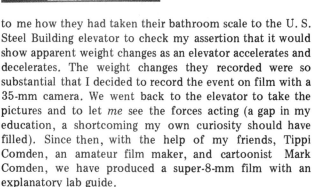

to me how they had taken their bathroom scale to the U. S. Steel Building elevator to check my assertion that it would show apparent weight changes as an elevator accelerates and decelerates. The weight changes they recorded were so substantial that I decided to record the event on film with a 35-mm camera. We went back to the elevator to take the pictures and to let *me* see the forces acting (a gap in my education, a shortcoming my own curiosity should have filled). Since then, with the help of my friends, Tippi Comden, an amateur film maker, and cartoonist Mark Comden, we have produced a super-8-mm film with an explanatory lab guide.

As this project progressed I came to understand elevators much better. They do not simply pull with constant force on a passenger cabin and then brake at the desired floor. They are much more sophisticated than that. Westinghouse designed this elevator to follow a preprogrammed velocity schedule. For a given elevator, no matter what the load, the cabin speed is nearly identical every trip. This is achieved by continuously monitoring the speed and using a feedback loop to regulate the acceleration produced by the drive motor. When the force is greater than the cabin weight, the cabin accelerates, when less, the cabin decelerates. As a result, braking is never necessary. Occasionally, when a rapid deceleration is necessary, a back-emf is introduced to reduce the angular momentum of the armature quickly. To reduce the strain on the motor, the cabin is counterweighted at 40% of the capacity load.

The acceleration adjustments appear on the accompanying graphs — Upward Acceleration near frames 45 and 100 (Fig. 5), and Upward Deceleration near frames 45, 105,

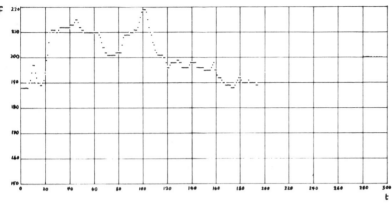

Fig. 5. A graph showing the change of force during upward acceleration. Data were taken from a motion picture of the scale dial, and were plotted frame by frame.

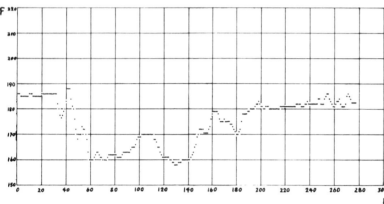

Fig. 6. Approaching the 64th floor the elevator loses speed. The upward force on Ron is less than his actual weight.

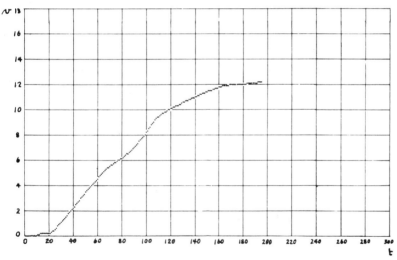

Fig. 7. Velocities were calculated by the trapezoidal rule after first determining the accelerations from the force data. This graph shows velocity versus time as the elevator accelerated upward.

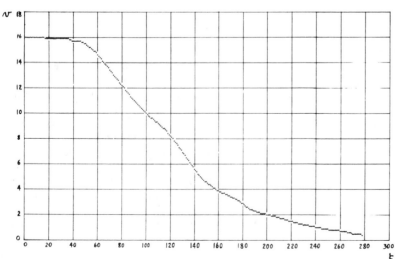

Fig. 8. Velocity versus time as the elevator approached the top floor. Note that Figures 7 and 8 show data for different elevators.

and 160 (Fig. 6). Notice how quickly these adjustments occur and how little fluctuation occurs in the corresponding speed graphs. Not all elevators in the building were identical as shown by the speed graphs, one accelerating to 12 mph (Fig. 7) and the other to 16 mph (Fig. 8).

For readers who wish to investigate this situation themselves, the following information may be helpful. The color filming was done in super-8-mm with Ektachrome 160G film. The cameras were a Minolta XL-400 and a Kodak XL-10 with a wide-angle lens. Low light levels in the elevator were exceptionally hard to deal with. For this we used Kodak 4-X, 16-mm film with a Beaulieu R16 camera. Later we refilmed in Kodak Tri-X, super-8-mm with a Nikon 8X. It may be possible for us to supply a print of our film for interested readers.

Finally, in closing, I would like to acknowledge the considerable and important assistance and advice from Tippi Comden and Mark Comden. To Ron Mikrut who initiated this project and to Ted Shlanta who helped finish off the film, a hearty thanks.

Using a laser to investigate free fall

Edward V. Lee
Landon School, Bethesda, Maryland 20034

The acceleration due to gravity (*g*) is measured by marking a falling meter stick at regular intervals of time. To the stick is taped a long strip of photographic film, which is exposed to a narrow beam of light that flashes on and off frequently. The developed film shows a pattern of spots whose spacing increases with the speed of the falling object. A straightforward analysis of this pattern shows that the acceleration is a constant and yields the value of *g*.

The flashing light is produced by directing the beam of a He–Ne laser onto a rotating shutter — a slotted wheel attached to the shaft of an electric motor (Fig. 1). As the meter stick falls, a spot is exposed on the film whenever the light passes through the shutter. With four narrow slots, evenly spaced around the wheel, a motor speed of 1500 rpm will produce about 45 spots in a one-meter strip of film (corresponding to a flash of light every hundredth of a second). The spots are recorded on 16-mm movie film, whose emulsion side faces the light. The falling stick is kept in a vertical orientation by two sets of wire guides, one above the other. These may be readily fashioned from a ring stand with two rings; wire is wound on each ring to form a rectangular opening that constrains the horizontal movement of the meter stick sufficiently to keep the film always in the path of the light beam. With the light flashing and the room darkened, the meter stick is slowly lowered into the beam; it is then released from rest. The initial motion is vividly displayed by the resulting pattern of spots.

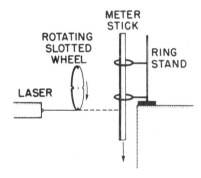

Fig. 1. *Falling meter stick with film attached is marked by flashing laser beam.*

To obtain the velocity *v* of the falling object at the midinstant of time interval Δt we use the formula for the average velocity during that time

$$v = \Delta x/\Delta t. \qquad (1)$$

Here Δx is the separation (easily measured with a ruler) between adjacent spots on the film, and Δt is the time between flashes of light. If the motor frequency (measured with a stroboscope) is multiplied by the number of slits, the result is the frequency of the flashes. The reciprocal of this frequency is Δt, the time between flashes.

The velocity is plotted as a function of time. It is immediately apparent from this linear relationship that the acceleration is a constant, and the value of *g* is simply the slope of the line.

APPARATUS-LECTURE DEMONSTRATION AND LABORATORY

Physical Effects of Apparent "Weightlessness" • Haym Kruglak Western Michigan University Kalamazoo, Michigan

The effects of "apparent weightlessness" can be best shown by allowing a body or system of bodies to fall freely vertically or to follow a parabolic path in free fall. The demonstrations described below can be performed readily in any classroom with simple and inexpensive equipment.

1. Freely Falling Objects

Tie or tape an object, such as a laboratory weight, to the hook of a spring balance with a large dial (Welch No. 4075 or No. 4091). When the balance is held up, the position of the pointer on the scale is interpreted as the "weight" of the object. Actually, the reading of the balance is the reaction force to the gravitational pull of the earth on the object.

Drop the balance and the object from a height of about 10 feet onto thick sponge rubber or other padding on the floor. During the free fall the reading of the balance will quickly return to zero. The balance and the object soon fall with the same acceleration, and the spring in the balance soon pulls the object and balance together so that the spring is no longer extended under tension. Question for student: What would the balance read if it and the object were arranged so that they fell with the same constant velocity?

In place of the balance, one may use a rubber band and a second object from which the rubber band hangs. When the assembly is held up, the gravitational pull on the lower object is counterbalanced by the elastic force in the rubber band. Again, drop the system onto some padding. During the free fall the elastic force of the rubber band pulls the bodies together. To one riding downward with the system, the only apparent force is that produced by the stretched rubber band.

Fasten a rubber band or a spring to the bottom of a container. Stretch the free end of the rubber band over the lip of the container and attach a small object to it so as to counterbalance the elastic force. Drop the system. During the free fall the object and container have the same downward acceleration and the object apparently becomes "weightless." It is pulled into the container by the unbalanced elastic force. A convenient container is a tin can open at one end. If an opening is made in the bottom with a "beer-can" opener, the depressed metal provides a hook for holding the rubber band or an eyehook may be soldered to the bottom of the can.

2. "Buoyant" Force

Place a small floating object such as a cork in a tall glass or plastic cylinder. A long cylinder (20 inches) is preferred. Fill the container with water and seal it. Turn the cylinder upside down to show that the object will rise to the top. Turn the cylinder over once more, but when the object has risen about half-way up, quickly allow the cylinder to fall to a catcher below, or toss it to an assistant several feet away. The object will appear to be at rest relative to the tube. The effective buoyant force on the object while in motion will be zero because the difference in the pressure at the top and bottom of the object is zero. Repeat the demonstration using a sinking object such as a rubber stopper. Fill the cylinder with water and an air bubble and repeat the experiment. Plexiglass or other transparent plastic tubing may be used to make permanently sealed demonstration cylinders. Convenient dimensions are: length 18-20 inches; diameter 1.5 inches.

3. Elastic and Buoyant Force

Attach a lead sinker to one end of a spring or rubber band; attach the other end to a large cork. Place the system in a tall cylinder with the sinker at the bottom. Fill the cylinder with water. Choose a spring such that the tension will allow some of the cork to be above the water surface. Drop the cylinder to a catcher. The buoyant force during the free fall becomes zero, but the elastic force of the spring will pull the cork under the water surface.

4. Convection Currents

Attach a candle to the bottom of a cork or cap which fits a gallon bottle. Light the candle and push the cork firmly into the neck of the bottle. Turn the bottle upside down and drop it to an assistant from a height of 8 to 10 feet. The candle will go out. Repeat by tossing the bottle straight up. Repeat by throwing the bottle in a parabolic curve to an assistant. During the motion there are no convection currents inside the bottle to remove the products of combustion and to bring oxygen to the wick. Heated air and cold air in the bottle fall at the same rate and there is no pressure gradient within the bottle. To the system of the bottle, the gas has no "apparent" weight. The air in a gallon bottle will support combustion for about a minute. The candle can be pushed onto a nail or sharp wire soldered to a metal cap.

5. Efflux from an Orifice

Drill or punch a small hole near the bottom of a tall tin can. Stopper the hole and fill the can with water. Hold the can several feet above a basin and remove the stopper so that water can flow out of the hole. Drop the can so that an assistant can catch it above the basin. During the free fall the flow of water will cease. The water and the container will be falling together and the water does not leave the container.

6. The Pendulum

Attach a pendulum to a clamp or to a short rod. Hold the pendulum support in your hand a few feet above ground or table. Let the pendulum swing. Drop the pendulum and its support while the pendulum string makes a fairly large angle with the vertical. The string will go slack; the pendulum will maintain the same position relative to its support. In other words, the pendulum stops swinging, that is, its period becomes infinite.

In deriving the formula $T = 2\pi(L/g)^{1/2}$ for the period of a pendulum, we assumed an inertial system, one in which Newton's laws of motion were valid and also assumed a gravitational force. If we ride downward with the falling pendulum and use a falling frame of reference, the apparent weight is zero (or $g = 0$). In this frame of reference, we must also be very careful about using Newton's laws of motion in the usual form. If we describe the situation from an inertial frame of reference, (say, the surface of the earth in which Newton's laws are valid) the tension in the cord attached to the bob of the pendulum becomes zero and the only force acting on the pendulum bob is its "true" weight. It will be accelerated vertically downward and will not swing.

7. Simple Harmonic Motion

Suspend an object from a vertical spring attached to a rigid stand. Pull the object down so that the spring is stretched several centimeters and let it oscillate up and down freely. The motion will be simple harmonic. If the system, rigid stand and all, is dropped, the object will be pulled up by the elastic force of the spring and the simple harmonic motion will stop. In the freely falling system, a relative displacement between the object and the coils of the spring does not change the "gravitational" potential energy of the object relative to the falling system. Consequently, there is no gravitational potential–kinetic energy conversion.

8. Buoyant Force and Centripetal Force

Another experiment in which an

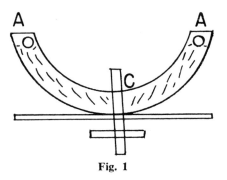

Fig. 1

"artificial gravitational field' is involved is described by J. S. Miller (*1*). A clear plastic tube (1-inch bore, 36 inch long) is bent into a circular arc and placed on a support which can be rotated about a vertical axis (Fig. 1). The closed tube is filled with water. If two small wooden blocks are in the tube they rise to A at the highest point in the tube when the tube is at rest, but "fall" to C, the lowest point when the tube rotates with sufficient speed about the central vertical axis. Try the experiment with steel balls in place of the wooden blocks.

It is a good exercise and may clarify thinking if these observations are accounted for by two "explanations,"

---

1. J. S. Miller, Am. J. Phys. **30**, 385 (1962).

one that uses the term "centrifugal" force and another that does not use this term. Note how in the rotating frame of reference, centrifugal force may be thought of as produced by an artificial gravitational field.

9. Other Possible Demonstrations

The demonstrations described here are not new. Interested readers are invited to think about and develop other demonstrations of "weightlessness." One possibility is to show that the force of friction vanishes during free fall if the normal force between surfaces is produced by gravity. It should also be possible to show that viscous friction is independent of "weight." You might ask your students: "Does the pressure of a gas in a closed container depend on the weight of the gas? Is it modified when a closed container filled with gas falls freely?" Devise an experimental test to support your answer. The writer and the editor of this Journal will be glad to hear of any suggestions for additional demonstrations and problems on the subject of weight and "weightlessness" as well as the "best" way to "explain" the observations in experiment (8) described above.

A center of gravity demonstration

Terrence P. Toepker
Department of Physics, Xavier University, Cincinnati, Ohio 45207

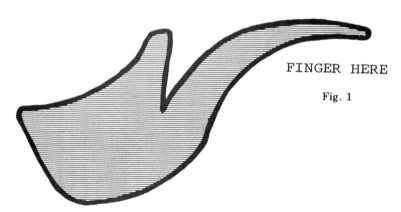

FINGER HERE

Fig. 1

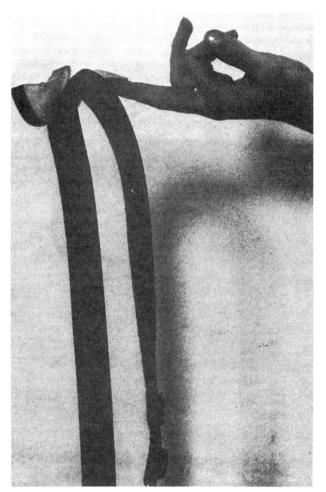

Fig. 2.

The recent article by Vincent Buckwash[1] on the center of gravity prompts me to show the readers of *The Physics Teacher* how we have used the same principle for a public relations purpose.

Our "Belt Hook" is both inexpensive and easy to make. It can be cut from 1 in. x 2 in. wood strips or from any scrap lumber that is available. Figure 1 shows the exact size that we use. After cutting it out, we stamp *Xavier Physics* on it and dip it in varnish for a finish coat.

To use it, try to balance it from the thin end on the tip of your finger as shown in Fig. 1. It obviously won't work. Now, by placing a belt in the slot and allowing it to hang over on both sides, the Belt Hook can easily be balanced on your finger as shown in Fig. 2.

While the practical uses of the Belt Hook are limited, it does provide an interesting demonstration of a fundamental physical principle.

Anytime we present a guest lecture at a high school, we give away a few of our Belt Hooks or our Tippe Tops.[2] We feel that this helps to keep our department in the minds of area teachers and students.

We hope that the readers will find the Belt Hook both entertaining and instructive.

References

1. Vincent Buckwash, *The Physics Teacher*, 14, 32 (1976).
2. Letters, *The Physics Teacher*, 5, 182 (1967).

Ways to demonstrate center of gravity

Vincent Buckwash
Unionville High School, Unionville, Pennsylvania 19375

In our solutions of problems concerning parallel forces and rotational equilibrium we eventually arrive at the point where we must include the torque produced by the weight of the bar, tree, bridge, etc. Hence, it becomes necessary to introduce the concept of center of gravity. I do this with several demonstrations.

First, I take a leather belt and place it in the slot of an ordinary clothespin, so that the clothespin divides the belt in half. The "system" can now be balanced by placing the open end of the clothespin on the end of your finger (see Fig. 1).

The second demonstration, which I call balancing a pin on a needle, requires a more difficult setup. It consists of a piece of soft wood approximately 2½ in. x 1¾ in. x 3/8 in. In the center of one of the 1¾ in. x 3/8 in. sides, insert a large brass straight pin.

The base of the setup is a 500-ml Florence flask half full of water. Push a steel needle through a cork that will fit your flask. The needle should protrude through the cork by about 3/4 of an inch. Insert the cork into the flask with the point of the needle up.

Finally, stick the blades of two pocket knives of approximately equal mass into each of the 2 ½ in. x 3/8 in. sides of the block and fold the knives so the handles are at right angles to the blades. (If the knives are of unequal mass, you can add paperclips to the lighter one to make them equal.)

To balance the pin, experiment with the knife-block-pin part of the setup by placing it on your finger. You will find a spot along the pin where the system will balance. Once you have visually located this spot, place the knife-block-pin system on the needle. It takes a little

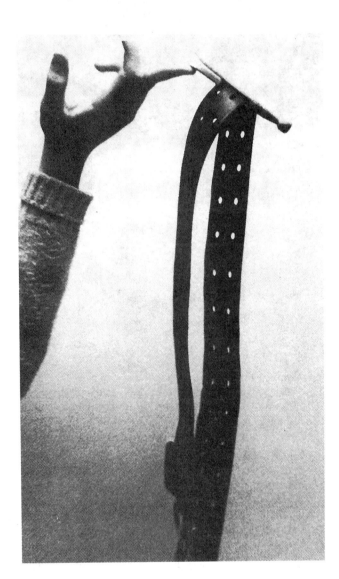

Fig. 1.

Fig. 2.

practice but it works (see Fig. 2).

The greatest interest, however, is created when a tall glass filled with water to about 2/3 of its height is balanced on its edge, on a coin. The coin provides a "heel" so the glass will not slide on the desk. For Fig. 3 we used a wide mouth 7-Up bottle.

Students have tried other shaped containers filled with water and have been able to balance a 250-ml Erlenmeyer flask, a wide mouth beer bottle, a 500-ml beaker and various types of drinking glasses.

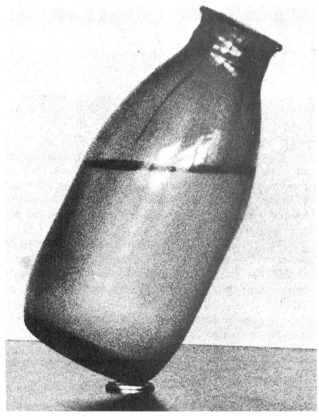

Fig. 3.

WITH A GRAIN OF SALT

Figure 1 shows a picture of a 2000 ml Erlenmeyer flask balanced on a few crystal grains of salt. We were intrigued by the balancing of a bottle on a coin as reported by Buckwash,[1] and succeeded in balancing a large flask on a penny. A passing student commented that they balanced glasses of water on salt grains in the dorm cafeteria. It obviously works with large flasks too. After the balancing is achieved, the extra salt grains can be blown away and the flask appears to balance without support.

For a finale, a little external water will trigger total collapse.

RICHARD MORANDI
University of California
Riverside, California 92521

Reference

1. Phys. Teach. **14**, 39 (1976).

Fig. 1.

A POTPOURRI OF PHYSICS TEACHING IDEAS—MECHANICS

Center of mass revisited

Ernie McFarland
Department of Physics, University of Guelph, Guelph, Ontario, Canada N1G 2W1

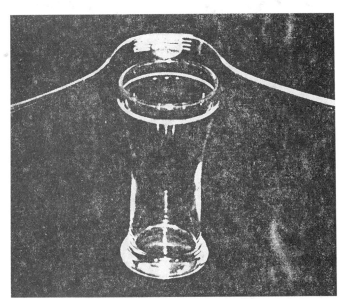

Fig. 1. Most students cannot explain this balancing "magic."

During the past few years, a number of notes[1-5] have appeared in *The Physics Teacher* which describe interesting demonstrations of balancing objects. In each case, stable rotational equilibrium is achieved because the center of mass of the object lies directly below the rotation axis. In this note some useful examples of unstable equilibrium involving the human body are presented, as well as a few more examples of stable equilibrium with inexpensive apparatus.

The stable equilibrium demonstrations will be mentioned first. One demonstration (Fig. 1) is similar to Buckwash's pin and pocketknife example[1] and Toepker's fork, spoon, and match.[2] The version shown in Fig. 1 requires two forks, a drinking glass (a beer glass gains plenty of student attention) and a quarter-dollar. The forks, with handles pointing in opposite directions, are positioned with their tines side by side; the quarter is placed in the middle gap between the tines and the fork-quarter system can be balanced on the edge of the glass. Because the fork handles curve down and in towards the glass, the center of mass of the fork-quarter system lies beneath the point of contact between the quarter and the glass; the result is stable equilibrium. With care the system can be made to oscillate about the equilibrium position, but beware — experimental evidence in support of Murphy's laws indicates that if an

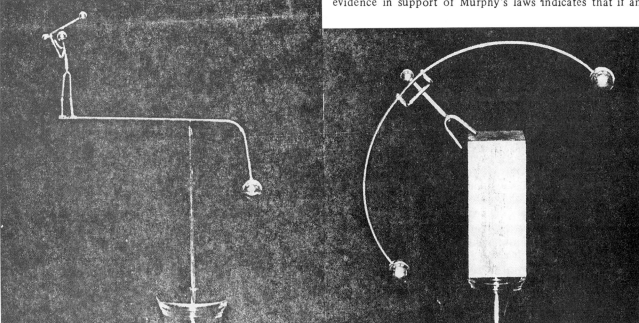

Figs. 2, 3. Commercially available balancers.

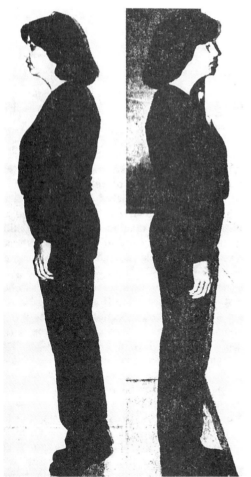

Fig. 4. Standing on your toes is easy . . .

Fig. 5. . . . unless you're facing a wall.

Fig. 6. I can balance on one leg . . .

Fig. 7. . . . sometimes.

Fig. 8. Some people can touch their toes . . .

Fig. 9. . . . but not here.

oscillation is attempted when the glass contains liquid (especially beer), the quarter invariably ends up at the bottom of the glass.

There are many examples of commercially available apparatus (Figs. 2 and 3) which exhibit stable equilibrium. Instead of purchasing them, the teacher can construct them easily and inexpensively. Alternatively, it is fun to have a student competition to design unusual balancing objects.

Four demonstrations related to unstable equilibrium (i.e., center of mass above rotation axis) will now be described. Since they all involve the human body, they can involve a great deal of class participation if desired.

In the first demonstration (Figs. 4 and 5) a student is asked to stand on her toes normally, and then with her toes against a wall. When standing on her toes normally, the student automatically maintains the position of her center of mass directly above a rotation axis through the toes. The torque due to the force of gravity acting on the student is then zero, and unstable rotational equilibrium results (the torque due to the force exerted on the student by the floor is of course zero as well). When the student stands against the wall, it is impossible to bring the center of mass far enough forward so that it lies above the toes. Hence, rotational equilibrium cannot be achieved, i.e., a person cannot stand on tiptoes with the toes against the wall.

Similar examples are shown in Figs. 6 through 9. In each case the student can maintain equilibrium when

Fig. 10. Positioning the matchbox.

Fig. 11. Most males cannot do this.

standing away from a wall by positioning his center of mass above a rotation axis through a foot (feet in Fig. 8), but cannot do so when standing adjacent to a wall.

A demonstration which is mentioned by Watson and Watson[5] is sometimes used as a "party trick," and illustrates the difference in position of the center of mass of a female and male student (Figs. 10 and 11). A kneeling student first places her elbows, arms, and hands together ("praying") with the elbows touching the knees and the forearms along the floor (Fig. 10). A matchbox or other object is placed at the student's fingertips. The student then clasps her hands behind the back (Fig. 11), and is instructed to knock the matchbox over with her nose without her entire body falling over. In general, female students can perform this task whereas males cannot. Equilibrium in the kneeling position can be maintained as long as the center of mass does not move forward beyond a point above the knees or backward beyond a point above the toes. Because the center of mass of a male is closer to the head than that of a female, a typical male cannot knock over the matchbox without moving his center of mass forward of the knees, thereby tipping over.

Students enjoy all of these demonstrations of equilibrium, both stable and unstable, and often are at a loss to explain them, especially when told that all the demonstrations can be explained by the same physics (rotational equilibrium of an object which is free to rotate about an axis occurs when the center of mass lies directly below or above the axis). The demonstrations involving the human body are particularly popular, and students have fun trying to amaze their friends with them.

Thanks go to students Wendy Collard and Jim Bennett who appear in the figures, and to Doug Fox, a teacher at Belle River High School, who provided some of the ideas.

References

1. V. Buckwash, Phys. Teach. **14**, 39 (1976).
2. T. Toepker, Phys. Teach. **14**, 499 (1976).
3. T. Toepker, Phys. Teach. **15**, 241 (1977).
4. T. B. Greenslade, Phys. Teach. **19**, 554 (1981).
5. J. Watson and N. T. Watson, Phys. Teach. **20**, 235 (1982).

CENTER OF MASS OF A ROTATING OBJECT

To show that the center of rotation of a rotating projectile is always located at its center of mass regardless of the trajectory or the rate of rotation, try this demonstration. From a 1-ft square of Masonite or heavy cardboard, cut a free form such as the one shown in Fig. 7. Find the center of mass by balancing the form on the point of a pencil and draw a series of heavy circles around the center of mass (Fig. 7A). On the other side of the form, draw a similar series of concentric circles around a center that is about 2 cm away from the center of mass (Fig. 7B).

Demonstrate the apparatus by flipping it so it travels in an arc across the classroom while it rotates rapidly with side A always facing the class. It will appear to the class that the target is not rotating. Then repeat the demonstration with side B facing the class and have them note the difference.

Marvin Ohriner
Elmont Memorial High School
Elmont, New York 11003

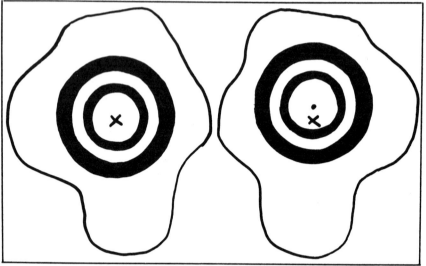

Fig. 7. Free form cut from Masonite has target around center of mass on side A. On the reverse side, B, the target is off center.

More center of gravity

Terrence P. Toepker

Dept. of Physics, Xavier University, Cincinnati, Ohio 45207

Recently Vincent Buckwash[1] presented a number of interesting demonstrations of the center of gravity. While the principle of the balancing clothespin-belt system has been used for a long time,[2] the simplicity of Buckwash's apparatus is greatly appreciated.

Since his note did not explain the phenomenon, I would like to pose the following questions and solutions in order to help the *understanding* of his clever demonstration.

Problem

A. Draw a free body diagram of just the clothespin, showing all of the forces acting on the clothespin. (Remember that the sum of the forces and the sum of the torques on the clothespin must be zero for equilibrium to be maintained.)

B. Draw a free body diagram of the belt showing all of the forces acting on the belt.

Solution

Once the clothespin-belt system has reached its equilibrium position, the upward force exerted by the finger (F_1) must equal the weight of the clothespin (W_p)

A POTPOURRI OF PHYSICS TEACHING IDEAS — MECHANICS

plus the weight of the belt (W_b) as seen in Fig. 1. Also the line of action of W coincides with the line of action of F_1.

$$F_1 = W_p + W_b = W \quad (1)$$

Clothespin

Now consider just the clothespin. Forces are exerted on it at four places:
1. The point of support by the finger.
2. Point 2 where the left side of the belt touches it.
3. Point 3 where the right side of the belt touches it.
4. The center of gravity of the clothespin (W_p).

Fig. 2 shows a free body diagram of the clothespin with the forces acting at the three contact points and at the center of gravity. Each of the four forces has been resolved into forces normal and parallel to the clothespin.

For clarity, the belt has been drawn as a rigid T-shaped object with the center of gravity of the object as shown in Fig. 2. (The "connecting rod" *could be curved* but is drawn straight and perpendicular to the rigid part which is in the slot of the clothespin. This is done merely for convenience.)

If you hold the belt between two fingers as shown in Fig. 3, you can feel both of the forces which the belt exerts on your fingers. Try it even if you believe it!

Linear equilibrium for the clothespin requires:

$$f_1 = f_2 + f_3 + W_p \sin \phi \quad (2)$$

and

$$N_1 + N_3 = N_2 + W_p \cos \phi \quad (3)$$

Neglecting the small torque from the friction forces, rotational equilibrium, calculated about the point of support, requires:

$$d_2 N_2 + d_4 W_p \cos \phi = d_3 N_3 \quad (4)$$

where d_2, d_3, and d_4 are measured along the clothespin. The presence of the force N_3 provides a counter-clockwise torque and cancels the clockwise torques from N_2 and W_p (and from the friction forces).

Belt

Figure 4 shows a free body diagram of the belt (in its simplified rigid form). The forces at points 2 and 3 (N'_2 and N'_3) are exerted by the clothespin on the belt. The weight vector is shown with its components parallel and perpendicular to the clothespin axis. The angle ϕ is the angle which the clothespin makes with the horizontal. Linear equilibrium for the belt requires:

$$W_b \sin \phi = f'_2 + f'_3 \quad (5)$$

and

$$W_b \cos \phi = N'_2 - N'_3 \quad (6)$$

By dividing (5) by (6) one obtains:

$$\tan \phi = (f'_2 + f'_3) / (N'_2 - N'_3) \quad (7)$$

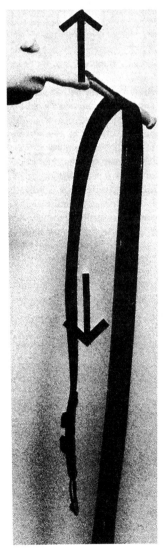

Fig. 1. The upward force on the clothespin exerted by the finger equals the net weight of the belt-clothespin.

Fig. 2. Free body diagram of the clothespin with the forces resolved into normal and frictional forces.

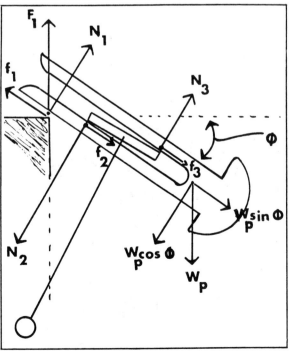

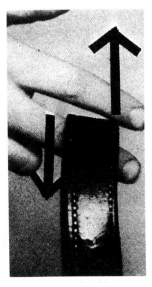

Fig. 3. Forces exerted by the belt on the fingers can easily be experienced.

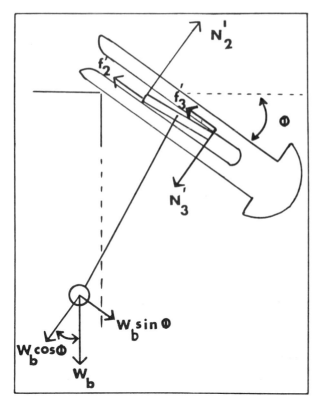

Fig. 4. Free body diagram of the belt with the weight vector resolved into components parallel and perpendicular to the clothespin.

Now using (1), (2), and (3) and the concept of action-reaction forces, it is shown that $\tan \phi = u$, the coefficient of friction between the clothespin and the finger.

Using (2) in the numerator and (3) in the denominator of (7) one gets the following:

$$\tan \phi = (f_1 - W_p \sin \phi) / (N_1 - W_p \cos \phi) \quad (8)$$

Now using

$$N_1 = W \cos \phi \quad (9)$$

Eq. (8) becomes:

$$\tan \phi = (uW \cos \phi - W_p \sin \phi)/(W \cos \phi - W_p \cos \phi) \quad (10)$$

Dividing numerator and denominator on the right side of (10) by $W \cos \phi$:

$$\tan \phi = [u - (W_p/W) \tan \phi] / (1 - W_p/W) \quad (12)$$

Cross multiply and add $(W_p/W) \tan \phi$ to both sides. This gives the desired result:

$$\tan \phi = u. \quad (13)$$

Thus the maximum angle of the clothespin below the horizontal is determined by the coefficient of friction between the clothespin and the supporting finger.

This can be observed by gradually moving the support finger out towards the end of the clothespin. As the finger moves back, the center of gravity of the belt-clothespin (W) must stay under the finger. However, for this to happen, the angle ϕ must increase. Thus beyond a certain point, the clothespin will no longer remain on the fingertip.

The result that the limiting angle is obtained from

$$\phi = \tan^{-1} (u) \quad (14)$$

should not be surprising. The same result is obtained for the angle where an object just starts to slide down an inclined plane!

In the analysis of this problem, the important insight is the realization that the stiff belt exerts *two* forces on the clothespin. If only one force were exerted (i.e., W_b directly) then rotational equilibrium could not be achieved. This also explains why a limp piece of rope will not work in the place of the belt.

Another demonstration which is somewhat similar can be shown by means of a fork, spoon, wooden match, and a glass. If the fork and spoon are forced together with the match placed between the tines of the fork, the entire "apparatus" can be balanced on the edge of the glass as seen in Fig. 5. The analysis of the forces on the match is essentially the same as the analysis for the clothespin. A dramatic conclusion to this demonstration can be achieved by lighting the end of the match which is extended over the inside edge of the glass. The wood will burn down until the glass acts as a heat sink and drops the temperature below the kindling temperature of the match. At this point the flame will go out leaving the fork and spoon balanced on practically nothing!

Special thanks to John B. Hart and Mike Gutzwiller for some interesting discussions concerning this problem.

References

1. Vincent Buckwash, "Ways to Demonstrate Center of Gravity," Phys. Teach. **14**, 39 (1976).
2. Dick Schnacke, *American Folk Toys* (Penguin Books, Baltimore, Maryland, 1974) p. 51.

Fig. 5. The match-fork-spoon system in equilibrium.

String and sticky tape experiments

Section Editor: **R. D. Edge**
University of South Carolina, Columbia, South Carolina 29208

Center of mass

There are a number of misconceptions about the center of mass of a body which can best be explained by simple experiments. For example, does the center of mass have to be within the object, and what happens to the center of mass of a flexibly jointed body — such as a person — when the body changes shape?

In most cases, we determine the center of gravity of a body, and not its center of mass. The center of gravity is the point of the body through which a single force, the sum of all the gravitational forces, may be said to act, whereas the center of mass is where the mass appears to be concentrated for kinematic purposes — for example, it is the point about which the body rotates when there are no external forces acting. Luckily, the two are almost identical, and only in rare cases does the difference matter — for instance, the LDEF satellite, a long cylinder, is stabilized pointing towards the earth because its center of gravity is slightly closer to the earth than its center of mass. This is because the earth's field is not uniform, but depends on $1/r^2$.

The following experiments require only paper, paper clips, scissors, pencil, string, marbles, and sticky tape.

When an object is supported from a frictionless pivot, the center of mass lies directly below the pivot. Cut out the shapes shown in Fig. 1 (and any others you would like). Open up a paper clip, tie one end of a piece of fine cord to the clip, and attach a marble, or other light object, such as a

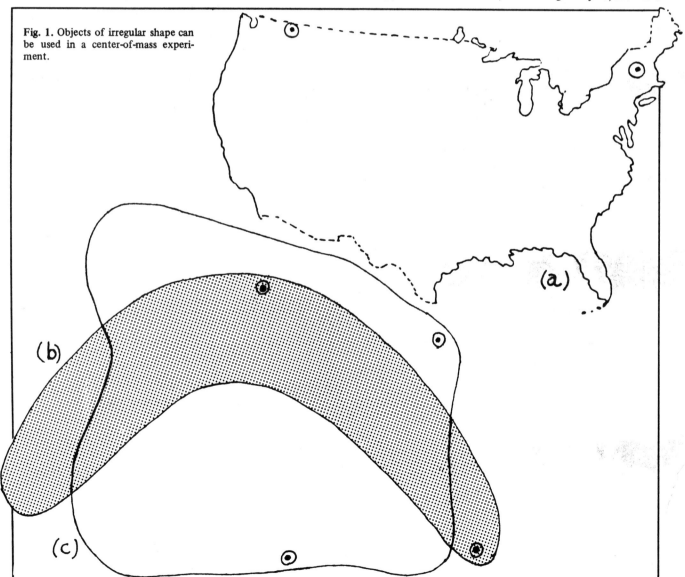

Fig. 1. Objects of irregular shape can be used in a center-of-mass experiment.

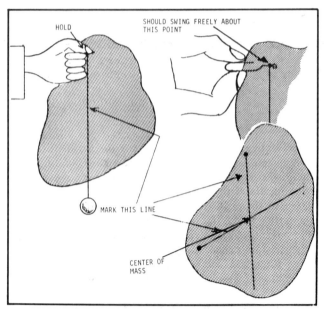

Fig. 2. The center of mass lies directly below the point of support.

Fig. 3. The stability of an object depends on the position of its center of mass.

bunch of keys, to the other end of the cord with sticky tape. Punch a hole, as shown in Fig. 2, through one of the points of the cutout, and hang it vertically. Mark where the string lies against the paper with a pencil. Now repeat about another point of suspension. Where the lines cross is the center of mass — does this, in fact, lie near the middle of the object? If you cut out a boomerang shape such as Fig. 1b, you will find the center of mass lies outside the object. Interesting shapes for the center of mass are, for example, your own state (Jerry Wilson, of Lander College likes to use this one) or the United States — Fig. 1a. The center of mass of the (continental!) USA lies a few miles into Kansas from Nebraska. Including Hawaii and Alaska would carry it much farther west!

It is possible to predict the center of mass of simple symmetrical objects, such as a rectangle or circle, by simply taking the geometrical center.

The stability of objects depends, of course, very much upon the position of the center of mass — this is why a race car, having a low center of mass, does not overturn on rounding a sharp corner or turning on a steep gradient whereas an old fashioned stage coach would (Fig. 3). Also, one can see why a tall lampstand is so easily knocked over. Your own center of mass is important — stand with your back against a wall, and try to touch your toes. You fall over because your center of mass lies ahead of your toes. The distribution of mass in men and women differs — a woman can stand with her back closer to the wall and yet touch her toes than can a man of the same height and weight. As we move about, our center of mass shifts. However, for certain classes of objects, their center of mass remains the same however they shift around. Mobiles come under this category. The reason is that each component of the mobile is supported at its center of mass, as shown in Fig. 4. The element A is supported at its center of mass from the beam (you can use a drinking straw) as is the element B, and the beam is supported at the combined center of mass of all three components, A, B and the beam — and so on.

Since each element rotates about its own center of mass, the whole device will always have the center of mass below the final pivot d. Mobiles also make interesting examples of moments, since, for example, in the figure $M_A g \times d_A = M_B g \times d_B$, $(M_A + M_B) g \, d_D = M_C g \, d_C$ assuming the mass of the beam to be negligible.

Another example of center of mass is a cantilevered pile of bricks — how far from the edge of the table can the top of a pile of bricks (Fig. 5) extend? To demonstrate, move the top brick first so that it does not fall off — it

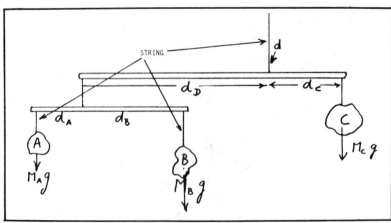

Fig. 4. In a mobile the center of mass of the whole device remains in the same place no matter how the positions of the parts change.

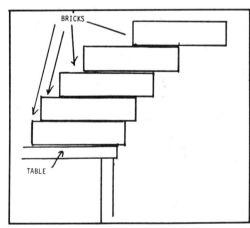

Fig. 5. A cantilevered pile of bricks.

A POTPOURRI OF PHYSICS TEACHING IDEAS — MECHANICS

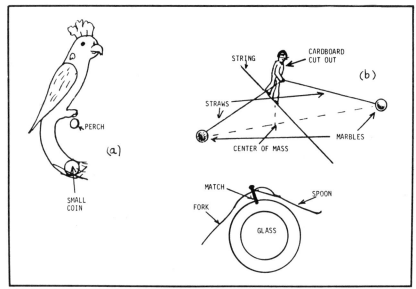

Fig. 6. Toys can be used to illustrate the idea of center of mass.

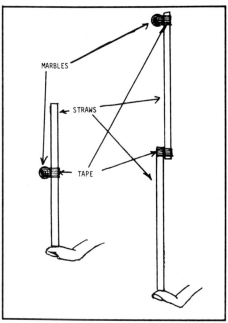

Fig. 7. Balancing a loaded straw.

will be half over the edge of the next brick down. Now, move the next brick out — it will extend 1/4 over the brick below, which will extend 1/6 over the brick below it and so on. Again, it is easy to show the nth brick extends 1/2n over the next brick down. By the fourth brick, the top brick will be clear over the edge of the table. Balancing toys are other good examples of center of mass — the familiar parrot (Fig. 6a) with the weight in its tail (this particular parrot is a modified version of that drawn by R. W. Wood, the famous optics researcher in "How to tell the birds from the flowers" — The Parrot and the Carrot — published by Dover) or the tightrope walker (Fig. 6b) with the long pole. Again, one can stick a spoon and a fork together, and push a match between the tynes of the fork to balance on a cup.

It might be thought easier to balance on short stilts, or a short unicycle, because the center of mass is lower. This is not so, because of the human element that comes in. Fasten a marble half way up a straw (Fig. 7a) and try balancing on one finger. Now tape two straws together, and attach the marble at the top end, balancing on one finger as shown. It is much easier, because the moment of inertia about the finger tip is so much larger, the marble accelerates slowly, and you can move your finger under the balance point more easily.

Another interesting balancing experiment is to support 12 large nails on the tip of another nail. First lay out one nail, and place the others as shown in Fig. 8a. Place a second nail on top (Fig. 8b), and the whole may be lifted and balanced on the blunt tip of the last nail, knocked into a block of wood (Fig. 8c). This was shown me by the inimitable Rae Carpenter and Dick Minnix, but at the annual meeting in San Antonio, the same experiment was demonstrated by someone else — a good experiment travels far.

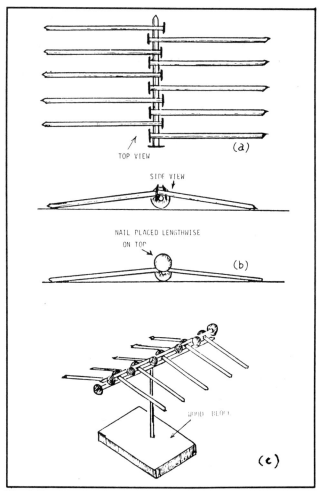

Fig. 8. Twelve nails can be balanced on the tip of another nail.

Center of gravity of a student

G. Stroink
Department of Physics, Dalhousie University, Halifax, N.S., Canada B3H 3J5

Several textbooks[1] written for students in the life sciences show, in one of their problems, how the center of gravity of a person can be determined by weighing the person on a board supported by wedges on two scales, one at each end of the board. This problem is the keystone of one of our lab experiments dealing with forces, torques, and the center of gravity. The preparation time to start this lab is relatively short, particularly when workshop facilities are available to cut wood. The students enjoy it, possibly because it requires them to participate actively, and it starts them thinking about the forces and torques acting on them in daily life. Two students require the following equipment:

- one board of 2 x 6 ft plywood, 3/4 in. thick
- one bathroom scale, flat on top (no handle)
- one meter stick
- two wooden wedges, each 2 ft in length, but of different heights.

To determine the location of the center of gravity the wedges are placed near to and parallel with the ends of the board. One rests directly on the floor, the other on the bathroom scale. The wedges differ in height by the height of the bathroom scale. This ensures that the board is horizontal when laid on top of the wedges. The student of predetermined weight lies down on the board and his or her center of gravity can be determined by calculating the sum of the torques around the point where one of the wedges supports the student (Fig. 1). In this approach the weight of the board is ignored. It is our experience that some students gain sufficient insight into the problem to include the location of the center of gravity of the board and its weight — as determined by the bathroom scale — to obtain a more accurate location of the center of gravity of the student.

One can simplify the lab by allowing the students to use two bathroom scales. In that case, one can compensate for the weight of the board by adjusting the scales to read zero with the board alone resting on the scales.

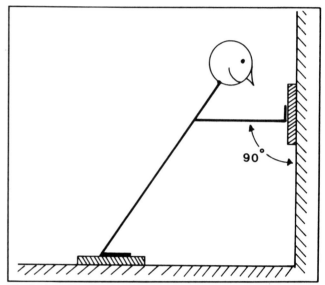

Fig. 2. A student leaning against the wall. Two bathroom scales are needed to determine all the forces that act on him in this position.

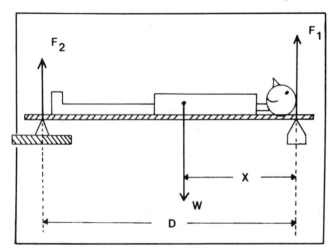

Fig. 1. The center of gravity of the student can be found from the equation $F_2 \cdot D = W \cdot X$. F_2 is read from the scale. W is the weight of the student, determined previously.

The experiment described above is a part of a more extensive lab dealing with forces.

In this experiment students study the forces while standing on one scale and pushing down on a second scale located on the bench, they check their measured center of gravity by balancing on a single wedge, and locate the center of gravity of a person lying on the board with his legs straight up.

In another part of the experiment, the student leans against a wall as far as possible, as indicated in Fig. 2. Two scales are used to determine the magnitudes of the forces.

In analyzing this situation we assume that the vertical component of the force at the wall is negligible.

If "momentum and impulse" have been treated already in the course then the following experiment can be added: Ask the student to stand very quietly on the scale and let him look carefully at the scale indicator. The periodic movement that can be observed on the scale corresponding with his heart beat is a nice demonstration of how a change in momentum (of the blood) can produce a force. This principle is applied in the Ballisto cardiograph.[2]

References

1. A. H. Cromer, *Physics for the Life Sciences* (McGraw-Hill, New York), 2nd ed., p. 63.
   J. Kane and M. Sternheim, *Life Sciences Physics* (John Wiley, New York), p. 78.
2. S. MacDonald and D. Burns, *Physics for the Life and Health Sciences* (Addison-Wesley, Reading, Mass.), p. 85.

THE WATER CAN PARADOX

Lester G. Paldy
Northwood High School
Silver Spring, Maryland

It is quite common for students studying fluids to be misled into accepting as fact a phenomenon which just does not occur experimentally. It is usual to introduce a water can with three holes in it, producing streams of water as indicated in Fig. 1. It is asserted that the increase of pressure with depth causes the horizontal distance traversed by the lowest stream to be larger than that traversed by the other two streams. However, if the situation is reproduced, (Fig. 2) it is found that the *middle* stream travels the greatest distance. This phenomenon is not at all in accord with most student's intuitions, and when observed for the first time, generates a lively discussion.

We can explain the observed result using simple kinematics, Torricelli's Law, and a simple graph. We will also show that the same result can be obtained analytically using only simple calculus.

It is easy to establish Torricelli's Law which states that the velocity of a stream of water exiting from a hole in the side of a can at a distance h beneath the surface may be found by equating the gravitational potential energy lost in falling through a height h to the kinetic energy of motion at the point of exit.

$$mgh = \tfrac{1}{2} mv_x^2 \text{ or, } v_x = (2gh)^{1/2} \quad (1)$$

The horizontal distance traveled by the stream will be given by

$$S_x = v_x t \quad (2)$$

where t is the time it takes for the stream to fall a distance $(l-h)$ as indicated in Fig. 3.

We know that for free-falling objects, $S = \tfrac{1}{2} at^2$ and we thus obtain $(l-h) = \tfrac{1}{2} gt^2$ or,

$$t = [(2/g)(l-h)]^{1/2} \quad (3)$$

Substituting Eqs. 1 and 3 in Eq. 2, we obtain,

$$S_x = [2gh]^{1/2} \cdot [2/g(l-h)]^{1/2} = 2[h(l-h)]^{1/2} \quad (4)$$

We are now in a position to plot S_x as a function of h (since l is a constant), and noting that $S_x = 0$ when $h = 0$, or $h = l$, we obtain the graph shown in Fig. 4. From this we can see clearly that S_x has a *maximum* that appears to be at approximately $h = l/2$. A careful plot would show that the maximum is exactly at $l/2$, and we can verify this analytically.

It is well known that at a maximum or minimum of a function, the slope at that point is equal to zero, and that the slope of the curve is represented by the first derivative of the function.

We see that Eq. 4 really expresses S_x as a function of h, and to obtain the slope, we need only take the first derivative of S_x with respect to h. This is done easily and we obtain,

$$\frac{dS_x}{dh} = \frac{(l-2h)}{[h(l-h)]^{1/2}} \quad (5)$$

Since this represents the slope of our function, we see that it is equal to zero only at $l-2h = 0$ or $h = l/2$ and that this is where the maximum occurs, explaining the observed phenomenon.

It should be noted that if the can is placed on a *raised* platform at a *sufficient* height as shown in Fig. 5, the flow will be in accordance with our intuition. The stream from the lowest hole will have the greatest horizontal displacement. It is an interesting exercise to find the height H of the platform for which the horizontal displacement of the stream from the lowest hole is equal to that from the hole half way up the can. If the platform has less than this height, the lowest stream will have a smaller displacement than that from the middle hole.

An elementary analysis of the situation shows that one should not have let his intuition lead him astray. The horizontal velocity of the stream from a hole just at the top of the can is zero, hence, although it takes a longer time to fall, its horizontal displacement is zero. The horizontal velocity of the stream from a hole at the bottom is large, but its time of fall is zero, hence the horizontal displacement is zero. The displacement for streams between the top and bottom varies from zero at the top to a finite value for intermediate values to zero at the bottom. Hence, we may expect to find a position of the stream that yields maximum displacement.

We feel that the experiment is an excellent one for the high school laboratory and classroom, for it can be handled with only simple kinematics and a graph by most students, while providing an opportunity for those students taking a concurrent first course in calculus to exercise their newly discovered tools.

In addition, we have been told by several students that it is an excellent party trick as well!

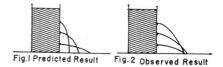

Fig.1 Predicted Result   Fig.2 Observed Result

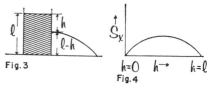

Fig.3   Fig.4

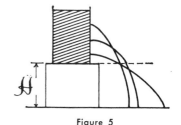

Figure 5

The Water Can Explored Again

Roy H. Biser
Lamar State College
of Technology
Beaumont, Texas

IN the September 1963 *Physics Teacher*, Mr. Lester G. Paldy described the water can paradox and showed the correct solution. Yet a recent physical science textbook for use in college arrived on my desk with the illustration shown in Fig. 1 A. The book has a 1965 publication date. I became interested in just how prevalent this misunderstanding was, and in a few days' time I was able to collect sixteen books with illustrations of this problem. These are shown in Figs. 1A to 1P. The diversity is amazing! For a while I was convinced that the only thing they had in common was that water squirts to the right, but in the fifteenth book (Figure 1 O) even this similarity was destroyed!

Out of the sixteen illustrations, eleven are completely wrong, since they show the water from a hole near the bottom having a greater horizontal range at the base of the can than that from the holes near the center of the can. Of the five that have the right idea, three are badly out of scale. The theory predicts that the water from the hole midway between the base of the can and the water surface will have a horizontal range at the base of the can equal to the height of the water in the can. Figures 2 and 3 show that water very nearly achieves this theoretical value. Only Figure 1 B seems to show the phenomenon well and to scale. Figure 1 E emphasizes the fact that two holes symmetrically spaced about the center will have the same range.

There are several very grievous mistakes in this collection. Fig. 1 F is from a book describing experiments for high school teachers to perform for and with their students. I wonder how many high school teachers have been frustrated by its not working this way. Figures 1 I and 1 N are from the same junior high school text, different editions.

Figure 1 P is from a junior high school workbook that purports to be encouraging thinking along

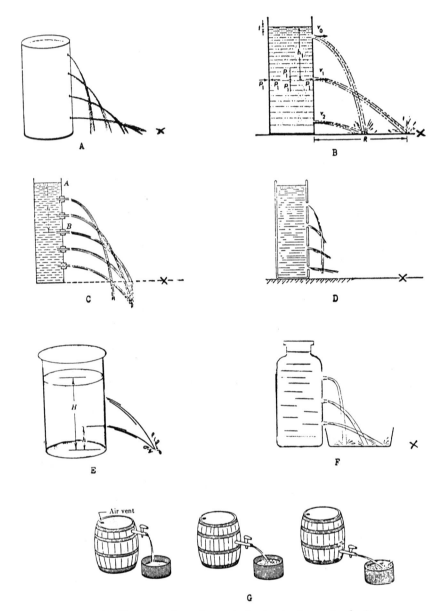

Figure 1. Illustrations from sixteen textbooks showing water flowing from holes at different heights in the side of a can filled with water. Horizontal base lines have been added to some and an X has been added to each to show the range equal to the height of the water in the can.

KEY TO FIGURE 1

A. Rosen, Siegfried, and Dennison, *Concepts in Physical Science*, Harper and Row, New York, 1965.

B. White, H. E., *Descriptive College Physics*, 2nd ed., D. Van Nostrand, Princeton, N. J., 1963.

C. Mendenhall, Eve, Keys, and Sutton, *College Physics*, 4th ed., D. C. Heath, Boston, 1956.

D. Gamow, G., *Matter, Earth and Sky*, Prentice-Hall, Englewood Cliffs, N. J., 1958.

E. Kingsbury, R. F., *Elements of Physics*, D. Van Nostrand, Princeton, N. J., 1965.

F. Joseph, Brandwein, Morholt, Pollack, Castaka, *A Sourcebook for the Physical Sciences*, Harcourt, Brace, and World, New York, 1961.

G. Bonner and Phillips, *Principles of Physical Science*, Addison-Wesley, Reading, Mass., 1957.

A POTPOURRI OF PHYSICS TEACHING IDEAS—FLUIDS AND HEAT

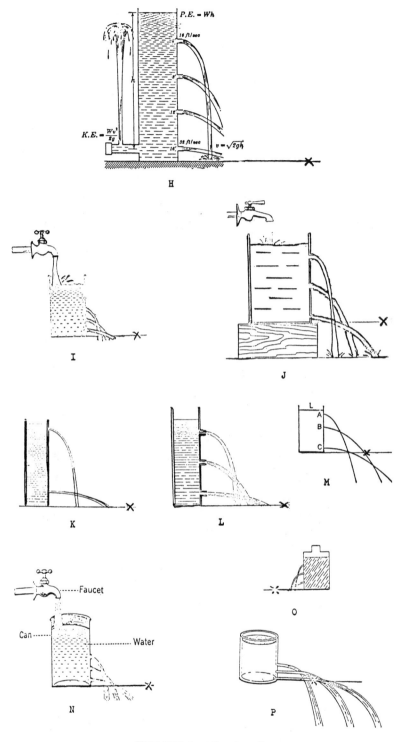

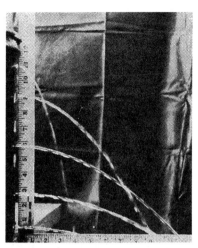

Figure 2. Photograph of the water can experiment. The can is filled with water to a height of 12 inches. The holes are at heights of 1 inch, 3 inches, 6 inches, 9 inches, and 11 inches. Note the parallax on the vertical scale.

the lines of cause and effect. It says that if the student has studied his lesson well he will be able to give the cause of the effect pictured. Pity the poor student who might have observed nature!

In the derivation, the horizontal range at the base of the can is independent of the acceleration of gravity. This means the range would be the same on the moon as on the earth. This comes about in the derivation because there is a g in the numerator and a g in the denominator. We should remind ourselves that g/g is not always one. For g = 0 this quantity is undefined, therefore the range is undefined. Looking back in the derivation we discover why; the water doesn't squirt out the hole at all. This case is not trivial in the age of satellites and "weightlessness." Maybe the moral of all this should be: If your can leaks, drop it.

FIGURE 1 (Continued)

H. Black, N. H., *Introductory Course in College Physics*, Rev. ed., Macmillan, New York, 1941.

I. Davis, Burnett, and Gross, *Science 3*, Henry Holt, New York, 1952.

J. Hogg, Cross, and Vordenberg, *Physical Science*, D. Van Nostrand, Princeton, N. J., 1959.

K. Stewart, Cushing, and Towne, *Physics for Secondary Schools*, Ginn, Boston, 1932.

L. Whitman and Peck, *Physics*, American Book Co., New York, 1950.

M. Hausmann and Slack, *Physics*, D. Van Nostrand, New York, 1939.

N. Davis, Burnett, and Gross, *Science 3*, Holt, Rinehart and Winston, New York, 1961.

O. *Ideas for Teaching Science in the Junior High School*, National Science Teachers Association, Washington, D. C., 1963.

P. Obourn, Heiss, Montgomery, and Lape, *Activities for Exploring the World of Science*, D. Van Nostrand, Princeton, N. J., 1963.

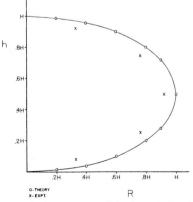

Figure 3. A comparison of the theoretical and experimental results. R is the horizontal range at the base of the can, h is the height of the hole, and H is the height of the water in the can. Experimental points were taken from experimental apparatus shown in Figure 2.

Floaters and Sinkers

By Terrence P. Toepker
Xavier University, 3800 Victory Parkway, Cincinnati, OH 45207

Fig. 1. The sealed cans with NutraSweet are floating while the old standards sink.

At a recent session of area physics teachers who were "Doing Physics" in Cincinnati, an interesting observation was made. (Attention, Earl Zwicker!)

The ice on our soft drinks had melted, and some of the full aluminum cans were floating!

Doing a Piagetian classification, we separated the "floaters" from the "sinkers." Diet pop floats!

In order to share with the TPT readers, we took a photograph of this phenomenon using an aquarium. The water temperature was 22°C. Figure 1 shows samples in various states of equilibrium in the aquarium. The drink called "Slice" (not shown) would sink or float, depending on which particular can was used.

Our conclusion is that aspartame (NutraSweet) makes the difference in the final density, since much more corn syrup is required in regular soda pop. Table 1 shows a small sample of data taken the day after the meeting.

Our next step is a graduate-level experiment using Budweiser, Strohs, Coors, Millers, and the comparable "light" counterparts!

Table I. Mass of Various Soda Pop Samples(g)*

	Classic Coke	Pepsi	7-Up	Diet 7-Up	Diet Coke
1.	399.0	393.5	391.3	379.5	374.5
2.	399.1	392.3	390.8	371.8	365.1
3.	392.1	392.5	392.1	378.1	368.1
4.	396.5	393.6	379.4	380.4	373.1
Avg	396.7	393.0	388.4	377.4	370.2

*This is the mass of the sealed container.
The fill volume for each can is listed as 12 fluid oz (354 ml).

How to cool a book by its cover

Robert Everett Vermillion

The University of North Carolina, Charlotte, North Carolina 28213

I accidentally discovered an armchair experiment using only a book with its paper dust jacket in place. The book was held in front of me, in a vertical plane, by pressing it between the palms and outstretched fingers of my hands. The book slipped *through* the dust cover as I relaxed the pressure somewhat, and I caught it before it fell all the way through by increasing the pressure of my hands on the cover. I was surprised to feel a cooling sensation in my hands. I subsequently found the cooling sensation to be maximum if a large book is initially held near the bottom, and allowed to slide as rapidly as possible through the dust cover until, when caught by increasing the pressure of the hands, the book is almost out of the cover and about to fall through to the floor.

I suggested that others try this experiment and describe to me their sensation, without telling them that I thought it would feel cool. In every case, after a few trials, they came to the conclusion that it felt cooler. This was surprising to everyone, in that we all felt that the friction associated with the book sliding inside the cover should generate heat, not the opposite. It seems a good experiment to leave students with for awhile to try to formulate their own hypothesis to account for the effect. I think the book sliding rapidly through the cover leaves a partial vacuum inside the cover at the top, and air from below expands through the region between the book and cover. The expansion-cooling of the air is momentarily discernable by one's hands.

Ed. note: What do you think?

HOW TO COOL A BOOK BY ITS COVER

Professor Vermillion has described an armchair experiment [Phys. Teach. 13, 165 (1975)] which seems to show that a book dropping through its dust jacket is cooled. He attributed the effect to adiabatic expansion of the air passing between the book and its cover. I think that a much more likely mechanism underlying the very real cooling effect described by Vermillion is that of simple heat transfer. In view of its very small thickness and relatively high density, dust jacket paper has a surprisingly small thermal resistance. The dust jacket and book cover warm quickly in the hands. Then, when the book slips down, fresh, cooler areas of the cover come into play and the enhanced heat transfer is sensed as a cooling.

The experiment was repeated in a laboratory "desert room," in which the temperature is controlled to 100°F — after the book had come to approximate temperature equilibrium (a few hours of standing vertically on a table, with its pages fanned out). A very distinct *warming* was sensed as the book slid down its dust jacket.

Vermillion's note should serve the excellent purpose of exciting the student's mind. For example, the student should try to find mechanisms by which there could be sufficient reversible adiabatic cooling to account for a temperature difference of, say, 1°C. Also, the student should examine the mechanical effect (on the falling book) of the pressure differential across the dust jacket-book cover combination that would be associated with an appreciable adiabatic cooling of the air: What would be the frictional force tending to slow the book's fall?

J.A. VAN DEN AKKER
*The Institute of Paper Chemistry
Lawrence University
Appleton, Wisconsin 54911*

Ed. note: Vermillion's description of his chance observation and our question, "What do you think?" brought several reactions from readers, of which Van den Akker's letter is typical. We do not have room to print the others but we acknowledge them here: The Physics Faculty of South Carolina State College collaborated in a letter agreeing generally with that presented above. Patrick J. Cooney of Middlebury College suggests the "desert room" experiment described by Van den Akker but says that the Vermont climate is working against him. Gerald Straks and Edsel Langdon of Freeport Senior High School, Freeport, Illinois made a solid state temperature measuring device and confirmed their suspicion that there is actual frictional heating of the dust cover. Gladys Luhman of City College of San Francisco, using reasonable figures, calculated that it would take 5.2×10^5 experiments, lifting the book each time, to cause the experimenter to lose 1.0 lb of body fat! Wayne Wright of Clackamas Community College points out that it works for a magazine, too. We are grateful to these correspondents for their interest.

"How to cool a book by its cover" revisited

Edsel M. Langdon
Gerald M. Straks
Freeport Senior High School, Freeport, Illinois 61032

After discussing Mr. Vermillion's book-cooling experiment,[1] we felt we could demonstrate experimentally a more reasonable explanation of the described phenomenon, as we explained in our letter.[2] We are convinced the only temperature drop occurs after the expected friction-heating and at no time is there any indication of temperatures below the original temperature of the system.

Before we had gone too far with our experiment, we verified our suspicion that the phenomenon is not unique to book/dust-jacket systems — the same effect can be achieved by sliding a piece of paper over a tabletop with approximately the same force and time interval needed to stop a book. Since this motion was more easily reproducible, we made most of our observations using this method.

For our temperature investigations we used the circuit shown in Fig. 1. This somewhat unusual use of a diode resulted when one of us (E.L.) heard someone mention that the forward voltage drop across a signal diode (such as a silicon 1N914 or 1N4148) was linear with temperature. He decided to design a circuit to use this characteristic. By placing the diode in a Wheatstone bridge and using the unbalanced potential to drive two differential amplifiers in series (a dual 741 or 558 works well), he produced a circuit with high sensitivity (breathing on it produced an almost immediate response), fairly high gain (up to 15 V per C°), and a means of controlling this gain without affecting the input characteristics, although at maximum gain the output voltage tended to drift slightly with time. Before he had time to eliminate this problem, we read Vermillion's article and saw a good opportunity to put the circuit to use. We merely conducted all our observations at less than maximum gain (1 V per C°). We were able to determine the time lag for the diode and circuit to be 100 μ sec or less by using a photographer's electronic flash unit as a heat and light source and a vacuum photocell to trigger the oscilloscope trace.

In establishing the parameters of our experiment, we assumed a 1.5 kg book, a drop h of 0.10 m before the "holding" force is increased to the "stopping" force (a distance which is also consistent with a reaction time of about 0.2 sec), and a stopping distance d of 0.05 m. Using familiar laws of dynamics and assuming that half of the work done in stopping the book would occur on each side of the book, we determined that the normal force N on one cover of the book should be

$$N = \frac{mg\,(h+d)}{2\,\mu d}$$

After using the inclined-plane technique to approximate μ (about 0.2), we calculated the needed force to be roughly 110 N (25 lbs), which seemed a little high until we had several people try the book-drop and then repeat the necessary squeeze with a bathroom scale. All readings were within 5 lbs of our calculated figure.

We placed the diode temperature detector on a sheet of paper of dust-jacket thickness, placed a piece of Styrofoam over it to prevent thermal interference from body temperature, and after several minutes of practicing on the surface of the bathroom scale to achieve the correct normal force with regularity, we slid them as a unit across the surface of the lab table.

With the output of the circuit connected to an

Fig. 1. The circuit used in measuring temperature changes.

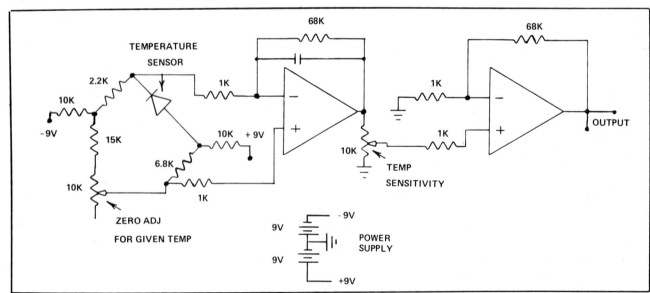

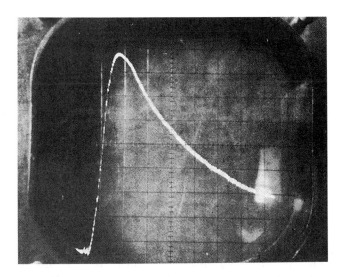

Fig. 2. Oscilloscope trace indicating warming of book cover.

oscilloscope, we photographed the results with a Polaroid model 110B camera using 3+ and 2+ close-up lenses held in front of its 127-mm lens. With the lens focused at infinity, the distance from the screen to the front of the lenses was 20.6 cm (8 1/8 in.). The oscilloscope trace was adjusted for ½ V per division vertically and ½ sec per division horizontally, while the circuit was adjusted for a 1 V per C°

output, giving us ½ C° per division vertically. The resulting photographic records were quite uniform in shape and extent. An example is shown in Fig. 2.

We feel that our hypothesis is substantiated by all of our photographs: There is about 0.25 sec temperature rise to a maximum of roughly 3.5 C° (3.5 V) above equilibrium, corresponding to the transfer of heat through the paper during the slide across the table top, followed by a slow cooling (lasting about 10 sec) as the heat is dissipated. The fast reaction time of the diode and circuit convinces us that the photographs are substantially accurate recordings of the actual temperature changes at the surface of the book jacket. A check with a stopwatch indicated that for most persons who tried the book-drop experiment or the table-slide experiment, the cooling sensation was noticeable after about 1 sec and lasted 2 sec or more. With the ½ sec/div horizontal scale, this would correspond to the portion of the curve in our photograph that occurs after a temperature drop of about 1 C° and includes the majority of the remaining 2–3 C° temperature drop back to the original temperature. At no point in this interval is there any indication of a bona fide "cooling of a book by its cover."

References

1. Phys. Teach. 13, 165 (1975).
2. Phys. Teach. 13, 324 (1975).

Newton's law of cooling
or is ten minutes enough time for a coffee break?

Chandler M. Dennis, Jr.
Pembroke State University, Pembroke, North Carolina 28372

A recent article[1] used a cup of coffee to gain some insights in cooling processes. The article gave some much needed attention to Newton's law of cooling. This interesting and, as we shall see, easily observed phenomenon is a much neglected topic in first year heat studies.

Newton found that, as long as the temperatures were not too extreme, the rate of cooling of an object is directly proportional to the differences between its temperature and the temperature of the surrounding region. In equation form, if

- $T(t)$ represents the temperature of an object at any time t,
- T_s represents the temperature of the surroundings,
- T_o represents the initial temperature of the object, and
- κ represents the temperature change constant (measured in reciprocal seconds);

then

$$T(t) = T_s + (T_o - T_s)e^{-\kappa t} \quad (1)$$

Newtonian cooling represents the combined effects of conduction, convection and radiation.

Unfortunately, the cooling law has a very low profile in college physics courses. Typically, it is either completely omitted[2] or appears in the problem section of physics texts.[3-6] Discussions on applications of ordinary differential equations[7-11] introduced me to the topic and these texts are still the best place to find a relatively complete discussion.

The rather simple experiment I am about to describe has at least two appealing aspects for first-year students (and their instructors): it works and the equipment is easy to manipulate. Each two-person group requires a stopwatch, two thermometers, a hot cup of coffee (the machine vendors around here love my physics sections) and a lid for the cup.

One thermometer is required to record the ambient temperature; the other monitors the coffee's slow slide to thermal equilibrium. When the thermometer is placed in the coffee (through a hole in the lid), start the stopwatch and (*important*) record the maximum temperature registered by the thermometer. Consider this temperature to be the coffee's initial temperature [T_o in Eq. (1)].

After exactly five minutes, the temperature of the coffee is recorded and Eq. (1) is then solved for κ, the temperature change constant.

Once κ is determined — and it usually takes about two minutes — the students are required to predict the coffee's temperature at the ten minute mark. The very accurate prediction appears on their calculator displays with about thirty seconds to spare. The quantitative sciences always win a convert or two after this experiment!!

The presence of a lid for the coffee cup appears to make the cooling conditions somewhat less than true-to-life. It is important for one to realize, however, that lidless heat loss is achieved through the action of two processes: Newtonian cooling (predominantly convection from the surface) and evaporation.[1,2] The rate of convective cooling is proportional to $(T - T_s)^n$, where T and T_s are as previously defined and n is *not generally* equal to 1. For natural convection (i.e., the situation in lidless cooling) n ranges from 1.3 to 1.6[13]; since our mathematical modeling assumes that the cooling is directly proportional to the temperature difference, that is $n = 1$ [see Eq. (2) below], it cannot accurately describe the temperature decrease. The lid insures that the coffee will *not* cool by convection or evaporation, and so Eq. (1) becomes an effective temperature predictor. [It should be pointed out that Eq. (1) can also describe the temperature slide of a lidless cup under *forced* convection, since in that case $n = 1$].

I have described an obscure yet potentially very flexible experience in first-year physics. This experiment can be beneficial to several different types of students. As described above, it allows students in an algebra-based course the opportunity to observe a decay similar to that of a radioactive substance and to "tone up their natural log muscles." The experiment could also serve as an introduction to differential equations for more advanced students.

Expressing Newton's cooling relation quantitatively we have

$$\frac{dT}{dt} = -\kappa(T - T_s) \qquad (2)$$

With this equation as a starting point, one could demonstrate the basic techniques for the solution of this typical first order differential equation. Solution of this equation results in the appearance of Eq. (1). Rush the students to the coffee machines and then into the lab where Eq. (1) can be examined empirically. The result of this attack gives your students a glimpse of the relationship of physics to reality. Finally, students with computer science backgrounds could probably devise some graphic and intriguing offshoots.

I should point out a possible "negative" aspect of this experiment. Students who eventually go from this lab experience into the American labor force, may very well form the tough core of a militant work-schedule revision movement. After all, they will be well aware that management has been short-changing labor for years: ten minutes is definitely not enough time for a coffee break!!!

References

1. R. Edge, Phys. Teach. **17**, 466 (1979).
2. A review of 12 standard first-year texts, which included the familiar books by Franklin Miller, Sears and Zemansky, and Harvey White, revealed that 8 texts did not mention the cooling law.
3. F. Bueche, *Principles of Physics*, 3rd edition (McGraw-Hill, New York, 1977) p. 267.
4. L. Greenberg, *Physics with Modern Applications* (Saunders, Philadelphia, 1978) p. 154.
5. D. Halliday and R. Resnick, *Physics*, 3rd edition (Wiley, New York, 1978) p. 471.
6. G. Shortley and D. Williams, *Elements of Physics*, 5th edition (Prentice-Hall, Englewood Cliffs, New Jersey, 1977) p. 328.
7. R. Agnew, *Differential Equations*, 2nd edition (McGraw-Hill, New York, 1960) pp. 77-79.
8. M. Boas, *Mathematical Methods in the Physical Sciences* (Wiley, New York, 1966) p. 357.
9. W. Boyce and R. DiPrima, *Elementary Differential Equations and Boundary Value Problems*, 2nd edition (Wiley, New York, 1969) pp. 59, 479.
10. E. Rainville, *Elementary Differential Equations*, 3rd edition (Macmillan, New York, 1964) pp. 47-49.
11. P. Ritger and N. Rose, *Differential Equations with Applications* (McGraw-Hill, New York, 1968) pp. 77-79.
12. G. Kell, Am. J. Phys. **37**, 564 (1969).
13. T. Brown, ed., *The Taylor Manual of Advanced Undergraduate Experiments in Physics* (Addison-Wesley, Reading, Mass., 1959) p. 115.

TENSIOMETER

The surface tension of water can be measured to a satisfactory degree of accuracy with the simple apparatus illustrated in Fig. 8. This equipment produces a force which is a factor of 8 greater than that one experiences with the commercial tensiometer. The equipment is readily available in any high school or college laboratory.

A styrofoam cup and counterweight are suspended from a triple beam balance, replacing the pan. A labjack supporting a shallow pan of water allows for making contact between the cup and water surface. The zero point of the scale can be maintained by adjusting the vertical level of the water surface with the labjack.

One finds the force necessary to break the contact between the water surface and the cup is 4900 dynes. Holes should be punched in the bottom or sides of the cup to prevent partial vacuum from altering the breaking force.

The equation for computing the surface tension by the ring method is:

$\gamma = fF/4\pi R$,

where f is the maximum force registered on balance scale and F is the correction factor determined by the shape of the liquid held up and the ring dimensions.[1] Substituting the above value of the force, the radius R(4 cm) of the cup, and a correction factor of 0.75, one obtains a value for the surface tension of 73 dyn/cm^2.

After the student has convinced himself of the reliability of the apparatus, he is asked to determine the surface tension of a liquid such as methyl or ethyl alcohol that (unknowingly to him) "wets" the cup. Of course, his calculations for this liquid do not yield a value that is in the ball park. Students offer many interesting explanations in an attempt to explain this apparent dilemma.

References

1. Harkins and Jordan, J. Amer. Chem. Soc. **52**, 1751 (1930).

JULIUS H. TAYLOR
Morgan State College
Baltimore, Maryland 21239

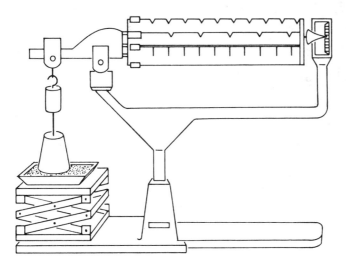

FIG. 8. *Beam balance measures the force to separate the styrofoam cup from the surface of the liquid.*

BROWNIAN MOTION

Many physics laboratories have the small Brownian movement apparatus[1-3] which is designed to be placed under a laboratory microscope to observe the Brownian motion of smoke particles which have been drawn into it by expanding a small rubber bulb when a lighted match is held at the opening of the chamber.

Two problems sometimes exist which make this a disappointing experiment. First, the objective lens of the microscope may have too short a focal length for focusing into the smoke chamber. This may be corrected by obtaining an objective lens of longer focal length, such as a 5X achromatic objective.[4]

The other difficulty is in finding the proper light source. The usual desk or microscope lamp is difficult to adjust and often produces a glare, and sometimes the extra light makes observing difficult.

Figure 8 shows a light source, with microscope and Brownian movement apparatus, that works and is easy to set up in the lecture room or laboratory. It consists of a 3 x 4 x 4 in. metal box with a 100-W projection bulb in it. A piece of ½-in. copper tubing is soldered into one side, and at the right angle to direct the light at the smoke particles in the chamber. Four legs are attached to the box so that it is always at the right height for the microscope and smoke chamber to be used.

It now takes less than two minutes to set up the equipment and the students have no difficulty in observing the Brownian motion.

WALLACE A. HILTON
William Jewell College
Liberty, Missouri 64068

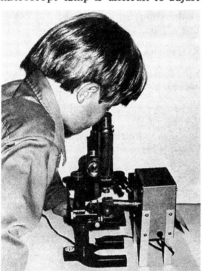

Fig. 8. *Student observes Brownian motion with specially constructed light source and microscope with 5X objective.*

References
1. Sargent-Welch Scientific Co. Catalog Item No. 4860.
2. Central Scientific Co. Catalog Item No. 71270.
3. H.E. White, *Modern College Physics*, 5th Ed., (Van Nostrand, New York, 1966) pp. 422-23.
4. Edmund Scientific Co., 5X Achromatic Objective. Item No. 30,045. ($6.15).

A Question of Air Pressure

Question:

In the course of discussing air pressure, many teachers introduce the following experiment: Take a glass and fill it to the brim with water. Cover the glass of water with a piece of cardboard and invert the glass. On inverting the glass, it is observed that the card stays in place and the water is contained.

The standard explanation given is somewhat as follows: *atmospheric pressure will support the column of water* (Fig. 1). There is nothing wrong with this explanation but some youngster might want to follow up this experiment by asking you to perform the same experiment with, say, half a glass of water (Fig. 2). You try it and find that the card does not fall off. The student may tell you that the pressure (atmospheric) of the air inside the glass plus the pressure due to the weight of the water is greater than the external air pressure, and, hence, one would not expect the observed result. What explanation can one now give the student?

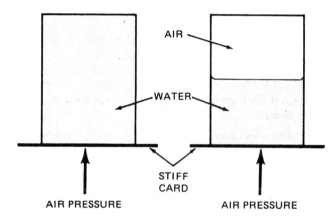

Fig. 1. *The inverted glass demonstration with a full glass.*

Fig. 2. *The inverted glass demonstration with a partially filled glass.*

Answer:

At least two hypotheses might be proposed:

Hypothesis 1: The adhesive forces between the card and the wet rim of the glass are sufficient to support the liquid column. But this seems highly unlikely in view of the relatively large weight of the contained water.

Hypothesis 2. There is a slight downward bulge in the card. The volume of air inside the glass is thereby increased and, in accord with Boyle's law, the pressure of the air is correspondingly reduced. This reduction in pressure is sufficient to produce the observed containment of the water. For this hypothesis to be tenable it is necessary to show that a very slight bulge suffices to produce the required pressure change in the captivated air. If a one centimeter bulge were indicated this hypothesis would certainly meet its demise. It turns out, however, that using the dimensions of a typical water glass calculations suggest that a bulge of the order of one millimeter would suffice to support this hypothesis. Its plausibility is established. (Experiments also tend to support this hypothesis since the half-full glass trick works with a flexible card, such as a file card, but fails with a stiff piece of plexiglass, which, however, is satisfactory with a full glass.)

Students might now be encouraged to suggest how one might confirm *Hypothesis 2* experimentally, what part the adhesive effect of *Hypothesis 1* might play, what the physical properties of the card must be, the effect of the weight of the card, etc.

For suggesting the treatment of this question I am indebted to one of my former students, Mr. Y. K. Tan, who is himself now teaching physics in Malaysia. Readers are invited to send us typical student questions for this column.

WALTER THUMM
*Queen's University,
Kingston, Ontario*

Boyle's Law Demonstration Using a Vacuum Gauge

William Carlson
Wilson Junior High School, St. Paul, Minnesota

A novel way to get good class results for a Boyle's law demonstration is to use a vacuum gauge (calibrated in inches of Hg) attached to the top of a 4-ft resonance tube about three-fourths full of water (Fig. 3). The air volume above the water level is varied by drawing water from the tube by some means such as a faucet aspirator. As the volume changes, readings of the pressure are taken.

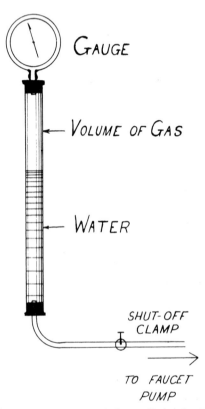

Fig. 3. Vacuum gauge connected to cylinder for Boyle's law demonstration.

Atmospheric pressure is recorded as well as the temperature. The pressure on the gas is the gauge reading subtracted from the barometer reading. Since the air above the water is saturated, a correction for the effect of vapor pressure must also be subtracted. Vapor pressures at room and water temperature are obtained from tables.

Too low a pressure should be avoided, because dissolved air in the water is released. Rubber stoppers can be greased to prevent leaks. Clamps stop the flow of water when readings are taken. The accuracy of the vacuum gauge can be checked against a column of mercury. To secure the vacuum gauge into a rubber stopper, ¼-in. copper or brass tubing is soldered to the appropriate fittings.

The results shown in Table I were obtained during a class demonstration. From these results, a graph is made showing the relation between the reciprocal of the pressure and the volume of the gas (see Fig. 4).

Table I. Air Volume, Atmospheric Pressure, and Temperature

Length of volume of air V	Gauge reading in. of Hg	Pressure (29.0 − gauge reading) in. of Hg	Corrected pressure p in. of Hg	$p \times V$
9.5 in.	2.0	27.0	26.2	248
12.2	8.0	21.0	20.2	246
15.2	12.0	17.0	16.2	246
19.7	15.5	13.5	12.7	250
24.0	17.6	11.4	10.6	254
29.2	19.4	9.6	8.8	256
33.0	20.5	8.5	7.7	254

Barometer reading 29.0 in.
Temperature reading 73 °F (22.8 °C)
Vapor pressure at 22.8 °C is 20.8 mm of Hg (0.8 in.)

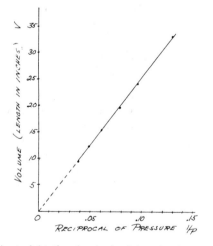

Fig. 4. A straight line is obtained by plotting the length of the air column vs the reciprocal of the pressure.

A POTPOURRI OF PHYSICS TEACHING IDEAS—FLUIDS AND HEAT

ARCHIMEDES' PRINCIPLE MEETS CHARLES' LAW

Clyde J. Smith
Physics Department, University of South Carolina, Columbia, South Carolina 29208

A lecture demonstration combining two concepts has been well received by audiences ranging from junior high school to adult level. A helium-filled balloon (preferably bearing some felt-tip message in praise of physics) is displayed while Archimedes' principle is explained. The class is led to realize that if the volume of the balloon were decreased sufficiently the buoyant force would no longer exceed the weight of the balloon. Then Charles' law is introduced, and a class member can often be induced to suggest cooling the balloon. It is pointed out that the helium acts essentially as an ideal gas down to liquid-nitrogen temperature, often most easily understood by the class as "three-fourths of the way from room temperature to absolute zero." The balloon is immersed in liquid nitrogen, and its corresponding decrease to approximately one-fourth its original volume gratifyingly vindicates Charles. Then the shrunken balloon is placed upon the lecture table, where it immediately begins warming and reexpanding. With appropriate gestures, the demonstrator intones "In the name of Archimedes, rise!" — just as the balloon lifts from the table. Audience response is customarily immediate and enthusiastic.

Notes for those attempting this demonstration: since rubber is quite porous to helium, inflate balloons as close to use as possible; the balloon may conceivably be buoyant even after cooling, necessitating the prior addition of ballast such as one or more paper clips; some practice is suggested to optimize the timing of the command to rise.

Elasticity shown with mirror blank and marble

Harry H. Kemp
269 West 1st South, Logan, Utah 84321

Fig. 1.

Students are amazed at this demonstration which I first saw done by Harvey White on a television program. A glass or flint marble, about one-half inch in diameter, when dropped from a height of one foot onto the concave side of a reflecting telescope mirror blank (Fig. 1) will bounce for about 25 sec. The blank I used was 8 in. in diameter by 1-3/8 in. thick. A larger size might be even better, but do not get a smaller one. No polishing was done on the mirror. Edmund Scientific Company[1] is a good source.

References

1. Edmund Scientific Co., Barrington, N.J. 08007

Bernoulli Demonstration

R. E. Worley
University of Nevada
Reno, Nevada

The spinning-ball demonstration, frequently cited to illustrate Bernoulli's principle, may be shown very effectively by means of a polystyrene ball and a simply-made wooden "thrower." The apparatus is shown in the accompanying photograph. The ball of 3-inch diameter is sold as a Christmas decoration or may be cut from a block of polystyrene. The thrower may be constructed by gluing or screwing together two strips of quarter-inch plywood, each about 2½ by 21 inches, to give an L-shaped cross section. Prior to gluing, the strips should be narrowed at one end to form the handle. Two strips of medium-grit emery cloth (100 to 150), each about 1 by 15 inches, are glued to the inner surfaces of the thrower so that the ball will roll along them.

With the thrower in the one hand, the ball may be held against the emery cloth with the other; or it may be held with the index finger. A swing of the thrower then ejects the polystyrene ball with an assured spin acquired by rolling along the emery cloth. Thanks to the lightness of the ball, its trajectory shows pronounced curvature. In addition, there is no danger of injury to students or equipment from impact.

I am indebted to Mr. Lewis Humason, formerly in charge of demonstration equipment at the University of California, Los Angeles, for showing me this device.

Repairing Thermometers with Split Mercury Columns

Charles L. Roberts
State University of New York,
Morrisville, N. Y. 13408

In almost every science lab you will find one or more thermometers that cannot be used because they have "gaps" in the mercury column. Most of these devices are eventually discarded because it is difficult to eliminate the separations once they occur. However, they may be repaired in a few seconds using a can of Freon that is purchased from your local refrigeration dealer for about the price of *one* new thermometer.

Invert the can of Freon and expel some of the liquid into a small beaker. Place the thermometers in the liquid Freon. The temperature will go down to about $-20°C$. At that point withdraw the thermometers and let the Freon on the bulb of the thermometer evaporate. This will lower the reading about 10 degrees more. Dip the thermometers in the Freon, withdraw and let evaporation occur again. Repeat the process until all the mercury is in the bulb of the thermometer. The job is done. When the mercury expands back into the column there will be no separations.

Two or three thermometers may be handled at a time and a can of the Freon will be enough to repeat the operation many times. Remember that Freon is a refrigerant and it is important to keep it away from your eyes.

THE CARTESIAN DIVER

An ideal bottle for a Cartesian Diver is a large one, such as the type used for "Micrin," a mouth wash. The usual demonstration is to sink the diver by applying pressure to the cork.

It is more interesting, however, and quite mystifying to the observers, to adjust the stopper until the diver is just ready to sink. Then hold the bottle in one hand and apply pressure front to back and watch the diver sink. If the pressure is correct the hydrostatic pressure will keep the diver submerged. Then, when pressure is applied to the sides, the diver rises. It is quite surprising that the volume of a glass bottle can be changed by applying pressure first front to back, and then sideways.

Pressure applied to the sides will raise a diver sunk by front-to-back pressure.

Pressure applied front to back when the diver is just ready to sink will keep him submerged.

—submitted by ROBERT N. JONES
Department of Mathematics and Physics
The Philadelphia College of Pharmacy and Science

THE RISING CARTESIAN DIVER

Professor R.N. Jones, [Phys. Teach. 11, 345 (1973)] describes a simple and effective demonstration for raising a sunk Cartesian diver. However, I would like to claim credit to a prior publication date for the method ["Cartesian Divers Designed by Pupils", *School Science and Mathematics*, 38, p. 141-2 (1938)]. Actually, the idea was submitted along with a working model by a former high school student of mine as a class project. My paper describes three other novel Cartesian divers constructed by students.

I am sure that Prof. Jones was not aware of my paper; with the information explosion, it becomes an almost impossible task to locate all the references dealing with a particular concept. Perhaps the AAPT or the Commission on Physics Education of IUPAP can indicate a computerized file of published topics on physics demonstrations and laboratory exercises to aid writers and editors in checking the originality or the source of an idea.

HAYM KRUGLAK
Western Michigan University
Kalamazoo, Michigan 49008

MORE ON THE CARTESIAN DIVER

Robert N. Jones's note on the Cartesian Diver (Phys. Teach. 11, 345 (1973)) brings back memories of when I first saw Prof.

V.E. Eaton of Wesleyan University squeeze a flat sided whisky bottle to make the diver perform. That was in the 1940's. Later the late Prof. John Albright at the University of Rhode Island showed me the north–south effect. With his hand on the bottle the diver only descended when the flat sides were in the north–south direction. If the bottle were turned 90 degrees the diver would rise.

FRANK T. DIETZ
University of Rhode Island
Kingston, Rhode Island 02881

Cartesian diver with pressure head

Edward V. Lee
Georgetown Day School, Washington, D.C. 20007

The usual Cartesian diver can nicely serve as a demonstration of the pressure head of a column of water (Fig. 1). The pressure head is maintained in a titration tube; stoppers and tubing are fitted in place of the stopcock to connect the tube with the large jug that contains the diver. The height of the pressure head is controlled by pouring water in at the top and draining it out at the bottom. After adding water to make the diver sink, it's quite instructive to ask students to predict the water level in the tube that will make the diver rise again.

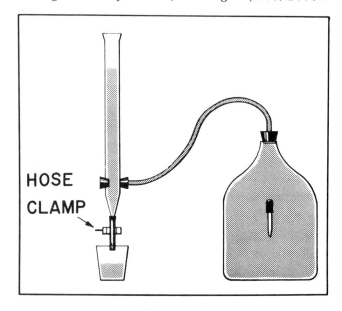

Fig. 1. The height of water in the titration tube controls the floating and sinking of the Cartesian diver.

ELASTICITY DEMONSTRATION

In discussion of the deformation of a body subject to various sets of deforming forces there may be difficulty in convincing the student that a particular elastic constant is involved. The model shown demonstrates the relationship between stress and strain vividly.

Figure 4 shows the unstrained model. It is made from polyurethane foam plastic with a grid ruled with a fine point marker. Wooden end blocks are cemented on with Elmers Glue.

The center picture shows the effect of a bend. It is obvious that what is actually involved is an extension along the upper layers and a compression along the lower layers. It is thus the "stretch" or Young's Modulus which is involved.

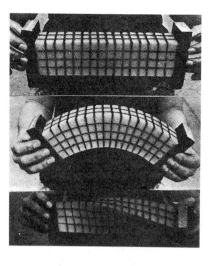

Fig. 4. Unstrained model at top, bent model in middle, and twisted model at bottom.

Furthermore, it is obvious that the maximum strain is along the edges and the strain along the center of the block is zero.

The lower picture shows the effect of a "twist." The squares are clearly deformed to a diamond shape which is just how the "rigidity" or shear modulus is defined. It is also apparent that the shear strain is a maximum at the centers of the faces and zero at the edges where the rulings remain perpendicular to the edge.

BERNARD SCOTT
*Behrend College
Pennsylvania State University
Erie, Pennsylvania 16501*

A POTPOURRI OF PHYSICS TEACHING IDEAS—FLUIDS AND HEAT

Elasticity of glass

Salvatore J. Rodano and James J. D'Amario
Department of Physics, Harford Community College, Bel Air, Maryland 21014

A demonstration to show the elasticity of glass can be accomplished as follows.

Fill a glass bottle completely with colored water. Then fit a glass tube (the smaller the bore the better) into a rubber stopper and press the combination into the bottle so that some water is forced into the glass tube and remains there.

Squeezing the bottle will cause the water in the glass tube to move, thereby demonstrating the elasticity of the glass. If an elongated bottle is used, then the change in volume will depend on the direction in which the force is applied to the bottle. Force on the short dimension of the bottle decreases the volume, whereas force on the long dimension increases the volume. Thus in the former case the liquid in the tube rises while in the latter case it falls. (See Fig. 1.) This effect can also be shown with a Cartesian Diver.[1]

The fact that glass is somewhat flexible and will bend slightly under an applied force is used by forensic scientists when determining the direction from which a window or pane of glass is broken. The glass first bends in the direction in which the force is applied, causing the glass on the opposite side to stretch. As glass will withstand more bending than stretching, it will break first on the side opposite that from which the force is applied.

Analysis of the stress lines which result from this bending and stretching enable one to determine the direction of the breaking force.

Reference
1. R. Sutton, *Experiments in Physics* (McGraw-Hill, New York, 1938), p. 124 (M-321).

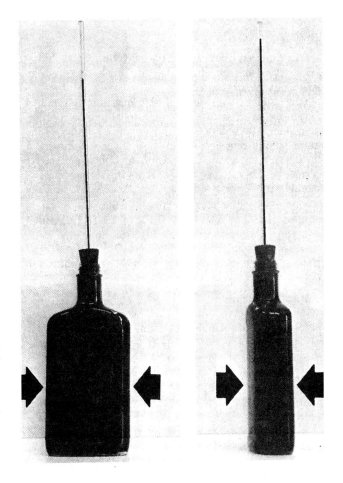

Fig. 1. Elasticity of glass: fluid level rises and falls depending on how glass is squeezed.

ANOTHER USE FOR A WHISKY BOTTLE

I have read with somewhat clouded interest the NOTE titled ELASTICITY OF GLASS – *THE PHYSICS TEACHER* DECEMBER 1979....

About 30 years ago I had a NOTE in *THE AMERICAN JOURNAL OF PHYSICS* titled AN ELEMENTARY DEMONSTRATION ON THE INCOMPRESSIBILITY OF WATER AND THE ELASTICITY OF GLASS – 18: 164 (L) 1950.

I have since extended this DEMONSTRATION and it is written up somewhere – I have forgotten ... but it goes like this:

Let us HEAT with a very localized flame and gently THE WIDE SIDE of the bottle ... WHAT DO YOU PREDICT?

NOW again heat the narrow side – after the system has cooled off ... WHAT DO YOU PREDICT?

Further to all this I had a letter from a physicist in Ljubljana (Yugoslavia) about this DEMONSTRATION and *he* extended it *again* with a paper in a Yugoslavia journal – which he sent me – which I could not read – which led me to the Yugoslav consul who could not read it because it was in *one* of the THREE languages in Yugoslavia which he could NOT read – Serbo-Croatian – Slovenian – Macedonian – which took me to a family he spoke of WHO COULD READ IT – and I had a Croatian dinner with them and the man of the house played music on a *gusla*... ALL a beautiful adventure arising out of play with a whisky bottle!

Julius Sumner Miller, *(Professor of Physics – Emeritus), 16711 Cranbrook Avenue, Torrance, California 90504*

Teachers' pets III: How thick is a soap bubble?

We invite you to contribute your own pet demonstration or experimental activity. Send them to the Editor, directed to this column.

Robert Gardner
Salisbury School, Salisbury, Connecticut 06068

The question above is one of many I presented to my physics class in an effort to get them involved in independent investigations.

One student who pursued this question determined the decrease in mass of some commercial bubble solution after making 100 bubbles. He also caught a number of the bubbles on the ring he used to blow them and measured them by sighting across the ends of the bubble's diameter to a ruler held behind the bubble. From the average mass of a bubble, the average diameter, and the density of the solution he determined the thickness of a bubble.

mass of 100 bubbles (m) — 0.43g (4.3 x 10^{-3} g/bubble)
average diameter of a bubble (d) — 6.0cm
density of solution (ρ) — 1.0g/cm^3

Surface area of bubble x thickness = volume of solution in bubble

$$\text{thickness} = \frac{\text{volume}}{\text{area}} = \frac{m/\rho}{\pi d^2} = 4 \times 10^{-5} \text{ cm}$$

Another student found that when he blew large bubbles the top of the bubbles became dark. Bands of color were clearly seen in regions farther down the bubbles' surface. He knew the dark area indicated a thickness less than 1/4 wavelength of light, but this was true for only a small portion of the bubble. Since the area of darkness at the top depended on the size of the bubble, and the number of colored bands varied, he decided that the thickness was on the order of one wavelength of light (5 x 10^{-5} cm) and that it was meaningless to try to be more accurate.

No one tried a method suggested by Malcolm Skolnick when we were colleagues at the Education Development Center about ten years ago. In this method you simply touch a capillary tube to the surface of a soap solution and measure the length of the liquid column that you carry away in the tube. Hold the tube in a vertical position and gently blow as large a bubble as possible on the lower end of the tube. Let the bubble fall onto a hard surface where it can be measured, or break the bubble with the tube and measure the diameter of the ring of tiny droplets that it leaves (Fig. 1).

When I tried this experiment the results confirmed the values obtained by my students.

Other similar questions can be used to encourage independent thinking in the laboratory. Here are a few I have used:

Is the index of refraction of a substance related to its density?
How powerful are you?
What is the temperature of a bunsen burner flame?
What makes a "dipping bird" dip?
How can you find due north in the daytime without a compass?

How does the surface area of an ice "cube" affect its melting rate?
Examine the shadow cast by an object illuminated by only a white light *and* a green light. How do you explain what you see?
How large are raindrops?
What determines how sensitive a balance is?
How do you explain the shadows cast by a fluorescent light bulb?

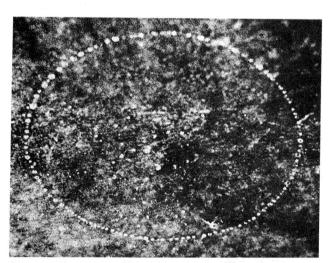

Fig. 1. Ring of droplets left by broken soap bubble. Photo by Henry Clark, Salisbury School, Salisbury, Connecticut.

A POTPOURRI OF PHYSICS TEACHING IDEAS—FLUIDS AND HEAT

Barroom physics, Part I[†]

James T. Schreiber
Trenton State College, Trenton, New Jersey 08625

Recently a student in one of my introductory physics classes posed what he called a barroom physics puzzle. He insisted he had won many free drinks by challenging bartenders and patrons to solve the following problem.

Three identical glasses* A, B, and C are arranged as shown in Fig. 1. A and B are filled with water** (or spirits if you prefer). B is supported on C by a few hollow stirrers. There are some additional hollow stirrers available on the table. The problem is to transfer the water from glass A to glass C. (Not all, but most of the water must be transferred.)

The following conditions are to be met. At no time may the experimenter physically touch or move the glasses or the hollow stirrers supporting glass B. The additional hollow stirrers, however, may be moved but may not touch the glasses or the stirrers supporting B. Fig. 1 shows the situation at the beginning of the "experiment," and Fig. 2 the situation at the end. It can be done!

In order to transfer the water from A to C (Fig. 3) without touching the apparatus, proceed as follows:

Use one of the additional hollow stirrers to blow air into glass A at any point where A and B glasses meet (see Fig. 3). Some of the air will enter the space between glass A and glass B and will rise to the top (bottom) of glass A and will exert enough pressure to force the water out of glass A, down the sides of glass B, and into glass C. This can be continued until all of the water has been transferred from glass A to glass C.

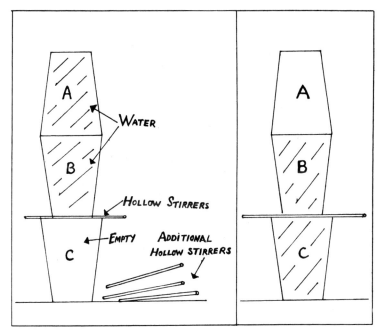

Fig. 1.   Fig. 2.

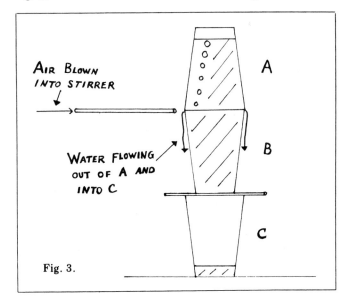

Fig. 3.

---

* Clear plastic cups do not seem to work as well due, I believe, to the fact that they are thin-walled.
** This can best be done by filling glasses A and B while they are submerged and then putting the mouths of the glasses together before removing them from the water.

[†] Barroom Physics, Part II will appear in a coming issue.

Barroom physics, Part II

James T. Schreiber
Trenton State College, Trenton, New Jersey 08625

Barroom physics Part I appeared in our September issue on page 361.

After presenting the first barroom physics puzzle to my classes, a second student approached me with another interesting problem which he claimed might also be used to win some barroom bets.

A glass of water is filled with water almost to the top. A small cork stopper (or any small object that will float high in the water) is made available. The problem is to place the cork in the water and have it stay floating in the middle of the glass (Fig. 1). The cork may be placed anywhere on the water surface and no other restrictions are placed on the "experimenter". The situation is diagrammed at right.

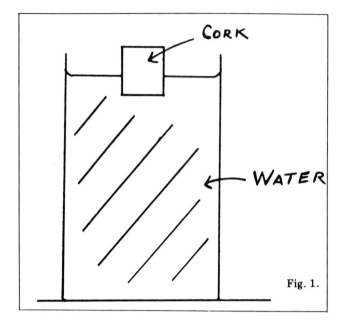

Fig. 1.

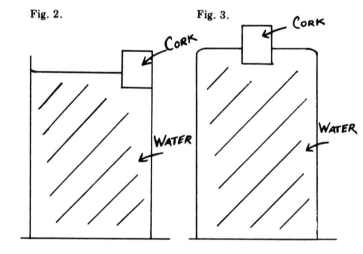

Fig. 2. Fig. 3.

No matter how hard you try, you cannot get the cork to float in the center with the setup described. The cork will always float off-center and adhere to the side of the glass (Fig. 2). If, however, the glass is filled above the brim (Fig. 3), the cork will float in the center every time. A quick review of the concepts of surface tension and angle of contact should make the explanation obvious.

Divergent barroom physics

George B. Barnes
Washington State University, Pullman, Washington 99163

James Schreiber has posed the problem: given a cork, or any other small object which will float high in water, and a glass of water which is nearly full, how can you make the object stay in the center of the glass? His solution is to fill the glass just past full.[1] I too was given this problem by a student. For several years now Schreiber's solution is the only one that my students have come up with on their own, but there are other solutions. In order to arrive at the first of these a closer look at Schreiber's solution is helpful.

Schreiber's solution relies on the fact that moving an object toward the edge of a very full glass stretches portions of the water's surface. Surface tension acts in opposition to this stretching, moving the object toward the center of the glass (Fig. 1). Similarly, when the glass is only partially full the surface area is decreased when the object moves toward the edge. In this case surface tension works to move the object toward the edge and hold it there (Fig. 2). It should be noted that when the curvature of the water's surface at the object and the curvature of the water's surface at the edge of the glass are in different directions (i.e., up and down) the object moves toward the center of the glass. Conversely when these two curvatures are in the same

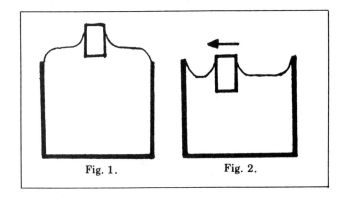

Fig. 1. Fig. 2.

direction the object moves toward the edge of the glass.

Some of the alternative solutions to this problem rely on making the floating object and the edge of the glass wet differently so that these curvatures are in different directions. For example, if the sides of the glass are coated with wax the object will stay in the center as shown in Fig. 3. Wax isn't wet by water while the object is. A bit more difficult, but similar, solution is to coat the object with wax

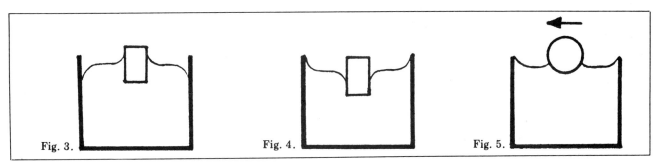

Fig. 3. Fig. 4. Fig. 5.

as shown in Fig. 4. Difficulties arise with this solution because most objects do not have the idealized shape shown in Fig. 4. The angle of contact between the water and the object's sides is primarily dependent upon the composition of the material, not the angle that the object's sides make with the horizontal. Thus, a well waxed ping-pong ball will produce a water surface like that shown in Fig. 5 rather than one like that shown in Fig. 4.

A different approach to the problem is to use a lack of surface tension to keep the object in the center. If soap is added to the water the object will wet better. Wetted objects drag water along with them when they move while objects which are not wetted move through the water with less obstruction. The friction that a wetted object experiences is sufficient to keep it from moving to the edge of a glass unless the object is already very near the edge.

Ignoring the whole question of surface tension there are other solutions. One is to stir the water. If the water is whirling around in the glass any floating objects will move toward the center. If it is not legitimate to stir the water the same thing can be accomplished by placing the glass on a turntable.

Another solution involves Bernoulli's principle. Simply pour water directly on the object while it is in the center of the glass. The stream of water will hold the object in the center. Ping-pong balls work especially well for this situation.

Still another, but probably not the final solution, is to hold a charged comb directly over the object. Here again ping-pong balls work well.

Reference

1. James T. Schreiber, "Barroom Physics Part II," *Phys. Teach.* **13**, 418 (1975).

An application of Archimedes' principle: Eureka! I'm 28% fat

Roland A. Hultsch
Department of Physics, University of Missouri, Columbia, Missouri 65211

One purpose of dieting and exercise is control or loss of weight. A less well known effect is the conversion of fat tissue to muscle without much change of weight.

Human-performance physiologists routinely measure the percent fat of subjects before and after a several-week exercise and diet program. The body is assumed to be fat and everything else: muscle, bones, and organs are collectively called muscle. From studies of cadavers, the densities of these are $d_F = 0.89 \pm 0.01$ g/cm^3 and $d_M = 1.10 \pm 0.01$ g/cm^3. From two measured values, total body mass and average body density, the masses of fat and muscle can be calculated.

The point of this note is to describe yet one more application of Archimedes' classic method for measuring the volume of an irregularly shaped object, in this case a human body. Figure 1 shows the ominous apparatus used to measure the buoyant force of water on an immersed body. Inside the tank is a 2-ft diam plastic dish (actually a sled) supported by the four chains to the spring scale. The subject sits on the dish, exhales as much as possible, and, if not rather fat, sinks. The operator then reads the apparent weight. Evidently it takes some getting used to, but after about five trials, consistent numbers are obtained. At least the water temperature is in a comfortable range, 30–36°C. Subjects who are too fat to sink are held under with the masses shown on the railing in Fig. 1. Of course, the weights of the immersed dish, chains, and added masses are subtracted from the scale reading.

Here are the steps in deriving the equation that physiologists use to calculate percent fat:

$$\text{density} = \frac{\text{mass}}{\text{volume}} \tag{1}$$

$$\text{volume} = \frac{1}{g}\left(\frac{\text{buoyant force}}{\text{density of water}}\right) - \left(\frac{\text{volume of}}{\text{exhaled lungs}}\right) \tag{2}$$

$$d_{\text{average}} = \frac{\text{mass (fat)} + \text{mass (muscle)}}{\text{vol (fat)} + \text{vol (muscle)}} = \frac{m_F + m_M}{\frac{m_F}{d_F} + \frac{m_M}{d_M}} \tag{3}$$

$$\% \text{ fat} = \frac{m_F}{m_F + m_M} \times 100 \tag{4}$$

Solve Eq. (3) for m_F and substitute into Eq. (4) to get

$$\% \text{ fat} = \frac{d_F}{d_M - d_F}\left(\frac{d_M}{d_{\text{avg}}} - 1\right) \times 100 \tag{5}$$

Fig. 1. Tank in which persons are immersed to measure the buoyant force of water.

The density of water is calculated from

$$d_{\text{water}}(\text{g/cm}^3) = 1.0055125 - 0.00032711\, T(°C) \tag{6}$$

Physics teachers may doubt the significance of all these digits. The volume of a person's exhaled lungs is obtained from

$$v_{\text{exh}} = [0.24 \text{ (male)}, 0.28 \text{ (female)}]\,(\text{vital capacity}) \tag{7}$$

Vital capacity is the difference of volumes of full and exhaled lungs, and is measured as part of the procedure. The subject exhales as much as possible into a 1-ft diam vertical graduated cylinder fitted with a low-weight piston.

In the case of the author, these values were obtained:
weight in air/g = 80.6 kg
weight in water/g = 1.6 ± 0.2 kg
vital capacity = 5.25 liter
water temperature = 32.5°C,
water density = 0.9949 g/cm^3
d_{avg} = 1.03 ± 0.01 g/cm^3

These yield

$$\text{vol}_{RAH} = 78.1 \text{ liters} = 2.8 \text{ ft}^3$$

and

$$\% \text{ fat}_{RAH} = 28 \pm 1\%$$

Few teachers will want to set up such an apparatus (though that shown in Fig. 1 is homemade), but there is one simple use of these ideas that amuses students. Most persons' densities are close to that of water, of course, so divide body weight in pounds by 62 lb/ft^3 to get body volume accurate to within a few percent. Then multiply by the density of air to get a buoyant force. I enjoy 0.2 lb of air support.

Ideal fat percentages for adults are considered to be 15% for men and 22% for women. Champion distance runners and bicyclists get down to 5%. Professional halfbacks and Olympic-class wrestlers have reached 2% fat. Young adults, for example high school wrestlers, may incur some health risk at these ultra-low levels.

Reference
1. Albert R. Behnke and Jack H. Wilmore, *Evaluation and Regulation of Body Build and Composition* (Prentice-Hall, Englewood Cliffs, 1974).

A large-scale electroscope

Sheldon Wortzman
Nassau Community College, Garden City, New York 11530

Are your students unimpressed with fleapower demonstrations of static electricity? Instead of picking up tiny bits of paper with a charged rod, why not turn them on with this forceful grabber!

Put a watchglass on a smooth table concave side up and balance a heavy window pole or any other wood pole or board on it (Fig. 1). Hold your favorite charged rubber rod near one end and watch the pole rotate! Can't find a watchglass? Then file a flat at the center of gravity of the pole and balance it horizontally on a soda bottle cap (Fig. 2).[1]

Reference

1. A similar demonstration, using a suspension, is described in R. M. Sutton, *Demonstration Experiments in Physics* (McGraw-Hill, New York, 1938) p. 249.

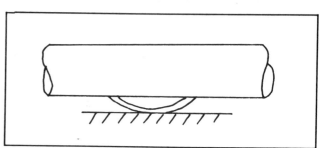

Fig. 1. Pivoting a pole on a watchglass . . .

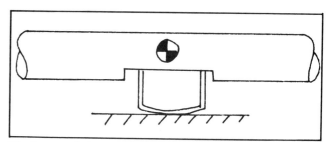

Fig. 2. or a soda-bottle cap.

NEGATIVE CHARGES FROM AN ELECTROPHOROUS

Mornings in Berkeley are usually damp, and developing a strong negative charge from cat's fur can be almost impossible (the fur readily absorbs moisture from the air). Strong positive charges, on the other hand, are generated and stored easily by rubbing cellulose acetate against a sheet of acrylic plastic (Lucite, Plexiglas, Polyglas, etc., 1/8 in. or greater in thickness). These observations have led us to use a positively charged electrophorous to induce large *negative* charges. The basic procedure is the same as for a conventional electrophorous:

1. Rub an acrylic plastic sheet with cellulose acetate to establish a positively charged surface.
2. Set an induction plate (from a conventional electrophorous) onto the acrylic sheet. A negative charge will be induced on the bottom of the induction plate.
3. Touch the upper surface of the induction plate with a finger. A spark will occur as your body donates electrons to the upper surface of the induction plate.
4. Lift the induction plate by the insulated handle. The plate will be found to possess a strong negative charge.

A word of caution: Do not rub any cat's fur against the positively charged surface; doing so will strip the cat's fur of free electrons and render it useless. However, the fur can be rejuvenated by rubbing it against the negatively charged plate of the electrophorous.

JOHN E. GIRARD
*University of California
Berkeley, California 94720*

Electroscope Shadowgraph

William J. Muha,
Notre Dame High School, Niles, Illinois 60648.

Reflections from the glass container of an electroscope make it difficult for pupils to see movements of the leaves. If projection of the image on a large screen is impractical, cover the front with a single sheet of facial tissue as shown in Fig. 4. A rubber band may be used to hold the tissue in place. When the electroscope is illuminated from the rear, the leaves will cast shadows on the tissue which are clearly visible to everyone in the classroom, even with the room well lighted.

Fig. 4. Illumination from the rear casts shadows of electroscope leaves on a sheet of facial tissue.

PITH BALL SUBSTITUTE

The repulsion of bodies carrying like electrical charges is demonstrated most directly when a light body is brought into contact with a charged rod, acquires charge from the rod, and then flies away from it. The usual light body for this experiment is a pith ball hung from an insulating thread. Since both the pith ball and the charged rod are poor conductors, it may be difficult to transfer the necessary charge to the ball without repeated attempts.

We have found that a pennant made from household aluminum foil gives definitely more dramatic effects than a pith ball. It is a good conductor, so that it charges up rapidly, and it is light enough to be repelled vigorously under the charge it acquires in a single contact. The speed of the interaction depends critically on the insulating properties of the thread from which the pennant hangs. Monofilament nylon, sold at notion counters as invisible stitching thread, makes an excellent insulator.

Monofilament nylon is uncomfortably stiff and wiry, but we have had good success with the following technique. The pennant is made from a 2-cm square of aluminum foil which has been creased on the diagonal. A blob of glue is placed inside the triangle which this forms; one end of a 40-cm length of the nylon thread is laid inside, and the triangle is pressed shut embedding the thread between its two folds. The other end of the thread is embedded in a cardboard square made by folding up and gluing together the two ends of a rectangular strip 1 cm x 2 cm. The cardboard square can be held by a laboratory clamp; it also makes a convenient handle if one wishes to tie the thread in a loose knot around a rod.

The only drawback we have found to this invention is that if the pennants are to be carried over from one year to the next, they must be stored separately. A group of them put into a single box will tangle rapidly and irrevocably.

ALFRED ROMER
St. Lawrence University
Canton, New York 13617

ELECTROSTATIC PONG

I have found the following demonstration to be quite effective in initiating discussion about electrostatic effects.

A grounded sphere (Fig. 1) is placed far enough away from a negatively charged Van de Graaff generator to prevent arcing. The field of the generator will force electrons off the grounded sphere, leaving it with a positive charge. A Ping Pong ball or Styrofoam sphere covered with aluminum foil is suspended from a support with thread. By induction, the small sphere will be attracted to the nearest large sphere. Upon touching, it will acquire a like charge, be repelled, and then attracted to the other large sphere. The small sphere will bounce back and forth until the Van de Graaff is completely discharged.

Robert J. Krohl, *Lafayette High School, Lafayette, New York 13084*

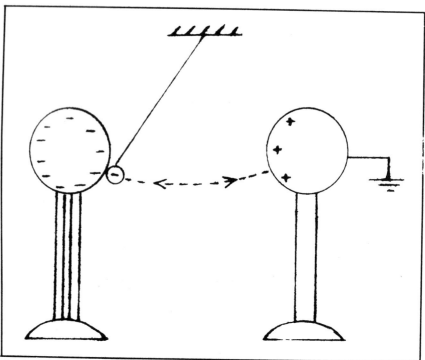

Fig. 1. A foil-covered Ping-Pong ball carries charge from a Van de Graaff generator to a grounded sphere.

ELECTROSCOPE DISCHARGE RATE

Almost every year when we were studying static electricity, a student would ask, "How long will the charge stay on the electroscope?" I usually replied that it will last at least a few hours and then discussed the reason for its eventual discharge. It finally dawned on me this year that this answer was not very scientific and that a more experimental approach would be better.

I placed a charged coffee can electroscope in a glass-fronted box on the teacher's desk so that no one would accidentally discharge it. After 36 h there was still a small amount of charge left on the electroscope.

The students were interested in the result, so I decided to extend the experiment. Since we had guessed that ions in the air were causing the discharge, I put two radioactive samples (one was a beta emitter and the other an alpha and gamma emitter) on the bottom of the electroscope. I charged the electroscope at the beginning of each physics class. For one class the charge lasted 75 min, and for the other class, the charge lasted 90 min.

I suggested next that we try putting a spark coil in the box with the charged electroscope to see how long the charge lasted (we had previously seen that a match flame will quickly discharge a charged electroscope). To the surprise of many students the electroscope discharged in seconds.

One student suggested that we try the laser beam next. So the next day I directed a laser beam on the top of the electroscope with no noticeable effect, although most of the students expected something to happen. This part of the experiment will be a useful reference when we later investigate the photoelectric effect by exposing the electroscope top to ultraviolet light.

After asking why the electroscope discharged when no obvious ionizing agent was nearby, one of the students said the air was ionized by radiation from outer space. To see if there were such particles around I put a geiger tube in the box and we counted about 15 counts/min, calculating then that there must be about 400 particles/min going through the box which could cause ionization. These ideas will be useful later in the year when we go more deeply into radioactivity.

The investigations described could be further extended intensively and extensively by testing both positively and negatively charged electroscopes in air and in a vacuum under varying conditions, reflecting the historical investigations leading to the discovery of cosmic rays. However, my purpose was only to answer a particular question. I spent about five minutes at the beginning of class for a week in this investigation, satisfying the students' curiosity and my own at the same time.

BRO. JAMES MAHONEY C.F.X.
Malden Catholic High School
Malden, Massachusetts 02148

ELECTROSTATIC LOBBY DISPLAY

The Kelvin electrostatic generator, so aptly described by Professor A. D. Moore,[1,2] when connected to a bell and chime arrangement such as Sargent-Welch No. 2039, becomes a new attraction and provokes interest even in the nonscience student (Fig. 3). We placed ours beside the Christmas tree allowing the spectators to recycle the water as it flowed from the top reservoir to the lower one. The pleasing tintannabulations of the bells, mixed with the blinking lights of the Christmas tree, served to renew the Christmas spirit in many a student and awakened a new interest in the science of electrostatics.

This apparatus was constructed from various sizes of tin cans. Tops and bottoms were removed from four cans in the central part of the apparatus; two of these have metal funnels inserted to catch the water droplets and conduct the charges to the outer surface of the cans (Fig. 4). The fact that the water is discharged from a neutral zone at the center of a can insures that the droplets will not carry away the charges as they leave the funnels.

Medicine droppers make excellent water jets and their flow can be controlled by pinch clamps connected to rubber tubing. If the bell arrangement is homemade, care must be taken to mount it on an adequately insulated stand or suspend it by a silk thread. Bare aluminum is recommended for the leads from the oppositely charged cans to the bells. Aluminum allows itself to be coiled neatly yet supports its own weight thus eliminating the possibility of short circuits which usually plague dangling wires.

Sometimes the apparatus does not work instantly but one can easily detect when the apparatus is accumulating charges by the fact that the water droplets begin to fall in a spray rather than a stream as droplets with like charges repel each other. In fact, the magnitude of the forces exerted by these charges is sufficient to deflect them considerably from their normal vertical path. Observations will be made as students reflect on the operation of this apparatus.

Many departments may already have a Kelvin water drop apparatus but perhaps have not thought about coupling it to the bell and chime system. We owe our apparatus to the efforts of Professor T. Jorgensen.

References

1. A. D. Moore, *Electrostatics* (Doubleday, New York, 1968), pp. 175-177.
2. Harry F. Meiners, *Physics Demonstration Experiments* (Ronald, New York, 1970), pp. 847-848.

MENNO FAST

Beblen Laboratory of Physics
University of Nebraska
Lincoln, Nebraska 68508

Fig. 3. Kelvin electrostatic generator made from tin cans. Water dripping from the top reservoir generates static charges that ring bells and spark student interest.

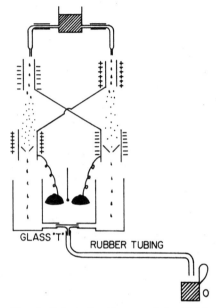

Fig. 4. Schematic diagram of the Kelvin electrostatic generator showing placement of funnels and other modifications to make an effective lobby display.

Kelvin water dropper revisited

Lester Evans
Tates Creek Senior High School, Lexington, Kentucky 40502
J. Truman Stevens
University of Kentucky, Lexington, Kentucky 40506

Innovative physics teachers are always looking for interesting ways of presenting new concepts and stimulating discussions in the classroom. In the search for new ideas for teaching, we don't want to overlook useful ideas of the past. With that in mind, let's dust the apparatus and take a look at a new/old device for observing phenomena related to electrostatic charges.

The Kelvin water dropper [1,2] is easy to construct and will pay large dividends for your time in the form of student curiosity and participation. To construct the water dropper, remove the tops from four large cans. Pound coffee cans work well for this purpose. (It may be helpful to use two different colored cans and position similar cans diagonally from each other.) Remove the bottoms from the two top cans and punch a hole in each of the lower cans. Attach the cans to an insulated stand as shown in Fig. 1.

Fig. 1.

Fig. 2.

Fig. 3.

Scrap wood and clear plastic covers for fluorescent fixtures are excellent materials for preparing the stand. Fasten the cans to the insulating arms with four-way binding posts to facilitate external connections. The bottomless cans are positioned directly above the ones with holes punched in the bottom. With electrical connecting wire, connect the top left can with the bottom right can. Using a separate insulated wire, connect the remaining two cans. A plastic tube from the water faucet leads to two finely drawn glass tubes (see Fig. 2) via a T-joint. With the addition of two containers to catch the running water, you complete the construction of the apparatus.

As you turn on the water, you get a stream through both sets of cans as shown in Fig. 3. In a few seconds an unusual phenomenon is observed. Water drops become charged by friction as they pass from the glass tubes. Through induction and contact, opposite electrostatic charges are found on the two sets of cans. An observer will find that the water no longer runs in a stream, but a mist of water is sprayed around the bottom sets of cans. The charged water droplets "feel" the presence of the electrostatic fields. Unfortunately this phenomenon could not be photographed clearly enough for reproduction in this note.

Electroscopes or other electrostatic devices can be used to demonstrate the presence and strength of the charges. A visible spark occurs when the cans are discharged.

One way to present this activity to a class is to show it with no explanation. Through questioning, the teacher can then encourage the students to discuss their observations and develop hypotheses. Be prepared for a lively discussion with lots of creative explanations.

Questions which might be used to lead the discussion follow.
1. What are your observations?
2. How do your observations differ from what you might have predicted?
3. Develop an hypothesis to explain the cause(s) of the phenomena which you observed.
4. How do your observations support or fail to support your hypothesis?
5. Assuming we accept an electrostatic hypothesis, would you expect the same set of cans to always have the same charge? Why? What variables might influence the type of charge on a given set of cans?
6. How can you determine whether a set of cans is positively or negatively charged?
7. How can you determine the quantity of charge on the cans?
8. Identify places where this phenomenon occurs naturally.

Obviously, these are only a few of the possible questions. The point is to get students involved in doing physics rather than just hearing or reading physics.

References

1. R. M. Sutton, editor, *Demonstration Experiments in Physics*, (McGraw-Hill, New York, 1938), pp. 261-62.
2. "Kelvin's water-drop electrostatic machine," *Scientific American*, p. 175, June 1960.

Questions students ask

Is a swimmer safe in a lightning storm?

Question:

What would be the area of a lake, say 10 m deep that would be affected by a bolt of lightning? If a swimmer were in the water, what would be a safe distance from the point of impact? After a storm, could one collect the fish that would be stunned by the lightning?

This question was submitted by **Lester Dwyer** *of Chaminade High School, Hollywood, Florida 33021 and also by* **Lee Ihlenfeldt** *of James Madison Memorial Senior High School, Madison, Wisconsin 53717. The answer was written by* **Captain Bobby N. Turman**, *who is assistant professor of physics at the Air Force Academy, Colorado 80840. Captain Turman received the Outstanding Achievement award from the 15th Annual Space Congress in 1978 for his studies of lightning observations from satellites. He has also done research on lightning activity in Florida thunderstorms at the Kennedy Space Center.*

Answer:

After you have watched a spectacular lightning display for awhile (and who hasn't?), you come to the conclusion that almost every lightning flash is unique. And this seems to be true because of the vast number of variables that are constantly changing as the atmosphere's electric charge is redistributed. To answer this question, we must discuss typical characteristics, and so the results will really mean something to you only if you are struck by lightning many times! The problem is divided into two parts: electrical and physiological effects. The first figure shows a typical lake with a lightning strike at the center. The typical lightning stroke has a peak current[1] of 20 kA. The typical lake has a resistivity of 200 Ω m, and the typical clay surrounding the lake has a much lower resistivity[2] of 50 Ω m. Thus we expect a nonuniform flow of current through the water, with more current taking the preferred path of lower resistance to the good earth connection at the bottom of the lake.

To study the effect on a swimmer, we need to find the amount of current flowing across the surface of the lake. This problem is most easily solved by using an analog approach, in which a scale model of the electrical system is tested. This technique is called the electrolytic tank method.[2] I used a cake pan as the scale model for the lake; the diameter was 25 cm and the tap water depth was 1 cm. The lightning flash was simulated by placing a wire at the center of the "lake," just making contact with the water's surface. A 10-V, 60-Hz ac current flowed through the partially conducting water from this wire to the pan. I first took voltage measurements across the surface of the water, and then calculated the radial electric field, which is the gradient of the voltage ($\Delta V/\Delta r$). Current density across the surface was found by dividing the radial electric field by the resistivity of the water. The surface current density for the lightning flash is scaled from these data and shown in Fig. 2.

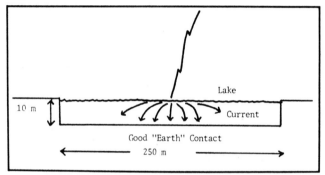

Fig. 1. Current flow from a lightning strike to a lake. A scale model of this lake was tested with a low voltage ac signal to deduce Fig. 2.

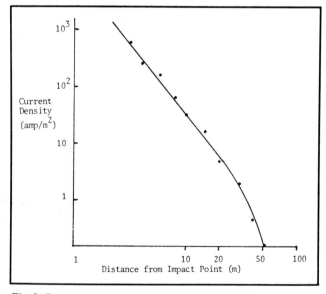

Fig. 2. Current density across the surface of the lake.

Suppose a hapless swimmer is in the lake when this lightning bolt strikes. This person has good electrical contact with the water and has roughly the same resistivity as the water. The current flowing into the body is the cross-sectional area (on the order of 0.1 m^2) times the current density. Various studies suggest that a current of 100 mA is sufficient to stun, if not kill, a person.[2] Thus the lethal current density is around 1 A/m^2. In this example, the lethal current density is found about 40 m from the point of impact.

Are fish also affected by this current? Yes. If we make the bold assumption that a fish will also be stunned by the same current density of 1 A/m^2, we can calculate the volume of water in the lake that will produce stunned, or dead, fish. In this example, the lethal volume is roughly 2% of the volume of the lake. Since it is hard to tell exactly where the lightning flash struck, and no one wants to venture on the lake until the storm has completely passed, there have been few reports of this unique fishing technique. There is a story, however, that fishermen collected almost a hundred stunned fish from a lake that had been struck by lightning.[3]

The final conclusion to all this is simply the warning to human and fish alike; stay out of the water when a thunderstorm comes!

References
1. Martin A. Uman, *Lightning* (McGraw-Hill, New York, 1969).
2. R. H. Golde, *Lightning*, Vol. 2: "Lightning Protection," (Academic Press, New York, 1977).
3. Peter E. Viemeister, *The Lightning Book* (Doubleday, New York, 1961).

Dissectible Leyden Jar

by G. Bradley Huff
Edison Computech High School, 540 East California, Fresno, CA 93706

Most of us discuss the effect of the presence of a dielectric on the capacitance of a capacitor. Some of us use a dissectible Leyden jar[1] (Fig. 1) to show that the charge on a capacitor actually resides on the surfaces of the dielectric and not on the metal plates.[2,3]

My demonstration consists first of charging the dissectible capacitor with a Wimshurst Machine or Van de Graaff generator, disconnecting the generator and discharging the capacitor. This produces a large, loud spark. Then I charge the capacitor a second time, disconnect it from the charging device, lift the dielectric jar out of the outer metal can, and then the inner metal can out of the jar using insulated tongs, and bring the two metal cans into contact. The students expect a large spark, but normally only a small spark, if any, is observed. The students usually suggest that "handling" the cans has discharged them. Using the tongs, I reassemble the capacitor and "just to be sure" reconnect the inner and outer cans. The resulting spark is almost as loud and bright as before. The students are quick to conclude that the charge resides on the inner and outer surfaces of the dielectric jar and not on the metal cans.

This demonstration served as background for an interesting question that arose this past year concerning the result when a dielectric slab is removed from a parallel-plate capacitor. Most textbooks, dealing with the topics of capacitance and dielectrics, present a series of problems involving parallel-plate capacitors with dielectric slabs which are inserted or removed. One particular problem[4] involves an isolated, charged capacitor with a dielectric slab of dielectric constant, k, filling the space between the plates. Initially, the charge is Q_0, the field between the plates is E_0, and the potential difference is V_0. *Question:* When the slab is removed, what are the new values of Q, E, and V?

This past year several of my high school students gave me answers far different from the "correct values." They said the values for all three quantities should be zero. They based their answers on the

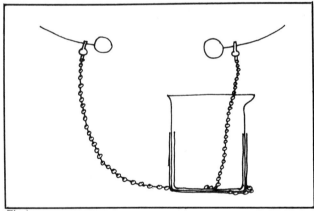

Fig 1.

demonstration I had done using the dissectible Leyden jar. They reasoned that if the dielectric is removed, so is all the charge.

The intent of this note is to bring the physical unreality of this fairly traditional problem to the attention of the physics teaching community. If you are not familiar with the dissectible Leyden jar demonstration, I strongly recommend it. You do not need to buy a commercial one. A beaker and aluminum foil cans can be used. The commercial ones do work better, though. □

References
1. Sargent-Welch Scientific Company catalog, 1983-1984, p. 644, item -1989.
2. R. M. Sutton, *Demonstrations in Physics*, (McGraw-Hill, 1938), p. 275.
3. H. E. Meiners, *Physics Demonstration Experiments*, (Ronald Press, 1970), p. 871.
4. Franklin Miller, Jr., *College Physics*, 4th ed., (Harcourt, Brace Jovanovich, Inc., 1977) problem -18-B18 (page 420).

"Dissectible Leyden Jar"; A Comment

A recent note[1] states that "the charge on a capacitor actually resides on the surface of the dielectric and not on the metal plates," referring to the fact that after a charged Leyden jar is disassembled and its plates shorted, it can still produce a healthy spark when reassembled.

As an apparent experimental contradiction of the usual introductory-level understanding of dielectrics in capacitors, this demonstration might seem to offer some deeper insight. But a search of *American Journal of Physics* cumulative indexes uncovered a rather thorough study done long ago which showed that the effect is mainly due to a simple transfer of charge from the plates to the dielectric surfaces. High charging potentials, over 2 kV, are required if the surfaces are dry. The apparatus thereafter works by induction, like an electrophorus, combined with charge transfer back from the dielectric to the plates by sparking when the charging potential is above 5 kV. The study concludes, "In view of the facts here presented, it is believed that the dissectible jar experiment is not a highly instructive experiment to perform in elementary classes."[3]

References
1. G.B. Huff, *Phys. Teach.*, **24**, 292 (1986).
2. J. Zeleny, Observations and Experiments on Condensers with Removable Coats, *Am. J. Phys.* **12**, 329 (1944).
3. Zeleny, p. 339.

Bruce H. Morgan, U.S. Naval Academy, Physics Dept., Annapolis, MD 21402-5026

More on "Dissectible Leyden Jar"

In *my* philosophy of **teaching physics**—which is always *abundant* with **demonstrations**—it is essential to our purpose that we get *all we can* out of a **demonstration**. Accordingly, every possible ingredient of it must be unveiled. This exposes the *beauty* and the *drama* of it and cultivates the *thinking* process. It is *now* an **intellectual process** and not just a passing "show". So I take the liberty of saying *more* on the "Dissectible Leyden Jar" [*Phys. Teach.*, **24**, 292 (1986)].

Alpha:
I have shown this enchanting thing at least 1000 times! It embraces the entire subject of electrostatics!

Beta:
The first thing we ask the class is HOW do we pronounce *dissectible*? (We write it on the board.) I have hardly *ever* heard it correctly! I might even put it in the Latin to suggest that I was schooled in another era! And *this* reads *dis-* apart + *secare* to cut.

Gamma:
The spark: What color is it and *why*? Does its length measure anything? What would its spectrum look like? Its temperature? What *is* a spark and *what* are we hearing?

Delta:
We have charged it—*what* does this mean? We now show ONE discharge. Instantly thereafter we show *another*—meaning that there is energy still there—and now again *another*—and still another! What's going on here?

Epsilon:
To add to the mystery, I secretly and surreptitiously—with furtive stealth—wipe the glass dielectric both inside *and* outside with a swift motion of the dry hand. Wipe it—*reassemble*—NO spark! NO wipe—a spark! T*his proves something.*

Zeta:
Let the spark pass through a sheet of paper, a pane of glass, or ignite a spoonful of alcohol.

Eta:
And what is the *capacitance* of this "jar"? How *big* would it be to have a capacitance of *one farad*?

Theta:
And of great importance: The story of Pieter van Musschenbroek *must be told!* Said he, playing with *his* electric jar: "I was struck so violently that all my body ached; I thought it was all up with me. I could not breathe. I would not take another such shock for all the Kingdom of France." How charming this little bit!

Iota:
Since little *history* of the subject is *ever* taught and none is *ever* learned, I have always "loaded" *my* classroom lectures with things biological and historical and humanistic and anecdotal—with pictures of men and events. This cultivates a new *feeling* for what **physics is**!

Julius Sumner Miller, 16711 Cranbrook Avenue, Torrance, CA 90504

A POTPOURRI OF PHYSICS TEACHING IDEAS—ELECTRICITY AND MAGNETISM

COULOMB'S LAW ON THE OVERHEAD PROJECTOR

Coulomb's law states that the electrical force between charged spheres depends inversely on the square of their separation and directly on the magnitude of the charge on each sphere.

Despite the efforts of the two national introductory physics programs, PSSC[1] and Project Physics,[2] a simple, reliable way of performing the Coulomb's law experiment has been elusive. It may not be possible to overcome all of the problems of sensitivity but the following arrangement minimizes them by combining larger apparatus, especially larger and heavier spheres, with the overhead projector. The heat from the projector lamp provides dryness to help maintain the spheres' charge. The projector magnifies the apparatus for a small laboratory group or a large lecture class. The complete apparatus (Fig. 5) is a modification of the PSSC arrangement and, therefore, the PSSC procedure may be used in performing the experiment.

The apparatus to be used with the overhead projector should be easy to collect and construct. Prepare a transparent displacement scale for the overhead projector stage using a series of fine, parallel lines, 1 cm apart. A penny may serve as a starting reference marker. Make a fence or corral for the stage to eliminate air currents, especially those created by the projector's cooling system. Cut a manila file folder into four strips, 2½ in. x 10 in. So that the corral can be folded easily for storage, leave a little space between the ends and tape them together with electrical tape.

Select two smooth, ½-in. diam polystyrene spheres[3] and coat them with graphite or other conducting paint. Prepare a small wooden block with a sawed slot to tightly hold an electrically clean polyethylene strip having a protruding, sharp tip. Carefully press one of the spheres onto the sharp tip and adjust the strip until the sphere is about 2 cm above the film scale. Secure a piece of 3/8 in. diam wood dowel, 16 in. long, and drill a small hole through each end. Loosely tape the dowel onto the center of the back of the projection head.

Measure out an extra long piece of electrically clean nylon or polyester thread, stranded not solid, for the pendulum. Take one end of the thread and insert it through the drilled hole in one end of the dowel. Tie the

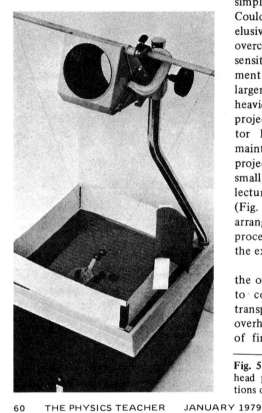

Fig. 5. Arrangement of apparatus on overhead projector for quantitative determinations of Coulomb's law.

thread securely around half the dowel. Take the remaining end of the thread and thread a small needle with it. Carefully push the needle and thread through the remaining sphere. The sphere will slide down the thread to the lowest point in the pendulum. Insert the needle and thread through the remaining drilled hole but do not tie the thread immediately. Run the projection head up as far as it will go and adjust the thread until the pendulum sphere is centered about 1 cm above the stage. Now tie the thread securely around half the dowel and cut off the excess.

Turn on the projector and focus the sphere mounted on the block sharply on the screen. Loosen the dowel from the tape on the projection head. Rotate the dowel between your fingers so that it winds up the extra thread, and bring the pendulum sphere into focus, the same distance above the stage as the other sphere. Tape the dowel securely. Charge up the spheres, and the apparatus should perform well. Excessive humidity, however, can interfere. The thread and polyethylene strip which insulate the spheres should never be handled to that they remain electrically clean. The thread and pendulum sphere may easily be wound up on the dowel for compact storage. All parts should be stored in a box or other dust-free environment.

One mechanical problem may arise. Not all projection heads are of the same design. Therefore, the dowel may not be so easily taped fast in all cases. The popular 3M projector (Fig. 5), for example, requires some modification. One solution is the use of a wood block with an enlarged 3/8-in. diam hole drilled through it for the dowel and a set screw. The block raises the dowel above an obstruction and is secured with a 2-in. C-clamp.

I wish to thank David B. Scott of Seattle for the original idea and his expertise. I also wish to thank the Union College Physics Department in Schenectady, New York for the use of their facilities in preparing this article.

JOHN B. JOHNSTON
Nanuet High School
Nanuet, New York 10954

References

1. Physical Science Study Committee, *Physics Laboratory Guide* (D.C. Heath and Co., Boston, 1965) p. 63.
2. Gerald Holton, F. James Rutherford, and Fletcher G. Watson, *Project Physics Handbook* (Holt, Rinehart and Winston, Inc., New York, 1970) p. 244.
3. Electrostatics Kit (PSSC) #1000, Eduquip-Macalaster Corp., 1085 Commonwealth Ave., Boston, MA 02215.

The Oersted effect on the overhead

It is well known that the effect on a magnetic compass needle of being deflected when placed near a current-carrying wire was discovered by Hans Oersted in 1820. An elementary demonstration of this effect is usually presented in any course dealing with electricity and magnetism, and it is a very convincing proof that moving electric charges produce magnetic fields. Several apparatus manufacturers* sell a simple device to demonstrate the Oersted effect to small classes. The apparatus consists of a metallic bar bent into a rectangular loop and mounted on an insulated base with a compass needle suspended at the middle of the loop. When a large current is sent through the loop the compass needle will deflect and line up perpendicular to the loop; i.e., tangent to the magnetic field line at that position. Reversing the current direction results in the needle reversing its direction, showing how the magnetic field direction is related to the current direction (right-hand rule).

In a large or auditorium-size lecture class it is difficult for all the students to see the effect demonstrated by this small apparatus. Since the overhead projector is used extensively in such situations it is natural to try to adapt this demonstration to the overhead. This is simply accomplished by replacing the opaque base with one made of Lucite and securing to it an inverted-U-shaped metal bar with screw terminals at each end for connection to a current source. The same compass needle that is used in the commercial apparatus is suspended under the bar by a needle point in the same manner as is found in the commercial device (see Fig. 1). When the apparatus is operated on the overhead the compass needle deflection is easily viewed by all. A small piece of paper can be taped to one end of the compass needle as a visible reference. A further modification (not shown in the figure) uses a smaller raised Lucite platform to place the compass needle above the metal bar for demonstrating the circular symmetry of the magnetic field.

Fig. 1. The Oersted effect demonstrated in place on the overhead.

SAM J. CIPOLLA
Creighton University
Omaha, Nebraska 68178

---

*For instance, Oersted's Law Apparatus, manufactured by the Sargent-Welch Company, Skokie, Illinois.

Teachers' pets:

Lenz's law

Harry H. Kemp
269 West 1st South, Logan, Utah 84321

The chances are good that one of your students can get a burned-out kilowatt-hour meter from the power company. The picture (Fig. 1) shows one that has had all the electrical parts taken off leaving the rotating disk and the permanent magnets. A stand was made for the device. Remove the two screws which hold the magnets on and give the disk a push. It should rotate easily with little friction. Now attach the magnets. With the model I used this was easily done in a few seconds with no clearance problem. The same push as before will cause the disk to move only a fraction of a turn. Students may need some help to find an explanation. When the aluminum disk moves it cuts magnetic lines of force and according to Lenz's law the current produced in the disk is in such a direction that its magnetic field interacts with that of the magnets to oppose the motion.

Electrostatic charges and copying machines

Robert P. Bauman
University of Alabama in Birmingham, Birmingham, Alabama 35233

It is well known that copies from xerographic copying machines tend to stick together, and nearly everyone recognizes the cause to be electrostatic charges accumulated during the copying and paper-handling process. It is somewhat less obvious why, if each page has picked up a charge, the pages attract each other, rather than repelling.

A quick experiment not only confirmed a suspected explanation but also provided a convenient cure. If one page is taken from the stack and turned over, there is a weak repulsion, showing that the effect arises from an electrostatic dipole induced across the sheet, front to back. Passing the inverted sheet across the stack, in contact, leaves the sheet with little or no residual attraction for other pages, greatly facilitating further handling. Similarly, with our IBM copier, after extended feeding there is a tendency for pages to stick to the input tray. By putting the page down upside down first, then right side up, the difficulty is removed.

It is often difficult to produce static charges in the lecture hall, so we seldom attempt to produce electrostatic dipoles. Perhaps we have overlooked a significant use for the copiers.

Electrical figures

Colin Pounder
161 Cotmanbay Road, Ilkeston, Derbyshire, England

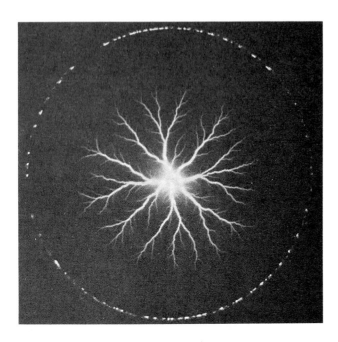

Fig. 1. *Lichtenberg's figures.*

Although the title could imply abstract calculations in the theory of electrical science, the opposite is in fact the case.

The following is a brief outline of two possible ideas for practical demonstration by a teacher to present an opportunity to experience some of the lesser-known phenomena found among early investigations into electricity.

Georg Christoph Lichtenberg (1744-1799), German satirist and physicist, showed how a capacitor discharging from a point to a plate can produce flower-like patterns of considerable beauty.[1]

Using a Leyden jar, "Lichtenberg's figures" (Fig. 1 may be demonstrated by connecting a metal plate to one terminal, covering the plate with a thin sheet of glass, sprinkling this with Lycopodium powder, and then discharging the other terminal via a sharp point held over the plate. The path of the discharge is made visible by the powder.

A more impressive approach was suggested by Saaski[2] in which a capacitor is charged to about 8 kV from a power source. One terminal is connected to a circular metal dish that holds water and the other capacitor terminal is connected to a wire which is carefully lowered to the surface of the water in the center of the pan. The results are magnificent figures which may easily be photographed by carrying out the experiment in the dark with a camera located over the pan of water.

In 1863 Dr. Strethill Wright demonstrated his less well-known "cohesion figures."[3] These may be demonstrated by the use of a small induction coil producing a spark about 0.5 cm long (I use a Model T coil). The diagram shows how the apparatus may be set up (Fig. 2). A sheet of glass is placed over a metal plate which is connected to one terminal of the induction coil. A drop of liquid, e.g., sulphuric acid, is placed onto the

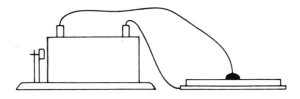

Fig. 2. *Wright's apparatus.*

glass near to the center and a wire from the second terminal of the coil is dipped into the drop. Arms will be seen to develop from the drop to form a figure not unlike a Lichtenberg discharge. Different liquids produce different patterns or figures.

In Ferguson's *Electricity*,[3] it is suggested that the plate be blackened to show up the figure. An easier possibility, particularly if the glass is to be retained for a time, is to sprinkle the figure with fine powder. The surplus is blown away and the figure is easily observed; if graphite powder is used, this may be set up on an overhead projector.

It was claimed that Wright produced figures by cleaving mica, placing a liquid drop on the clean surface and breathing on to the drop which spread into a flower-like figure. This is not easy to achieve and is perhaps worthy of further investigation.

References

1. H. Pupke, *Ostwalds Reprints of Classical Writings No. 246* (Lichtenbergsche Figuren), Leipzig, 1956. (This is a translation of G. C. Lichtenberg's original, "Novo methodo naturam ac motum fluidi electrici investigandi," Societatis Regiae Scientiarum Gottingensis 1778.)
2. E. W. Saaski, Sci. Amer., **211**, No. 6 (Dec. 1964).
3. R. M. Ferguson, *Electricity*, (W. & R. Chambers, London, 1871), pp. 193-194. Details of the practical application of Lichtenberg figures are found in M. A. Uman, *Lightning* (McGraw-Hill, 1969), p. 116.

Demonstration of Gauss' law for a metal surface

T. W. Haywood and R. C. Nelson
The University of North Carolina at Wilmington, Wilmington, North Carolina 28401

An important prediction of Gauss's law, that an excess charge resides entirely on the outer surface of a conductor, can be dramatically demonstrated with a Van de Graaff generator[1] and a few of the Styrofoam buttons commonly used as packing material.

A metal cup filled with the Styrofoam buttons is placed on the top of the generator by means of a banana plug attached to the bottom of the cup. With the generator running, large sparks can be drawn from the can, showing that a considerable excess charge resides on the can; yet the Styrofoam does not move, as no charge reaches the inside surface of the can (Fig. 1). If the metal cup is then replaced by a plastic cup (an empty cream whip container is excellent), charge will be transferred into the Styrofoam buttons when the generator is turned on. After a few seconds, the Styrofoam buttons will begin to fly out of the cup like popcorn as they collect enough charge to lift them in the high electric field (Fig. 2). Many of them may be left sticking to the walls and ceiling after the generator is turned off.

Reference

1. The unit shown in Fig. 1 is model number 27-4100 of The Ealing Corporation.

Fig. 1. Styrofoam buttons responding to the absence of charge inside their container.

Fig. 2. Styrofoam buttons fly out of their container in response to charge.

A motor is a generator and vice versa

John A. Johnson and Franklin Miller, Jr.
Kenyon College, Gambier, Ohio 43022

A motor and a generator have the same basic construction — a coil (armature) free to rotate in a magnetic field. A simple way to demonstrate this equivalence is to connect two identical student galvanometers together. A generator converts mechanical energy into electrical energy. If you pick up one galvanometer and wiggle it, you are providing mechanical energy. As the coil of the galvanometer rotates through the field of its permanent magnet, an emf is induced in the coil. The second galvanometer (the motor) converts electrical energy into mechanical energy and the needle of that galvanometer moves when the first galvanometer (the generator) is wiggled. The equivalence between the two devices is shown by wiggling first one then the other galvanometer.

We use two Weston Model 375 galvanometers which have full scale deflection of about 0.5 mA; any inexpensive galvanometer of the type commonly used in Wheatstone bridges would be suitable. The damping is small enough so that several oscillations of the motor can be observed before the energy of a single wiggle of the generator is dissipated. A continuous agitation of the generator will produce a sustained wiggling of the motor coil. In order to obtain a large amplitude of vibration, the demonstrator will tend to agitate the generator at its resonant frequency. Since the galvanometers are identical, this frequency is also the resonant frequency of the motor, and a large response is obtained. It is desirable to use several meters of connecting wire to avoid any suggestion that the two galvanometers have purely mechanical coupling.

FORCE BETWEEN PARALLEL CURRENTS ON THE OVERHEAD PROJECTOR

A straightforward demonstration of the force of attraction or repulsion between parallel currents can be made with an overhead projector, using the apparatus shown in Fig. 7. A 7 in. x 11 in. piece of 1/8 in. Plexiglas is mounted on insulated legs. Four 2-56 screws 1 in. long are used to support two parallel wires made from No. 18 magnet wire which has varnish insulation. An L is formed at the end of each wire, the insulation is removed from about the last 1/2 in. and the wires rest in the screw slots. Any power supply capable of delivering 5 to 10 A, momentarily connected with a tap switch, will furnish plenty of current to make the forces clearly apparent. This demonstration, like that recently reported by Kruglak[1] et al., avoids the use of mercury contacts needed for similar demonstrations previously suggested by Sutton[2] and Meiners.[3] It also has the advantage of small size.

R. C. NICKLIN
Appalachian State University
Boone, North Carolina 28608

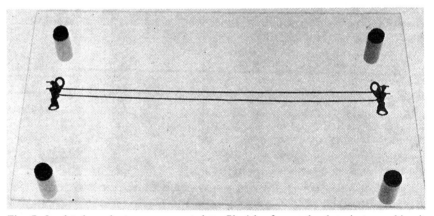

Fig. 7. Insulated conductors are mounted on Plexiglas for overhead projector making it possible for large groups to observe effects of forces generated by current passed through the conductors.

References:

1. H. Kruglak, W. Merrow and P. Rood, Phys. Teach. 14, 454 (1976).
2. R. Sutton, *Demonstration Experiments in Physics* (McGraw-Hill, New York, 1938) p. 310.
3. H. Meiners, *Physics Demonstration Experiments* (Ronald Press, New York, 1970) p. 927.

How things work

Sealed batteries

Fritz G. Will

General Electric Corporate Research and Development, Box 8, Schenectady, N.Y. 12301

The two most commonly used rechargeable batteries are the lead-acid and the nickel-cadmium battery. Every automobile uses a lead-acid battery to start its engine; the more durable but also more expensive nickel-cadmium battery is used in aircraft, flashlights, pocket calculators, portable electric tools, toys, and many other applications.

When these batteries are recharged, hydrogen and oxygen gases are usually formed by the electrolysis of water. In the past, the gases had to be vented to prevent the buildup of gas pressure inside the battery. This meant, however, that electrolyte could spill through the vent openings and water was lost by electrolysis and evaporation. Furthermore, the battery had to be kept in an upright position and water had to be added regularly.

Only a few years ago, one learned how to react the gases inside the battery and, therefore, how to make a sealed and maintenance-free battery. Actually, the first sealed nickel-cadmium batteries were in use in the late thirties while the first sealed lead-acid batteries became available only in the early seventies.

When researchers first experimented with sealed batteries, they made several key findings: During charging, the gas pressure inside the cell first increased and then leveled off. However, after completing the charging, the gas pressure started decreasing slowly. Analysis of the gases inside the cell revealed that oxygen was gradually being consumed by some reaction whereas the amount of hydrogen stayed essentially constant. These initial findings were the beginning of many detailed studies of how hydrogen and oxygen react inside a sealed battery. Today it is well-known that oxygen is readily reduced on the negative electrode, that is, the lead electrode in a lead-acid battery or the cadmium electrode in a nickel-cadmium battery. Aqueous electrolyte is reformed during that reduction. Hydrogen gas, on the other hand, reacts exceedingly slowly inside the battery and its formation must be prevented if the battery is to be sealed safely.

Sealed batteries, which are now being used increasingly in many applications, employ negative electrodes with a larger equivalent weight than the positive electrodes and contain uncharged negative mass at all times. Furthermore, a limited amount of electrolyte is used and held by capillary forces in the fine pores of the electrodes and the battery separator. During charging, the positive electrode evolves oxygen gas at a time when the negative electrode is only partially charged. The oxygen gas can readily reach the negative electrode through the porous separator and through the gas-filled space above the electrodes. The oxygen oxidizes — or discharges — the negative electrode at the same rate as the current attempts to charge it. On balance, a negative electrode remains in a partially charged state. Under these conditions, little if any hydrogen gas is formed and the oxygen is continuously reduced; therefore, no abnormally high gas pressure is formed. Battery chargers and alternators are designed such that the battery is charged with desirable and safe current levels and overcharging is kept at a minimum. Charging equipment designed for one type of battery must not be used with other types of batteries. In the event of accidental charging with higher than recommended currents, however, a safety vent guards against the buildup of excessive pressure inside the battery.

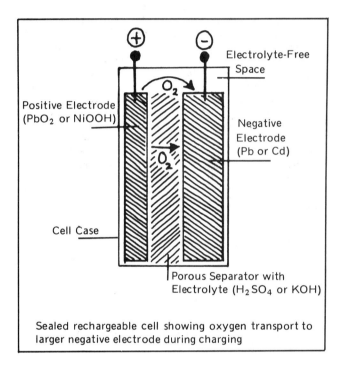

Sealed rechargeable cell showing oxygen transport to larger negative electrode during charging

How things work

C. H. ANDERSON, RCA Laboratories, Princeton, New Jersey 08540

The smoke detector

J. R. Young
General Electric Research & Development Center

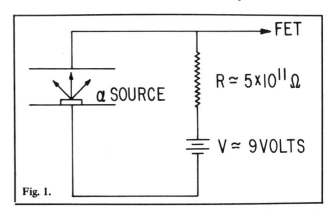

Fig. 1.

In most commercial smoke detectors the smoke sensor is an ionization chamber (photoelectric smoke detectors represent only a small fraction of the home smoke detectors). The simplest smoke detector ionization chamber consists of two parallel metal plates a few square cm in area, mounted in air and separated by about 1 cm. A small radioactive α-particle source is attached to the center of one plate of the ionization chamber. The α particles are ejected into the air between the plates and produce ionization (negative and positive ions) along their path. By applying a small dc voltage of about 5 V across the two plates, a small current will flow ($\sim 10^{-11}$ A). This current remains almost constant under normal conditions even when there are large changes in humidity and temperature. However, when smoke particles enter the ionization chamber, the current decreases. This lower ionization chamber current produces a voltage change across a large series resistor which turns on a field-effect transistor (FET) or similar device and activates an alarm indicating the presence of smoke (Fig. 1). The ionization chamber current decreases when smoke is present because the smoke particles provide surfaces for neutralization of the negative and positive ions. In most cases when a positive or negative ion strikes a solid surface, the ion will be neutralized. This neutralization results in a lower density of charge carriers in the ionization chamber which in turn produces a lower current.[1]

The radioactive α source is quite weak ($\sim 2\mu$c) and since the α particles have a range of only a few cm in air, simple baffle shielding reduces radiation exposure to the user far below normal background levels.

Reference

1. J. J. Thomson, *Conduction of Electricity Through Gases*, (Cambridge University Press, New York 1928) Chapter II.

Ohm's law mnemonic

Carl H. Hayn
University of Santa Clara, Santa Clara, California 95053

Fig. 1. Ohm's law mnemonic

In the teaching of high school and college noncalculus general physics courses, every teacher has the optimistic hope of making his presentation of Ohm's law so clear and logical that all of his students will understand and easily remember. Considering the relationships involved is the most attractive approach, indicating how the flow of charge through a resistor is influenced by the potential difference impressed across it and also by the magnitude of the resistance itself.

Despite the most painstaking efforts, some students exercise very little reasoning and merely memorize equations. For this type of student I have drawn Fig. l(a) on the blackboard and have asked what it represents. Without exception they have replied that it is an Indian teepee. I then suggest that they invert the teepee, place an I on one side of it and an R on the other side as shown in Fig. l(b). They can think of the I and R as indicating "Indian Reservation," where teepees may be found. This figure gives them a mnemonic for remembering Ohm's law. The V-opening in the teepee represents the voltage obtained by the product IR. The current I is found by dividing V/R and R is similarly found by V/I.

The Omega competition

Robert P. Lanni
Department of Physics, State University of New York, Albany, New York 12222

ohmbrella

coolohm*

ohm fries

Recently, a student who had spent a weekend at West Point told me that the cadets had introduced her to a game that involves using the symbol Ω to be equivalent to ohm and modifying it in some way or incorporating it into a picture that would then represent a common word or phrase. For example  becomes mobile ohm, $\Omega C_6 H_{12} O_6 \Omega$ becomes ohm sweet ohm. The student showed me these and three others, two of which are printable.

Feeling challenged, I tried to see what I could ⌠⌡ (cohm) up. In an hour I had fifteen. The thought occurred to me to start a competition in class. The rules were announced and examples given. The students were given a week. The contest would be judged by a colleague and the prize would be an omega on a silver chain (for a girl) or an omega on a pin (for a boy). We had a ball doing it and I heartily recommend it as one way of humanizing the beginning course. Some of the entries are shown. I'd like to thank Roslyn Weinstein, one of the students, for doing the art work.

*Contest winners

stockohm

$\Omega = \vec{F} \cdot \vec{d}$
ohm work

$\Omega E \; \Omega I$
oh me oh my

a house is not a ohm*

$\frac{\Omega}{1} \; \frac{\Omega}{2} \; \frac{\Omega}{3} \; \frac{\Omega}{5} \; \frac{\Omega}{6}$
ohmission

The Volt competition

Donna A. Berry
Shaker Heights High School, Shaker Heights, Ohio 44120

While teaching high school physics, I incorporate competitive contests to humanize the course of study (cartoons, Physics Olympics, Bridge Building, essays and energy projects). It is a challenge to motivate individual creativity.

I suggested the word "VOLT" to my Project Physics classes as a sequel to "The Omega Competition."[1] The object of the contest was to make up "puns" using the word VOLT. Some students even drew their ideas and the best entries are shown here.

Reference
1. R. Lanni, Phys. Teach. 16, 483 (1978).

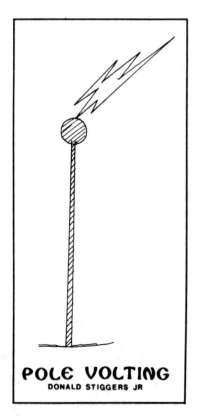

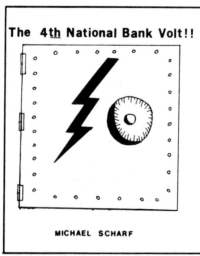

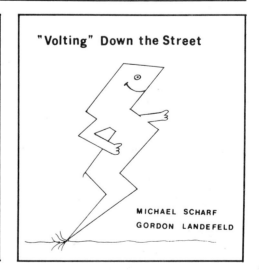

Questions students ask

Why is the ac power line grounded?

QUESTION:

Courtney Lantz, *physics teacher in Pacifica High School, Garden Grove, CA 92667 writes:*

In my physics classes I spend considerable time discussing typical household electrical circuits. I make a point of going over both grounded and ungrounded circuits and the nature of the third-wire ground. I cannot give a really good reason to students, however, for grounding our ac in the first place. Can you clear this up?

After calling the writer's attention to a note on grounding by the late Walter Thumm in our issue of February 1974, we asked **Charles H. Anderson** *for a more detailed explanation. The answer that follows was written by Dr. Anderson, who is Head of Applied Mathematics and Physical Science at the RCA Laboratories, Princeton, NJ 08840 and* **S. Bozowski** *who is Senior Technical Associate at the same laboratory.*

Answer:

It is illegal and unsafe to install electrical power circuits without a third grounding wire in a home. In the ideal situation the ground wire looks superfluous, but in the real world with circuits in hundreds of millions of homes things can and will go wrong that create potentially lethal conditions that are invisible. The primary purpose of the ground wire is to prevent this from happening. All electrical appliances sold in this country are required to have their metal cases connected to the ground. When a wire in an appliance happens to touch a grounded metal case, a fuse blows when it is turned on rather than having the case sit at a high voltage.

To fully understand the problem one has to start with the power lines that come into a home. These originate at a center tapped transformer with 220 V across the outer two lines; the center line is called neutral. The 110-V circuits are derived from the neutral and one of the other two lines. Electricians try to balance the load so that electrical outlets in different parts of a home can be derived from both halves of the circuit. Since the neutral is grounded this means the hot sides of two different outlets can differ by 220 V! This however does not have anything to do with why the neutral is grounded. Consider a kitchen that is wired from only one side and the neutral is not grounded. If the case of a toaster happens to touch one side of the 110, it by itself is not dangerous. However, if the case of the refrigerator happens to touch the other side of the line then a lethal situation is created. Or if you and your neighbor are on the same transformer circuit and he decides to ground his neutral to a water pipe, then your toaster and sink are a dangerous combination. Grounding the neutral establishes a common ground so that ambiguous situations like this cannot occur. The third ground wire in the circuit ensures that when the metal case of an appliance happens to touch the hot side, a fuse blows and a dangerous situation is prevented. The third ground wire helps prevent mistakes by electricians who could wire a two-socket outlet backwards. This could cause no problem for years and then suddenly a defective appliance could be plugged in and create a dangerous situation with no warning. In conclusion it is recommended that one read the Underwriters Code on electrical wiring that provides very sound rules that should be strictly followed at all times.

ELECTRIC FIELD USING AN OVERHEAD PROJECTOR

We teach that the electric field, $\vec{E}$, between two oppositely charged parallel plates is uniform (except for the fringing effects) and that a charge q, placed between the plates will experience a force, $\vec{F}$, which is equal to the product of the charge and the electric field intensity, and which is constant for any position of the charge between the plates. Here is a method of displaying this fact to a large class using an overhead projector.

Mount two metal plates (about 20 cm square) in an insulated stand made of wood or polystyrene, and connect the plates to a source of about 5000 V dc. Place the metal plates on edge on the overhead projector. Suspend a metallized pith ball from a thread attached to an insulated arm clamped to a retort stand and mount the stand on a dynamics cart or a small wheeled stand. Place the stand supporting the ball near the projector so that the stand is free to move parallel to the plates using a meter stick on the bench as a guide. The pith ball should hang centrally between the plates.

Charge the pith ball, either positively or negatively, and switch on the projector. When focussed, the image position of the charged ball may be used to draw an equilibrium line on the chalk board. When a potential is applied to the plates the ball will move in a direction perpendicular to the plane of the plates to a new equilibrium position determined by the weight of the ball and the electrical force exerted by the field on the charged ball. By tracking the wheeled stand supporting the charged ball parallel to the plates, the displacement of the charged ball from the initial equilibrium line can be shown to be constant, and so the deflecting force, $\vec{F}$, is constant and in consequence so is the electric field, $\vec{E}$, causing the force. (The ball is following an equipotential line.) At any time the potential to the plates may be switched off and the ball will then return to the initial equilibrium line.

If desired, the ball may be moved to another position between the plates and the procedure repeated. It is also possible to show the fringing effects at the plate extremities by tracking the ball fully out of the region between the plates.

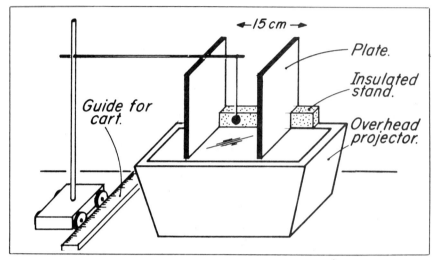

Fig. 4. Method of mounting insulated metal plates on overhead projector stage.

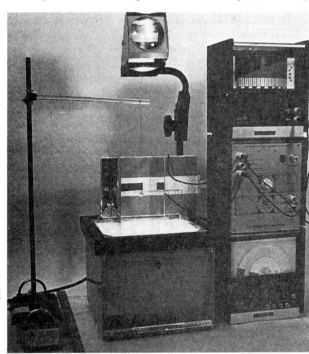

Fig. 5. Electric field apparatus connected to 5000 V power supply.

IAN MENNIE
CYRIL SNOOK
Memorial University of Newfoundland
St. Johns, Newfoundland, Canada

A POTPOURRI OF PHYSICS TEACHING IDEAS—ELECTRICITY AND MAGNETISM

TURN-BY-TURN TRANSFORMER DEMONSTRATION

Several aspects of transformers are made particularly clear to students who see the act of winding a secondary coil on the core of a primary which is already connected to an ac source. It is easy to demonstrate the transformer by placing a coil-bulb combination over the core of a solenoidal ac electromagnet. The solenoid is part of the widely-used Elihu Thompson Apparatus (Fig. 3). The usual procedure is to use a prewound secondary. However, the importance of the number of turns on the secondary is demonstrated much more vividly if the secondary is wound turn-by-turn on the core of the ("plugged-in") primary. A bulb connected to the two ends of the growing coil will become brighter with each new loop.

This demonstration can provide data for a sample calculation using the turns ratio. In one approach, the secondary voltage may be estimated to be approximately 6 Vrms when a 6-V flashlight bulb appears to be as bright as it does when connected to a 6-V battery. From this voltage, the known primary voltage and the (counted) number of secondary turns, the number of primary turns can be computed (assuming an ideal transformer). A more careful analysis using voltmeters and ammeters will reveal that this transformer differs considerably from ideal behavior. For example, the primary current may be several amps when the secondary current is zero. This non-ideal behavior can lead to valuable discussions of several concepts including energy conservation and eddy currents.

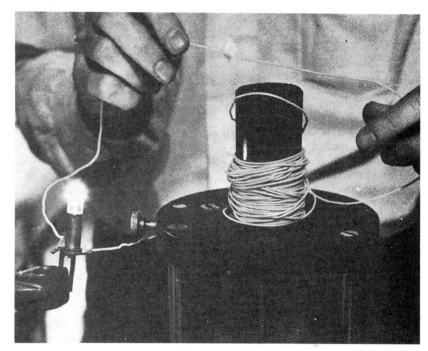

Fig. 3. Bulb gets brighter as more turns are wound on transformer secondary.

JOE L. FERGUSON
Physics Department
Mississippi State University
Mississippi State, Mississippi 39762

Static Electricity Demonstration

Philip E. Highsmith
Converse College
Spartanburg, S. C. 29301

Many aspects of static electricity can be shown with equipment readily obtainable in almost any school. For this demonstration you only need a dark room, a fluorescent tube, a generous piece of saran wrap, and a piece of flat conductor. The following can be shown:

(1) Stroke the tube vigorously with a large piece of crumpled saran wrap. The tube will glow.

(2) After stroking the tube with the saran wrap, touch one end of the tube with the part of the saran wrap that has been rubbed. (Actually, a jabbing motion with the saran wrap is most effective). Short blips of light can be seen illustrating charge does not tend to flow from the saran wrap.

(3) To illustrate electrostatic induction, rub the tube vigorously with the piece of crumpled saran wrap to produce a charge on the saran wrap. Place the crumpled saran wrap on a table with the rubbed side up. Next, place a conductor on top of the saran wrap, being careful to hold the conductor at its insulated edges (Fig. 1). When the tube is touched to the conductor, a strong flash of light is seen. When the conductor is removed and the end of the fluorescent tube is touched to the conductor another strong flash is observed.

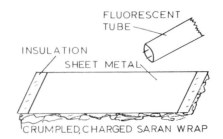

Fig. 1. Touching the fluorescent tube to the metal produces a bright flash of light.

If needles are taped on the end of the fluorescent tube and the tube brought near the conductor, a series of small blips will be seen rather than a strong flash which is an excellent illustration showing how charge leaks from a point. This demonstration is also effective with a neon spectra tube.

Lenz's Law Demonstration

Thomas D. Miner
Yeshiva University
New York City, N. Y.

The inclusion of Lenz's Law in the elementary physics course is probably best justified by its aspect as a consequence of the law of conservation of energy. Students can see the connection most easily by considering the force between, say, a moving magnet and the coil carrying the induced current. Thought experiments involving directions of fields show that this force always opposes the relative motion of coil and magnet, but may leave students with an unhealthy doubt about the reality of a force they cannot feel. A simple and easily-assembled piece of apparatus can show that this force really exists.

A ring of heavy copper wire hangs by two threads from a horizontal rod. The dimensions are not critical. A 3-in. ring of No. 16 wire on a suspension about a foot long works well. The ends of the wire should be joined by soldering. A strong alnico bar magnet (Welch 1812c or equal) when thrust into the ring causes the ring to swing back slightly. By moving the end of the magnet back and forth in the ring at the same frequency as the natural frequency of the ring-pendulum, large oscillations can quickly be generated. Show that the ring is repelled by the approaching magnet, attracted when the magnet is withdrawn, and is unaffected by the stationary magnet.

To satisfy skeptics who may think the effect due to air currents, and as a little puzzle for all, have ready an apparently identical ring with a small gap hidden by one of the suspension threads. Of course it fails to respond to the same motion of the magnet.

VOLTAGE SURGE PROTECTION

A simple over-voltage protection has been developed for integrated components. Voltage surges often occur when electronic equipment is interconnected. The circuit in Fig. 5 has no appreciable effect on the input wave as long as the voltage stays below the operating voltage, usually 5 or 12V. The Zener diode in the circuit clips any voltages down to the breakdown voltage preventing I.C. component burnout. The Zener diode is usually selected by considering the operating voltage plus one volt.

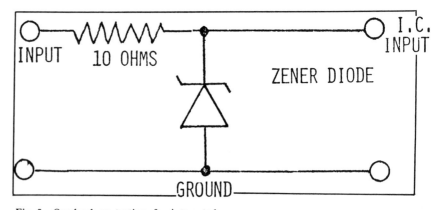

Fig. 5. Overload protection for integrated circuits, schematic diagram.

ROBERT P. BARRETT
Messiah College
Grantham, Pennsylvania 17027

A circuit demonstration

Renato Lichtenstein
Rua Oswaldo Cochrane No. 49, Santos-11100-S. P. Brasil

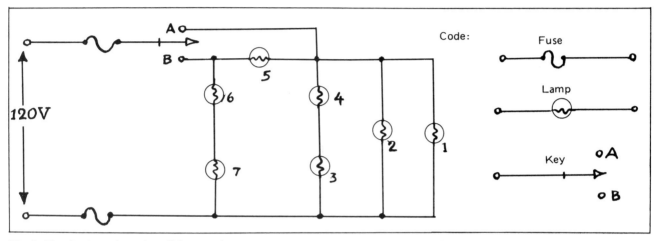

Fig. 1. The circuit - series and parallel connections

Figure 1 shows a circuit I have used to teach electricity. The circuit permits students to learn about electric connections in series and parallel in a simple way. It consists of seven lamps of 40 W and 120 V, a single-pole, double-throw switch and two fuses. It is also possible to use lamps with other power ratings. To connect and disconnect a lamp it is enough to turn it in its socket.

NEON LAMPS AND STATIC ELECTRICITY

When one terminal of a small neon lamp is touched to a charged object and the other held in the hand a bright flash occurs which is frequently more impressive than the sight and sound of a spark. The neon lamp may be used in addition to, or in place of, an electroscope.

A neon glow lamp consists of two closely spaced electrodes immersed in a neon gas at moderately low pressure, forming a gas discharge device.[1] Some neon glow tubes have a photosensitive cathode with photo electrons contributing to improved lamp operation. The response of such tubes tends to become erratic when operated in the dark. To counter this effect, other tubes contain minute quantities of radioactive material which provide a continual source of ions. The experiments reported were conducted with Ne-23 and Ne-88 tubes. The Ne-23 is a low cost lamp with radioactive material added, and is available through local electronic supply houses.

Because the glow is confined to the cathode, when one lead of the lamp is held in the hand and the other touched to a charged metal object, the lead associated with the glowing terminal will identify the negative surface. This allows immediate determination of polarities without reference to ebonite, catfur, etc. A second important feature of the glow lamp is its firing potential. It requires approximately 70 V to cause the lamp to flash. This allows the student to have a rough quantitative feeling for the voltage involved in his experiment.

The neon lamp may be used when working with a Wimshurst machine. The lamp may be fastened at the end of an insulating handle and used between the machine and objects being charged to show the polarity of charge transfer. For this purpose, a lamp with a higher current rating such as the 6 AC (Ne-67) is suggested.

With a Van de Graaf generator, the field may be studied by connecting one of the wires of a neon lamp to a ground wire and moving the tube in the vicinity of the sphere. The brightness of a continual glow (with the generator running) will vary nicely with distance.

Whatever the teaching condition, the addition of a neon discharge lamp can do much to liven the study of static electricity. Earlier references to the use of a neon wand have been made[2,3] but these lamps did not have visible terminals and the polarity effect so important to this suggested use of the lamps may not have been observed.

References

1. *Glow Lamp Manual*, General Electric, 1966, pp. 1, 2.
2. Richard M. Sutton, *Demonstration Experiments in Physics* (McGraw-Hill, New York, 1938), p. 255.
3. D. S. Ainslie, Amer. J. Phys. 30, 69 (1969).

JOHN W. LAYMAN
University of Maryland
College Park, Maryland 20742

DELBERT J. RUTLEDGE
Oklahoma State University
Stillwater, Oklahoma 74074

A simple transistor demonstration

David L. Mott
Department of Physics, New Mexico State University, Las Cruces, New Mexico 88003

In the normal transistor circuit, the collector current is under the control of the base current. Figure 1 shows a very simple circuit that effectively demonstrates this control process to students of general physics. The few parts can be wired together directly without need for a circuit board. Three flashlight batteries in series can provide the necessary power. A photoresistor costs less than a dollar at Radio Shack. The one I use has a resistance of several hundred ohms under ordinary room light, and megohms of resistance in darkness.

In this circuit, the pilot lamp is on when the photoresistor is illuminated, and goes out when the room lights are turned off or when the photoresistor is shielded from light with the hand. This clearly demonstrates the concept: Control of one current by another in a transistor.

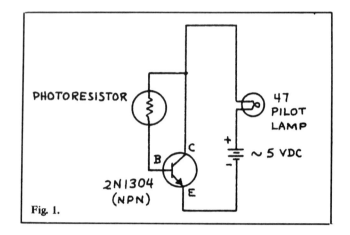

Fig. 1.

ac made visible

Lloyd Harrich
San Ramon High School, 140 Love Lane, Danville, California 94526

In a well known demonstration a neon lamp, powered by alternating current is swung in a large circle at the end of 3-foot connecting wires. In a darkened room the ac is "made visible" by the dark intervals when the voltage is too low to light the lamp causing the circular trace to be a dashed line. Here is an improved version.

Instead of a neon lamp, use a bi-colored light-emitting diode (LED).[1] When powered by dc the LED glows either red or green depending on the polarity. Use a step-down transformer to convert line voltage to about 6 V ac. Connect a current-limiting resistor (100 Ω to 330 Ω) in series with the LED and attach the combination to the transformer through 6 to 8 feet of flexible wire (Fig. 1). The resistor and the LED can be combined as a unit by coating with silicone or using heat-shrink tubing.

When the circuit is plugged in the LED looks yellow, but when it is twirled in the air the resulting circle is made of red and green segments showing the alternating polarity of the ac.

The diffused bi-color LEDs can be obtained at most electronics stores. In the event you cannot find one, the members of my electronics club can supply one for one dollar plus a stamp.

Reference

1. Radio Shack Stores

Fig. 1.

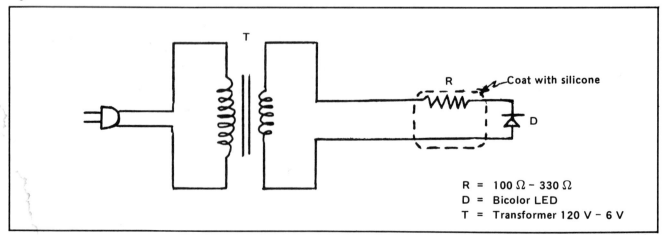

Mysterious lights in series and parallel

Clifton Keller
Department of Education, Andrews University, Berrien Springs, Michigan 49104

Many science experiments and demonstrations take advantage of student curiosity. The sealed black box which contains an unknown object has been popular. In a similar manner a sealed black box with six light bulbs on top has been used effectively in my science classes to improve questioning skills and to test student understanding of parallel and series circuits. At times the brightness of the bulbs is inversely related to their size. For example, a 25-W bulb may be brighter than the 100-W bulb. This makes for true mystery.

The sockets are mounted at the points of a six-pointed star, colored for attractiveness. This arrangement does not give away the circuitry, and the colors make it possible to refer to certain sockets without referring to the size of the light bulb. This is important if the positions of the bulbs are changed. The sockets are connected and the bulbs are located as shown (Fig. 1). Students are allowed only to loosen or tighten bulbs to determine the exact wiring of the circuit. The varying intensities of the lights as they are in series or in parallel with other lights are intriguing and make this a very popular learning center activity. Since the positions and sizes of the bulbs can be changed and other circuits can be designed, interest can be maintained for extended periods.

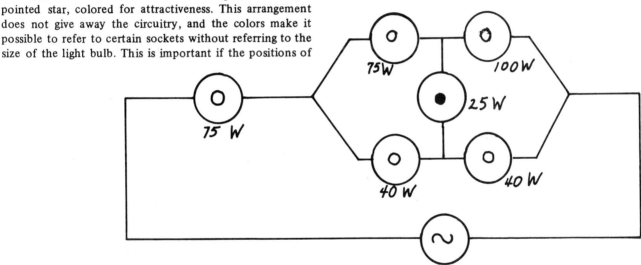

Shape of an Electric Field

Mike Weiss
Brooklyn Technical High School, N. Y.

Electric force field patterns may be shown with the aid of small objects which line up to form a picture of the field. Various materials such as needle crystal epsom salts, dried pig hair, and bent grass seed have been recommended by scientific supply companies and the Physical Science Study Committee.* Using carbon tetrachloride with bent grass seed is no longer recommended for educational purposes because the inhalation of the CCl_4 fumes is hazardous. An alternate material for the same purpose is fibers of flock suspended in clear mineral oil. (Flock is a fiber about 1/32 in. long which is sprayed on wallpaper to give a velvet effect.†) The apparatus employs the use of an overhead projector and is similar to a setup shown at the AAPT Meeting in New York City about two years ago.

Obtain some light mineral oil from a drug store and pour it into a transparent tray on the stage of an overhead projector to a depth of ½ in. Mix in a quantity of flock. (Black is the most desirable color.) Attach a pair of electrodes to a Wimshurst machine and immerse the electrodes in the mineral oil. The electrodes may be small rods, parallel plates, or various combinations of these shapes. When the machine is operated, the flock will line up with the configuration of the electric field present. The oil mixture may require stirring from time to time as the flock will eventually settle to the bottom of the container. A two-dimensional pattern is clearly projected on a screen; if it is possible for a group of students to observe the apparatus at closer range, they may even observe the field in three dimensions.

---

* PSSC, *Physics* (D. C. Heath and Co., Boston, 1965), p. 488.
† If any reader has difficulty in obtaining these fibers, send a self-addressed envelope to Mr. Weiss.

Force on Current Carrying Aluminum Foil

Unusual effects may be demonstrated by placing a strip of *lightweight* aluminum foil between the poles of horseshoe magnets and observing the reaction of the foil as increasing amounts of current are applied. That aluminum foil used for covering foods is much too stiff to react in this demonstration, but an ideal source of lightweight foil may be obtained by unrolling a 0.02 microfarad paper-dielectric capacitor. These capacitors are available in electronic supply stores and repair shops for 10¢ or 15¢. When the energized foil is placed in a magnetic field, as in Fig. 2, the foil arches in two places. Reversing the

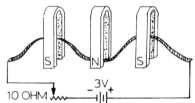

Fig. 2. Strip of lightweight aluminum foil arches dramatically as current is applied.

magnets or the current direction also reverses the crests and troughs in a manner that the class should be able to predict by using one of the hand rules. If desired, a variable low-voltage dc power supply may be substituted for the battery and potentiometer shown in the diagram.

Suggestions for many additional experiments, using similar apparatus, have been made by Siddons.*

---

* J. C. Siddons, The School Science Review **44**, 74 (1962).

Standing Waves by a Current Carrying Conductor

Marvin Ohriner
Elmont Memorial High School, New York

Application of ~ 3 A from a low-voltage ac source to the ends of a suspended wire will produce standing waves with 3–5 nodes after slight adjustments of the tension of the wire and the proper positioning of a magnet (see Fig. 2). A 0.5-mm diam copper or nichrome wire was found to work equally well with a large magnetron magnet. The choice of a fine and preferably light-weight (low-density) wire is dictated by the need for large accelerations while still having adequate tensile strength. To obtain the low voltage, a suitable step-down transformer (or Variac) may be connected across the ends of the wire, and the voltage adjusted so as to produce 2–4 A.

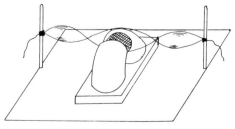

Fig. 2. Standing waves shown with four nodes.

How things work

Section editor: H. Richard Crane, Department of Physics, University of Michigan, Ann Arbor, Michigan 48109

Touch-panels in elevators, and idiosyncrasies of gas tubes

Physicists often while away the time of an elevator trip speculating as to what is behind the little square that lights up when it is touched. But the trip is over before the answer is found. The part that is touched is solid: it is not a mechanical push-button. Does it depend on conduction through the finger? Or is it capacitive coupling? It works through a leather glove, but not a mitten — unless you have come in out of the rain. Only one guess seems safe: the color indicates a neon discharge.

Having heard that many years ago Peter Franken[1] had taken the direct approach and disassembled a live touch panel, I tried the same. But everything I did called the elevator — sometimes with people. So I quit. Rescue came when Dave Shalda, the University's expert on elevators, supplied a circuit diagram and a spare "glow tube" with which to experiment.

The glow tube proved to be just a modified version of a neon lamp. The characteristics of it that are essential for touch operation are present also in the garden variety neon lamp, and that makes it easy to study and demonstrate touch operation with materials readily at hand. I connected up the circuit of Fig. 1a, using a variable dc power supply. The lamp was an NE45, Fig. 1b. The resistor shown in the diagram is the one that is inside the metal base of the lamp. (Most neon lamps with screw or bayonet bases have built-in resistors.) As the voltage of the power supply was turned up, the discharge did not "strike" until about 80 V was reached. Once started, it did not go out until the voltage was turned back to about 60. Thus in the interval between 60 and 80 V, the discharge is capable of continuing *if started*, but it will not start by itself, and that is the first requirement for making a touch switch possible.

The second, and only other, requirement is that there be a simple way to trigger the discharge, when the voltage is set within the interval 60 to 80 V. This happened very reliably with the NE45 when the glass bulb was touched, the area of greatest sensitivity being that nearest the anode. (The anode is the piece that stays dark; the cathode the one that becomes sheathed in the orange glow.) I tried a few other things. With ac instead of dc voltage the discharge could be triggered, but could be kept going only by holding the finger on the glass. That's because it goes out and has to be restarted every time the voltage passes through zero, 120 times a second. The familiar "wheat seed" neon lamp, Fig. 1c, could be triggered, but not as sensitively as the NE45. An external resistance of 20 000 Ω was used. *Caution*: The above makes a good demonstration, but fingers will wander! Make sure the high-voltage metal parts and wires are taped.

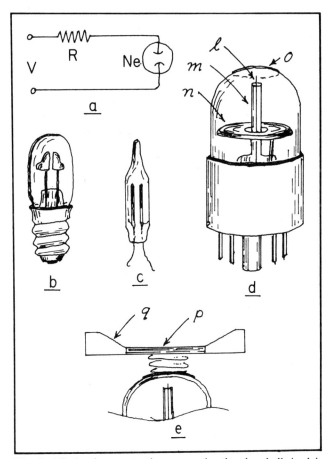

Fig. 1. All sketches are to the same scale: the glass bulb in d is 1 in. in diameter.

Moving from here to controlling an elevator is engineering. The current is routed through the coil of a relay, which starts the chain of events that calls or stops the elevator. Switches activated by the elevator interrupt the circuit to extinguish the discharge when the elevator arrives. The special design of the gas tube increases the sensitivity, the light output, and the interval between the spontaneous striking and extinguishing voltages. The specimen at hand is sketched in Fig. 1d. The anode, l, is a wire that extends to within a millimeter or two of the inner surface of the glass. It is shielded, except at the tip, by a glass tube, m. The cathode, n, over which the orange glow resides, has a large area, annular in shape. On the outside of the glass there is a spot of a transparent conducting film, o. (There is a third electrode, a grid, consisting of a ring of wire just above the cathode, not used in systems having only one elevator, and not shown here.) In operation the supply voltage is 135, which is midway between the spontaneous striking and extinguishing voltages. The external resistance is such that the current is about 30 mA, enough to give a bright glow.

Of course the elevator rider does not touch the gas tube itself. The inner square, p, in Fig. 1e is of conducting material, and is connected to the conducting patch on the outside of the gas tube by a coiled wire. The surrounding frame, q, is translucent plastic, through which the orange light comes.

Unanswered questions remain. How does touching the outside disturb the electric field inside enough to start the discharge? Is it the capacitance-to-ground of the body? The static electricity we always carry around? Or the 60-Hz ac we pick up by being antennas? If an elevator is handy, experiment. Touch the panel with a fine wire that is held in the hand, or grounded, or connected through a small capacitor – or a resistor. Stick thin plastic on the touch area. But don't leave it there!

More questions (Send yours today.)

Air purifiers for rooms.[2] What do they remove and how?
Light pens that work on a computer's CRT.[3] How does the computer know what the pen is pointing at? What if it is pointing at a dark area?
Non-mechanical phonograph or video disk pickup. No record wear.
Cordless telephone extensions.

Notes

1. Chm., Optical Sciences Center of University of Arizona
2. Sent by Glen L. Green, physics teacher, Mundelein High School, Mundelein, IL 60060.
3. Sent by John S. Wallingford, Pembroke State University, Pembroke, NC 28372.

Send questions, answers or comments to the Section Editor.

A magnetic tripole

Michael Davis
Clemson University, Clemson, South Carolina 29631

While rummaging through our demonstration equipment recently, I found a solid steel bar magnet that was, innocently enough, marked "N" on one end and "S" on the other. However, a quick search with a small compass and subsequent mapping revealed the field in the figure.

How does a bar magnet get this way?

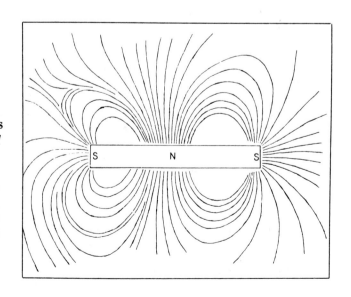

A MAGNETIC TRIPOLE — WHAT CAUSED IT?

In response to Michael Davis ("How does a bar magnet get this way?")[1], I'd say the key word in his note is "rummaging." If a bar magnet is in a box of equipment with no other magnet to "pair with" and is subjected to jostling in the process of rummaging, demagnetization is hastened by the jarring. If there is in the box a second magnet which in the shaking-up process winds up more or less perpendicular to the first magnet, with the S-pole of the second touching the center portion of the first, then in due time the ends of the first will be S-poles and its center an N-pole (of pole strength equal to the sum of the pole strengths of the two S-poles). Stroking the first magnet with the appropriate pole piece of another (untampered with) magnet will remedy the situation. Of course the condition sketched in Davis' note can be produced by dc in a helix wrapped around the bar, with the direction of wrapping reversed at the center (or at whatever point is chosen for the "odd" pole).

GLADYS F. LUHMAN
City College of San Francisco
San Francisco, California 94112

Reference

1. Michael Davis, Phys. Teach. 14, 34 (1976).

The note from Michael Davis [Phys. Teach. 14, 34 (1976)] on the magnetic tripole is presumably somewhat "tongue in cheek" but nevertheless deserves a response. Two dipoles arranged back to back do indeed make an interesting quadrupole field!

It could be argued on symmetry grounds that a dipole magnet left to itself cannot develop a pole in the middle since that change makes an asymmetric distinction between north and south polarities. There was a time,

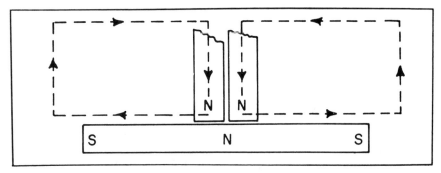

Fig. 1.

however, in the teaching of introductory electromagnetism, when a little effort was put into the study of so-called consequent poles [see for example Poynting and Thomson's "Electricity and Magnetism," page 174]. The double touch method of magnetizing iron is attributed to a Dr. Knight ca. 1750 for making a strong dipole. Reversing one of the pair of stroking magnets produces the desired consequent poles (Fig. 1).

One is left wondering why are they called consequent poles and who was Dr. Knight?

ARTHUR R. QUINTON
University of Massachusetts
Amherst, Massachusetts 01002

In response to the question about a "tripole" magnet by Michael Davis [Phys. Teach. 14, 34 (1976)] such a magnet can be produced with a magnetizer. These devices are sold by scientific supply houses to remagnetize bar and horseshoe magnets. They contain two coils into which two bar magnets are inserted with a piece of iron placed across the top to connect the two magnets. A short pulse of unidirectional current in the coils remagnetizes the bars.

If a third bar magnet is used in place of the metal connector and placed so that the magnets being remagnetized contact the third magnet at one end and in the middle, the third magnet will become a "tripole" magnet. If the desired polarity does not result simply repeat the process.

DEAN HARTMAN
Grant Wood Area Education Agency
Cedar Rapids, Iowa 52401

In reference to Mr. Davis' question on how a bar magnet could have three poles, my teacher read the article and asked me to see if I could come up with the answer. Earlier in the year some students and I were playing around with a large horseshoe-like magnet. The distance between the two poles is about two inches. When we put a six-inch bar magnet on the larger magnet, one end of the bar magnet hung out farther than the opposite pole on the large magnet. Since the pole was in the middle of the bar magnet, the strength of the magnetic field pulled and rearranged the atoms in the protruding end of the bar magnet. The large magnet will then create a new pole in the middle of the bar magnet and make the opposite pole at the end of it. We have many magnets like this in our lab.

ROD FISHER (student)
Orem High School
Orem, Utah 84057

A three-pole bar magnet?

Jerry D. Wilson
Lander College, Greenwood, South Carolina 29646

Bar magnets are commonly used to demonstrate the so-called "law of poles" when studying magnetism; and even though students are familiar with the attracting and repelling properties of such magnets, they can rarely resist bringing the ends of the magnets together for a hands-on confirmation that like poles repel and unlike poles attact. In the physics and physical science classes at Lander College students are given pairs of bar magnets with unmarked poles. They quickly find that two ends repel (or attract), and upon turning both magnets around, the opposite ends are also observed to repel (or attract). However, if only one magnet is turned around, the ends still repel! (or attract!).

This raises some eyebrows. Bar magnets with like poles at both ends? With further investigation, students find an unlike pole in the center of the bar. Now a three-pole magnet?[1] When sprinkled with iron filings, it would appear so (Fig. 1).

Of course, the magnets are not common bar magnets, but have been specially prepared. One half of an Alnico bar is magnetized, then the bar is turned around and the other half magnetized with the same polarity. In effect, the bar is two common bar magnets with two like pole ends "glued" together. The student response and interest have been found to be well worth the cost of the special preparation. Perhaps you would like to obtain some of these segments that were introduced to me by a friend in science education, Dr. Lester Mills. They may be specially ordered from:

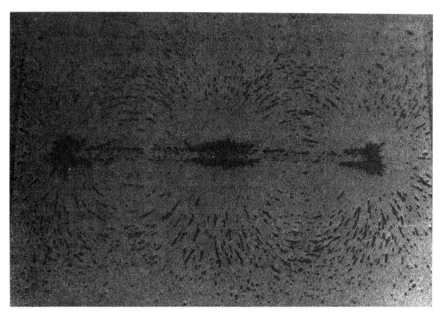

Fig. 1.

Permag Dixie Corp.
1919 Hills Ave., N.W.
Atlanta, GA 30318

The cost of an 8-in. pair of specially prepared "three-pole" cast Alnico 5 magnets is currently around $30. One-pole magnets are not available at the present time.

Reference

1. A question about the magnetic "tripole" was raised by Davis in this journal [Phys. Teach. 14, 34 (1976)] and elicited a number of letters in response which appeared in the letters section in our May issue.

THERE'S STILL A LOT WE DON'T KNOW

After reading *The Physics Teacher* for May 1976 I wished I had written a letter more promptly, discussing Michael Davis' note on the magnetic tripole [Phys. Teach. 14, 34 (1976)].

Many years ago, inspired by the discussion of consequent poles in Charles E. Dull's high school text *Modern Physics* I had one of my students at Kelvyn Park High School in Chicago make a three-pole magnet. He entered it in a science fair and I think I remember him winning a prize.

Later I had the Welch Scientific Company make this magnet and it was listed in their catalogue as the "Mystery Wobbly Bar" (Fig. 1). Many of these were sold and must still be on the shelves of physics departments today. I am amazed that no one mentioned this in the Letters column of May 1976.

The magnet was made at Welch by the method shown in Michel Henry's letter [Phys. Teach. 14, 318 (1976)]. Mr. Welch made me keep one of the magnets for a whole year before he consented to list it in the catalogue. He feared that it would not retain its

When demonstrating the Mystery Wobbly Bar to anyone it was fun to watch his reaction. At first he would say, "Oh, I've seen that before," but when you turned the floating magnet end-for-end and it was still repelled, he was really surprised.

Years ago Bell Labs sent out free demonstrations like the one described in Elizabeth Wood's letter [Phys. Teach. 14, 317 (1976)]. It used a cobalt alloy magnet and was the first producing sufficient repulsive force to "defy gravity." Alnico, invented later, was a marked improvement and revolutionized the loudspeaker and small motor industries.

Consequent poles are not really mysterious — they are easily understood, but there's still a lot we don't know about magnetism. Heussler's alloy (a magnetic alloy made of nonmagnetic materials) seems to me to be really mysterious. And, of course, the real baffler: the truly independent north or south pole existing alone [see article by Richard A. Carrigan, *The Physics Teacher*, 13, 391 (1975)].

DWIGHT L. BARR, Sr.
Retired High School Physics Teacher
P. O. Box 87
Lake Geneva, Wisconsin 53147

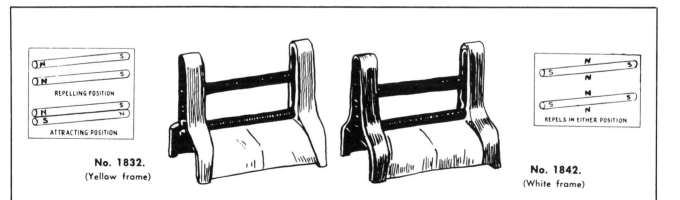

Ask your students to explain why the white one behaves as it does!

1832.   WOBBLY BAR. Two alnico magnets are held in a tough yellow plastic frame, the lower one permanently in place, the upper one removable for reversing. With like poles adjacent the upper magnet floats well above the other. Over-all length is 2½ inches.

1842.   MYSTERY WOBBLY BAR. Identical in appearance to the Welch Wobbly Bar except that the frame is white. This device challenges the student to explain why the upper magnet floats regardless of which way it is oriented with respect to the lower one.

Fig. 1.

RECYCLING A MAGNET
or, The Little Magnet That Could*

When is an accelerator magnet obsolete? Not for a long time according to a note in *The Magnet,* a publication of Lawrence Berkeley Laboratory.

When the Bevatron was being designed in 1948, a quarter-scale magnet was built to facilitate studies of the predicted magnetic performance. After completing construction of the Bevatron, the smaller test magnet was put into dead storage and it seemed like the end of the line. But in 1954 scientists at Caltech asked for it; it was transported to Pasadena and modified to become the Caltech Synchrotron.

In 1966 synchrotron operation was discontinued at Caltech. Again it seemed like a death sentence for the old magnet. But meanwhile at Los Alamos, scientists started the design of a proton storage ring for use as a high-intensity neutron source. In a "bull-session" someone remembered the quarter-scale Bevatron magnet. As a result, it has been dismantled again and shipped to a new home.

Los Alamos scientists believe that recycling the old magnet will result in savings of about a half million dollars, and nine months of time.

Not bad for an obsolete magnet!

The ring-shaped quarter-scale magnet and auxiliary equipment. Dr. Edward Lofgren, lower right, was in charge of development. (Photo: Radiation Laboratory, University of California, Berkeley.)

*Condensed from *The Magnet*, **16,** 9 (1972).

CONSTRUCTION OF A SIMPLE COMPASS

A compass, 3 cm long and 1.2 cm wide (Fig. 2) was built employing a discarded tin can (made of iron and tin), a pin, a snap fastener (of which only the bottom half was used) and a piece of Lucite. The procedure was as follows:

1) The compass was cut according to the specified dimensions, and a hole was made of a diameter just large enough for the male connector of the snap fastener to fit. If the compass is going to be used on the overhead projector then another small hole should be made to identify the north pole.

2) The compass, so prepared, was made red hot, and allowed to cool in water (quenching) so that the hardening of the iron enhanced the retentivity of the magnet.

3) The snap fastener bottom was put in the hole by simple pressure.

4) The compass was then magnetized by using a coil and direct current. The compass can be magnetized also by induction with a strong magnet.

Fig. 2. Magnetic compass made from discarded tin cans has a transparent base for use with overhead projector.

5) A small Lucite base was built and a 2.5 cm pin was inserted in it. The Lucite base is specially good when one uses the compass on the overhead projector. In other cases a wooden or cardboard base is good enough and makes the construction cheaper.

6) The compass was mounted on the pin, and was ready for use.

We consider that this method can be employed with great advantage in developing countries for the manufacture of school compasses.

LIBERIO MAR
Universidad Nacional Agraria
Lima, Peru

CARLOS HERNANDEZ
Universidad Nacional de Ingenieria
Lima, Peru

A POTPOURRI OF PHYSICS TEACHING IDEAS — ELECTRICITY AND MAGNETISM

CERAMIC MAGNETS

Small cylindrical or ring ceramic magnets can be obtained inexpensively from scientific supply companies. These magnets have a lifting force of approximately 1 N and retain their magnetism for a relatively long time. When several are placed in a test tube with like poles opposing (Fig. 5), their motions can be studied as a wood or glass piston is moved back and forth to produce longitudinal wave motion. Ring magnets mounted on a length of lubricated glass tubing also provide a handy demonstration apparatus (Fig. 6).

One of the most useful applications is to have the magnets represent neutrons, protons, and electrons in atomic diagrams on a metallic chalkboard. Particles, identified with different colors of spray enamel, can be added, deleted, or moved to different positions to show fission, fusion, transmutations, and energy level transitions without the necessity of continual erasures on the diagrams.

We purchased a large quantity of surplus magnets (15 mm diam x 7 mm high) and would be happy to share them. For 10 magnets, send $1 and a self addressed, stamped (2 stamps) envelope to the address below.

HERBERT H. GOTTLIEB
*Martin Van Buren High School
Queens Village, New York 11427*

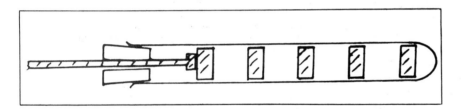

Fig. 5. With like poles facing each other, magnets space themselves uniformly along a horizontal test tube. Operation of the plunger produces a longitudinal wave.

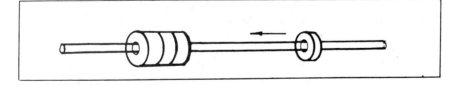

Fig. 6. Ring magnets sliding along a length of lubricated glass tubing serves as a "poor man's" air track.

Three Dimensional Views of Magnetic Effects

Phillip E. Miller
Western Kentucky State College
Bowling Green, Kentucky

As Dyer and Powers* have indicated, there is great need for three dimensional experiments or demonstrations of magnetic force field effects. The effects they obtain by suspending iron filings can also be attained without using a gel. The following method can also be easily used by a full laboratory of students, with fast recovery of all materials used without the problem of temperature control.

Arrange any number of magnets in any desired pattern on a level surface. Next place a glass plate over the magnets (if the magnets have a flat surface to rest the glass plate upon, this procedure is simplified), and sprinkle iron filings evenly on the plate. Now stack two microscope slides (or other uniform support) near each end of the glass plate in order to support a second glass plate. Sprinkle filings on the second plate as on the first, and continue this stacking procedure until you get a three dimensional effect of the magnetic force on the filings.

For a more complete view of the "magnetic field" (realizing we are seeing only the "misleading" result of magnetic force), the "level surface" can also be a glass plate, with the necessary number of glass plates previously stacked beneath the magnet or magnets to be studied.

---

* *The Physics Teacher*, May, 1965.

Questions students ask

Why are so few substances ferromagnetic?

Question:

This question was asked by **Jean Frazier,** *a student of* **Joseph Mosca** *at the Wheatley School, Old Westbury, New York 11568. The answer was provided by* **Henry H. Kolm,** *one of the founders and now Senior Scientist of the M.I.T. Francis Bitter National Magnet Laboratory. Dr. Kolm has written a number of articles on magnetism for encyclopedias and periodicals, in addition to one in this journal, "Tackling the sixth state" [Phys. Teach.* **13,** *73 (1975)] on using magnetic effects to reduce water pollution.*

Answer:

Magnetic behavior of certain substances is caused by the fact that they contain atoms or molecules which have an inherent magnetic dipole moment caused by a circulating current in the form of outer electrons which orbit and also spin. We shall neglect, for the present explanation, the interaction between such electrons, and assume that each dipole behaves independently of its neighbors except for its alignment. The lines of force which characterize each magnetic dipole are shown in the sketch, and it will be noted that the lines of force, or the field direction next to a dipole is *opposite* to the direction in which the dipole is pointing. Thus, if all the dipoles in a material were perfectly aligned, the field between atoms would be pointing in the opposite direction. The existence of this "backward" or "demagnetizing" field is crucial to the rarity of ferromagnetism.

Substances containing magnetic dipoles in moderate concentration, namely paramagnetic substances, are not rare at all. In such substances the dipoles are far enough apart to behave in essence independently. What this means is that each dipole is affected by an applied magnetic field, but not by its neighbors. When no external field is applied each dipole is subject to violent thermal motion and random orientation prevails. An applied external field competes with the thermal agitation and causes a net magnetization, which increases more or less linearly with the strength of the applied field, the ratio (or slope of the magnetization curve) being known as the *magnetic susceptibility*. The lower the temperature, the easier it is to align dipoles. The fact that the susceptibility of paramagnetic materials is inversely proportional to temperature is known as *Curie's law*.

What happens when the density of magnetic dipoles in a substance becomes high enough for neighbors to affect each other? One might expect neighbors to align each other, but one would be wrong! Only neighbors in a head-to-tail configuration would tend to align each other. Side-by-side neighbors will have the opposite effect on each other, for a reason which becomes clear if one examines

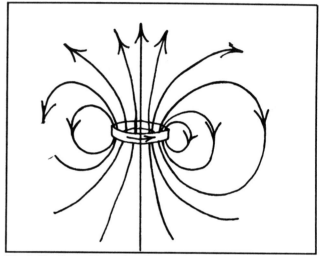

Fig. 1. The lines of force that characterize a magnetic dipole.

Fig. 1: the field next to each dipole points in the direction *opposite* to that of the dipole, and is therefore called the "demagnetizing field." A second dipole brought alongside the first will therefore tend to point in the opposite direction! What will actually happen in a crystal is that every other dipole in each layer of crystal, like the white squares on a chess board, will point in the direction opposite to the dipoles on the black squares. We then say that the crystal has two sublattices, and that it is "anti-ferromagnetic." In other words, the two sublattices interact strongly or ferromagnetically, but they cancel each other's magnetization, so that the overall effect is just as if the crystal were nonmagnetic, or only very slightly magnetic. This phenomenon of antiferromagnetism is actually quite common, and explains why ferromagnetism is as rare as it is. In other words, when atoms having magnetic dipole moments are crowded together, they are more likely to disalign their nearest neighbors than to align them.

The miracle therefore is not that ferromagnetism is rare, but that it exists at all! How can a high concentration of magnetic dipoles ever result in a "cooperative effect"? The answer is that certain "exchange forces" begin to intervene when dipoles are very close to each other, and the dipoles no longer behave independently. Under those conditions a state of lowered overall magnetic energy can be achieved if groups of dipoles align each other into "magnetic domains," with neighboring domains closing the magnetic circuit, so to speak. As the applied field is increased, domains aligned with the field will increase in size at the expense of domains aligned opposite to the field. The motion of domain walls as a ferromagnetic material is magnetized was discovered by Barkhausen, and domain walls were first explored visually by Francis Bitter, who placed a drop of colloidal ferromagnetic suspension over a polished iron surface and looked at it through a microscopy cover glass. The cooperative effect, i.e., domain formation, no longer exists above a certain temperature known as the "Curie temperature," above which a ferromagnetic material becomes paramagnetic.

In a certain sense, therefore, ferromagnetism represents a "phase change" similar to the solidification of a liquid. It is a cooperative effect which becomes possible only when thermal agitation is reduced to such an extent that interaction forces between neighbors begin to dominate.

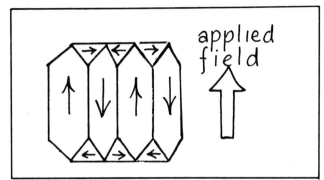

Fig. 2. ↓ domains must be smaller than ↑ domains.

Questions Some Students Ask...

A Question of Magnets and Keepers

Question: From elementary school on, students have been cautioned to store magnets using the proper keepers. So it is not surprising that some student asks you, *"What is the purpose of keepers on magnets and how do these keepers fulfill their purpose?"*

Answer: First of all, as most students (we hope) know by the time they get to high school, the purpose of the keeper is to help the "permanent" magnet retain its magnetism. How the keeper achieves this purpose is a somewhat more difficult question to answer honestly in elementary terms.

One might try this (assuming one has dealt with the concept of magnetic poles): The free poles at the gap (Fig. 1) exert a demagnetizing force on the magnetic domains in the iron. This is so because the magnetic forces of these poles tend to oppose the desirable magnetic alignment of the domains. Consequently, in the presence of free poles disalignment of the domains is relatively readily achieved through the agency of thermal and mechanical shocks.

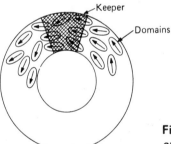

Fig. 2. *The magnet of Fig. 1 with a keeper.*

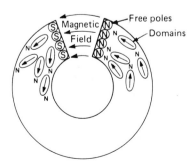

Fig. 1. *Schematic representation of a permanent magnet showing free poles, magnetic domains, and the direction of the magnetic field in the air gap. This diagram and Fig. 2 should be viewed as they are meant, that is, schematic representation, and should consequently not be taken too literally.*

(A caution is in order at this point. Single free magnetic poles do not really exist, at least under these circumstances, and hence any description employing the magnetic pole concept is somewhat of an artifact. Nevertheless the pole concept is often quite useful if one realizes its shortcomings. The so-called free poles in Fig. 1 can be considered representative of the net magnetic effect of the magnet; they are not free in any rigorous sense since they are associated with the so-called free poles of the opposite kind at the other end of the magnet.)

If, on the other hand, a soft iron keeper is used to close the gap, then, as is shown in Fig. 2, there are no free poles. There is no net magnetic force at either end of the magnet and hence no force which could cause disalignment of the domains.

For a more rigorous explanation, employing the vectors **B** and **H** the reader is referred to the author's note on this subject in *The Physics Teacher*, Vol. 6, No. 1 (1968), pp. 38-40.

WALTER THUMM
Queen's University, Kingston, Ontario

String and sticky tape experiments

Section Editor: R. D. Edge
University of South Carolina, Columbia, South Carolina 29208

Experiments with nickels and magnets

Some time ago, I described an experiment in which a nickel is held horizontally between two quarters by the finger and thumb, as shown in Fig. 1 [Phys. Teach. **18**, 309 (1980)]. You release the lower quarter, and catch it a foot (30 cm) or so below, to find, paradoxically, that the nickel is under rather than over the quarter. Since it takes a quarter of a second for the coins to fall a foot, the angular velocity of the two coins is 720° per second. If we assume the quarter is 2.5 cm in diameter, the velocity of the center must be about 15 cm/s at the point when the edge of the coin ceases to stick to the finger to provide the correct angular velocity. Kinematically, this occurs after the center has fallen little more than a millimeter! When I mentioned this experiment to John Webb, at the University of Sussex, he said "Oh yes, I remember that very well, in fact, you can win quite a lot of money in pubs with that experiment." It appears that there is a further extension of which I was unaware. If, instead of simply dropping the coins, you give them a slight twist as they are released, shown in Fig. 1, they do not turn over at all! Of course, this fact is not mentioned in making the bet — but why do they not turn over when given this twist? The secret is in the gyroscopic forces involved. The twist makes the coins spin about a vertical axis. The small torque given the coins as they fall from the fingers, which previously turned them over, now merely gives a slight precession about the vertical axis! Voila!

Magnets

Experiments with magnets give students a lot of physical insight, with very little equipment, since about the only primary prerequisite is a magnet! These may be purchased in the dime store, but if you are going to buy them

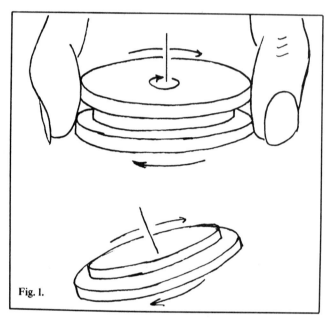
Fig. 1.

in bulk, it is probably better to get several of the chubby alnico bar type, such as are sold by suppliers like the Edmund Co. of Barrington, N.J.[1]

An obvious experiment is to use the magnet as a compass, which can easily be done by hanging it by a thread, or floating it on a piece of Styrofoam on water in a Styrofoam cup.

We can check the attraction of unlike poles by magnetizing a paper clip and using it as a compass. Take the clip, and, holding it as shown in Fig. 2, stroke it from end

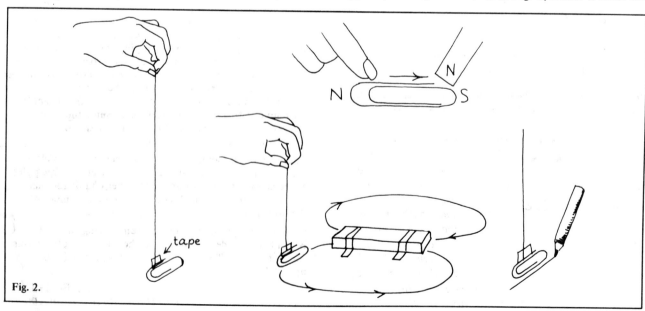
Fig. 2.

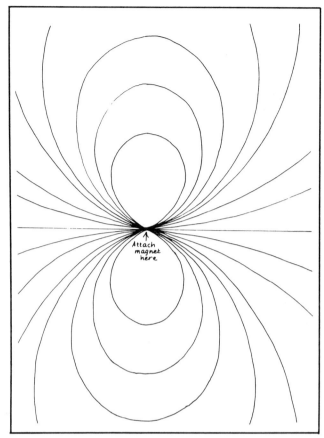

Fig. 3.

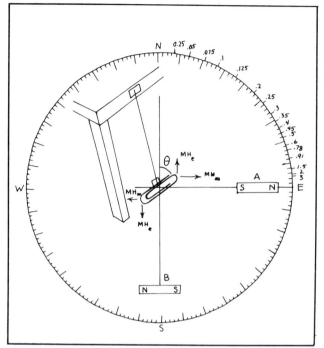

Fig. 4.

to end with one pole of the magnet a large number of times, always in the same direction. Hang the clip from a thread, as shown, so that it balances horizontally and see in which direction it points. The thread should be as long as is convenient so that the twist of its fibers supplies a minimum torque. Notice that the end of the clip where the stroke finishes has opposite polarity to the pole doing the stroking. Bring up the north pole of the magnet, and see that it repels the like pole of the clip, and attracts the opposite end. The stroking magnetizes the clip because the strong magnetic field causes the magnetic domains of which the clip is composed to align with the field.

We can plot the lines of force around our magnet using the paper clip hung on the thread. The clip orients itself along such lines, the poles at opposite ends being forced in opposite directions to provide the torque. Hold the thread supporting the paper clip as far up as possible to minimize the thread's restoring torque. A small piece of tape on the clip air-damps vibrations.

Tape the magnet to the center of a sheet of paper, support the clip somewhere near the magnet, and lower it gently to the paper. Draw a pencil mark along the line where the clip touches the paper, as shown in Fig. 2. This is a field line. Now, move the clip a little farther along, and try to make a continuous field line. Do this all around the magnet, holding the thread closer to the clip as you approach the magnet, to prevent it being drawn in. For a very small magnet, the lines of force follow curves of the form drawn in Fig. 3. Compare them with the lines you obtained. A larger magnet tends to make the loop flatter.

So far, the experiments have been somewhat qualitative. However, we can construct a tangent magnetometer which will measure the strength of our bar magnet by comparison with the earth's field.

The magnetized paper clip is hung by a long thread from a table or chair, as shown in Fig. 4, to swing just above the floor. Make sure the thread is unwound so that it provides the minimum torque, and that the piece of tape attached to the clip acts as an air damper. Place the graduated circle of Fig. 4 beneath the clip so that 0° is orientated north.

Bring up the magnet with its axis along the 90° or E.W. line, as shown. Two couples act on the clip. One arises from the horizontal component of the earth's magnetic field H_e, and the other the field at the clip produced by the magnet H_m. When the paper clip is sitting at an angle of θ to the north, as shown, the two couples are in equilibrium. That due to the earth's field is $mH_e \ell \sin \theta$, where m is the pole strength of the paper clip, and ℓ the separation of the poles. The magnetic moment, $M = m\ell$, and the couple due to the magnet is

$$H_m M \cos \theta$$

Then

$$H_e M \sin \theta = H_m M \cos \theta$$

or

$$\tan \theta = H_m / H_e$$

The horizontal component of the earth's magnetic field varies with geomagnetic lattitude. It averages about 0.2 G for the United States, varying smoothly from 0.25 G near Miami to 0.15 G in Northern Maine and the Great Lakes. On the West Coast it is about 0.25 G in Los Angeles and 0.18 G in Seattle. There is also a slow secular variation of this field. Using these values, it is possible to calculate the field strength produced by the bar magnet a distance d from its center. The edge of the circle has been graduated in units of $0.2 \tan \theta$ so that you can read the values of a field

H_m directly off the circle, for a place where the earth's field is 0.2 G.

Far along the axis of the bar magnet, the field is given by M/r^3, where r is the distance from the center of the magnet to the paper clip. It is simple to check on this relationship, by plotting the measured field strength against ℓ/r^3. The most accurate readings of H_m are obtained when θ is approximately $45°$, so it is a good idea to start by placing the magnet at a distance giving this deflection. With the magnet placed at position B, rather than A, the field at the clip is again perpendicular to that of the earth, and is given by $2M/r^3$, which can also be checked. This experiment makes the point that we can get a reasonable quantitative value for the magnetic moment of a magnet using the earth's field for calibration. Most experiments do the inverse to use a calibrated magnet to obtain an accurate value for the earth's field.

Reference

1. Edmund Scientific, 101 E. Gloucester Pike, Barrington, NJ 08007.

THE FIELD STRENGTH OF A PERMANENT MAGNET

In the PSSC physics text there is a problem describing an experiment which measures the magnetic field of a permanent magnet.[1] I thought that this problem would be an interesting investigation to follow Experiment 42 in the PSSC laboratory manual—The Measurement of a Magnetic Field in Fundamental Units.[2]

Using a large permanent magnet I set up the apparatus as illustrated in the PSSC text. I used a large demonstration Newton scale, from which I hung a block of styrofoam packing material (any insulator which will fit between the poles of the magnet is satisfactory) with 20 turns of insulated wire wound around it (Fig. 7). The coil was hung so that the direction of the current causes a downward force on the coil. Any forces on the side wires of the coil are directed to the right and to the left so that these forces do not contribute to the downward pull.

The weight of the coil and block was 0.4 N and a current of 8 A caused a force of 2.7 N. For a current of 6 A, 2.1 N; and for 4 A, 1.5 N. I ran the current only long enough to read the force. If smaller currents are used, one has to carefully adjust the height of the coil to keep it in the same part of the magnetic field. Also the error in measurement will be more significant with a small force.

Using the definition of magnetic field strength, $B=F/(IL)$, and subtracting the weight of 0.4 N from the measured forces (the L measurement for the part of the loop between the magnet poles was 0.085 m), the class calculated field strengths of 0.17, 0.17, 0.16 N/A-m. In the calculations, one must remember to multiply the current by the 20 turns of wire. By coincidence, these values are close to the value of 0.15 N/A-m for the magnet in the PSSC problem.

This demonstration experiment is interesting to the students for 3 reasons: (1) they can actually see that there is a magnetic force pulling on a spring scale; (2) they get some idea of the approximate field strength of a large permanent magnet (approximate, because the bottom of the loop is not entirely between the pole pieces); and (3) they can compare the permanent magnet field strength to the field strength of the solenoid in PSSC experiment 42 which was about 0.01 N/A-m for a current of 2 A.

I wish to thank student Carl Hindy of our Camera Club for the photograph used in this note.

BROTHER JAMES MAHONEY
Malden Catholic High School
Malden, Massachusetts 02148

References

1. Haber-Schaim et al, *PSSC Physics*, 3rd ed. D.C. Heath, Lexington (1971), p. 517, problem 15.
2. Haber-Schaim et al, *PSSC Physics Laboratory Guide*, 3rd ed., D. C. Heath, Lexington (1971).

Fig. 7. Scale indicates force of magnetic field on current carrying conductors.

A POTPOURRI OF PHYSICS TEACHING IDEAS — ELECTRICITY AND MAGNETISM

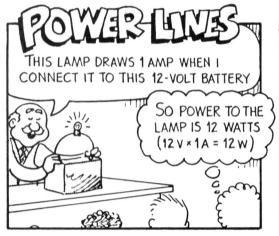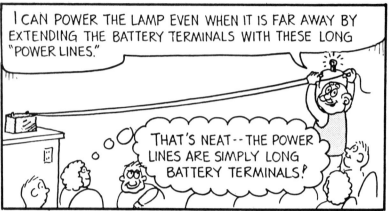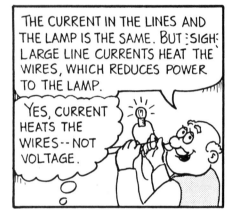

A choice observation

Ronald A. Brown
SUNY at Oswego, Oswego, New York 13126

Some years ago, a colleague[1] informed me of an interesting observation. Consider the words:

CHOICE OXIDE MATERIALS   (1)

If you hold a cylindrical glass rod *above* (not touching) the words in (1), they will appear as:

CHOICE OXIDE WVLEBIVrS   (2)

The effect is even more dramatic if CHOICE OXIDE is in one color, say red, while MATERIALS is in another color, such as blue. This coloration adds a bit of misdirection, as if to imply that the difference in the index of refraction for the two colors has something to do with the effect (it hasn't).

Upon mulling over this curiosity for some time, I wondered if the effect could be developed further into a useful pedagogic technique. I therefore offer the following suggestions to those readers who would like to pose a puzzling situation to their students in order to get them to think a bit more critically concerning the way that lenses and mirrors form images.

1A. Turn this page upside-down, holding it in front of a mirror. Then (1), viewed thus upside-down and reflected, will look like (2) does normally, and vice versa. The reason for this is clear: turning the page upside-down is equivalent to rotating it by 180° about a horizontal axis perpendicular to the plane of the mirror. Also, reflection in a mirror corresponds to a 180° rotation of the object about a vertical axis in the plane of the mirror. The combined effect of these two successive rotations is thus equivalent to a *single* 180° rotation about a horizontal axis. However, letters such as B, C, D, E, H, I, K, O, X are symmetric under this latter single rotation, while all other letters are not, so that we can see at once why (1) becomes (2) when viewed upside-down in a mirror. One can easily think of a variety of symmetric words, such as ICE BOX, HIDE, DICE, HICKOCK, CODE, BIKE, HEX, CHOKE, in addition to CHOICE OXIDE.

1B. Next, a glass rod, if held sufficiently *above* (so that the object falls outside of the focal point) the words in (1) will act like a one-dimensional lens which, of course, inverts the image by means of a single 180° rotation about the horizontal axis of symmetry of the rod. Accordingly, the viewing of (1) through a glass cylinder, held at length, is equivalent to viewing (1) upside-down in a mirror. In a recent talk, J. Watson, Jr.[2] discussed a variation of the viewing procedure, using the words CHICK TURK and viewing them through a water-filled glass tube.

2. If one holds the glass rod directly on (touching) the paper, over the words in (1), then an enlarged, erect image of (1) is seen. This is because the object now falls within the focal point of the rod. It is a surprising and mystifying effect to first view this enlarged, erect image and then, by simply moving the glass rod a small distance away from the paper, to see one of the words appear to "magically" turn upside-down and backwards, while the other words appear to remain unchanged.

3. In Fig. 1 there is seen a set of two silvered 45° prisms, of the type that are commonly used in periscopes, arranged one above the other as shown. A semi-transparent piece of onionskin paper, with (1) written on it, is placed in position between the two prisms in order to see (1) and (2) simultaneously. It is left to the reader to explain this combination effect. The apparatus also provides a simple way to view both sides of an object, such as a coin, simultaneously (although mirror-reversed).

Fig. 1. Double-prism apparatus showing an apparent optical paradox.

References

1. D. H. Frisch (private communication, 1966)
2. J. Watson, Jr., "The Impact of Physical Science Training on Prospective Elementary School Teachers," AAPT Announcer, 7, 76 (1976). For other versions of this demonstration see J.S. Miller, Amer. J. Phys. 22, 343 (1954); J.S. Miller, Amer. J. Phys. 23, 71 (1955); W.A. Hilton, *Physics Demonstrations at William Jewell College* (Wallace A. Hilton, Liberty, 1971) p. 90.

BLUE SKY AND RED SUNSETS

The blue color of the clear daytime sky and the redness of celestial objects close to the horizon are terrestrial events resulting in large part from selective Rayleigh scattering of the sunlight by particles in the atmosphere. According to Minnaret,[1] molecules of oxygen and nitrogen are probably the most effective scattering agents in our atmosphere being less than 1/1000 the size of a wavelength of violet light. Since Rayleigh scattering varies as $1/\lambda^4$, more violet light is scattered than blue light by our atmosphere; however, the human eye is not very sensitive to violet and therefore we perceive the sky as the hue of the next greatest amount of scattered color which is blue. Celestial objects, like the sun and the moon, near the horizon appear redder in color since, as shown in Fig. 4, light from the horizon travels through a greater thickness of the earth's atmosphere. The more atmosphere the light must penetrate, the greater the number of scattering centers which remove the blue color, so the transmitted color of the object will appear redder than normal.

A description of a simple experiment used to demonstrate these effects has previously been published with various amounts of detail (Refs. 2 and 3). Essentially, the phenomena are demonstrated in a dark room by using a tank of water with a suitable suspension (which simulates the earth's atmosphere) through which a colimated beam of white light shines onto a white screen. (See Fig. 5.)

First measure a predetermined quantity of water into your glass tank. Then dissolve the $Na_2S_2O_3$ (sodium thiosulfate) in the water. When the demonstration begins, add the HCl and stir — the reaction begins slowly. The proportions are not critical: 2-gal (7.6-liters) H_2O; 40-g $Na_2S_2O_3$; 18-ml HCl (concentrated).

The reaction is:

$$Na_2S_2O_3 + 2HCl \longrightarrow 2NaCl + SO_2\uparrow + H_2O + S\downarrow. \quad (1)$$

Sulfur precipitates slowly and there is time to observe the blue coloring of the water as the white light is scattered by the precipitating sulfur. The transmitted beam will slowly change from its white color to a yellow-orange and then red.

As a side benefit, those interested in plane polarization of light through scattering can demonstrate that the scattered light in the water tank is plane polarized and has a maximum polarization at right angles to the direction the beam of light is traveling. This plane polarization can also be observed by carefully studying the daytime sky with polaroid sun glasses or films.

MARLA H. MOORE
Montgomery College
Takoma Park, Maryland 20012

References
1. M. Minnaret, *The Nature of Light and Colour in the Open Air* (Dover, New York, 1954).
2. "Polarization by Scattering," in *Handbook of Physics* (N.Y. State Dept. of Educ., Albany, NY, 1970), p. 88.
3. E. G. Savage, "Blue Sky and Sunset Effects (Tyndall's Experiment)," *Science Masters Book* (John Murray, London, 1931), Series 1, p. 101. Available from Transatlantic Arts, Inc., North Village Green, Levittown, NY 11756.

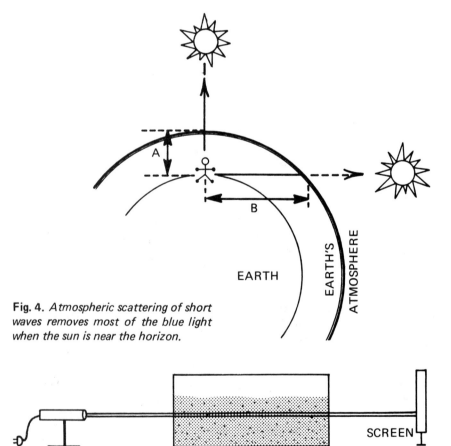

Fig. 4. *Atmospheric scattering of short waves removes most of the blue light when the sun is near the horizon.*

Fig. 5. *Sulfur, precipitated in a tank of water, simulates atmospheric scattering of blue light and a red sunset appears on the screen.*

A SIMPLIFIED SUNSET DEMONSTRATION

A light beam passing through a colloidal suspension is frequently used to demonstrate qualitatively the inverse relationship between the intensity of scattered light and wavelength (Rayleigh Scattering).[1] The transmitted beam is orange or red; the scattered light from the beam appears blue. This selective scattering by air molecules and other impurities in the atmosphere causes our blue sky and red sunsets.

Directions for carrying out the demonstration with colloidal sulfur are also described in Sutton's book.[2] The method requires the preparation of two solutions; this may be a deterrent to many instructors particularly in schools which have no lecture setup technicians. The reaction is also dependent on temperature and of limited duration.

The apparatus shown in Fig. 1 was developed for an "instant" demonstration of light scattering. A Plexiglas tube 40-cm long and of 6-cm diam with a section removed along its length was cemented between two rectangular plexiglass uprights. The height of the uprights was selected so that the light beam from an available 35-mm slide projector would pass through the center of the tube. The tube is filled with water and the projector beam is focused on a screen which may be a piece of white cardboard.

An "instant" colloidal suspension can be produced by stirring into the water a very small quantity of instant nonfat dry milk. The light projected onto the screen changes color or hue, depending on the concentration of the suspension. The scattered light appears blue. The chemical reaction involving colloidal sulfur is over in a few minutes; the milk suspension will maintain a fairly constant concentration for several days. Thus, the "instant" solution may be used for a continuous demonstration in the hallway or during the laboratory period.

A better control of the colloidal concentration in the water may be achieved by making "thick" milk out of the powder and then adding it to the suspension a few drops at a time.

Polarization by scattering can be demonstrated by rotating a Polaroid disk in the light beam.[3] The Polaroid material may also be placed in front of the tube.

The Plexiglas tube holds about one liter of liquid. The apparatus is easy to set up, clean, and store. A packet of instant dry milk should last for many years.

If a projector is not available, a flashlight which has a reflector may be used effectively.

HAYM KRUGLAK
Western Michigan University
Kalamazoo, Michigan 49001

References
1. Marla H. Moore, Phys. Teach. 11, 436 (1973).
2. "Scattering of Light — Sunset Experiment" in *Demonstration Experiments in Physics*, edited by R. M. Sutton (McGraw-Hill, New York, 1938), p. 387.
3. See Ref. 2, p. 424.

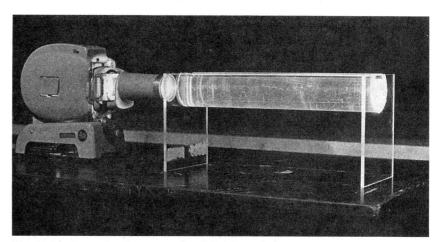

Fig. 1. *Apparatus for demonstrating light scattering.*

Dispersion and Inversion

Samuel Hirschman
Forest Hills High School
Forest Hills, New York

Puzzles can make even a reluctant pupil do extra work, especially if he thinks that the teacher is trying to put one over on him. After teaching about lenses and dispersion, the pupil is ready to be hooked.

On the unlined side of an index card, very carefully print the following sentence *exactly* as shown in Fig. 7. Use red ink for the first part of the sentence and blue ink for MARY'S HAT. (The peculiar ampersand does not seem to bother even the most suspicious student.)

The fun begins when the student is asked to view this sentence through a lens consisting of a small piece of thick glass rod or a test tube filled with water and held with the axis along the sentence.

Despite themselves, even the laziest students find their thought processes stimulated and many are even moved to consult texts to try to explain what they seem to observe. Of course, their first impulse after critically examining the lens, is to say that the index of refraction of the lens medium is greater for blue than for red light.

After the truth dawns on them, and it will, discussions can follow on the difference between spherical and cylindrical lenses and the importance of symmetry in physics and mathematics.

Fig. 7. When this message is viewed through a cylindrical lens, the red letters are readable but the blue letters are inverted.

ATMOSPHERIC REFRACTION

Atmospheric refraction causes the continuous bending of starlight, especially when stars are located near the horizon, due to the changing density of the earth's atmosphere. It has long been described in most high school physics and astronomy textbooks but has not been simple to demonstrate in the classroom. Here is a rather simple way to do it. A liquid (a variable sugar solution) is used because it is much easier to handle than a gas (air). Layers of different sugar concentration must be established; this, of course, creates a vertical gradient of index of refraction, similar to that in the earth's atmosphere.

Start with a clear, transparent tank such as American Optical's demonstration tank (2 x 8 x 20 cm) or a utility tray (7 x 6 x 22 cm). If such a tank is not available, one can easily be made out of clear Plexiglas and acrylic cement. In a beaker, make up a concentrated sugar solution using 50 ml of warm water and 80 g of sugar with vigorous stirring, at 40-45°C; let it cool. For final display, place the tank on something stable where it will not have to be moved again. Fill it to a depth of 3 cm with tap water a few degrees warmer than the sugar solution. Then add all of the sugar solution to the bottom of the tank by gently pouring it down a stirring rod to minimize mixing as the rod is slowly moved from one end of the tank to the other. Line up a HeNe laser so its beam enters the solution about 1 to 2 cm above the tank bottom. Using a darkened room, the simulation of atmospheric refraction should be clearly visible (Fig. 4.).

For many, an even more interesting effect will be noted. After the first refraction occurs, the beam can be made to "bounce" at least three times. This demonstrates total internal reflection at the plastic-air interface of the outside surface of the clear plastic tank bottom and total internal refraction, the reverse deflection of the upward, reflected beam (Fig. 5.). By adjusting the height of the laser and its beam and changing the beam's angle of incidence with the tank and solution, still other interesting phenomena can be observed.

Once the gradient has been established it will remain stable for several

Fig. 4. Laser beam bends in nonuniform sugar solution to simulate atmospheric refraction.

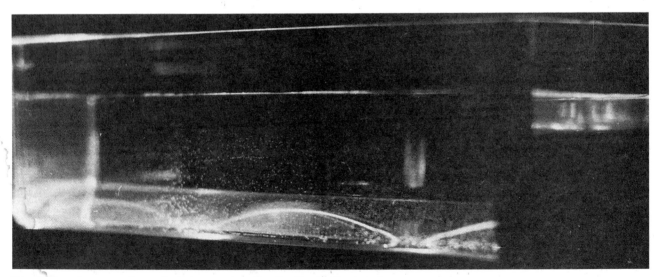

Fig. 5. Laser beam can be made to bounce by adjusting the incident angle.

days unless the solution is disturbed.

I wish to thank Mr. William M. Strouse of Research Laboratory, American Optical Corporation, Framingham Center, Mass. 01701 who kindly granted me permission to use information from his article, "Bouncing Light Beam," *American Journal of Physics*, Vol. 40 (June 1972), pp. 913-914. I felt that it was applicable to introductory physics education and received his enthusiastic support. Mr. Strouse's article should be consulted for a more complete discussion of the "bouncing light beam."

I also wish to thank Professor Christopher C. Jones of the Union College Physics Department in Schenectady, New York for his advice and expertise in preparing the photographs for this article.

JOHN B. JOHNSTON
Nanuet High School
Nanuet, New York 10954

Colored lights and shadows

Robert Gardner
Salisbury School, Salisbury, Connecticut 06068

The colors produced when the primary colors of light (red, green, blue) are mixed can be demonstrated very nicely with three overhead projectors.[1] The stage of each projector is covered with a cardboard sheet that has a hole cut in the center. A sheet of blue, red, or green plastic filter is taped over the openings in the cardboard sheets.[2] By adjusting the projectors so that the colored beams cast on a white screen partially overlap, the additive nature of color mixing is readily evident.

An exciting extension of this common demonstration can be done with no additional equipment. A few questions will help you to assess how well students have mastered the concept of color mixing.

"What will be the color of my hand's shadow when I hold it in a beam of blue light?" (Don't be surprised if a few say "Blue.")

"If the green beam is turned on too, how many shadows will be cast when I hold my hand near the screen?"

"What will be the color of each shadow?"

"What will be the color of each shadow if blue and red beams are used? How about red and green beams?"

"Suppose all three beams pass by my hand to the screen, how many shadows will you see? What will be the color of each shadow?"

Once students can explain the cyan, magenta, and yellow shadows obtained with the three colored beams, ask what seems to be a simple review question: "If I hold my hand near a screen that is illuminated with a red beam and an ordinary beam of white unfiltered light, what will be the color of each of the two shadows?"

As expected, the shadow illuminated by the red light will be red, but the shadow cast by the red light on which white light falls will *not* be black or even gray! You may be as surprised as your students to see a cyan shadow.

Similarly, beams of green and white light will produce a magenta shadow as well as a green one! A yellow shadow as well as a blue one can be obtained by using blue and white beams.

If you think these strange colored shadows are some kind of optical illusion, photograph them using colored film. You will see the same colors on the prints that you see with your eyes.

Can anyone explain these odd shadows?

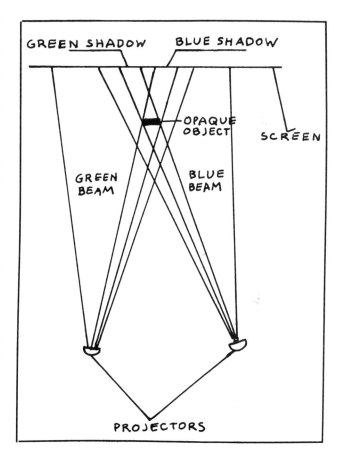

Fig. 1. An opaque object such as your hand will cast a green and a blue shadow when held close to a screen and illuminated by blue and green beams of light.

References

1. Students can carry out their own investigations of color addition using the light box described by the author in an earlier article. See Phys. Teach. **13**, 50 (1975).
2. Satisfactory filters can be obtained from Rosco Laboratories, Inc., 36 Bush Avenue, Port Chester, NY 10573. The filters come in 20 x 24 in. sheets. Order medium red No. 823, medium blue No. 863, medium green No. 874.

Making rainbows in the classroom

Fred B. Royalty

Science Education Department, University of Hawaii at Manoa, Honolulu, Hawaii 96822

Rainbows are a daily occurrence here in Hawaii, but this does not lessen the grandeur of the spectacle. Whitaker[1] has provided this magazine with a definitive treatment of the topic, and Robinett's follow-up article[2] provides an excellent method for producing "rainbows" outside the classroom. But what about making rainbows within the classroom? I have developed an easy way to produce "rainbows" in the classroom and an easy way to draw them on the chalkboard.

"Rainbows" can be produced within the classroom using a clear plastic box (the size of a shoebox) and an overhead projector. Center the overhead projector in the front of the room. Tape a cover over the upper mirror unit. Place the clear plastic box on the stage of the overhead projector. Fill the box three-quarters full with water. Dim the room and turn on the overhead projector. An arced spectrum should appear on each side wall (Fig. 1).

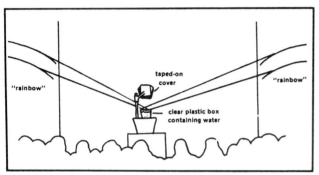

Fig. 1. Producing "rainbows" in the classroom

The Fresnel lens of the overhead projector causes the effect. The lens projects a cone of light from the 10 in. x 10 in. stage to the upper mirror unit. The light enters the sides of the container of water and is dispersed into colors by refraction (Fig. 2a and b). The Fresnel lens provides a curved source of beams at the proper angle. That is to say, the locus of beams that can be dispersed by refraction form a curve and the spectrum is, therefore, curved like a rainbow.

You can draw a rainbow on the chalkboard simply and quickly by using what I call a rainbow-maker. Remove the colors — red, orange, yellow, green, blue, indigo, and violet — from a box of colored chalkboard chalk (please do not use pastels on the chalkboard as previous teachers have done) and wrap them in that order with some corrugated cardboard. This should then be taped, tied, or wrapped with Velcro as I have done (Fig. 3). (By the way this makes a great give-away item for your next meeting.) A steady sweep of the arm, while holding the rainbow-maker with red topmost, produces a beautiful primary rainbow. Inverting the rainbow-maker enables you to draw the secondary bow.

References

1. Robert J. Whitaker, Phys. Teach. **12**, 283 (1974).
2. R. W. Robinett, Phys. Teach. **21**, 388 (1983).

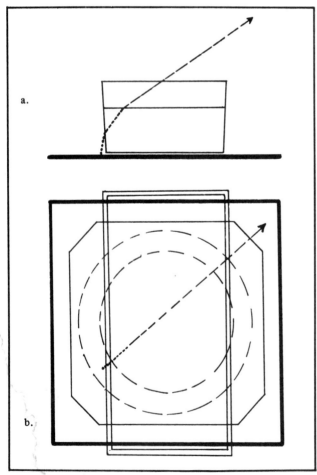

Fig. 2. A side view and top view of one ray.

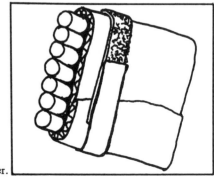

Fig. 3. Rainbow maker.

SOLAR SPECTRUM PROJECTION

A simple and inexpensive arrangement for observing the spectrum of sunlight can be assembled from a pan of water and a pocket mirror. The mirror should be completely submerged in the water and should rest against the side of the pan at an angle of about 45° (see Fig. 1).

When sunlight falls on the mirror, a bright spectrum is seen on the ceiling or on the wall opposite.

The demonstration is most dramatic when done in a relatively dark room where the light is allowed to enter through a small opening.

SAMUEL DERMAN
Trenton State College
Trenton, New Jersey 08625

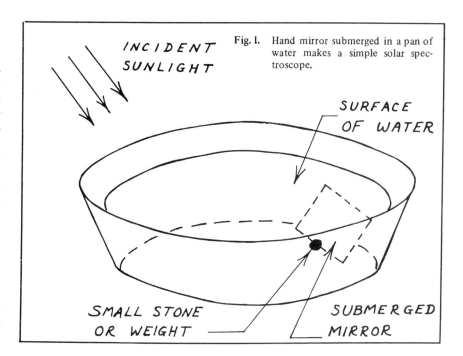

Fig. 1. Hand mirror submerged in a pan of water makes a simple solar spectroscope.

MAKING RAINBOWS WITH A GARDEN HOSE

In "The Physics of the Rainbow" [Phys. Teach. 12, 283 (1974)], R.J. Whitaker mentions the existence of higher-order rainbows which are produced by three or more internal reflections, and the difficulty of observing such bows. I would like to point out that "higher-order" bows are also produced by an interference effect and that these are both readily observable and a quite beautiful phenomena. Wood[1] in *Physical Optics* mentions the phenomena of a plane wave after two refractions and one reflection forming a peculiar cusp wave which is really two sets of wave surfaces and which then produce interference maxima and minima, according to an explanation that was first developed by Airy.

These higher-order or supernumerary bows are seen inside the primary bow and have the same order of color distribution as in the primary bow. Some overlapping of the bows is observed and "mixed" colors are seen at the edges of the different orders.

I first observed these supernumerary bows quite accidently while using a garden hose spray and looking closely at the primary and secondary rainbows which were produced when the spray was illuminated by the sunrays coming from behind my shoulder. When a spray is directed so that the primary bow is in front of a dark background or shadowed area, several of the supernumerary bows are readily apparent.

Wood also mentions that supernumerary bows can be seen in the laboratory by looking into a minute drop of water suspended from a greased glass thread while standing with one's back to an arc light. I have had little success with this technique, but have felt that the use of the garden hose to produce the ordinary and supernumerary bows is a useful and enjoyable observation exercise for students.

HELENE F. PERRY
Loyola College
Baltimore, Maryland 21210

Reference

1. R.W. Wood, *Physical Optics*, (Macmillan, New York, 1936) p 388.

A PINCH OF COFFEE-MATE

I very much enjoyed reading [Phys. Teach. 11, 436 (1973)] Marla H. Moore's description of a very interesting demonstration on Rayleigh scattering (i.e., the blue-sky—red-sunset demonstration). While most physics teachers have access to sodium thiosulfate and hydrochloric acid, many will not bother to perform this beautiful demonstration if they have to search for chemicals. However, salvation is at hand for the chronic coffee drinker since a small pinch of Coffee-mate or other non-dairy creamer in water will provide the necessary suspension for the effect.

BEN M. DOUGHTY
East Texas State University
Commerce, Texas 75428

Using a video projector for color-mixing demonstrations

Richard A. Bartels
Department of Physics, Trinity University, San Antonio, Texas 78284

The large screen size available with color television projector systems makes them very useful for showing videotapes in the classroom. These systems are increasing in video quality and decreasing in cost and thus should become more common in classrooms. The main purpose of this note is to suggest how these projector systems, if available, can be used for color-mixing demonstrations.

When discussing color in the classroom, color television is often mentioned as an application of additive color mixing. The effectiveness of this example could be increased if the instructor had available a television receiver with which to manipulate the separate red, green, and blue images. Also, with control of the separate color images, the receiver could be used to demonstrate color mixing. With a television projector, manipulation of the three primary colors can be done by simply covering and uncovering the three separate beams.

Though some color-mixing effects could be demonstrated while viewing normal programming, a more controlled video display is desirable. After some experimentation I have found the following video tape to be very useful: one minute of a solid white background followed by five minutes of a National Television Systems Committee (NTSC) color-bar pattern. The white background is useful for an introduction; for example, the red, green, and blue colors can be projected separately onto the screen, then the blue beam can be covered resulting in a yellow display, the green beam covered resulting in a magenta display, and the red beam covered resulting in a cyan display.

The NTSC color-bar pattern contains seven vertical bars: white; the three additive primaries, red, green, and blue; and the three subtractive primaries, yellow, cyan, and magenta. The arrangement of colors is shown in Fig. 1. Television broadcasting stations have color-bar generators and it may be possible to get a station to make a videotape of the pattern. Alternatively, an off-the-air tape can be made since these patterns are often broadcast by commercial television stations or commercial cable services (such as Home Box Office) when they are not broadcasting their normal programming. Several demonstrations can be made while running the color-bar pattern; for example, suppose the blue beam is completely covered, then, referring to Fig. 1, bar 7 turns black, bars 2, 4, and 6 are not affected, bar 1 becomes yellow, bar 5 becomes red, and bar 3 turns green. Analogous demonstrations can be performed by

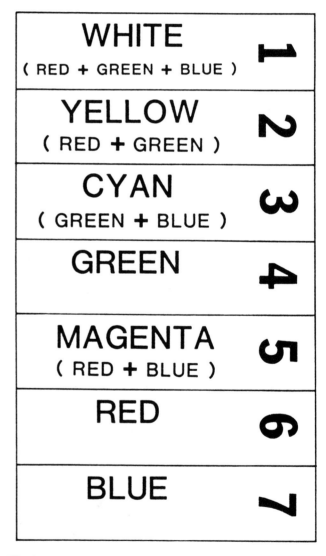

Fig. 1. Arrangement of colors in the NTSC color-bar pattern.

covering the other two beams. It is probably a good pedagogical technique to ask the class to predict what will happen when a particular color is absent.

An alternative to the color-bar pattern is simply to make a videotape of patches of color. I have found this to be less satisfactory but after considerable experimentation have found colors that, for the type of demonstration mentioned above, approximate the additive and subtractive primaries. Useful here are the packages of about two hundred different color patches commercially produced[1] to aid artists and designers in color matching. These can be found in art-supply stores or perhaps can be borrowed from academic art departments.

Of course, the type of demonstration discussed here could be done with an ordinary television receiver if the individual color guns could be switched. Most professional monitors have this feature and it would probably not be too difficult to modify other receivers to allow for this possibility.

I would also like to add that if a television projector is available in the classroom, it can provide a good example in geometrical optics. The main features of the projector's optical system can be seen by looking into the beam tubes and the several brands that I have examined all appear to use optics similar to those found in Schmidt-Cassegrainian reflecting telescopes.

Acknowledgment

The author would like to acknowledge the aid he received from William Bristow of Trinity University's Department of Art and from Ronnie Swanner and his staff in Trinity's Department of Instructional Media Services.

Reference

1. "Color Match Designer's Color-Pak" available from Craftint Mfg. Co., 18501 Euclid Avenue, Cleveland, Ohio 44112.
   or
   "Color-aid Packet" available from Color Aid Corp., 116 E. 27th Street, New York, New York 10016.

A simple reflection experiment

T.T. Crow
Mississippi State University, State College, Mississippi 39762

In a physical science survey course with no laboratory, involvement of the students in the classroom is particularly important. To this end each student can be supplied one microscope slide and deduce the law of reflection. The slide is used as a mirror, straight edge, and protractor.

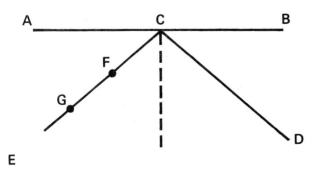

Fig. 1. *A simple reflection experiment.*

Follow these operations in sequence:

1. Lay the slide on a sheet of paper and draw a line *AB* along one edge.

2. Lay the slide across *AB* and draw a second line *CD*.

3. Hold the plane of the slide perpendicular to the paper and along *AB*.

4. Look into the slide from approximately point *E* and the image of *CD* is clearly visible.

5. Along the continuation of the image line into the "front" half-plane, put two dots *F,G* and connect these to line *AB* by a straight line.

6. Use the slide as a protractor and draw a dotted line perpendicular to *AB* and passing through the crossing point of *CD* and *FG*.

7. Now by folding the paper along the dotted line, *CD* and *FG* will be found one on top of the other thus proving the angle of incidence equals the angle of reflection.

Recombination of spectral colors

Fred T. Pregger
Department of Physics, Trenton State College, Trenton, New Jersey 08625

Newton and many others since his day observed that if a glass prism is properly placed in a narrow beam of white light, the light will be broken up into the various colors of the spectrum. They also showed by interposing another prism, a converging lens, or a set of mirrors set at the proper angles, that the colors can be mixed again to produce white light. I discovered a simpler way to show that the recombination of the colors will give the impression of white light.

Set up a slide projector using as the slide a piece of cardboard with a thin slit no wider than a millimeter and couple of centimeters long. A piece of oak tag or file card will do. You can cut this slit with a razor blade or hobbyist's knife. Put the slide in the slide holder with the slit in the horizontal position. (You can do the experiment with the slit in either position, but I think the results are easier to see if the image is in the same plane as that of the students' eyes.)

Stand a flat screen on the floor a couple of meters from the projector at an angle of about 60° above the horizontal as shown in the diagram. The screen I use is a scrap of hardboard about 18 in. by 8 in. painted white held up by two ring stands and burette clamps. Clamp an equilateral prism just in front of the projector lens and adjust it and the projector so that you get a sharply defined spectrum on the screen. This works best if the projector is focused at the distance of the screen and the prism is adjusted so that the spectrum appears at or near the angle of minimum deviation. In a dark room under these conditions the continuous spectrum is spectacular if you have a bright lamp and very thin cleanly cut slit.

Look at the spectrum from a high angle such as Position 1 in the diagram. Of course you will see the array of colors. No white light will be apparent.

Now look from Position 2, a very low angle, as near to grazing angle as you can get. If your eyes are far enough away so they can't resolve the colors in the narrow band of light that you will see from this angle, you will observe a thin band of white light, not the colors of the spectrum. Your eyes will react to the melange of colors and you will see the mixture as white.

To test the idea that your eyes are reacting to the mixing of all the colors, cut out a larger slit in a piece of cardboard such as a file folder. A slit of about 5 mm by 10 cm works fine. Place the cardboard in the spectrum beam not far from the screen, permitting only one "color" to pass through. From grazing angle, you will see that color, not white. By passing the slit across the beam, you can observe all the colors in sequence. Take away the card, and you will see white again.

It is apparent that when your eyes see but cannot resolve the colors in the spectrum, you interpret the mixture as white.

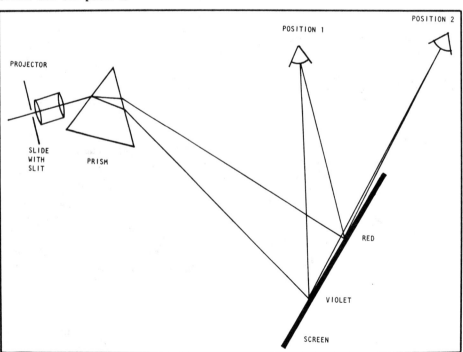

Fig. 1. Apparatus to show recombination of spectral colors. *Notice that this is a side view.* The students are looking down.

Color mixing for a large audience

William A. Butler, A. Douglas Davis, and Charles E. Miller, Jr.
Eastern Illinois University, Charleston, Illinois 61920

To demonstrate color addition to large audiences, we have found it quite successful to use three slide projectors, each with a slide of a different primary color acting as a filter. Light from the projectors is directed onto a screen where the beams overlap, revealing the three primaries, the resulting secondary colors, and a white area in the center. The slide projectors are readily available and the resulting demonstration is large enough and bright enough for everyone to see clearly — even with very large audiences.

We anticipated two problems with this scheme. First, we expected some difficulty and expense in obtaining "just the right" colors of filters to use in the projectors. But this is not critical at all. Readily available, inexpensive color filter sheets available from photography supply companies proved excellent.[1]

We also expected that the intensities of the different primary colors would have to be adjusted carefully. Adjustable iris diaphrams would allow us to control each intensity. Again, this problem proved of less consequence than expected. The high and low settings on the slide projectors provide a coarse adjustment in intensities which usually proved sufficient to give quite acceptable results.

Since we use slide projectors supplied by our audio visual department, the projectors are often different each time we use them. The intensities of their lights do vary. Nonetheless, the use of the high/low switch has usually proven sufficient to correct this. If the intensities varied enough that this did not give reasonable colors — if the area where all three overlapped was not white — then a strip or two of masking tape in front of the lens proved more expedient than the iris diaphragms and just as effective.

References

1. We have had excellent results with filters obtained from the American Science Center, 5700 Northwest Highway, Chicago, IL 60646 (catalog number 40676) for less than $2.00.

A different way to use Newton's color wheel

Gordon R. Gore
Westsyde Secondary School, Kamloops, British Columbia, Canada V2B 6P1

A rapidly rotating Newton's color wheel[1] appears off-white. Darken the room and illuminate the rotating wheel with a stroboscope. Vary the frequency of the strobe flashes until you obtain a fixed pattern of colors which are not on the disc at all! Beautiful shades of brown, purple and light green can be seen, apparently because of the automatic color-mixing that occurs at the correct relative frequencies of color wheel and stroboscope, which would not occur were it not for persistence of vision. A student looking for an interesting Science Fair project could have fun with this phenomenon.

Reference

1. For example Cat. No. 0955, Welch Scientific Co. or S40954, Fisher Scientific Co.

LIGHT BOX, INEXPENSIVE BUT VERSATILE

There are a number of commercial devices currently used to demonstrate light phenomena. The Klinger Blackboard Optics set or the Hartl Disc are used in many classrooms; however, these devices are expensive and generally are not used in such a way that students get "a piece of the action". Welch does make an individual Ray Box (Welch No. 3665A) for $25 that enables students to investigate reflection and refraction experimentally; but for less than $5 you or some of your students can build a light box that will accommodate six students and allow them to trace rays as they investigate reflection and refraction. They can use the same light box to produce pinhole images, shadows, and colored light beams for investigating the mixing of colored light and the formation of colored shadows. With a diffraction grating they can use the box to analyze colored light produced by various filters and colored solutions.

The light box (Fig. 2) includes: a cardboard box 8 to 12 inches on a side; a clear 150 watt bulb with a straight filament perpendicular to the bulb's axis, (a PSSC light source is good); a light socket and cord; and six oak tag or light cardboard strips, 1 in. x 4 in.

Colored plastic strips — red, green, and blue, 1 in. x 3½ in. can be cut from 20 in. x 24 in. colored sheets. Sheets can be purchased from: Rosco Laboratories, Inc., 36 Bush Avenue, Port Chester, N.Y. 10573. To make covers for light box order medium red #823, medium blue #863 and medium green #874. Filters for color analysis are: medium magenta #837, straw (yellow) #809 and light green blue (cyan) #858.

Figure 3 shows the various covers that you can make to produce rays, beams, and colored beams with your light boxes.

To produce sharp single rays we make two single slit covers for each opening. One is used to cover the opening; the other to collimate the narrow beam. Should you cut the slits too wide use a strip of black tape to make them narrower.

You will be able to design a great variety of experiments and activities to use with your light boxes. Interested readers may obtain copies of some of the experiments by writing to the author.

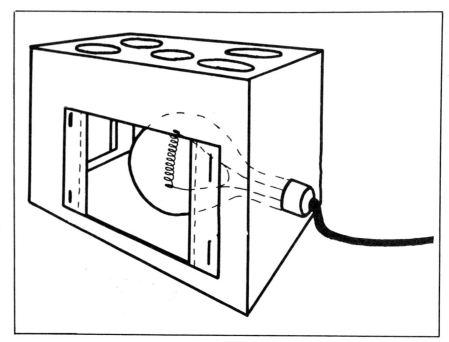

Fig. 2. Light box has ventilation holes in the top and a cutout in the side to mount slits and other accessories.

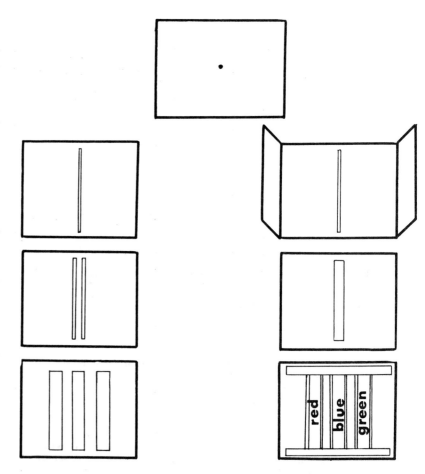

Fig. 3. Suggestions for cardboard panels to mount over cutout in light box.

ROBERT GARDNER
Salisbury School
Salisbury, Connecticut 06068

Questions students ask

Why is the string colored?

Question:

When standing waves are generated in a white string at 120 Hz and the string is illuminated by light from fluorescent fixtures and viewed against a dark background, the string looks pink on one side of a loop and blue on the other. In the adjacent loop the colors are reversed. Why do we see these colors?

This question was asked by students of **Sue Gray Al-Salam** *at Christopher Newport College, Newport News, Virginia 23601. The answer was supplied by* **Ronald B. Edge** *of the University of South Carolina, Columbia, South Carolina 29208. Professor Edge is Editor of our section "String and Sticky Tape Experiments."*

Answer:

Transverse standing waves on strings are often generated by a tuning fork resonating with the 60-Hz ac mains, one prong being attached to a taut cord. The fork vibrates at twice the mains frequency, since it is powered by an electromagnet between its tips, which attracts the prongs when the current peaks twice per cycle. If the string is in line with the prong of the fork to which it is attached, it is easy to generate 120-Hz standing waves on the string by applying the correct tension. Observed under fluorescent light, against a dark background, the color of the string on one side of the loop is quite definitely different from that on the other side. The string, in fact, acts as a stroboscope for the light.

Most hot-cathode fluorescent tubes, used for room lighting, have two components to their light output. A typical spectrum for such a tube is shown in Fig. 1. The sharp spikes are the emission lines of the mercury vapor filling the tube. The tube is powered by ac and these lines are excited when the voltage applied to the tube is highest, which is sixty times per second in one direction, and sixty times the opposite: 120 Hz. You can see from the figure the line spectrum peaks in the blue, most radiation coming from the 435.8 nm line. The continuous part of the spectrum arises from fluorescent materials deposited on the glass wall of the tube. These materials absorb the high-frequency mercury radiation, and reemit it at a lower frequency, peaking toward the red. They are chosen so that the net effect of the fluorescent material together with the mercury discharge simulates daylight. Since the mercury light only occurs at the peaks of the cycle, it illuminates only one side of the string loop, and this side appears blue. No colors will be seen if the fork vibrates in the same direction as that in which the string is stretched, for then the string is vibrating at half the fork frequency.

You can see the spectral lines by looking at the string through a diffraction grating. Now block off the blue string with a piece of dark card; the light coming from the red side appears more or less continuous. The fluorescent material takes at least a cycle for the light it emits to decay, so the side of the string's oscillation where it looks red is illuminated only by the fluorescent powder in the lamp.

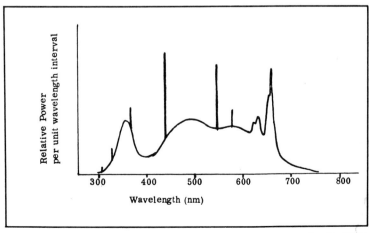

Fig. 1. Spectral power distribution of a fluorescent tube (Artificial daylight, Thorn Lighting Ltd.)

Reflection on the study of flat mirrors: Two demonstrations

Jim Nelson
Harriton High School, Rosemont, Pennsylvania 19010

I.

When students do an experiment on reflection of light from a flat mirror they are often confused by the different pins used. To help students sort this out, we give the pins names:

1. Object pin
2. Image pin
3. (Parallax) Test pin
4. Sighting pin

To aid the discussion of the experiment, a large demonstration model of the student experiment can be set up (Fig. 1). If the model is placed on a rotating platform, it can be turned to permit students to view the demonstration from any angle regardless of their position in the room. To set up the demonstration, photocopy the signs in Fig. 2, place the "IMAGE" and "OBJECT" signs back to back, attach signs to one stick, and place sticks at the position of the object and image. The two sticks should be identical. I also have a third stick of a different color to be used to show the parallax when the image is assumed to be at the wrong location.

When adding sighting sticks to the demonstration (or pins in the experiment) I suggest that you use completely different types of sticks so students do not confuse the object with the sighting sticks. At this point you could use strings to represent rays of light as suggested by John Layman.[1]

The fact that a flat mirror has a magnification factor of 1, can be directly observed by comparing the width of the object stick with the width of the identical image stick placed at the position of the virtual image.

If you have a large mirror set up in front of the classroom, you might cover it with a cloth so it will not draw student attention until you are ready.

Fig. 1.

II.

You can easily involve students in a debate concerning the smallest size mirror which will show the entire face. After letting students debate this for a few days, you can immediately resolve the issue with the following demonstration:

1. Tape a mirror firmly against a flat wall or chalkboard so the student can see his face.

2. Give the student a felt marker and ask him to carefully outline his face on the mirror while keeping one eye closed. Even if you have done the typical chalkboard proof for this, I think you will be surprised at the small area outlined.

3. For a quantitative result have the student grip a pencil in his teeth (like a pirate's knife!) and draw the apparent length of the pencil and then measure and compare this distance with the length of the pencil.

Be sure to point out that the image is still behind the mirror and not "on" it.

Reference
1. Phys. Teach. 17, 253 (1979).

Fig. 2.

A favorite experiment

R. Kennedy Carpenter
Butler High School, Butler, New Jersey 07405

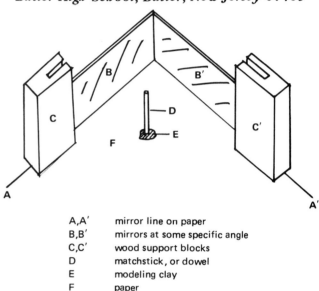

A,A'	mirror line on paper
B,B'	mirrors at some specific angle
C,C'	wood support blocks
D	matchstick, or dowel
E	modeling clay
F	paper

Fig. 1. *The experimental setup.*

This is a good experiment for several reasons. Economically, it employs low-cost re-usable materials and apparatus which take up little storage space and can be used for different experiments. Educationally, it allows a student to collect his own data, and, with graphical analysis plus teacher help, derive a mathematical formula from which he can make predictions. The more accurately he works, the easier it is for him to arrive at his conclusions.

The experiment, done early in the school year in the unit on light, relates the number of images seen in two facing plane mirrors opened at specific angles apart and these angles. Although each pupil collects his own data and creates his own graphs, the analysis, used as a learning process, is done as a class project under teacher guidance.

Two small plane mirrors, about 2 in. X 6 in., are the heart of the study. They can often be bought at small cost from some glass or mirror shop, cut from leftovers. Mine were made by the school custodian from a broken mirror I supplied. The two support blocks which hold the mirrors erect were cut from scrap in the school wood shop. Their size depends upon the choice of mirror dimensions, but their essential element is the deep saw groove into which the mirror can slide snugly without damaging its reflecting surface. A straight edge is used to draw the guide lines on the paper upon which the reflecting surface of the mirror rests (usually the back of glass mirrors). If no commercial rules are available, folding a sheet of paper into several thicknesses gives a rigid edge that can be used as a pencil guide. A protractor, or precut angle template, establishes the angular position of the mirrors. A matchstick or 2-in. length of 1/8 dowel held

Table I. *A sample completed data chart for this experiment.*

Trial	A	1/A	1/A × 10³	Observed	Calculated
1	180	0.0056	6	1	1
2	120	0.0083	8	2	2
3	90	0.0111	11	3	3
4	60	0.0167	17	55	5
5	45	0.0222	22	7	7
6	30	0.0333	33	10	11
7	15	0.0667	67	22	23

erect by pushing it into a small ball of modeling clay or other supporting medium serves as a convenient optical object.

There are seven trials to the experiment, one for each of the angles listed in Table I. A student draws a straight line about the length of his mirror and from it a second line at the desired angle between the two. He then supports both mirrors on the lines so they touch at the apex. At some convenient spot about midway between them he presses the object and clay into position. By moving his head along the lengths of the mirror fronts he counts the number of images of the object appearing in the total reflecting surface. He records this in his data chart and proceeds to the next trial.

The first step in the analysis is the plotting of the number of images, I, vertically, against the angle in degrees, A, horizontally, on a graph. His result, even if some of his counts are incorrect, will resemble Fig. 2. This indicates that an inverse proportion is involved, but does not designate the exponent of the horizontal variable. That is $I \propto 1/A^x$ in which x is a numerical exponent of A.

The second analytical step is to plot the number of images vertically against the reciprocal of the angle. To avoid complications in a plot using a decimal scale the student converts these values into a factor of ten notation, $1/A \times 10^{-3}$, and employs only the factor. Thus, by multiplying each decimal value by 1000 and rounding off, he converts them all to whole numbers for easy locating on the graph. This plot resembles Fig. 3 and is the key to the derivation of the formula.

The fact that the resulting line is straight indicates that the numerical exponent of A is unity; that is, there is no square, cube, or root function involved. The slope of the line (rise/run) turns out to be very close to 360. Individual values will vary somewhat, but by using two significant figures, a class average comes out to that value. A little intuition will lead to the fact that 360, the number of degrees in a whole circle, would be a logical theoretical choice for the proportionality constant.

The line intercepts the vertical axis at negative unity. Thus, if a positive unity were added to each of the image counts, the resulting line would then shift vertically to the origin of the graph and indicate a simple direct proportion between the variables plotted. Figure 3 indicates that $(I+1) \propto 1/A^x$.

Therefore, $(I+1) = k(1/A^x)$. But $x = 1$, and $k = 360$, so that $(I+1) = 360/A$, or

$$I = \frac{360}{A} - 1.$$

The formula is now checked three ways. First by substituting all the angle values used experimentally in the formula; the resulting I values tell the number of images which should have been seen. Recounts on disagreements can be made. The second check is to calculate the angle which should give 17 images. The answer comes out 20°. Students now draw this angle on their paper and make a physical confirmation count. Lastly, the number of images one should see if the mirrors were set at 72° apart is computed. The formula indicates 4. Students set up a 72° angle and make a direct image count.

Allowing for the "personal equation," "experimental error," "individual difference," or whatever, everything checks out nicely; so it is with a high degree of confidence that each student has discovered the relationship.

The same formula can also be derived using the horizontal intercept on Fig. 3. It measures about 2.8×10^{-3}, or 0.0028. To move the graphed line horizontally to the origin, therefore, a correction factor of -0.0028 would have to be added to all horizontal scale values. This correction factor when multiplied by the proportionality constant of 360 gives a negative unity within about 1%. Figure 3 indicates

$$I \propto \frac{1}{A^x} + c.$$

Therefore,

$$I = k(\frac{1}{A^x} + c).$$

But $k = 360$, $x = 1$, and $c = -0.0028$ (horizontal correction factor), consequently

$$I = 360\,(\frac{1}{A} - 0.0028).$$

Therefore,

$$I = \frac{360}{A} - 1$$

My experience, however, has convinced me that correcting the number of images is easier for the average pupil to visualize and understand. A completed data chart for this experiment might resemble Table I.

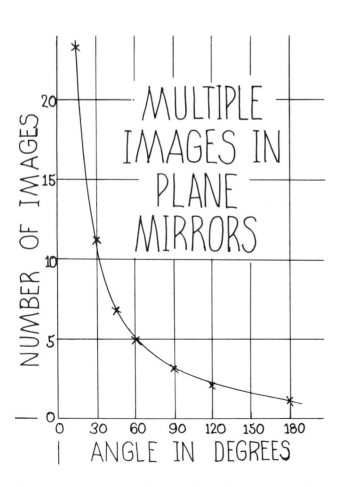

Fig. 2. *A student graph of the number of images versus the angle.*

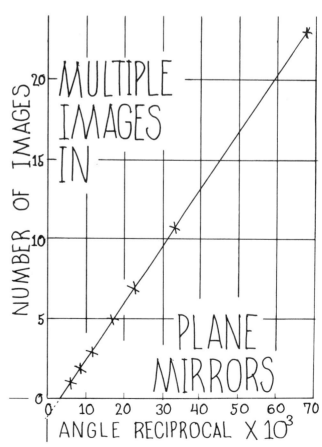

Fig. 3. *A student graph of the number of images versus the reciprocal of the angle.*

An introduction to pinhole optics

Wallace A. Hilton
William Jewell College, Liberty, Missouri 64068

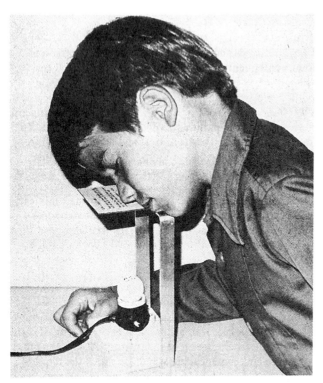

Fig. 1. *A student observing the words on a light bulb through a pinhole at a distance of 4 inches.*

Recent papers[1,2] have indicated the importance of the study of pinhole optics. A simple experiment for the beginning student is to assemble the apparatus shown in the figure. A small 7.5-W bulb has the words "Physics is Fun" printed on the top side. The student is first told to place his eyes at a distance of 4 in. from the bulb and try to read the words. Most persons will find it difficult, if not impossible, to focus their eyes on the letters. The student is then asked to look with one eye through a very small pinhole placed at this same distance from the words. He will see the letters in sharp focus and almost as clearly as if he were using a small magnifying lens. A piece of cardboard with a very small pinhole is satisfactory for the experiment.

The figure shows a paperclip holding the card with pinhole to a 4-in. x 4-in. metal sheet with a half-inch hole at its center. The pinhole is centered over the metal hole, and the metal sheet is hinged to the support which is 4 in. above the words on the bulb. This makes it easy to move the pinhole in and out of position so that the student may quickly compare the difference between the two observations.

References:
1. M. Young, Appl. Opt. **10**, 2763 (1971).
2. M. Young, Amer. J. Phys. **40**, 715 (1972).

That can't be: a virtual comment

John W. Layman
Physics Department, University of Maryland, College Park, Maryland 20742

"The image can't be behind the mirror because most plane mirrors are hung on walls" — quotation from a young lady in an introductory physics course on light and color, even after viewing the following lecture demonstration.

A relatively thin plane mirror with a reflecting surface on both sides is mounted on a rotatable platform as shown in Fig. 1. Across the top is a board with nails in various positions. Attached to each nail is a string of 10-m length on one side with a smaller length behind the mirror. An object taller than the mirror (ring stand as shown in Fig. 1) is placed in front of the mirror. The strings are passed to students in the class who are asked as a group to shift the strings until each is aligned with the image of the ring stand seen in the mirror. All students along a particular string are asked to confirm its alignment. After this certified alignment, other students are instructed to go behind the mirror to pick up the short extensions of the strings and align these with the rest of each string. The students holding the string extensions are seen behind the mirror in Fig. 1, along with the image of at least one of the strings extended into the classroom. The camera angle makes the strings look like they are at different levels, but because of the size of the lecture hall they are in nearly the same plane.

Lo and behold, these string extensions all cross at a particular point behind the mirror, a point corresponding to the location of the virtual image of the front ring stand. If a second ring stand, identical to the first, is now placed in this position its top will be aligned with the virtual image of the first stand thus offering physical evidence for the actual position of the virtual image of the real object. The alignment of the virtual image of the front ring stand and

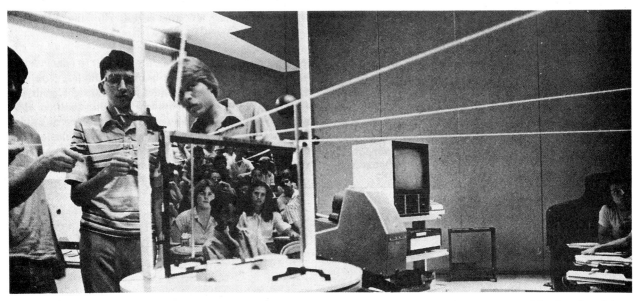

Fig. 1. Locating the virtual image of a ring stand.

the top of the back ring stand placed in the virtual image position is clearly shown in Fig. 2.

Having the mirror mounted on a rotatable platform allows the apparatus, strings now removed, to be rotated to show that the original image ring stand becomes the object stand and the original object is now in the position of the virtual image. It is easy to show that any other position for the ring stand marking the image produces parallax with the virtual image as the mirror is rotated, thus strengthening the physical evidence for one unique virtual image location. It is a most effective demonstration.

The young lady's response quoted at the beginning of this note was blurted out in class after the demonstration, illustrating conceptual difficulties of the nonscience student with the idea of a virtual image having a position behind the mirror. Without this demonstration to return to for further discussion matters would be even worse.

Fig. 2. The virtual image of the ring stand, marked by a second stand behind the mirror, and puzzled students.

Why is your image in a plane mirror inverted left-to-right but not top-to-bottom?

Kenneth W. Ford
University of Massachusetts, Boston, Massachusetts 02125

The title of this note is a popular brain teaser often asked of students during their study of geometrical optics. An excellent discussion of the question appears in Martin Gardner's *Ambidextrous Universe*.[1] Gardner points out that the mirror in fact produces neither a left-to-right inversion nor a top-to-bottom inversion, but a front-to-back inversion.[2] He argues that we interpret this as a left-to-right inversion because of our own (approximate) bilateral symmetry. This is, I think, only a partial explanation. Here I want to focus on the important psychological (perceptual) aspect of the question. It may be that our illusion of left-right inversion is attributable more to a social habit, the way we turn around, than to our bilateral symmetry. This "social habit," in turn, stems directly from the fact that we live in a gravitational field.

The *physics* of mirror reflection is not in dispute. We physics teachers can all demonstrate with ray diagrams that a plane mirror produces an image that is erect, virtual, and unmagnified. To demonstrate the front-to-back inversion, it is probably best to work with the mathematical transformation from object to image. This transformation can be defined as inversion through a plane; all the lines drawn from points in the object to corresponding points in the image have the mirror as their perpendicular bisector. Alternatively, the transformation can be described as inversion through a point in the mirror plus a 180 deg rotation (Fig. 1). None of this answers the left-right vs. up-down question, however. The girl in Fig. 1 is standing on her right foot with her right arm raised. In the mirror she sees a girl who is *upright* but appears to be standing on her left foot with her left arm raised. Why the asymmetry between left-right and up-down?

Consideration of the physics only leads us to conclude that there *is* no asymmetry. In your image, your left shoulder is at your left, your right shoulder is at your right, your head is up, and your feet are down. Yet the question about asymmetry seems to make sense, and it deserves an answer. Obviously the answer must be in more psychological than physical terms. Here is my proposition: We call the image "inverted left-to-right" because of the way it differs from

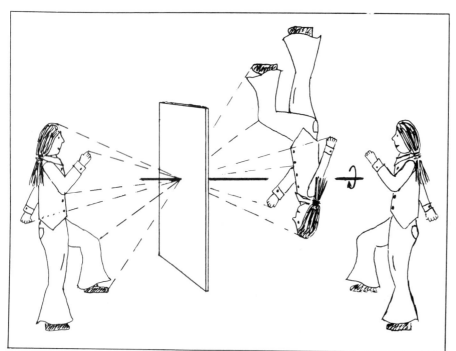

Fig. 1. The transformation from object to image is an inversion through a point, plus a rotation through 180 deg about an axis that passes through the point and is perpendicular to the mirror (the only direction defined by the mirror!). The point can be located anywhere in the plane of the mirror. The order of the inversion and rotation can be interchanged. In the figure, the image is artificially displaced to the right. The young woman is an object only in the traditional usage of geometrical optics. (Drawing courtesy of Caroline A. Ford.)

our customary view of another person whom we meet face-to-face. Relative to oneself, that person is turned through 180 deg about a *vertical* axis [see Fig. 2(a)]. Her left shoulder is at our right, her right shoulder is at our left, her head is up, and her feet are down. Imagine a world in which people customarily meet face-to-toe instead of face-to-face [see Fig. 2(b)]. In that world, one person "facing" another would be turned, relative to the other, through 180 deg about a *horizontal* axis.[3] If you lived in that world, you would expect to see a person's left shoulder at your left, his right shoulder at your right, his head down, and his feet up. Looking into a mirror, you would see an image that differed from your customary view of other persons; you would say that your image was inverted top-to-bottom but not left-to-right and you might wonder why.

The apparent left-right inversion produced in our world by a mirror applies, of course, to things as well as people. If the left side of an automobile is viewed in a mirror, it appears to be the right side of an automobile. In the mirror view of a printed page, the lines run from right to left but suffer no up-down inversion. An automobile viewed from the side has no bilateral symmetry; neither does a printed line. This suggests that the sensation of left-right inversion is produced not so much by bilateral symmetry as by our customary way of looking at things in our gravity-dominated world. If we walk around an object to look at its other side or if we turn it around, almost always we are dealing with rotation about a vertical axis. Therefore we interpret what we see in a mirror by comparing the image with the result of such a vertical-axis rotation.

Here are some further questions that might be posed to students: 1. Suppose you were the only living creature in the world. What kind of an inversion, if any, would you attribute to your mirror image? 2. If we lived in a gravity-free part of the universe, would we say that mirrors produced left-right inversion? 3. Write a short essay answering the question "Why is a mirror image inverted top-to-bottom but not left-to-right?" from the standpoint of creatures who habitually turn themselves and their possessions around horizontal axes.

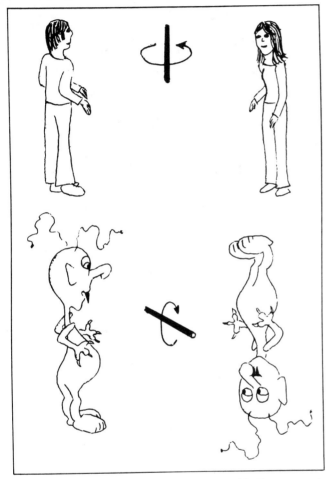

Fig. 2. (a) A normal encounter in our world. One person is turned about a vertical axis relative to the other. These people consider their mirror images to be turned left-to-right but not top-to-bottom. (b) A normal encounter in another world. One person is turned about a horizontal axis relative to the other. When one of these people looks in a mirror, he wonders why his image is inverted top-to-bottom but not left-to-right. (Drawings courtesy of Sarah E. Ford.)

References

1. Martin Gardner, *The Ambidextrous Universe* (Basic Books, Inc., New York, 1964) pp. 22-24.
2. This part of Gardner's explanation has been summarized by Walter Thumm, Phys. Teach. **10**, 346 (1972).
3. We should inquire of astronauts which method of turning around they find more congenial when they are weightless.

Physics at home and in the back yard (Part III)

George Ficken
Cleveland State University, Cleveland, Ohio 44115

Spooky "pinhole" camera image

A year ago a full page article, with pictures, appeared in the Cleveland Press[1] on the "pinhole" camera effect (camera obscura). Some high school students had been exploring a vacant house with boarded up windows on a bright sunny day. They got really spooked when they saw the inverted images[2] shown in Fig. 1 of the house next door (Fig. 2) on two darkened interior walls. There were several small openings in the boards (Fig. 3) through which the light reflected from the other house could pass and thereby produce the ghostly images. The newspaper was called; the science editor, David Dietz, immediately recognized the effect, sent out a photographer, and the article resulted.

Reference

1. Photos supplied by *The Cleveland Press*, Cleveland, Ohio 44114.
2. Careful inspection of the photo reveals at least two fainter images in addition to the brightest one caused by the largest hole. Gross imperfections in the interior wall are apparent.

Fig. 1. "Pinhole" camera images appearing on darkened interior wall of house.

Fig. 2. "Pinhole" camera source: The brightly lit house next door.

Fig. 3. Pointing out largest of "Pinholes" in boards covering window.

"HIS SPECS - USE THEM FOR BURNING GLASSES"

During the past decade or more, numerous groups of high school and college students have participated in discussions of Golding's "Lord of the Flies," either in formal literature courses, college orientation programs, leadership retreats, or other programs. The tale concerns life for a group of young boys marooned on an island without the adult supervision they had previously always had. A key element is the ignition of fire—for signaling possible rescuers, for cooking, and for comfort.

Can you spot in the following quotations the physical inconsistency of which the author is apparently unaware?

From Chap. 2 when it is discovered that no one has a match, attention turns to Piggy, the one boy wearing glasses: ". . . 'His specs—use them as burning glasses!' . . . . Ralph moved the lenses back and forth, this way and that, till a glossy white image of the declining sun lay on a piece of rotten wood. Almost at once a thin trickle of smoke rose up. . . .".

Much later Jack and his band of hunters steal Piggy's glasses. Early in Chap. 11, the author writes: ". . . The twins watched anxiously and Piggy sat expressionless behind the luminous wall of his myopia. . . ."

Answer: Myopia, or nearsightedness, denotes an eyesight abnormality in which the point of most distant vision is too close to the viewer. Rays from points on more distant objects are converged "too strongly" by the eye so that the real image is formed in front of, instead of at, the retina. The condition is corrected by inserting a *diverging* lens in front of the myopic eye so that rays from distant objects are first diverged from their original incident directions so that the eye now converges the rays to a real image on the retina. Only a *converging* lens could be used as a "burning glass," since it must focus the parallel incident rays from the sun into a real image. The *diverging* lens used in glasses to correct myopia would only spread the incident parallel rays from the sun apart, rather than producing a "glossy white image."

C. BOB CLARK
University of North Carolina
Greensboro, North Carolina 27412

Pinhole glasses?

Russell Patera

Physics Department, University of Miami, Coral Gables, Florida 33124

Few people realize that glasses are not the only way to correct poor vision. There is an extremely simple yet very effective method for improving vision. Make a pinhole in an opaque material and place it over the eye so that one can look through the pinhole. We found that a pinhole in a pop bottle cap serves this purpose well. When placed over eyes suffering from nearsightedness and farsightedness a remarkable improvement in vision occurs. This pinhole monocle is most useful when glasses are not at hand. For example, while swimming the pinhole monocle can be produced by the fingers of one or both hands. It is good to know that one can improve ones' own vision without the use of glasses. A discussion of this effect provides motivation for studying aberration in lenses.

The pinhole monocle works because it converts the eye into a pinhole camera which has infinite depth of field. The eye itself can automatically produce this same effect. In very bright light the pupil decreases in size thus cutting out some of the aberration and improving vision. Squinting is based on the same principle of reducing the amount of poorly focused light but it is not as effective as the pinhole technique.

Pinhole glasses have two disadvantages. First, the field of vision is narrow because the pinhole is not very close to the eye. Second, the intensity of light striking the retina is reduced. Therefore, bright lighting is necessary for pinhole glasses to be effective.

A discussion of the pinhole monocle adds some flavor to the study of geometrical optics. This whole discussion takes on added meaning for those students who need corrective lenses because they can experience the effect themselves.

To make a Camera Obscura

For observation of the sun, a simple device as suggested in Fig. 11 is satisfactory. The "pinhole" may be of the order of 1/16-in. in diameter. To get the brightest image attach a sheet of white paper to the end of the box opposite the pinhole to serve as a screen. Of course, the longer the box, the larger the image of the sun.

An instrument similar to that shown in Fig. 5 can be constructed by using a convex lens, a plane mirror, a sheet of glass and some cardboard boxes.

The lens should have a focal length of at least 15 in. so the image will be of adequate size, and a diameter of several inches to give a sufficiently bright image.

Instead of having one box slide inside another as shown in Fig. 5 the design uses a single box. Focusing is provided for by mounting the lens in a tube that slides in a second tube attached to the front of the box. (See sketch.)

The tube for a 5-in. diameter lens is a large oats box with the ends removed. (For smaller lenses, a salt box or a smaller oats box will do nicely.) Mount the lens close to one end of the tube by means of six small wood blocks (three on either side of the lens) glued inside the box. The second tube is half of another large oats box, cut parallel to its axis and opened to form a collar in which the lens tube fits snugly but can slide freely.

Choose a rectangular carton whose length is approximately equal to the focal length of your lens. Cut a hole in one end of the carton to accommodate the collar and glue the collar in place, taking care that its axis is perpendicular to the end of the carton.

When the glue is dry, put the lens tube in the collar and point it at a window. Slide the lens tube in or out until there appears on the back surface of the box an inverted image of what you would see if you look out the window. You now have a simple camera obscura with lens.

To complete the camera obscura as used by artists, make some 45-degree triangles of cardboard and glue them inside the box at the bottom corners furthest from the lens. These will support the plane mirror (M in Fig. 5). With the plane mirror in place, put a sheet of glass on the open top of the box directly above the mirror. On the glass place a piece of tracing paper. Point the lens at a window and adjust the lens position until there is an image on the tracing paper. The image will be right side up but reversed left-to-right.

If the sun is shining, the image will be bright and you can sketch in the outline of the scene on the tracing paper. However, you will find that different lens positions are necessary to obtain sharp images of objects in the foreground, in the middle distance and in the background. You probably will find, too, that even with the best focus, an image is not as sharp as you might like.

To improve the definition, cut a hole about an inch in diameter in a piece of cardboard and mount the cardboard in front of the lens with the hole coinciding with the center of the lens. There will be improvement both in the sharpness of the images and the "depth of field," i.e., the ability of the instrument to focus images of objects at different distances with a single lens setting. However, there also will be a marked decrease in the brightness of the image.

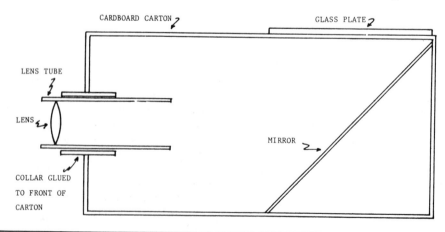

INVERSION OF AN IMAGE ON THE RETINA

Sutton's book *Demonstration Experiments in Physics*,[1] includes a delightful little experiment to demonstrate that an image on the retina of the eye is inverted in relation to the object which is being observed. However, the work involved in making the apparatus is likely to deter a busy physics teacher. Two simpler ways of producing the small piece of equipment are described here. Since the book is out of print and may be inaccessible to many physics teachers, the experiment will be described in enough detail here so that it can be reproduced without the reference. The easiest approach is to use a new spool of thread (or a spool, even without the thread), whose paper end covers have not been punctured. Three holes in the form of an equilateral triangle (about 1 or 2 mm on an edge) are needed in one end and a single hole in the other. These holes can be made by an ordinary pin or needle. By holding the end with the three holes very close to the eye with one base of the triangle horizontal, the image of the three holes appears to be upside down. The fact that the object distance is less than the focal length of the eye lens prevents the lens from being able to invert the image as it does when the object distance is greater than the focal length, as is usually the case. Of course, the image formed cannot be in sharp focus, but it can easily be recognized as being upside down.

Another form of the apparatus can be made from a plastic 35-mm film container, such as is used to enclose Kodak film. The same size needle or a very small drill is used to make the holes. Other arrangements of the holes are also possible. Consider the letters of the alphabet, for example. Holes arranged in the form of any letter which is not the same when inverted can show the effect. For example, the letters A or P could be used, but O or H could not.

George Barnes
Department of Physics
University of Nevada
Reno, Nevada 89557

Reference
1. Richard M. Sutton, *Demonstration Experiments in Physics* (McGraw-Hill, New York, 1938) Experiment L-59, p. 393.

INVERSION OF SHADOWS ON THE RETINA

Barnes[1] has described simple ways of demonstrating that images on the retina of the eye are inverted in relation to the object being observed. One of his methods involves the use of a plastic 35-mm film container, such as is used to enclose Kodak film. A pattern of pinholes is made in the bottom of the container and appears inverted when viewed from the open end. A slight modification of this setup gives an even more interesting effect:

A single hole is made in the center of the bottom of the container using a straight pin. Then the pin is stuck through the side of the container (from the inside), about 5 mm from the open end, so that the shaft lies along a radius of the container and the head is located on the axis. When the open end of the container is held up to the eye the pin is so close that an image cannot be formed. Light coming through the pinhole, however, will cast a relatively sharp shadow of the pinhead on the retina. The shadow is, of course, right-side-up but it appears up-side-down. If several pinholes close together are used, multiple shadows may be seen. This demonstration is similar to one described by Bragg[2] but is a bit simpler and more reliable since the hole and pin are held fixed with respect to each other rather than manipulated independently.

References
1. George Barnes, Phys. Teach. **19**, 499 (1981).
2. Sir W. H. Bragg, *The Universe of Light* (G. Bell and Sons, Ltd., London 1936) p. 50.

Karl C. Mamola, *Appalachian State University, Boone, North Carolina 28608*

BEHIND THE EYE

In the October issue, George Barnes described some nice ways to demonstrate the fact that retinal images really are inverted.[1] However, his explanation of the theory behind what is observed cannot be correct. The triangular shaped pattern of object holes does not have to be closer to the eye lens than its focal length. Moving the holes to a distance several or more times the diameter of the eyeball from the lens will not destroy the effect. I stumbled onto this myself, years after having many times given the same incorrect explanation, exactly as I first heard it in 1957.

Walker[2] describes the effect as "Inverted shadows" cast by a nail held close to the eye, and gives several references. One of these, by Hirsh and Thorndike,[3] includes ray diagrams to explain the shadow formation and its interpretation by the eye. However, they do not show the bending of the light rays in the cone of light formed by the single (source) hole as they pass through the eye lens. Figure 1 is a sketch of the setup. The way the demonstration is usually done, the tiny source hole S is well within the near point of the eye. This is to insure a large enough bright background circle within which to view the pencil point customarily used as an object.

This causes the real image S' of S to be located behind the retina and produces a fuzzy circle of light on the retina if S is the only hole used. The diverging pencils of light rays from the "point" source S, defined by object holes 1 and 2 of the triangular pattern, are refracted by the eye lens and hit the retina at locations 1' and 2'. Note that 1' and 2' are not inverted relative to each other, even though hole 2 lies on the axis and location 2' is below the axis. The diagram was purposely drawn with holes S and 1 aligned parallel to the axis, so that one set of the construction lines was known. This causes 1' to be located at the focal point F of the eye lens, on the retina for a relaxed eye. The separation of holes 1 and 2 is exaggerated so that locations 1' and 2' can be easily distinguished.

George W. Ficken, Jr., *Department of Physics, Cleveland State University, Cleveland, Ohio 44115*

References
1. G. Barnes, Phys. Teach. 19, 499 (1981).
2. Jearl Walker, *The Flying Circus of Physics with Answers* (Wiley, New York, 1977), No. 5.23, p. 121.
3. F. R. Hirsh, Jr., and E. M. Thorndike, Am. J. Phys. 12, 164 (1944).

IMAGE FROM A PINHOLE

I think the December issue is an unusually good one and, as one particularly interested in light, I greatly enjoyed the article on the camera obscura by Alley.[1] Much of the material in it is also applicable, of course, to the making of pictures by the use of a pinhole camera. A "pinhole" image of the sun can also be formed by use of a very small reflecting surface, as I mentioned in a 1955 note to the *American Journal of Physics* (vol. 23, p. 544). For a number of years I had one mounted so that the sun's image would come to my attention at my desk when it was time to go home — though on rainy days my system broke down.

By the way, how many people notice the images on the sidewalk formed by pinhole images of the sun formed by light from the sun passing through the foliage? It always catches my attention in the summer time. Many people who see them for the first time are astonished that all of the images are of nearly the same size and, of course, round.

E. Scott Barr, *Box 3174, East Side Station, Tuscaloosa, Alabama 35404*

Reference
1. Phys. Teach. 18, 632 (1980).

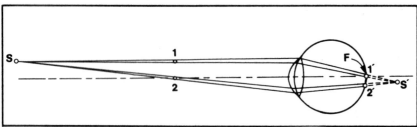

Fig. 1. Pencils of rays through holes 1 and 2 form fuzzy images 1' and 2' of source hole S on retina.

Optics of the Rear-View Mirror: A Laboratory Experiment

By Edwin R. Jones and R. D. Edge
Department of Physics and Astronomy, University of South Carolina, Columbia, SC 29208

The number of optics experiments performed in class which are relevant to everyday life are few and far between. The day-night rear-view mirror found in most automobiles provides a meaningful example of applied geometrical optics. The device is simplicity itself—the mirror is a back-silvered glass wedge.[1] For day vision, it works like a regular rear-silvered mirror (Fig. 1a). Light enters from the front, is reflected from the back, and exits back through the front. The mirror behaves as a glass prism of small angle. The thin end of the wedge is at the bottom, so that a bright source of light will have a fringe of red at the bottom and blue at the top. This dispersion is not usually noticed in the daytime but is readily observed in the reflection of headlights in the dark. At night, the mirror is turned up as shown in Fig. 1b. The image of the rear window is then seen reflected from the front surface of the wedge, and the light reflected from the back of the mirror is deflected above the driver. Only about 5% of the light is reflected from the front surface, as opposed to approximately 80% from the back. The mirror is used in this position to reduce the glare from headlights of cars behind. Because the reflection is only from the front surface of the glass, the image is sharp and shows absolutely no dispersion. These are not the only images provided by the mirror. Figure 2 shows some other possibilities. In addition to the two reflections mentioned above, further reflections produce images labeled 3, 4, 5,—all of which can be seen. They arise from successive reflections between the front and back surface.

If the mirror is improperly positioned, the image labeled 3 may be used for the reduced glare image at night. In this position, about 4% of the incident light will be reflected. Because the light traverses the prism four times before it emerges, the dispersion is equivalent to that of a prism with an apex angle four times that of the wedge. Consequently, the dispersion will be quite pronounced.

We devised a novel student laboratory experiment in which the student examines the reflection of light from an auto mirror. The light source may be either a small laser (which provides the simpler and more convenient method) or a small bright source of light (such as an incandescent lamp with a clear envelope).

Mirrors may be obtained from a junkyard and removed from their housing to expose their wedge shape. The angle of the wedge should be measured using a micrometer or vernier caliper to determine the thickness at the thin and thick edges. The wedge angle is then given in radians as the difference in thickness divided by the separation between the edges. In our case, the angle was 3.2°.

The laser beam is reflected off the mirror to the wall behind the laser as shown in Fig. 3. A protractor or set square is used to align the laser beam normal to the wall. The mirror is then turned so that the back surface reflection returns along the incident beam as shown in Fig. 4a. The angle of the front surface reflection is computed from measurement of the position where the reflected beam strikes the wall (the distance from 1 to 2 in the figure), and the distance between the mirror and the wall. Half of that angle is the angle of incidence. Snell's Law may then be used to find the index of refraction n of the glass. The angle of transmission is equal to the wedge angle (in our

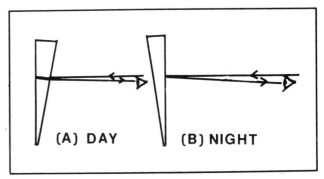

Fig. 1. (a) In daylight use, the auto mirror is tilted so that the rear view is reflected off the silvered back surface to the driver's eye. (b) At night, the mirror is repositioned so that the rear view is reflected off the front surface of the glass to the driver's eye.

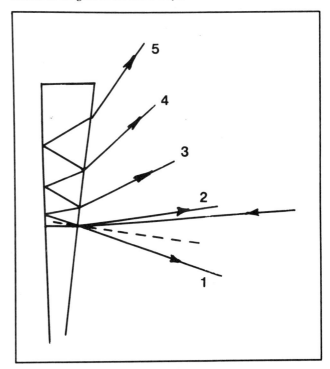

Fig. 2. In addition to the front surface reflection, ray 1, and the back surface reflection, ray 2, three other reflected rays can also be seen. Rays 3, 4, and 5 correspond to 1, 2, and 3 internal reflections from the glass surface.

case, 3.2°) in order for the light reflected from the back surface to come back along the path of the incident ray. For our mirror, the index was found to be n = 1.54.

The emerging angles can then be calculated for the three beams labelled 3, 4, and 5 which correspond to 1, 2, and 3 internal front surface reflections. The positions where they strike the wall can be measured and the values used to compute the measured angles of emergence of the three beams. The results obtained in our trials are shown in Table I where the measured and computed angles agreed to better than one tenth of one degree.

In a second experiment, the mirror may be aligned as shown in Fig. 4b so that the beam reflected from the front surface is directed back along the incident beam, i.e. perpendicular to the wall. The positions of the other four beams can be measured and the angles computed. These measured angles may then be compared with the values computed directly from Snell's law. Our results for this case are shown in Table I, also.

In addition, the beam intensities may be calculated assuming a front surface reflectance of 0.05 and a reflectance of the silver surface of 0.90. The values, which we confirmed experimentally, are listed in Table II.

The experiment can also be conducted without a laser. Instead, a clear envelope lamp can be used as the source of light. A line can be marked on the face of the mirror, using a china marker or felt tip marker. The reflected images can then be seen directly by looking into the mirror. By moving the eye until the lamp filament is visually aligned with the line marked on the mirror, the direction of the reflected beam can be measured. The mirror and lamp may be arranged on a tabletop so that lines may be drawn along the directions of the incident, and reflected rays and their angular separations easily measured. This technique yields the added bonus of directly showing the dispersion.

The measurements required for this experiment take only a few minutes, using a meter stick and a protractor to provide very good results. Thus, the auto mirror provides the basis for an excellent experiment in geometrical optics for beginning physics students. □

Reference

1. H. R. Crane, *Phys. Teach.* **23**, 238 (1984).

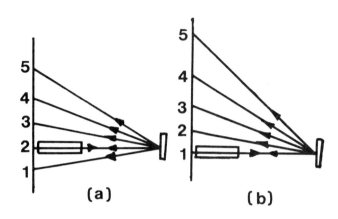

Fig. 3. The laser beam is reflected back to the wall where it grazes the protractor or set square. The laser and mirror can then be adjusted to align the beam perpendicular to the wall.

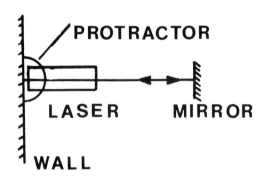

Fig. 4. (a) In the first experiment, the strong ray reflected from the back surface is returned along the incident ray. (b) In the second experiment, the ray reflected from the front surface is returned along the incident ray.

Table I. Angles of Reflected Rays Relative to the Incident Ray

Ray	Angle Θ in First Experiment		Angle Θ in Second Experiment	
	measured	calculated	measured	calculated
1	-9.83°	-	-	-
2	0.0°	-	9.89	-
3	9.96°	9.97°	20.2°	20.0°
4	20.2°	20.2°	30.4°	30.4°
5	30.9°	31.0°	41.7°	41.7°

Table II. Calculated Intensities of the Reflected Rays from an Automobile Day/Night Mirror. Incident Intensity of I_0.

Ray		Intensity
1	External front surface reflection	$0.05\ I_0$
2	Back surface reflection	$0.81\ I_0$
3	One internal front surface reflection	$0.037\ I_0$
4	Two internal front surface reflections	$0.0016\ I_0$
5	Three internal front surface reflections	$0.000074\ I_0$

Mirrors in air and water

Robert Gardner
The Salisbury School, Salisbury, Connecticut 06068

Physics students often investigate the reflection of light from a plane mirror by standing a mirror on a sheet of paper, using two pins to determine an incident ray, and locating the reflected ray by sighting along a ruler, or aligning two other pins. By measuring the angles between each of these lines and the normal to the mirror at the point where the rays meet, students find that the angle of incidence is equal to the angle of reflection.

The procedure can be modified to introduce students to the refraction of light. In this experiment the mirror is placed in a transparent rectangular box that is nearly full of water. A line is drawn along the rear edge of the box, and the box moved so that the reflecting surface of the mirror is above the line. Again, pins are used to establish an incident ray that must pass through the water to reach the mirror. (See Fig. 1.) Students look through the water to find the images of the object pins, and determine the path of the reflected ray with another set of pins. The box of water is removed so that the incident and reflected rays can be drawn. Where do these rays appear to meet?

Because the rays, when extended into the area previously occupied by the water, meet in front of the mirror, many students think they have "done it wrong" and will often repeat the experiment only to find similar results. When they realize that others are getting the same data, they will begin to generate explanations. Some actually believe the light is somehow reflected at the point the rays intersect in the water. After some discussion, someone usually suggests that perhaps the light rays are bent when they enter and/or leave the water. This hypothesis can be tested readily if you have a light box[1] or some way of producing a narrow beam of light.

As an alternative method, or as an extension of this experiment, use a rubber band to hold a mirror against the back of a clear plastic block as shown in Fig. 2.

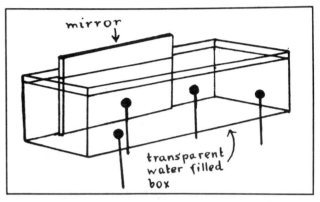

Fig. 1. The light ray located by two pins enters the water before reflection by the mirror.

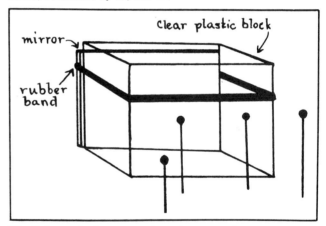

Fig. 2. Instead of water, a plastic block can be used.

Reference

1. Robert Gardner, "Light box, inexpensive but versatile" Phys. Teach. 13, 50 (1975).

An optical puzzle that will make your head spin

Samuel Derman
Department of Technology, City College of N.Y., New York, N.Y. 10031

An optical puzzle, easy to construct and demonstrate, is the following.

A rectangular strip of polished metal foil is bent into the shape of an arc, parallel to the long dimension (Fig. 1). If one looks at one's reflection in the concave surface of this arc, as indicated, the image will appear inverted. But if the strip is now slowly rotated through 90° about the line of sight (keeping the reflection in view all the while) the image will do a peculiar thing. It will flip over and turn right side up. That is, a 90° rotation of the reflector will result in an image rotation of a full 180°.

Why should that be?

Here's the answer. By bending the strip into an arc as described, one forms (approximately) a segment of a right circular cylinder. This cylinder has a fairly well defined focal length.

A tall skinny object that is more than one focal length away from the metal surface and is parallel to the long dimension of the strip (i.e., perpendicular to the cylinder axis) will indeed form an inverted image (Fig. 2). But once the strip is rotated through 90° — the object distance remaining fixed — the tall skinny object now finds itself *parallel* to the cylinder's axis. In this direction the surface performs no optical inversion. So the skinny object must appear noninverted — i.e., right side up. In the experiment the observer's face acts as the tall skinny object.

The crucial (and perhaps obvious) point is that such an effect could not occur if the strip formed a surface of revolution.

To construct the puzzle one needs highly reflective metal foil, which may not be readily available. Thin aluminized mylar, however, is.

The mylar should be cut to shape (any convenient size) and glued onto a stiff piece of cardboard. The type of cardboard commonly used for file folders possesses the necessary "springiness" and rigidity.

Some commercial suppliers of aluminized mylar are:
1. Arex Graphics, 256 East Third Street, Mt. Vernon, N.Y. 10550
2. Vacumet Corporation, 14 East Wesley Street, South Hackensack, N.J. 07606
3. AIN Plastics, 300 Park Avenue South, New York, N.Y. 10010

In addition, small sized cylindrical mirrors are available from Sargent Welch Corp.[1] (No. 3524) and from CENCO[2] (No. 85440).

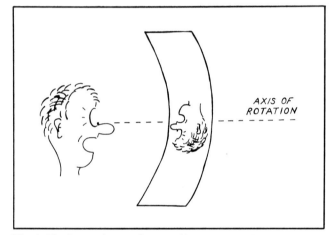

Fig. 1. The image changes from upside down to right side up when the reflector is rotated 90°.

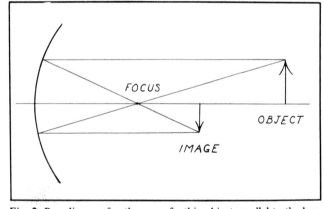

Fig. 2. Ray diagram for the case of a thin object parallel to the long dimension of the reflecting strip.

References
1. Sargent Welch Co., 7300 Linder Ave., Skokie, IL 60077.
2. Central Scientific Co. (CENCO), 2600 S. Kostner Ave., Chicago, IL 60623.

MARKERS FOR YOUNG'S EXPERIMENT (PSSC VERSION)

Paper markers on the meter stick are all right if you keep the distance between the slits and the light source small, but if you use the full length of the lab table, it is difficult to see the markers against the bright diffraction pattern. Using the arrangement in Fig. 6 and allowing a small bit of the filament to shine below the meter stick, a sharp image of the filament may be focused on the meter stick by means of any concave mirror mounted on a movable pedestal and the "mark" can be seen easily.

WILLIAM H. PORTER
Gilman School
Baltimore, Maryland 21210

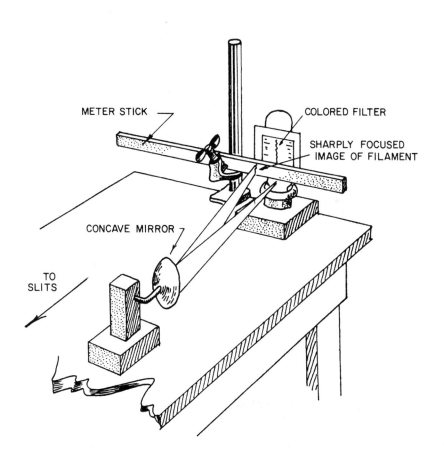

Fig. 6. *Observer uses mirror to project marker on meter stick.*

CYLINDRICAL MIRRORS

Samuel Derman's article, "An optical puzzle that will make your head spin," [Phys. Teach. 19, 395 (1981)] suggests the use of Mylar or aluminum foil to construct a cylindrical mirror. A cylindrical mirror having a variable focal length can be found in most dark rooms. It is the chromium-plated squeegee plate used for drying glossy prints. Hold the squeegee plate in front of your face, with your left hand on the top and your right hand on the bottom of the plate. Now gradually bend the plate into a concave shape (the axis of the cylinder being formed is horizontal), and observe your reflection. You will observe the transformation of an upright virtual image into an inverted real image. Now rotate the mirror through an angle of 90° (such that the axis on the cylinder becomes vertical) and observe the flip-over of the image as described by Derman.

An additional interesting phenomenon can also be observed with the mirror held with the axis vertical. When the mirror is slightly concave, wink your left eye and observe that your left eye is looking into your blinking left eye and your right eye is looking into your nonblinking right eye. You are seeing your reversed virtual image. Now bend the mirror more and again blink your left eye when the curvature is such that your face is beyond the focal point of the mirror. Your right eye will now be looking into your blinking left eye and your left eye will be looking into your nonblinking right eye. This is the way others see you when you face them. You are looking at your upright real image.

Paul E. Wack, *Department of Physics, University of Portland, Portland, Oregon 97203*

Recording timer tape for interference demonstrations

FR. EARL R. MEYER
Thomas More Prep, Hays, Kansas 67601

One of the important lessons in the PSSC program is the section on interference effects of wave motion. The treatment of physical optics and many other topics in the later part of the text depends on an understanding of the principles of interference.

Students may find this important lesson difficult and can become confused about the geometry of interference from two point sources. The text books traditionally illustrate this effect by a series of arcs from each point source, overlapping the arcs to indicate interference. The complex pattern of many intersecting arcs confuses some students. The following design (perhaps not original) is an attempt at a simple explanation of interference from two point sources.

The demonstration material consists of two strips of recording timer tape, each about 1 m long. Make a series of dots on each tape at some fixed distance, e.g., 10 cm. The dots will indicate the crests and troughs of a wave and must be distinguished—alternate colors or solid dots and circles.

The two strips can be attached to the frame above the chalk board with thumb tacks. The tacks represent the two point sources, and the distance between them is d. The strips of tape may now be arranged to show the nodal lines at each point where a solid dot coincides with a circle. The nodal lines can be recorded directly on the chalk board (cf. Figs. 1 and 2). If you make such a setup for your bulletin board, you will find the students doing this exercise and explaining it to each other.

These tapes work wonders in explaining interference from thin films. A section of the tape is used to represent the incoming wave. For the section of the tape inside the film, the dots are marked for a wavelength of $\lambda/2$, where n is the index of refraction. At the upper surface where there is a phase shift of 180°, a second piece of tape is used to indicate the half-wavelength shift. This works nicely for the standard thickness films (cf. Figs. 3–5). A set of these tapes can be fixed into a single unit and moved along a wedge to show positions of positive and negative interference.

A further use of these tapes is to demonstrate the integral units of DeBroglie wave lengths as the quantized orbits of the Bohr atom.

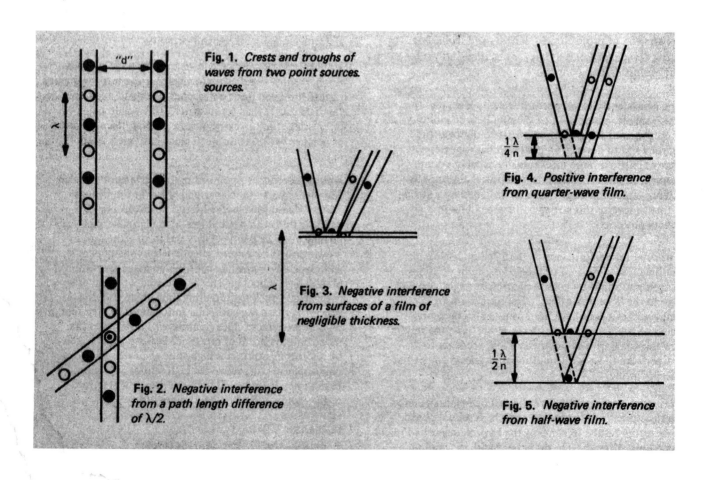

Fig. 1. *Crests and troughs of waves from two point sources.*

Fig. 2. *Negative interference from a path length difference of $\lambda/2$.*

Fig. 3. *Negative interference from surfaces of a film of negligible thickness.*

Fig. 4. *Positive interference from quarter-wave film.*

Fig. 5. *Negative interference from half-wave film.*

Drawing wave diagrams for the interference of light

Frank A. Anderson
The Bolles School, Jacksonville, Florida 32217

Most students find it difficult to grasp the idea of the interference of light because it is difficult to "see." Presented here is a more effective means of explanation — a means by which drawings of light waves may be constructed so as to show the interference of light.

"Fixed" diagrams are drawn in order to represent a picture of a wave train at a particular instant in time. There is no attempt to show the wave at t_1 and then again at t_2, a short time later.

The nomenclature and sign convention are as follows:

Crest and Trough — A crest is located to the left of the axis as one looks along the direction of propagation of the wave. A trough is located to the right of the axis as one looks along the direction of propagation of the wave. These are illustrated in Figs. 1 and 2.

Destructive and Constructive Interference — Interference is a result of adding two or more waves. Destructive interference occurs when the crest of one wave and the trough of another are in the same space at the same instant of time. Constructive interference is a condition in which the crests of two waves or the troughs of two waves are in the same space at the same time. This is shown in Figs. 3 and 4.

Starting Crest and Starting Trough — A starting crest is a condition such that if one moves a pencil along the wave in the direction of propagation of the wave, the pencil would move across the axis and start to form a crest. Point *B* in Fig. 1 is such a condition. The same applies to point *B* in Fig. 2. Henceforth this condition will be called an SC (Starting Crest). A starting trough is a condition such that if one moves a pencil along the wave in the direction of propaga-

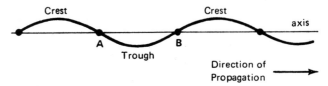

Fig. 1. *The crest is located on the left of the axis and the trough is on the right side of the axis when one looks in the direction of propagation of the wave. Point A is where a trough is starting and point B is where a crest is starting.*

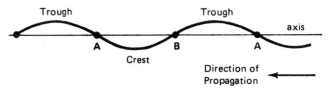

Fig. 2. *As one looks along the direction of propagation of a wave, crests are located on the left and troughs are located on the right. Points A are ST's and point B is an SC.*

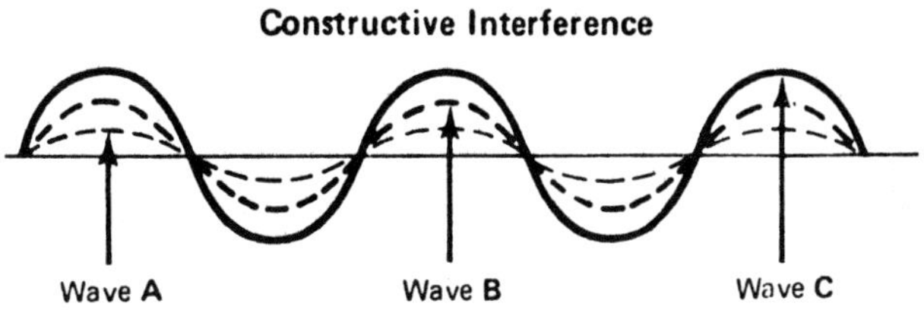

Constructive Interference

Fig. 3. *Wave C is the resultant obtained by adding wave A and Wave B. The crest of wave A is on the crest of wave B. The same applies to the troughs. This is constructive interference.*

tion of the wave, the pencil would move across the axis and start to form a trough. Point A in Fig. 1 and point A in Fig. 2 indicate this condition. Henceforth this condition will be called an ST (Starting Trough).

When reflected light is turned at a boundary it may be reflected with a change in phase or it may be reflected without a change in phase. When light is reflected from a boundary going from a more dense optical medium to a less dense optical medium, there is no change of phase in the reflected wave. In order to show this with a wave diagram, draw the incident wave so that at the surface of reflection the wave is starting to cross the axis. It makes no difference whether this intersection is an SC or an ST. Refer to Fig. 5. Point A represents an ST. Since the wave is being reflected as it goes from a more dense optical medium to a less dense medium, there is no change in phase. The reflected wave is drawn as if it were a continuation of the original wave. Since the incident wave is starting to form a trough, then the reflected wave should also be starting to form a trough.

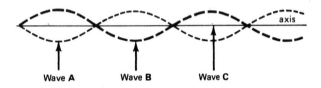

Fig. 4. *The crest of wave A is on the trough of wave B. Both wave A and wave B have the same amplitudes. When the two waves are added the result is total destructive interference.*

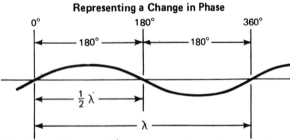

Fig. 6. *An SC and an ST are separated by a distance of one-half wavelength. They are also separated by 180°. A change from SC to ST can be equated to a change in phase of 180° or a path difference of one-half wavelength.*

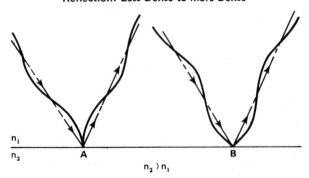

Fig. 7. *When light is reflected as it ravels from a less dense optical medium to a more dense medium a change of phase is observed. This change of phase is depicted by changing an ST on an incident wave to an SC on the reflected wave (point A). At point B the SC on the incident wave connects with an ST on the reflected wave.*

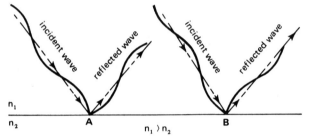

Fig. 5. *When reflection occurs as light goes from a more dense to a less dense optical medium, no change of phase occurs. At point A both waves are in the same state, an ST condition. At point B both waves are in an SC condition.*

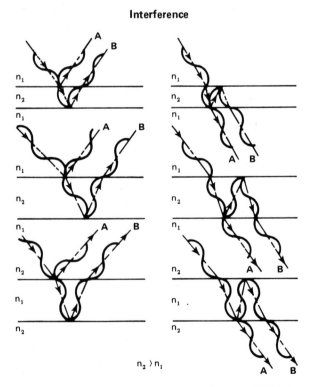

Fig. 8. *The left column shows the interference of light that is reflected in a thin film. The right column shows interference of light transmitted through a thin film. In practice waves A and B would be close enough together to cause interference.*

Starting at point A in Fig. 5 it can be seen that the reflected wave is starting to form a trough according to the definition previously stated.

Consider point B in Fig. 5. The wave at this point is in a condition of an SC. The reflected wave should also be in a similar condition since there is no change in phase. Point B on the reflected wave is an SC point.

When light is reflected at a boundary going from a less dense optical medium to a more dense medium, there is a change in phase. This condition is usually equated to a path difference of one-half wavelength or to an angular change of 180° (see Fig. 6). In this system such a change of phase is represented by the reflected wave in a condition opposite to the incident wave. If an incident wave were starting to form a crest the reflected wave would be in a state where it is starting to form a trough (see Fig. 7). Consider point A on the incident wave. This corresponds to an ST. Point A on the reflected wave corresponds to an SC, or just the opposite condition. The difference between an SC and an ST is one-half wavelength. Thus, a change in phase has taken place. Just the opposite is true of point B. An SC on the incident wave is an ST on the reflected wave.

This method of drawing waves has been most useful to show constructive and destructive interference when light is reflected in various thin films. In the examples drawn in Fig. 8 it is assumed that the incident waves of light are normal to the surface. In order to make the drawings more easily understood, the angle of incidence has been exaggerated. Not indicated in the diagrams is a change in wavelength when the light enters a substance of different optical density.

The reflected rays in the diagrams are too far apart to interfere with each other. Again, this is done in order to make the diagrams more understandable. In actual conditions the two waves would be close enough so that the addition of the waves would take place.

This model for the reflection of light is very useful in the explanation of such phenomena as interference in thin films, interference due to double and single slit diffraction, Lloyd's mirror, and other phenomena of interference due to either a change in phase or a difference in path.

Long Lasting Soap-Gelatin Films
Paul A. Smith
Coe College, Cedar Rapids, Iowa 52402.

With the solution given in the fourth paragraph below soap-gelatin bubbles have been blown which have lasted up to ten days while resting on untreated 3 × 5 in. file cards. Portions of bubbles have been mounted in 1 in. diam holes in 3 × 5 in. file cards, and when mounted between pieces of flat glass (with cards having 1.5 in. holes on each side to keep the film away from the glass—slots from hole to edge for ventilation) have survived more than five years of general handling.

Flat sheets of the long lasting thin film may be formed on a wire (or model airplane rubber band) ring by dipping it in the solution. A rubber band ring may be stretched out (in a square frame by four strings to the corners) before the film dries in order to get extra large uniform sheets; up to 2 ft square may easily be made.

Both bubbles and flat sheets of film may be made with a thickness less than a quarter wavelength of visible light, as indicated by the loss of interference colors on drying. Multi-MΩ resistors of about 5% stability in resistance value may be made from strips of flat film mounted on black photo album pages.

To prepare the solution, mix the following in the order indicated: (1) 60 cc cold water (tap water works satisfactorily); (2) one tablespoon Knox pure gelatin; (3) when thoroughly mixed, heat to about 90 °C *double boiler style* to dissolve gelatin and let bubbles rise; (4) add 9 cc glycerin and 3 cc Joy liquid detergent and stir *gently* to avoid making a pesky foam. Keep warm for use and dilute with water as you wish. Reduce glycerin content to make hard dry films.

Bubbles blown with this solution need to be kept suspended in air for a few minutes to let them dry out before mounting them on cards or letting them roll across the floor. If they bounce off the card on which you want to mount them, put a damp spot on the desired mounting location. A flat film in a frame should be allowed to dry only partially before blowing it gently against a sheet of black photo album paper to mount it. The glass of an eye dropper put in the end of a short piece of rubber hose will be helpful in blowing the periphery of sheets of film against their paper mounts.

This formula was developed by the author while a high school student and used as an entry in a science fair in Illinois in 1952. A simpler formula has been given by Sharkey[1].

Reference
1. B. Sharkey, The Physics Teacher 3, 285 (1965).

SOAP FILM INTERFERENCE PROJECTION

This note is concerned with one of the classics of lecture demonstrations: the projection of beautiful colors due to the interference of reflected light from the two surfaces of soap films. This demonstration has been in the literature for many years and therefore the content of this note concerns technique rather than subject.

The method used is basically outlined in Sutton and somewhat improved by Highsmith[2] and others.[3] However, because of the dual problem associated with air conditioning (a low relative humidity and air currents) Dr. Dudley Williams of Kansas State made the following suggestion: The reliability of the demonstration would probably be increased by constructing a protective Lucite box in which the wire frame for the soap film could be suspended.

The Lucite cube offers the following advantages: With the transparent cube serving as the frame support, there is increased student visibility with the elimination of a ring stand, rod, and clamps. The walls of the cube completely eliminate troublesome air currents. The relative humidity of the atmosphere inside the cube can be increased by placing water in shallow dishes in the bottom of the chamber, thereby increasing the evaporation time. The soap film is illuminated by an ordinary slide projector from the front (thereby protecting the student's eyes from the intense light). When the thin-film frame is aligned along the diagonal of the cube, the reflected light coming out the side of the cube can be focused (a 33-cm focal length plano-convex lens was used) on a projection screen at one side of the lecture hall. The students can easily see the brilliant colors as different orders of interference occur while the vertically suspended soap film drains down.

One can also arrange the apparatus so that both the incoming light and the reflected light pass through the same side of the cube. This alternate arrangement cuts down on the focusing difficulty. However, some shielding is necessary to protect the student's eyes. While the room arrangement might determine which set-up method to use, one would normally have to choose between the following: an easily-seen apparatus arrangement with an adequate display of the desired effect; or a slightly better display with partially hidden apparatus.

One can, in addition, by utilizing these controlled conditions, obtain with repeated success a soap film with the thickness of only one-half wave length. It is very dramatic to see the brilliant colors change to black along the definite boundary where the thin film is one-half lambda thick. These nearly-ideal conditions allow the soap film to drain off to the point where the thickness of the entire film is one-half lambda—with all the reflected light appearing black due to destructive interference.

Our Lucite box is ten inches on all edges. If the back wall of the cube is made flat black (by either painting or inserting black cardboard), an appreciable amount of undesirable reflected light is eliminated. The framework for the thin film is made of brass welding rod and is suspended from a Lucite plug in the top of the box. With this particular arrangement using no clamps, the demonstration can be repeated quickly and easily after the soap film breaks. The frame is approximately two-inches square. Larger frames were tested, but focus problems occurred as a result of the large depth-of-field. The soap solution, called Wonder Soap Bubbles, is

WATER LENS

Sometimes the most simple demonstrations are the best. Make a water lens by placing a penny in a test tube with the Lincoln head up. Fill the test tube about half full with water and look down into the tube at the coin. Because the water miniscus has a concave shape, the coin looks very small and far away. Now fill the test tube to the top and add an extra few drops of water to overfill the tube and obtain a convex surface. The coin will now appear much larger and closer.

I would like to thank David B. Scott of Seattle, Washington, for showing me this simple apparatus.

JOHN B. JOHNSTON
Nanuet High School
Nanuet, N. Y. 10954

STANDING WAVES ON THE OVERHEAD PROJECTOR

Standing waves may easily be demonstrated using ropes, springs, etc. The explanation of this phenomenon in terms of interference between two waves traveling in opposite directions is more elusive. Introductory texts[1,2] usually employ rather complicated multi-image, time-sequenced graphs which many students have difficulty following. Moiré patterns analogous to standing waves may be produced using pocket combs.[3] In this note, a method to show clearly the superposition of two waves traveling in opposite directions is described.

An overhead projector equipped with two rollers for moving an acetate strip across the stage of the projector is needed. Cut two pieces of the acetate from the roll, each about one meter long. Draw or trace an accurate sine curve along the center of each piece using a different colored felt tip pen for each. The sine curves should have equal wavelength (about ten centimeters) and equal amplitude (about two centimeters). Tape one end of each strip to one of the rollers and overlap them on the stage of the projector (Fig. 1). By turning the roller knobs slowly in opposite directions, alternative constructive and destructive interference of the two moving waves can be seen on the screen.

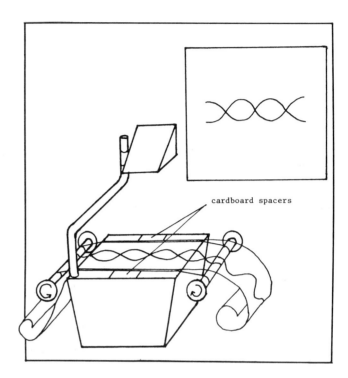

Fig. 1. Overhead projector with two sine curves moving in opposite directions.

An additional overlay can be used to show more clearly the formation of the nodes and antinodes. It consists of a graph grid transparency and two transparencies (in a third color) showing the sums of the two moving waves when they are in phase (Fig. 2). Each of the summation curves has the same wavelength as the moving waves and twice the amplitude. They are one-half cycle out of phase with each other. These are taped onto opposite edges of the graph grid which is placed over the moving strips. Cardboard spacers (Fig. 1) taped to the top of the projector stage allow the moving strips to slide beneath the overlay and keep the moving strips aligned. Initially, the moving strips should be adjusted to give antinodes at the same points on the graph as either of the summation transparencies. Then, as the moving strips are advanced in opposite directions at the same rate, the summation transparencies can be flipped into position alternatively at half-cycle intervals to show the nodes and antinodes.

I have found this visual aid to be quite helpful to students in understanding the formation of standing waves. They also see easily that the distance between adjacent nodes (or antinodes) is one half the wavelength.

References

1. F. Miller, *College Physics* (Harcourt, Brace Jovanovitch, New York (fourth edition), 1977) p. 248.
2. D. Halliday and R. Resnick, *Physics* (John Wiley and Sons, New York (combined edition) 1963) p. 413.
3. R. Prigo and R. Wormsbecher, Phys. Teach. 15, 187 (1977).

William Warren, *St. Christopher's School, Richmond, Virginia 23226*

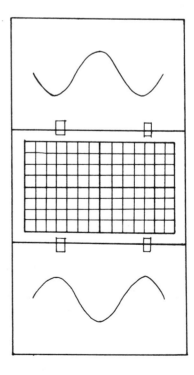

Fig. 2. Overlay transparencies.

SINE WAVE ANALOG

One of the items developed and constructed at the Ontario Science Centre for large audience demonstrations is a large sine wave analog.

The mechanism is housed in a box about 1 ft-8 in. deep, 2 ft high and 4 ft wide. The face consists of a black painted surface with three horizontal rows of verticle slots through which protrude the ends of wooden bars covered by brightly colored discs. At the left end of the apparatus (Fig. 3), are three hand wheels and two levers for operating the mechanism.

Figure 3A shows the machine in a storage condition with levers A and B in the up position, thereby raising all bars off their respective cams. When lever A is turned down, the bars in the center row are lowered and rest on their cams. The first and second rows assume a wave-like pattern as indicated in Fig. 3B. If lever B is turned down instead of A, the lower row of bars will rest on their cams and the first and third rows assume wave-like patterns. In either of the above conditions a traveling wave may be created by turning cranks 2 or 3. Crank 1 reverses the direction of travel of the wave in the center row. When levers A and B are turned down as in Fig. 3C the top row assumes the pattern of the sum of the two lower patterns. A traveling wave, the sum of the two lower waves, may be obtained by turning cranks 2 and 3. To create an undulating standing wave,

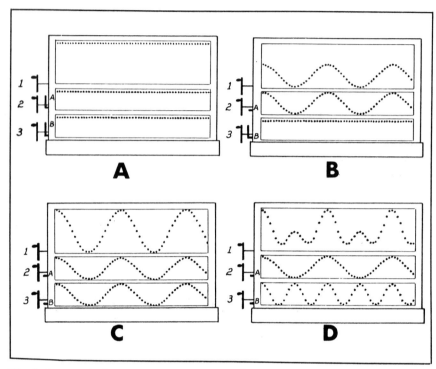

Fig. 3. Large scale demonstrator represents traveling waves and wave interference by turning cranks.

created by waves moving in opposite directions, cranks 1 and 3 are turned in the same direction.

Another wave length may be obtained by turning crank 3 clockwise and crank 2 counterclockwise at the same time (Fig. 3D).

Although a set of drawings does not exist for this machine, information for its construction may be obtained from the author. The whole mechanism is constructed of wood with the exception of three shafts and a few very simple metal parts which may require the use of a machine lathe and a drill press.

DONALD H. TROLLOPE
114 Douglas Avenue
Toronto, Ontario, Canada

Images from a Piece of a Lens

Samuel Hirschman
Forest Hills High School
Forest Hills, New York

Lens light ray diagrams show image formation by means of the principal rays (i.e., the one parallel to the principal axis and the other through the optic center). Many students then get the idea, from the ray diagram, that the whole lens participates in the image formation, but they arrive at the right conclusion for the wrong reason. When we draw rays to the lens parallel to the principal axis from the extremes of the subject, it appears that both the "top" and the "bottom" halves of the lens are employed.

We should ask students to predict the description of the image formed by half of a lens. "Will we get half of an image or a complete image but half the size?" Or, "Describe the image if the lens were smashed and we used only a large sliver."

To cut a large lens in half or smash it is a needless waste. Use a large magnifying glass (about 4-in. diam) to project the image of a carbon filament lamp onto the wall of the darkened classroom. Project the image again after masking half of the lens. Finally, use a cardboard mask which only exposes a jagged sliver of the lens.

The Psychedelic Student-Getter

Struck by the beauty of cellophane tape patterns viewed through crossed polarizers, and aware of the difficulty class officers were having in obtaining decorations for a school dance entitled "Kaleidoscope," I hit upon a happy idea. I brought cardboard slide mounts and cellophane tape to school and passed these out so that each student could create his own design. Eighty-one of these slides filled the tray of the school's Carousel slide projector. One polarizing sheet was placed between the condensing lens and the slide, the second polarizer consisted of eight small 2 in. × 2 in. squares mounted on a wheel which could be rotated in front of the lens by a small motor. The projector automatically changed the slides every 15 sec, while the rotating polarizer changed the colors several times a second—in time with the beat of the band, yet! This was projected on one wall of the gym, and the place really *jumped*. The fascination of this display, especially when timed to the deafening roar of a teen-age combo, has to be experienced to be appreciated; and the interest it generated in physics is quite incredible.

William R. Franklin
Westchester Senior High School
Spring Branch Independent School District
Houston, Texas

Optics in a fish tank

Richard Breslow
The University of Connecticut, Stamford, Connecticut 06903

The refracting properties of lenses are most always investigated using lenses of glass or plastic. Most students are taught that a double convex lens is convergent and a double concave lens causes divergence of light.

The lens maker's equation

$$\frac{1}{f} = (n-1)\left(\frac{1}{R_1} - \frac{1}{R_2}\right)$$

Fig. 1. Unaided view of object.

Fig. 2. Object viewed through double convex air lens.

Fig. 3. Object viewed through cylindrical air lens.

in which f is the focal length, n is the relative index of refraction and R_1 and R_2 are the radii of curvature of the surfaces indicates that the characteristics of a lens depends not only on the curvature of the two surfaces but is also dependent upon the relative index of refraction of the lens with respect to the surrounding medium.

It is interesting to observe the change in the image position that occurs when one immerses both an object and lens in water. If this can be demonstrated it may lead into a discussion as to why under water the human eye cannot focus sharply on an object. Some student will recall that most of the focusing of the eye takes place at the corneal surface when the eye looks out into air.

What is even more interesting is to observe the refracting effects of lenses made of air immersed in water. Pairs of watch glasses will act as air lenses under water if taped together to form a waterproof trapped air space.

However, an excess fish tank from the biology laboratory, a few flasks from the chemistry stockroom and a bell jar from our own stockroom quickly illustrates the interesting focusing effects shown in the accompanying figures.

Figure 1 shows the "object" viewed directly while Fig. 2 shows how the double convex air lens behaves as a diverging lens. In Fig. 3 the results of viewing the object through a cylindrical air lens shows the astigmatism reversed from what it would be for a cylindrical glass lens. For comparison, Fig. 4 shows the magnification obtained when the object in air is viewed through a water lens.

My thanks to Bruce Cuseo for taking the photographs.

Fig. 4. Object viewed through water lens.

Optical effects in a neutral buoyancy simulator

Van E. Neie
Physics and Education Departments, Purdue University, W. Lafayette, Indiana 47907

The accompanying photo (Fig. 1) from NASA has some rather interesting optical effects. The astronaut, mission specialist Dr. Shannon Lucid, is working on the Space Telescope in a Neutral Buoyancy Simulator. Note the size of her head relative to the rest of the body. The plastic helmet (which contains air for breathing) forms a spherical air-water interface with the air being on the concave side of the interface. Because the index of refraction of air is less than that of water, the image formed is virtual, erect, and has a magnification less than one. This reduction is quite evident in the photograph. If we assume that the head occupies a space within 2/3 the radius of the helmet, the magnification would vary from about 0.6 on the far side to about 0.9 at the near side (based on formulas related to refraction at spherical surfaces).

If the astronaut were outside the tank, where air (or a vacuum) made up the external environment, no such effect would be observed. Moreover, if the helmet were filled with water (momentarily, of course!) the air-water environments would be reversed from that of the photograph and we would observe an image having a magnification greater than one, a kind of fish-bowl effect. In the present case, the effect appears unusual because we aren't accustomed to looking from water into air.

This phenomenon can be used as a basis for fruitful classroom discussion. A line of questioning might include:

1. Do you see anything unusual in this photograph? (Many of my colleagues did not, at first.)
2. How do you explain the smallness of the head compared to the rest of the body? — or — What does the smallness of the head suggest in terms of the environment in which she is working? (If the bubbles are not noticed, the student may be unaware that the astronaut is working underwater.)
3. Can you draw a ray diagram that at least qualitatively illustrates the observed effect? (Fig. 2)

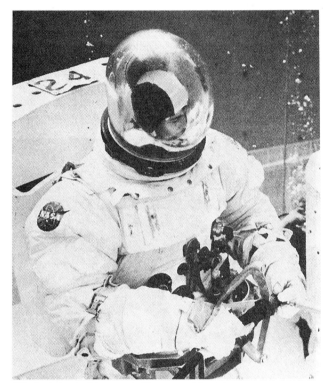

Fig. 1. Astronaut in a neutral buoyancy simulator.

4. From the point of view of Dr. Lucid, would the external surroundings be distorted, e.g., would the telescope on which she is working appear larger or smaller than its actual size?

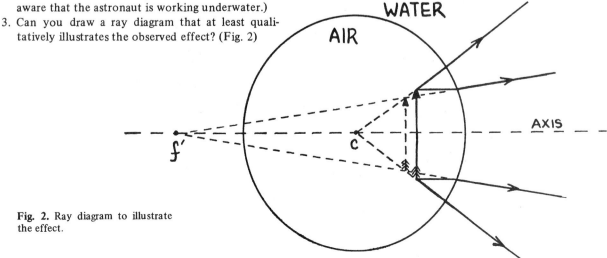

Fig. 2. Ray diagram to illustrate the effect.

A Problem in Image Formation

Mr. Hyman A. Cohen of H. Frank Carey High School, Franklin Square, L. I., New York, has submitted a note on a demonstration in optics as follows:

"In the course of teaching 'Optics' to my high school physics class, I demonstrate the formation of real images by projecting the image of a lighted bulb in the direction of the class—through a large reading glass lens. It occurred to me to use a 'water lens,' made by pouring water into a 500 ml round flask, to show that water can be used as the refracting medium in a lens.

"I noticed that, from my side, I could see two images being formed by each lens (as seen in the photograph)."

He then describes the images seen from *his side* of the arrangement.

Four images are observed as shown in the photograph below. The photograph—not the easiest to obtain—was made by a student, John Meyer of H. Frank Carey High School. The photo was "retouched" by the editor.

Mr. Cohen, in the note, gave a partial explanation of the images. It appeared to the editor that the complete explanation of this demonstration required rather full understanding of the principles of geometrical optics and some ingenuity in their application to a real situation. Hence, the partial explanation given by Mr. Cohen is omitted and the problem is left as an "exercise" for our readers.

The high school teacher who shortly sends us the explanation will be given a free 1966 subscription to *The Physics Teacher* or, if he is a member of A.A.P.T., his prize will be a book costing $5.00 (or more). The explanation should account for the type of image (real or virtual), the size, the relative location with respect to the observer and its orientation (erect or inverted). Some numerical values in an assumed arrangement would be in order.

The curvature of the flask surface is greater than that of the convex lens surfaces. The lens is relatively thin.

Note the reversed orientation of the large and small images as seen "in" the flask and lens.

Ray models of concave mirrors and convex lenses

Robert Gardner
Salisbury School, Salisbury, Connecticut 06068

A light box such as the one described in an earlier article[1] can be used to demonstrate how concave mirrors and convex lenses form real images or virtual images depending on the distance between the object and the lens or mirror.

A four-slit mask (Fig. 1) can be made by using a razor blade to cut narrow slits in a piece of black construction paper. One pair of slits is covered with a small sheet of colored plastic or cellophane.

The mask, when placed over an opening in the light box, produces two pairs of rays; one pair will be colored. Two mirrors, held in position by small lumps of clay, can be used to deflect one ray of each pair so that it crosses the other member of the pair. These crossed rays may now represent two points of light — one at the top and one at the bottom of an "object" as shown in Fig. 2.

Upon striking a cylindrical mirror, the rays reflect to form a "real image" as you can see in Fig. 3. It is clear that the image is inverted and real. The size of the "image" can be controlled by moving the mirror closer to or farther from the "object." When the mirror is one focal length from the object, the reflected rays will be parallel, indicating that the image is at infinity. As the mirror is moved still closer to the object, the reflected rays diverge suggesting a virtual but erect image that appears to be behind the mirror.

A clear water-filled plastic jar can be used as a cylindrical lens. In this case the "lens" can produce a real image as shown in Fig. 4. Again, by moving the lens closer to or farther from the object, the nature of the image (real or virtual) as well as its size and position can be changed.

Reference

1. R. Gardner, "Light box — versatile but inexpensive," Phys. Teach. **13**, 50 (1975).

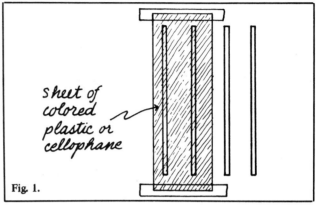

Fig. 1.

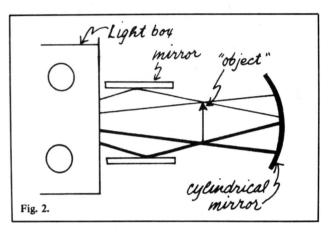

Fig. 2.

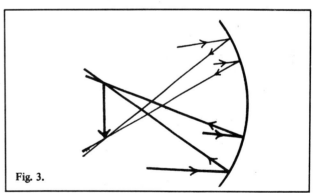

Fig. 3.

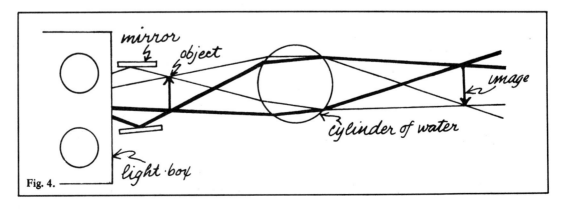

Fig. 4.

Binocular vision – A simple demonstration

Richard Didsbury
Science Department, King's Edgehill, Windsor, Nova Scotia B0N 2T0

The usefulness of two eyes for the quick and accurate determination of distance is well understood.[1]

A rather simple, yet convincing demonstration of this is to toss, without warning, small objects (pieces of chalk are very convenient) at various members of your class. Most students have no trouble at all in grabbing the object out of mid air. Next, have each student close one eye and repeat the experiment. The students are usually quite amazed that such a seemingly simple task has now become rather difficult.

Reference

1. Uri Haber-Schaim, Judson B. Cross, John H. Dodge and James A. Walter, *PSSC Physics* (D. C. Heath and Company, Lexington, Massachusetts, 1971), Third Edition, pp. 19-20.

Physics for automobile passengers

Elizabeth A. Wood
New Providence, New Jersey 07974

While driving along a dark road in a lightning storm, I observed several discrete images of the windshield wiper during each flash of lightning that appeared single to the eye. Moving-film photographs have shown that lightning flashes are composed of multiple strikes, but the windshield-wiper observation is one that anyone can make. Ambient darkness is essential so that the wiper is not visible between strikes.

This is an amusing reversal of the usual stroboscopic experiments in which the experimenter provides the stroboscopic lighting to observe rapid motion. Here the experimenter provides the rapid motion to observe Nature's stroboscopic lighting.

THE CHESHIRE CAT[1]

The disappearing dropper described earlier[2] can be followed by a demonstration of the delightful scene described by Lewis Carroll in *Alice's Adventures in Wonderland*. In Chapter Six, "Pig and Pepper," Alice was upset because the cat kept appearing and disappearing suddenly. "'All right,' said the cat; and this time it vanished quite slowly, beginning with the end of the tail, and ending with the grin, which remained some time after the rest of it had gone."

"'Well! I've often seen a cat without a grin,' thought Alice; 'but a grin without a cat! It's the most curious thing I ever saw in all my life.'"

Using a high-speed hand grinder I formed a resemblance of the Cheshire Cat on a piece of window glass cut to fit the bottom of a glass dish (Fig. 3). The shallow dish can be placed on an overhead projector so the cat figure can be shown to a large audience. With the glass tilted at a small angle, Anisole[3] or a mixture of carbon tetrachloride and bromobenzene[4] is slowly added to the dish. [Editor's note: Carbon tetrachloride and bromobenzene have been labeled carcinogenic and banned from school use in some states.] As the liquid rises over the cat figure the ground portion becomes transparent and disappears on the screen.

In order to keep the grin from disappearing I outlined the mouth of the cat figure with a soft lead pencil. As the liquid with an index of refraction very close to that of the glass covers the cat figure only the pencil marking shows on the screen and hence we have a grin without a cat.

Walter C. Connolly, *Appalachian State University, Boone, North Carolina 28608*

Fig. 3. Photo courtesy of John Dinkins, University Photographer.

References

1. Lewis Carroll, *Alice's Adventures in Wonderland* (Random House, New York, 1946).
2. Walter C. Connolly and Thomas L. Rokoske, "The Disappearing Dropper," Phys. Teach. **18**, 467 (1980).
3. Anisole (Methyl Ether – $CH_3OC_6H_5$)
4. A mixture of CCl_4 and C_6H_5Br to give an index of refraction of 1.526.

THE DISAPPEARING DROPPER

The disappearing dropper in a dropping bottle[1] is one demonstration that has been popular with all audiences since we introduced it 15 years ago. This is probably because it borders on the unbelievable and until explained or thought about, it might as well be magic.

The dropper in the flint glass dropping bottle is made of borosilicate glass and has an index of refraction of about $n = 1.517$. Anisole[2] has an index of refraction of $n = 1.517$ and is a clear colorless liquid. Therefore with a dropping bottle filled with anisole the dropper can be made to disappear by forcing the air from the dropper and then releasing the bulb. If one views the bottle by transmitted light one can still see the outline of the dropper because of dispersion. However, if one views the bottle by reflected light the outline of the dropper is very difficult to see and so appears to vanish.

The method we use for this demonstration is to hold the bottle with one hand such that the top is held by the thumb and forefinger (Fig. 7). This places the bottle in front of the palm of the hand and prevents transmitted light from dominating the reflected light. Now when the bulb is squeezed and air is forced out of the dropper the disappearance of the end

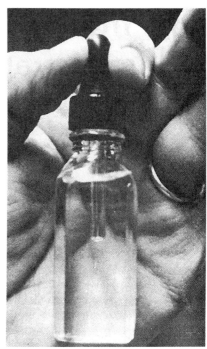

Fig. 7. Dropper becomes invisible as liquid enters.

of the dropper is very striking.

After being in the bottle for some time, the anisole will become discolored due to a reaction with the rubber bulb of the dropper. This discoloration does not reduce the effectiveness of the demonstration but can be prevented by storing the anisole in a separate container between demonstrations.

The pictures are due to the courtesy of John Dinkins, University News Photographer.

Walter C. Connolly
Thomas L. Rokoske
Appalachian State University
Boone, North Carolina 28608

References

1. Dropping bottle – Curtin Matheson Scientific Co., P. O. Box 43528, Atlanta, GA; Catalog Number #034-793; 30 ml $11.04/dozen.
2. Anisole (Methyl Phenyl Ether – $CH_3OC_6H_5$) – Fisher Scientific Co., P.O. Box 11666, 3315 Winton Road, Raleigh, NC 27604; Catalog Number A-834 $22.95/Qt.

RIPPLE TANK PROJECTION WITH IMPROVED CONTRAST

Ripple tanks which use overhead projectors can have greatly improved contrast if a vertical mask is placed in the center of the projector lens towards the screen.

A couple of fingers placed across the lens gives the same result and may be used to verify the effectiveness.

It seems that the information that we want most is contained in the scattered light which is gathered by the edge of the lens. The light that travels straight through the center of the lens only bleaches out the image.

Before and after pictures (Figs. 5 & 6) clearly show the results.

Fig. 5. *Typical image of double slit interference in a ripple tank lacks contrast when lens of the overhead projector is left wide open.*

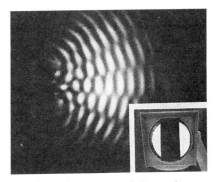

Fig. 6. *Covering part of the projector lens with a piece of masking tape greatly improves contrast.*

ROBERT W. SMITH
Georgia Institute of Technology
Atlanta, Georgia 30332

How the world looks underwater — A demonstration for nonswimmers

Samuel Derman
Department of Technology, City College of New York, N.Y. 10031

How does the world look to an underwater swimmer — or to a fish? The answer can be found in many basic physics texts. But it is possible to get a firsthand look yourself, and a bathing suit won't be necessary.

The apparatus to be described, although trivial to construct, nevertheless permits a variety of optical experiments to be performed, not only in class, but also by students at home.

Lay a pocket mirror face up at the bottom of a flat pan full of water (Fig. 1). Poke your finger into the water and the mirror will show by reflection how your finger appears to a submerged fish. The arrangement of Fig. 1 shows what the fish would see looking straight up, perpendicular to the surface.

But suppose the fish decides to direct its view at some arbitrary angle to the surface. That situation can be duplicated too.

Simply lean the mirror against the side of the pan (Fig. 2). As you change the angle of inclination of the mirror you simulate the variation in angle between the fish's line of sight and the water's surface. Looking into the mirror you see just what the fish sees as it looks up towards the surface (at some angle).

In fact, at a particular angle, total internal reflection will occur. The surface of the water (viewed from underneath) will become mirror-like and will reflect any object situated below. Poke your finger through the surface again and the mirror will show you the odd-looking, symmetric object drawn in Fig. 3. Or place a coin at the bottom of the pan (Fig. 2), and you will be able to see it by looking into the mirror.

It's important to realize that you can see the top of the coin even though the mirror is tilted upward — an indication that light rays from the coin first traveled *up* towards the water's surface, then were totally reflected *down* towards the mirror, and finally were reflected *up* again, towards the eye.

To view the coin by total internal reflection it's best to proceed as follows. First, lay the mirror flat, near one wall of the pan. Next, lay the coin down flat also so that the distance between coin and mirror is about 3½ inches (center of coin to nearest edge of mirror). Lastly, slowly tilt the mirror upwards until the coin comes into view.

Many interesting effects are possible. For example, once you've determined the proper angle for total internal reflection, set the mirror at that angle by leaning it against the side of the pan. Use a small stone or weight to prevent the mirror from slipping, as shown in Fig. 2.

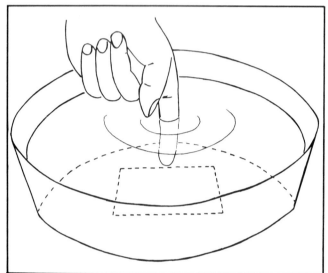

Fig. 1. Poke your finger into the water and look down into the mirror. You will see what a submerged fish would see — a finger pointing at you.

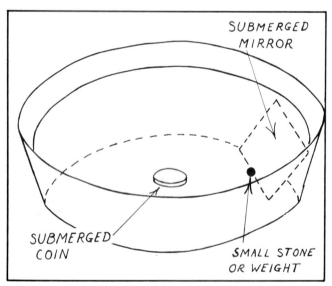

Fig. 2. Lean the mirror against the side and you can simulate what the fish sees when it looks obliquely towards the surface. At a particular angle of inclination total internal reflection occurs.

Next, put your hand into the water and slowly raise the submerged coin until it's near the surface. To keep the coin in view while it is being raised, the angle of inclination of the mirror may have to be varied. The mirror will show you two coins, and two hands, because now, two reflections occur simultaneously (Fig. 4). One is direct reflection by the mirror. The other is the double reflection described earlier, that is, from the coin to the surface of the water, to the mirror.

Note that the setup of Fig. 2 can be used as a simple solar spectroscope.[1] If sunlight from a window is allowed to fall obliquely on the water's surface, a full spectrum of colors will be observed on the ceiling or opposite wall of the room. The darker the room, the more dramatic is the effect. By varying the angle of inclination of the mirror you can shift the position of the image on the ceiling or wall.

Reference
1. S. Derman, Phys. Teach. **16**, 58 (1978).

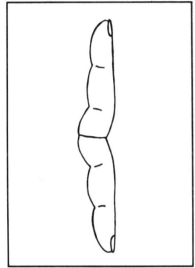

Fig. 3. Total internal reflection prevents the mirror from showing you anything above the surface. Poke your finger into the water, and this time you see only the part beneath the surface, together with its reflection.

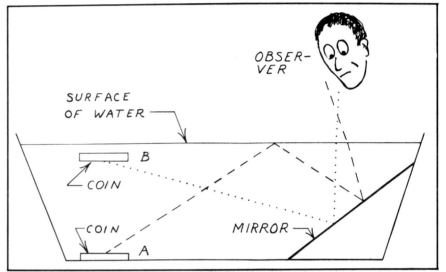

Fig. 4. When the coin is in position A (on the bottom), light rays (dashed lines) from its upper face undergo total internal reflection at the surface before striking the mirror. The rays are then mirror-reflected out to the observer. When the coin is raised to position B, rays leave from both its upper and lower faces. Rays from the upper face (not shown) are totally internally reflected as before, and reach the observer. However, now the lower face is also visible (rays shown dotted).

The physics of visual acuity

Michael J. Ruiz

Department of Physics, The University of North Carolina, Asheville, North Carolina 28814

A very interesting optics experiment can be performed using an eye-test chart, a ruler, and a calculator. A satisfactory eye chart can be obtained from Sargent-Welch.[1] The objective of the experiment is to determine an estimate for the limit of visual acuity.

The student measures the length of the largest letter on the chart and notes at which distance the letter should be discernible. This minimizes experimental uncertainty which increases when measuring the smaller letters. For a letter which should be read at 50 ft by a normal eye, one finds that the linear dimension of the letter is 22 mm. Figure 1 shows how a simple proportion can be used in order to obtain the length of the real image of the letter on the retina. The length of the eye can be taken to be approxi-

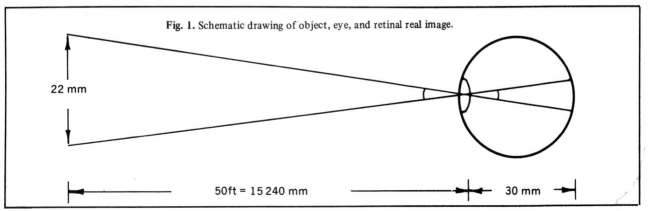

Fig. 1. Schematic drawing of object, eye, and retinal real image.

mately 30 mm. We immediately find the following proportion:

$$\frac{22 \text{ mm}}{15 \ 240 \text{ mm}} = \frac{x}{30 \text{ mm}}$$

$$x = 40\mu$$

where 1μ is one micron, which is one thousandth of a millimeter.

In order to distinguish a P from an F the observer must see clearly one half of the letter. The image of this portion of the letter has a linear dimension of 20μ. Since the mean spacing between the centers of the light receptors is about 3μ, one half of the letter is approximately 6 receptors in length. The student can now draw an array of circles as a model to represent the array of receptors in the retina. The student should recognize two points. First, for 20/20 *normal* vision, an individual is required to discern letters when the image of one half of the letter spans 6 receptors. Second, for the *best possible* vision the minimum number of receptors, in principle, needed to distinguish a P from an F is 3 x 3 receptors (Fig. 2). This problem is analogous to representing letters or numbers on the minimum number of light-emitting diodes in a calculator.

Most people cannot resolve a P or F when half of the letter falls on the minimum 3 x 3 receptors and the test for 20/20 vision does not require them to do so. However, some individuals have better vision than the normal 20/20 and the aim of the experiment is to place a limit on the best vision possible.

A reasonable estimate of the limit of visual acuity corresponds to the illustrated case where the upper half of the P or F falls on 3 x 3 receptors, i.e., the retinal image is one half of the size it would be for 20/20 vision. Therefore, the limit is realized by standing 100 ft from the chart rather than 50 ft in an attempt to identify a 22 mm letter. The visual acuity for an eye that can read the 20/a line of a Snellen chart at b feet is 20/20 (a/b). Our rough calculation

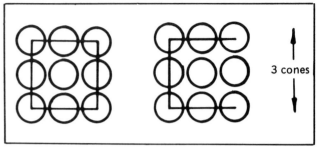

Fig. 2a. Part of the letter P.   Fig. 2b. Part of the letter F.

indicates that the limit of visual acuity is approximately 20/10. It is interesting to allow the students to test each other. From time to time a case of about 20/10 vision will be found.

The limit of resolution based on diffraction can be found using the Rayleigh criterion

$$x_d = 1.22 \frac{\lambda}{d} f = 1.22 \frac{(500 \text{ nm})}{2 \text{ mm}} (30 \text{ mm}) = 9\mu$$

using a wavelength of 500 nm (green light) and a pupil diameter of 2 mm (small opening for high light level). This surprisingly spans 3 cones. It is remarkable that the limit for visual acuity corresponds to the limit of resolution obtained from diffraction arguments.

The author is indebted to David S. Falk for sharing his ideas on resolution of the human eye and diffraction which he discusses in a very popular nonscience course "Light, Photography, Perception, and Visual Phenomena" at the University of Maryland. The author would also like to thank Mark Clark (MD) of Anderson, South Carolina for discussions on the physiology of the eye and visual acuity from a medical viewpoint.

Reference

1. Cat. No. 3539. Sargent-Welch Scientific Company, 7300 N. Linder Avenue, Skokie, IL 60077.

VISUAL ACUITY REVISITED

It occurred to me that some readers might be concerned about my rough estimate of f = 30 mm for the focal length of the human eye[1] since the light rays of Fig. 1, Ref. 1 should cross at some point behind the cornea, resulting in a better estimate of f = 20 mm.[2] If we tighten up on our estimate of the focal length, we need to be more precise in our estimate of the minimal length of a discernible retinal image. Carefully taking the linear dimension of the letter in Fig. 2, Ref. 1 as the distance between the centers of the extreme cones, we find an image 6μm in length (2/3 smaller than the rough estimate of 9μm in Ref. 1). This factor of 2/3 cancels the factor of 20/30 = 2/3 introduced by the improved focal length (see the diffraction equation), leaving the conclusion of Ref. 1 essentially the same: the limit of visual acuity is approximately 20/10.

Michael J. Ruiz, *Department of Physics, University of North Carolina, Asheville, North Carolina 28804*

References

1. M. J. Ruiz, "The physics of visual acuity," Phys. Teach. 18, 457 (1980).
2. T. N. Cornsweet, *Visual Perception* (Academic Press, New York, 1970) p. 40.

PYREX "VANISHING SOLUTION"

In studying large scale optical phenomena it is often desirable to eliminate the interface effects of a Pyrex container: for example,[1] a round-bottomed flask used as a double convex air lens. If one were to fill the flask with a liquid whose average refractive index (for white light) matches the refractive index of the glass, all interface (glass-liquid-glass) effects are minimized (Fig. 6). Such a solution is easy to prepare and use; in fact, glassblowers commonly make use of it to determine if glass stock is Pyrex. In such a solution the Pyrex is invisible but common soft glass is quite visible. This effect itself makes a striking classroom demonstration.

One such solution may be prepared by mixing approximately 2.3 volumes of 2-propanone (acetone) (CH_3COCH_3) with 10 volumes of benzene (C_6H_6). The most rapid method of preparation is to add less than the recommended volume of acetone, use a Pyrex tube to stir the solution, and then add the acetone by drops until the rod is invisible. "Overshooting" can easily be corrected by the addition of a small amount of benzene.

For two miscible compounds, a and b, the refractive index of the mixture (n) may be written

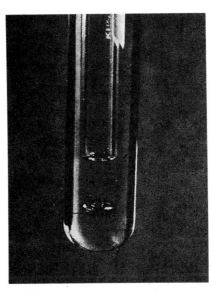

Fig. 6. Inner test tube partially filled with refractive solution has drop of mercury at the bottom. When it is immersed in a larger test tube with the same refractive solution, a portion of the inner test tube disappears.

$$n = x_a n_a + x_b n_b$$

where x_a and x_b are the mole fractions of a and b and n_a and n_b are the refractive indexes of a and b respectively. The compounds a and b are chosen such that $n_a < n < n_b$. One can then easily show

$$\frac{V_a}{V_b} = \frac{M_a \rho_b (n_b - n)}{M_b \rho_a (n - n_a)}$$

where M is the molecular mass (grams/mole), ρ is the density (grams/cc), V the volume (cc) and the subscripts again denote compounds a and b. Caution should be exercised whenever one is using volatile organic compounds; most organic chemists classify all such compounds as either food or poison, and inhalation is a most rapid means of ingestion.

F. J. WUNDERLICH
D. E. SHAW
M. J. HONES
Villanova University
Villanova, Pennsylvania 19085

Reference

1. *Optics in a Fish Tank*, Richard Breslow, Phys. Teach. 14, 234 (1976).

A SAFE PYREX "VANISHING SOLUTION"

I was surprised to see *The Physics Teacher*[1] publish an article suggesting the use of such an acutely toxic substance as benzene. I realize that the implication was that the benzene would be used only by the instructor, but unfortunately, clever demonstrations have a way of becoming lab experiments and student projects. Although the article contained a paragraph pointing out the need for caution when using benzene, the nature of the compound was not made clear; e.g., the fact that benzene is readily absorbed through the skin.

In the earlier note 2-propanone (acetone) (CH_3COCH_3) and benzene (C_6H_6) were suggested as index-matching liquids for Pyrex glass. Since these two organic compounds have indexes of refraction for the Sodium D line of 1.3588 (acetone) and 1.5011 (benzene) they bracket the index of Pyrex and fused silica ($n_D \simeq 1.46$). Unfortunately, both benzene and acetone are highly volatile, and in addition benzene presents a severe biological hazard and is a known carcinogen.[2-4]

The purpose of this note is to point out the easy availability of nonvolatile, nontoxic organic compounds with appropriate indexes of refraction. All of the compounds listed below are available from Cargille[5] and may be ordered by specifying "Index of Refraction end members" and the code listed below.

Code S10 and Code 50 are miscible and are therefore suitable for index matching for n from 1.399 to 1.458. Code 50 and Code 40 are useful for matching from 1.458 to 1.570. Code S10 and Code 40 are not miscible although they may be in the presence of the Code 50 oil. Since these oils are relatively viscous, molecular mixing takes place as much as a minute after mechanical mixing, so a slow approach is recommended.

When mixing liquids to match the index of refraction of a given type of glass, a quick test of the match can be made with the aid of a low-powered laser. Pass the laser beam through a glass rod, lens, or prism that is immersed in the liquid. Any deviations of the beam from a straight line indicate a mismatch. When the match is perfect, the laser beam will remain perfectly straight through any portions of the glass and liquid.

DONALD K. DAY
Montgomery College
Rockville, Maryland 20850

References

1. F. J. Wunderlich, D. E. Shaw, and M. J. Hones, "Pyrex Vanishing Solution," Phys. Teach., 15, 118 (1977).
2. L. J. Casarett and J. Doull, editors, *Toxicology: The Basic Science of Poisons*, (Macmillan, New York, 1975).
3. N. Irvine Sax, *Dangerous Properties of Industrial Materials*, 3rd Edition, (Van Nostrand Reinhold, New York, 1975).
4. C. Gosselin, et al, *Clinical Toxicology of Commercial Products*, 4th Edition, (Williams and Wilkin, Baltimore, 1976).
5. R. P. Cargille Laboratories, Inc., 55 Commerce Road, Cedar Grove, New Jersey 07009.

Code S10	Silicon oil	n = 1.399	$12.00/16 oz.
Code 50	Mineral oil	n = 1.458	$ 8.00/16 oz.
Code 40	Mineral oil	n = 1.570	$15.00/16 oz.

OPTICAL ASTIGMATISM MODEL

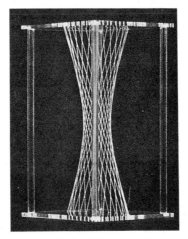

Quarter view.

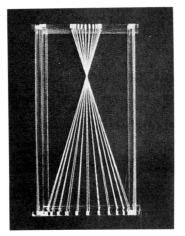

Front view.

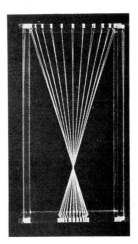

Side view.

Visualizing optical astigmatism has always been difficult for the student and worker in geometrical optics. There are few effective drawings in text books, and demonstration devices are generally inadequate.

A manufacturer of educational visual aids is offering a unique aid for visualizing astigmatism. It is a string model of a quartic surface which accurately demonstrates how light rays focus in horizontal and vertical directions in planes at different distances from the light source. It also clearly shows the circle of least confusion between these two planes. The model shown in the illustrations is made of clear plexiglas so that the "light rays" can be viewed from any direction. Educational Products Corporation (Box 9632, Station A, Colorado Springs, Colorado 80909)

WAVE MOTION DEMONSTRATOR

Devices for demonstrating the phenomena of one-dimensional wave propagation have been known at least since the time of Lord Kelvin. The complexity and expense of many of these wave machines has discouraged their widespread use. We describe here the construction and use of an unusually simple and inexpensive device suitable either for lecture demonstrations or an introductory laboratory course.

The wave machine consists of a length of Mylar tape recorder tape, one end attached to a support and the other to a removable weight. Along the tape at roughly 3-cm intervals are arrayed a series of colored plastic drinking straws. If the weight at the bottom is quickly twisted around a vertical axis and returned to its initial position, a wave pulse will propagate slowly up the tape and be reflected from the top (Fig. 1).

Best results have been obtained with a wave machine roughly 2 m in length. Obviously, considerable flexibility in materials is possible.

The first stage of construction is to cut two small slits in the center of each straw for the tape to pass through. This is best done with a sharp scalpel or X-acto knife. The slits should be 3/8-in. to 5/8-in. long, straight, and as parallel as possible to the axis of the straw.

A small paper clip, or other small mass, should then be inserted into both ends of each straw to increase the moment of inertia. A bit of colored adhesive tape may be wrapped around one end of each straw to make the wave motion more visible.

To thread the Mylar tape through the straws, simply squeeze each straw near the slits to open them slightly and pass the tape through

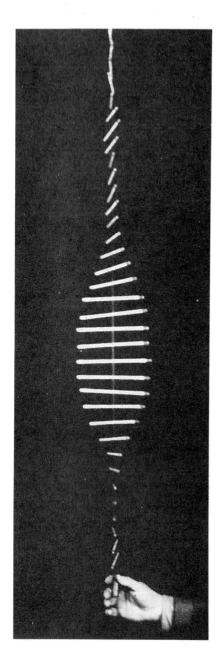

Fig. 1. *A wave pulse viewed edge on.*

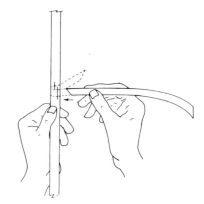

Fig. 2. *Threading the tape through a straw.*

in the manner shown in Fig. 2. If the tape is inserted in a direction tangent to the side of the straw, it will slide around the inside and exit through the other slit. A little practice will make the operation almost automatic. The Mylar tape should be wrapped around the end straws rather than passed through them and then clamped with the binder clips as in Fig. 3.

The tape may then be hung from some support, a weight attached to the bottom, and

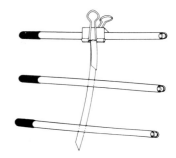

Fig. 3. *The upper end of the wave machine. The tape is wrapped once around the topmost straw and clamped tightly with a binder clip.*

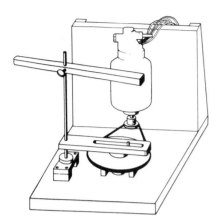

Fig. 4. Sinusoidal drive for generating standing waves.

the straws arranged at regular intervals. When this is done and the straws made approximately horizontal, the wave machine is ready for operation!

A few precautions should be taken to make the demonstrations of wave phenomena as dramatic as possible. Since the wave machine is so light, it is very sensitive to air currents and care should be taken not to hang it in a draft. The upper support should fasten to the binder clip securely so that the topmost straw cannot turn. A less rigid support will lead to poor pulse reflection properties. When undisturbed, the wave machine may hang in a "twisted" configuration; it may be straightened out by holding the weighted end in the proper orientation. The bottom of the wave machine will be essentially free if the weight has a small enough moment of inertia around its vertical axis. In practice, a 3/4 in. diam cylinder of brass attached *firmly* to the bottom clip will cause downward propagating pulses to be reflected without a change of sign. Finally, the wave effects are most visible when the wave machine is viewed edge-on from some distance.

To observe the propagation of a wave pulse, quickly rotate the weight around a vertical axis through 90° from its equilibrium position and then rotate it back. A distinct wave pulse will propagate up the wave demonstrator with velocity

$$V = W\sqrt{TD/12I}$$

where W is the width of the tape, T the tape tension, I the moment of inertia of each straw, and D the distance between straws. The propagation velocity is slow enough for convenient visual observation. The wave machine, constructed from the materials which are listed in Table I, had a wave speed of 23 cm/sec.

Many experiments showing the propagation and reflection properties of wave pulses require no additional equipment. For example, removing the weights from all of the straws in part of the wave machine introduces an impedance mismatch leading to visible reflections. The properties of standing waves may be demonstrated by applying a periodic displacement to the lowest straw either manually or with a simple sinusoidal drive (Figs. 4 and 5).

We hope that many physics teachers will find in this simple and inexpensive wave demonstrator an interesting way to enliven the discussion of some of the fundamental phenomena of physics.

MARC D. LEVENSON
Harvard University
Cambridge, Massachusetts 02138

Table I. *Materials required for wave demonstrator.*

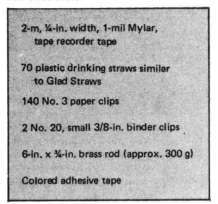

2-m, ¼-in. width, 1-mil Mylar, tape recorder tape

70 plastic drinking straws similar to Glad Straws

140 No. 3 paper clips

2 No. 20, small 3/8-in. binder clips

6-in. x ¾-in. brass rod (approx. 300 g)

Colored adhesive tape

Reference

1. R. Resnick and D. Halliday, *Physics* (Wiley New York, 1966), Chap. 19.

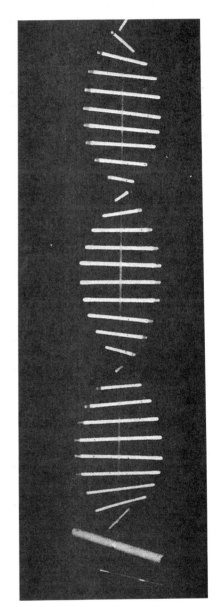

Fig. 5. Standing waves on the wave machine.

STANDING WAVE ANALOGY USING POCKET COMBS

In a past note in this journal,[1] Peter Melzer introduced a simple demonstration of beat production using two pocket combs. The overlay of two combs of slightly different teeth spacing produced the desired moiré beat pattern. A similar moiré beat pattern can also be seen in the overlay of two combs of identical teeth spacing inclined relative to each other. Furthermore, the number of beats varies with inclination angle. Quantitatively, the projection of the angled comb superimposed on the other produces a beat frequency f_b given by

$$f_b = f(1-\cos\theta)/\cos\theta,$$

where f is the comb frequency (the number of teeth/inch) and θ the angle of inclination, assuming an observation perpendicular to one of the combs.

Two combs can also be used to demonstrate standing waves. Superimposed combs of equal teeth spacing, if moved slowly over each other in opposite directions, will produce a "standing" moiré pattern. The frequency of the oscillating light and dark pattern matches the frequency of the individual comb teeth, as demonstrated by varying the relative speed of the combs. In fact, combs divided into two regions of different teeth spacing work best.[2] When moved across one another, each region "oscillates" at a visibly different standing wave frequency (Fig. 3).

When used in conjunction with Melzer's beating combs, these standing wave combs can help clarify the different conditions responsible for the production of each phenomenon as well as their common base in the superposition principle.

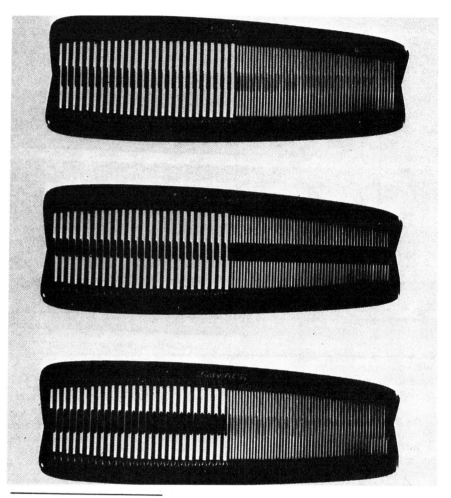

Fig. 3. Two standing moiré patterns oscillate at different frequencies as double combs are moved across each other.

References

1. P. Melzer, Phys. Teach. 14, 120 (1976).
2. For example, "Lady ACE" combs.

ROBERT PRIGO
RICHARD WORMSBECHER
*Department of Physics
University of California
Santa Barbara, California 93106*

*Present address: Quantum Institute and Department of Chemistry, University of California, Santa Barbara, CA 93106

A POTPOURRI OF PHYSICS TEACHING IDEAS—OPTICS AND WAVES

BEAT PRODUCTION ANALOGY USING POCKET COMBS

Two combs of equal lengths but of differing number of teeth make a useful apparatus for demonstrating the principle behind the production of beats. The moiré pattern obtained in superimposing the two combs yields a number of dark fringes equal to the difference in the number of teeth (Fig. 3). The teeth, of course, represent the number of waves in a wave train of one second duration. Two combs of differing lengths may also be used as long as the teeth spacings are also different. In this case it would be advisable to display the superimposed combs against a mask of arbitrary length. The mask would then represent a one second time interval.

A similar apparatus consisting of two combs was recently described by M.R. Wehr to illustrate the difference between group and phase velocities.[1]

PETER MELZER
Bronx High School of Science
Bronx, New York 10468

Reference

1. H. Meiners, Editor, *Physics Demonstration Experiments*, (Ronald Press, New York, 1970), p. 480.

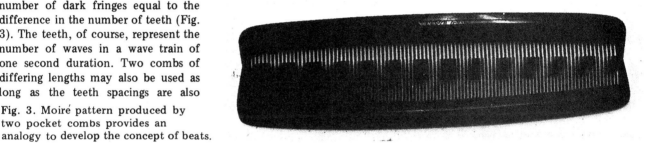

Fig. 3. Moiré pattern produced by two pocket combs provides an analogy to develop the concept of beats.

Hot Standing Waves

Harold C. Jensen
Lake Forest College, Lake Forest, Illinois 60645.

A versatile apparatus can be assembled using the basic idea that was recently suggested by Ohriner.* I used #24 (0.5-mm) nichrome wire. A variac was employed for power and adjusted so the wire glowed dull red when not vibrating (about 5 A). As can be seen from Fig. 1, approximately 200 g supplied the tension, and the length of the vibrating part of the wire was adjustable from 1 to 4 ft. Because the wire becomes hot when the alternating current is applied, it was necessary to suspend the wire a short distance above the magnet. The effect was dramatic as the wire vibrated and formed standing waves. Glowing continued at the nodes but cooling at the antinodes was extraordinary.

We demonstrated:
(a) the heating effect of electric currents;
(b) the effect of temperature on length (the wire expanded about ¼ in.);
(c) the effect of tension on a vibrating wire;
(d) the effect of wire length on the standing waves;
(e) the interaction between electric currents and a magnetic field;

Fig. 1. Vibration at antinodes cools red hot wire in contrast to the glowing nodes as wire vibrates in magnetic field.

(f) that there is cooling at the antinodes where vibration is at a maximum; and
(g) that there is no cooling at the nodes where vibration is at a minimum.

_____
* M. Ohriner, The Phys. Teacher 5, 287 (1967).

VERSATILE MOUNT FOR SLINKY WAVE DEMONSTRATOR

This Slinky wave-motion demonstrator has the advantages that it a) will demonstrate both longitudinal and transverse waves with either nodes or antinodes at the ends, and b) is easily stored by sliding the runners to the end of the rail and holding the spring in contact with the wall with a crocodile clip.

The suggestion by Kirwan and Willis[1] is useful but the suspended spring tension of a vertical Slinky will vary from ceiling to floor causing a change in the velocity of the wave as it proceeds along the spring. Since both ends are fixed it is not possible to use the spring to demonstrate antinodes at the ends of the spring. Reynold's[2] suggestion of bifilar suspension is fine for longitudinal waves but will not allow free-moving transverse waves.

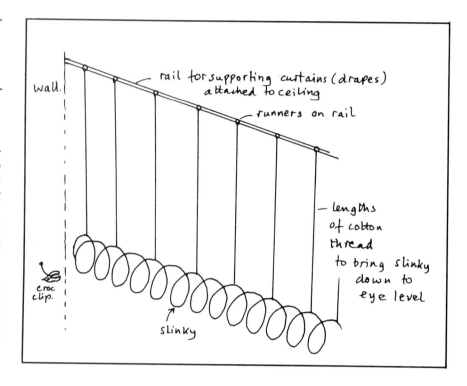

John F. Spivey, *Berry Stores Cottage, Thorverton, Exeter EX5 5NU, Devon, England*

References

1. D. F. Kirwan and J. Willis, Phys. Teach. **17**, 471 (1979).
2. Leon M. Reynolds, Phys. Teach. **16**, 652 (1978).

Cigar Box Spectroscope

Brother Shamus, C.F.X.
Notre Dame High School, Utica, New York

Cut slits at opposite ends of a cigar box and then glue a piece of diffraction grating over one of the slits, as shown in Fig. 3. For convenience, centimeter tape may be attached to the front end to measure the distance from the slit to the spectral line. With the slit aimed at a distant source, one student locates the spectral lines in the box and another student moves a wire indicator along the front edge of the box until the first student sees the indicator and the spectral line coincide. One can then easily calculate the $\sin\theta$ from the ratio X/L and substitute in the familiar equation $n\lambda = d \sin\theta$. Using this equipment, students measured the yellow line of helium (5876 Å) with the largest error in a class of fifty students being 4.5%.

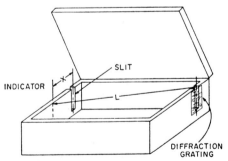

Fig. 3. Cigar box spectroscope.

Finding the principal focus with laser and thread

Kenneth Fox

Smoky Hill High School, Cherry Creek School District, Aurora, Colorado 80015

The principal focus of a parabolic reflector is an elusive point at best, and hard for some students to visualize. A previous article in *The Physics Teacher* by Layman showing students locating images with long strings[1] made me consider the use of a gas laser and threads to show the convergence of light rays (themselves an elusive idea!).

The laser was set on a dynamics cart which had books stacked on it. In a semidarkened room, the laser was directed toward a large spherical concave mirror,[2] bought for general demonstration, set on the far end of the table (Fig. 1). The beam should hit above the center of the mirror so that the reflected beam hits the table.

The laser is first aimed at the upper left side of the mirror, about 10 cm from the edge. A student takes a white thread and sticks it to the mirror with a small piece of clay so the thread just touches the red spot made on the mirror by the beam. A second student sets the thread's other end on the table where the reflected beam hits, pulls the thread taut (gently), and fastens it with another piece of clay. The thread will "sparkle" with reflected laser light.

The cart is now rolled to the right to a second and third position, being careful to stay clear of the mirror edges since the mirror is not parabolic. Again threads are stretched and to the amazement of all (especially me the first time), the threads really do cross at one point (Fig. 1).[3]

This was enough for me, but not for some of my

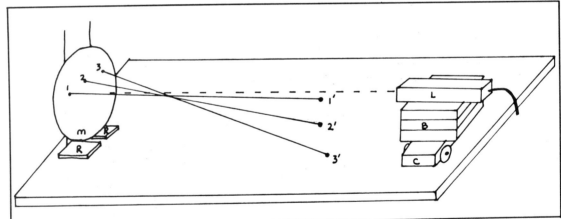

Fig. 1. Arrangement of apparatus.

L laser
B books
C cart
M curved mirror
R ring stand

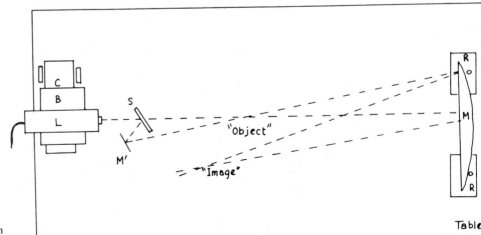

Fig. 2. Locating the image of an object.

C – cart; B – books; L – laser; M – curved mirror; R – ring stand; S – beam splitter; M' – flat mirror.

favorite skeptics. What about when the beam is on a second, lower parallel plane? A ring stand, with a strip of paper taped to it, was set so the paper was located at the point where the threads intersected. As books are carefully slipped out and as the cart rolls back and forth the spot on the paper stays remarkably motionless, considering the rough approximation of parallel rays. With this setup, the deviation from a parabolic reflector near the edges also becomes very apparent.

A similar setup with a beam splitter (Fig. 2) can be used to show that two rays that start at one point together (the "object") will indeed converge at another point (the "image") after reflection. A smoke cloud, or chalk dust, helps show this. I enjoyed this demonstration and so did my classes. Best of all, though, they seem to appreciate the existence of the principal focus as a "real" point.

References

1. Phys. Teach. 17, 253 (1979).
2. Sargent Welch Scientific Company 3525-A.
3. The threads converged 35 cm from the mirror's surface. The house across the street formed an image 36 cm from the mirror.

Paper waves

Richard A. Lohsen
Department of Physics, United States Military Academy, West Point, New York 10996

The picture shows how paper waves can be used to demonstrate Young's double-slit experiment. This three-dimensional model may help some students who have difficulty in visualizing the phenomena if they see only static text diagrams.

To make the paper waves, cut a sinusoidal wave form into the edge of each of two long strips of paper, giving both the same wavelength. This can be easily done if they are cut out like a long string of paper dolls.

To illustrate the double slit experiment, draw the two slits on the blackboard, and tape up the strips. Illustrate the central bright spot by pulling both strips out to a central point, which can be seen to be equidistant from the slits because it is the same number of wavelengths from each. The paper waves represent a snapshot of the two separate wave forms. Point out that at the central point, chosen so it occurs at peaks in both waves, the displacement of each wave varies in time, giving constructive interference. Then follow through by finding the first dark spot. Keeping the

paper strips taut, move to the first spot where a peak of one wave matches up with a trough of the other wave. Now, discuss how each wave varies in time at this point, and how the waves destructively interfere. Point out that the path lengths differ by half of a wavelength. Now the general relationship of the location of the bright and dark spots to the path length difference can be illustrated.

Traveling waves on a rope

Keith Fillmore
Department of Physics, Saint Mary's University, Halifax, Nova Scotia, Canada B3H 3C3

Propagation of a wave in one dimension is readily demonstrated on a rope under tension. The usual procedure is to attach one end rigidly, then pulling on the other end to provide tension (and at the same time keeping the rope reasonably horizontal), one gives a vertical flip and observes a pulse travel down the rope. Almost any rope will work if not too stiff; I have even made do with the power cord of an overhead projector.

One readily discovers, if one didn't know already, that the pulse is reflected from the fixed end and travels back toward the hand with reversed amplitude. While this reflection may be interesting in itself, it is rather a nuisance if one wishes to demonstrate the propagation of a continuous sinusoidal wave or "traveling" wave. The reflected wave is superimposed on the transmitted wave, and what one actually observes is a standing-wave pattern of nodes and antinodes. Although one can easily prove mathematically that a standing wave is the sum of two traveling waves, the fact remains that one does not *observe* a traveling wave.

Using a longer rope doesn't help much. To avoid excessive sag under its own weight, one must increase the tension, which in turn increases the speed of the waves, and the wave arrives at the fixed end just as quickly as for a shorter rope. The ideal "infinitely long" rope often shown in textbook diagrams is difficult to approximate in the laboratory.

Here is a simple method to show real traveling waves. Using a flexible rope about 8 to 10 m long, tie one end to a table leg *near the floor*. Apply a tension at the other end of the rope to keep it fairly taut, and hold the end just high enough above the floor so that about one quarter of the rope lies along the floor, and the rest is suspended in air. The floor should be quite smooth, such as tile or wood, but not carpeted. Now at the hand-held end apply a sinusoidal transverse motion in a *horizontal* plane. Waves travel down the suspended portion of the rope and are then damped out progressively in the portion lying along the floor, thus eliminating any reflected wave. A little experimentation will soon determine a suitable combination of tension and hand height above the floor. A rather vigorous motion of the hand is required since it must move through the full amplitude of the wave. In comparison, for standing waves, a small hand motion produces large amplitude because of resonance.

This demonstration nicely illustrates the propagation of energy in a wave. The source of the energy will be readily apparent to the person providing the aforesaid vigorous motion. The energy then propagates along the rope and is converted to thermal energy as the rope moves against friction on the floor.

I actually discovered this demonstration in a playground. Some kids were playing a game called "rattlesnake," in which one produced the kind of rope waves just described, and the others had to jump back and forth across the "snake" without getting "bitten." Might be a good way to introduce the subject.

LASER GIMMICK

A simple and inexpensive way to split a laser beam into several beams for mirror and lens demonstrations is to stack four or five microscope slides together with strips of cardboard sandwiched at one end to form

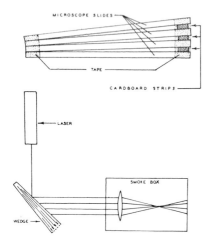

Fig. 1. Reflections from microscope slide wedge provide multiple rays for optics demonstrations. The slightly divergent rays are near enough to parallel for the purposes of demonstrations.

a compound wedge (Fig. 1). The *reflected* components give nicely separated beams sufficiently intense for clear demonstrations using an aquarium or plastic box containing a little smoke.

WILLIAM H. PORTER
Gilman School
Baltimore, Maryland 21210

LASER BEAM SPLITTING FOR THE STUDENT LAB

When an institution has only one laser but would like to run a number of optics experiments at one time it is very easy to use a series of beam splitters to accomplish this. The beam splitters do not have to be of exceptionally good quality. An ordinary piece of flat glass will do.

If the flat glass is fairly thick one gets a reflection from both the front and the back surface. A slit can be used to eliminate one of the beams. From there it is a simple matter of using diverging/converging lenses to expand the beam to a desired diameter.

While teaching an evening laboratory course in optics at Arizona State University, I found myself with only one operating helium-neon laser. We wished to run Lloyd's mirror, Fresnel's bi-prism, Fresnel's mirror, and other experiments during one night's laboratory. Using the idea above, I was able to do all the experiments at the same time with enough intensity to project the patterns on a white screen for measurements.

ROBERT J. JULIAN
Universiti Pulau Pinang
Palau Pinang, Malaysia

LASER SAFETY GOGGLES: AN UNNECESSARY EXPENSE

In response to current advertising, many students and teachers are purchasing goggles for use with 1-mW helium-neon lasers at $20 or more a pair. Although such goggles do not seriously impair the visibility in the classroom or the laboratory, they attenuate the laser beam (6328 Å by a factor of at least 1000. Thus unless a student has a special need to look directly down the barrel, the beam of a laser of 5-mW or less output power becomes completely invisible. The overall effect is identical to that of operating a laser with the power switch off (but at a much greater cost). The Radiation Safety Act of 1968 does not require, or even recommend, the use of laser goggles. Any advertisements that make statements to the contrary should be brought to the attention of the physics teachers' association.

HERBERT H. GOTTLIEB
Apparatus Editor

Long-playing diffraction grating

Scott R. Welty
Physics Department, Lewis University, Romeoville, Illinois 60441

When doing interference and diffraction experiments I find that it is helpful to be able to measure something of common interest. Usually the experiments involve finding the wavelength of light or the grating spacing. These are two quantities in which the students tend to have little interest. By using a 33-rpm record for a reflection grating, I find an automatic interest simply because of the familiarity of the record. Also, with the record there is an independent way to check the measurement of grating (groove) spacing.

We set up the record approximately 4 m from the blackboard with the laser on the front desk near the board (Fig. 1). With the room darkened and a white sheet of paper taped to the board, fine adjustments of the laser trained on a particular band of grooves will result in a fairly wide-spaced pattern of dots (about 1 cm apart) on the paper. Using the standard equation for diffraction gratings, one can solve for the grating spacing, d.

$$\Delta x = \frac{\lambda L}{d}$$

Where:

Δx = distance between successive maxima.
λ = wavelength of laser.
L = distance from record to screen.
d = grating spacing (distance between grooves).

This calculation can then be checked easily and independently by using the frequency of revolution (33-1/3 rpm), the width of the band used, and the time given on the label for that number. Then:

$$d = \frac{\text{(width of band)}}{\text{(no. of revolutions for band)}} \quad (2)$$

or:

$$d = \frac{\text{(width of band)}}{\text{(frequency)(time for band)}} \quad (3)$$

We consistently achieved agreement between these

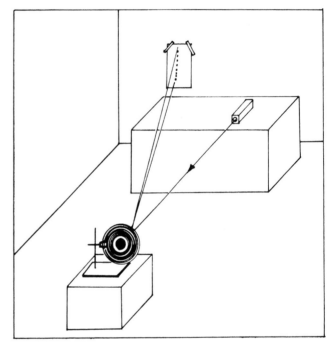

Fig. 1. Arrangement for measuring groove spacing of a long-playing record.

two measurements to about 5%. Care should be taken that the selection from which the laser beam is reflected is the same as the one whose groove spacing is measured using the second method. Also, there is quite a bit of "noise" in the interference pattern and some practice is advised to be able to pick out the regular sequence of dots that represent the diffraction pattern.

CALIBRATING AN INEXPENSIVE STROBE LIGHT

Inexpensive strobe lights are not calibrated accurately enough for photographic illumination when time measurements are desired. I discovered that these strobes can be calibrated within pencil line accuracy if one has access to a good audio oscillator. Connect a small loud speaker to the output of the audio oscillator and set the oscillator to the frequency that you wish to use for the strobe photography (such as 20 or 30 Hz). Put your ear to the case of the strobe and press the speaker to the case. The strobe may now be "tuned" to the audio oscillator by listening for the absence of any beat frequency between the two sources. A piece of tape on the strobe may now be marked for that particular frequency.

ALFRED M. EICH
C. F. Brush High School
Lyndhurst, Ohio 44124

MERCURY VAPOR STREET LAMPS

P. Wentworth reported in *The Physics Teacher* (May 1970) that interference experiments can be carried out with a mercury vapor street lamp. I should like to present some other experiments conducted with this kind of lamp.

The outer cover of the lamp was cut off with a diamond saw at a glass shop. The rim of the cover and the glass protruding from the metal base were care-

The lamp is connected to the mains in series with the ballast coil that was obtained with the lamp. It takes the lamp about 4 min to warm up. This lamp, as any other mercury light source, will not ignite immediately after it has been switched off because of the high pressure of the hot vapor.
fully polished. A cylindrical shield for the inner tube was made of iron sheet 1.5 mm thick with a window and exchangeable diaphragms. A round support for the shield or the outer cover was welded to the metal base of the lamp. As the temperature is very high when the lamp is switched on, both the shield and the support must be welded with brass.

Figure 1 shows the details of the modified lamp. The shield has small ventilation holes in the upper cover for cooling.

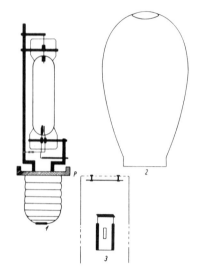

Fig. 1. Mercury vapor lamp intended for outdoor illumination adapted for classroom use as a spectral source. (1) Inner mercury lamp with welded base. (2) Outer globe cut with a diamond saw and removed from lamp. (3) Metal cylinder with window serves as an alternate cover.

The following experiments may be recommended:

1. The lamp with the cylindrical shield and a slit diaphragm is placed in front of the collimator of a prism spectroscope and the spectrum is projected on a screen. Nine mercury spectrum lines are clearly visible.

2. An historic experimental arrangement for demonstrating the photoelectric effect is shown in Fig. 2. The cathode is a freshly polished 10×10 cm Zn plate. The anode is made of a fine copper grid fixed to a wire frame and its surface area is the same as that of the cathode. Both electrodes are mounted on insulating stands. Any regulated dc power supply is suitable.

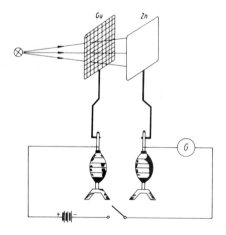

Fig. 2. Apparatus setup for photoelectric demonstration.

An experiment was carried out at our laboratory using a galvanometer with a sensitivity of about 3×10^{-9} A per scale division. With the electrodes 1 cm apart and the lamp 1 m away from the anode, the light spot was deflected up to 50 divisions when the voltage was varied from 0 to 320 V. A better visual effect may be obtained with a more sensitive mirror galvanometer whose light spot can be projected on a screen. Ultraviolet rays may also be demonstrated with a quartz prism. If this is not available, the light may be dispersed with a common glass prism and the ultraviolet radiation nearest to the violet end may still be detected. This ultraviolet region is assessed to extend from 4000 to 3150Å. A simple detector of this radiation is a sheet of blotting paper saturated with potassium iodide and a few droplets of one per cent starch solution. Good effect is obtained with the following solution: 25 cc of 1% iodide were mixed with 25 cc of dilute sulfuric acid (5.3g H_2SO_4 per 100 g of water) and a few droplets of 1% starch solution. If the treated paper is attached to the screen beyond the violet part of the spectrum, a strong bluish coloring occurs due to iodine freed by the uv photochemical dissociation. A fresh starched handkerchief placed in the ultraviolet region glows with a faint bluish light.

3. The near ultraviolet region is very effective in demonstrating luminescence. Many organic substances, e.g., hair, fingernails, teeth, etc. and fluorescent paints, glow when irradiated with these rays. Some invisible details of banknotes may also be shown in this light.

With an opaque screen to protect the students from uv rays, the inner tube may be lit. When it is in full glow, the teacher, wearing protective glasses, puts on the outer cover. The effect of the ultraviolet rays on the luminous substance of the outer cover is very impressive.

The light may once again be dispersed with one of the arrangements discussed above and an x-ray screen may be placed in the region next to the violet light. Two additional distinctly visible lines will be seen in the region of the near ultraviolet. In this experiment as well as in the one with the outer cover, we observe the transformation of the frequency of light.

ZYGMUNT PRZENICZNY
University of Poznan
ul. Grunwaldzka 6, Poland
Translated By Zenon Gubanski

DIFFRACTION OF LIGHT APPARATUS

The following is a simple and inexpensive method for demonstrating the diffraction of light to an entire class:

Each student is supplied with a variable width slit. The light source is on the table at the front of the room. The slit is formed by two unused, single-edged razor blades sandwiched between two microscope slides. The sandwich is held tightly together by plastic electrical tape wound around each end of the pair of glass slides (Fig. 4). The glass slides protect the student against the sharp edges and the tight friction fit permits the slit width to be varied easily, yet maintained at any desired setting.

An effective light source may be prepared by using a 25 W incandescent show case lamp (straight filament and tubular envelope) mounted vertically. Because the filament is not accurately straight, a good pattern can be obtained by masking the source except for, perhaps, the straightest 5-cm section. A 12 oz. soft drink "pop open" can serves as an excellent mask and filter holder.

First, remove the bottom of an empty soft drink can. The portion of the top that was torn open serves two purposes. The central round portion (where the tab was attached) becomes the suspension point to mount the can coaxially over the lamp. The remaining open sector serves for ventilation. A slit, 2 cm wide is cut up the side of the can and is masked by black plastic electrical tape to expose only the straightest 5-cm portion of the filament.

A filter set is taped over the slit to provide red, blue, and white light sources. Scraps of red and blue cellophane from candy wrappers proved very effective in producing

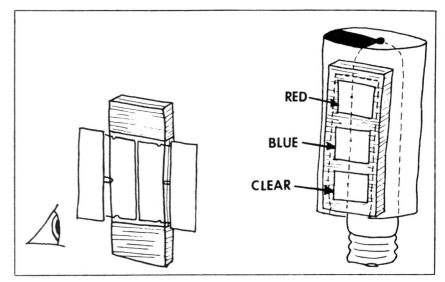

Fig. 4. Improved variable-width slits are adjusted by students at their seats while the instructor uncovers filters at the front of the room to demonstrate diffraction phenomena.

"monochromatic" light. Because the color densities of candy wrappers vary, it may be necessary to experiment by folding the cellophane until there are enough layers to filter the light while providing a diffraction pattern that can be seen from the farthest point in the room. The completed filter set, composed of a red, a blue, and a colorless area, is sandwiched between two glass microscope slides and sealed and masked by black plastic electrical tape.

A piece of black paper, held by a rubber band around the can, masks the filter so that different sections can be displayed independently or one above the other, for comparison.

The slit width can be adjusted easily by grasping the protruding spine of the razor blade on each side between the thumb and forefinger while pushing or pulling with the tips of these fingers against the edges of the slides. By first exposing only the red source, each student can see for himself how the width of the central maximum varies with slit width.

Sliding the mask down over the filter set exposes both the red and blue light sources and the diffraction pattern for each color appears to float, one directly above the other, so that the student can easily compare the width of the central maxima and thus see their dependence on wavelength for a given slit width.

Sliding the mask all the way down permits spectral fringes to be seen for at least three or four orders on each side of zero. This can lead to a discussion on the use of monochromatic light of short wavelength to improve resolution.

Though the diffraction patterns can be viewed under ambient light conditions, better results can be obtained extinguishing the room lights and drawing the shades.

SAMUEL HIRSCHMAN
Forest Hills High School
Forest Hills, New York 11375

Classical demonstration of polarization

Robert P. Bauman
University of Alabama in Birmingham
Dennis R. Moore*

One of the more puzzling demonstrations for students is the insertion of a polarizer, at 45°, between crossed polarizers. The transmission of light by parallel sheets of Polaroid and absorption by sheets placed perpendicular to each other represent phenomena that are well known and therefore "understood." Why the insertion of the diagonal sheet *between* the first two, but not in front or behind, should increase the transmission is not so readily grasped.

Dirac has shown that the phenomenon in question provides a good basis for introducing concepts of quantum states and the superposition of states. Thus the problem is not only interesting as an optical curiosity but is important for students going on to the study of quantum mechanics.

The classical basis of the effect can easily be shown with a length of rubber tubing, four ring stands and clamps, a long rod for spacing the ring stands, and three wave guides made of electrical conduit silver-soldered together at right angles as shown in Fig. 1. Clamp one end of the hose to the far ringstand. Slip the three wave guides over the vertical posts of the other ringstands as shown in Fig. 2, and pass

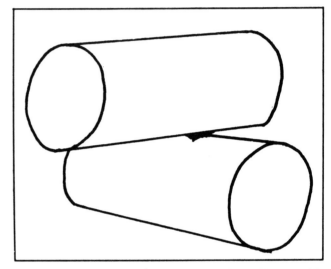

Fig. 1. A 2 in. length of 3/4 in. conduit is silver-soldered at right angles to a 1 in. length of 1/2 in. conduit.

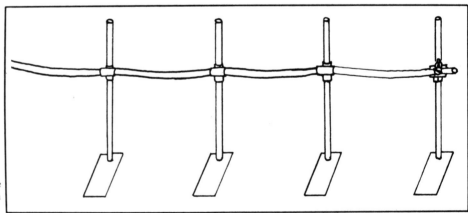

Fig. 2. The hose is clamped at one end and passes through guides, each of which can slide on its ring stand.

*High school student at Hueytown Resource Learning Center, Jefferson County, Alabama; currently an undergraduate at Vanderbilt University.

the hose through the guides. You can excite a wave in the vertical plane that will pass through the three guides. If one of the intermediate ring stands is now released from the clamp holding it in position and is turned to a horizontal position, perpendicular to the wave, the wave will be stopped at that point.

If, however, the first ring stand remains vertical, the second is at 45°, and the third is horizontal (Fig. 3), the wave will go from a vertical polarization to a somewhat weaker diagonal polarization, and will then pass beyond the horizontal ring stand, being transmitted as a horizontal wave. The classical model demonstrates the important concepts of the optical and quantum problems: a wave is transmitted only upon conversion (if necessary) to the polarization of the transmitting filter, and that conversion can only occur insofar as the incident wave has a component in the direction of the filter.

The initial stage of this demonstration was developed during the 1973 summer program on lecture demonstrations at the U.S. Naval Academy, supported by the National Science Foundation. Members of the team contributing to the preliminary model were Joel Blatt, Ray Dameron, William Harris, John Myers, Ronda Ryder, and Marllin Simon, along with one of the authors (RPB) and Professor Smithson of the Academy.

Fig. 3. A wave produced in the vertical plane produces a wave at 45° upon encounter with the diagonal rod. The 45° wave produces a horizontal wave at the horizontal rod.

Demonstrating interference

H. L. Armstrong
Department of Physics, Queen's University, Kingston, Ontario, Canada

Here is a way to show what happens when light from two coherent sources, e.g., two Young's slits, interferes. The rays of light from the two sources to the place in question (it is convenient often to think in terms of rays) may be represented by two cords. The heavy white cords often used for clotheslines are convenient. Alternate sections, of suitable length, are colored black and left white. The white sections may represent crests of the waves, the black, troughs.

Each cord may be fastened at one end to places a little apart. The cords are held by the other ends, stretched out, and let cross somewhere. The place at which they cross represents the place in question, at which the interference is being considered.

If there should be a white section of each cord at the place where they cross, that means, in terms of the waves, that crests from each source will meet there at a certain time. A little later, troughs from both sources will meet there, and so on. So the waves from the both sources will add in their effects; there will be what is called constructive interference. Likewise if the cords should cross where they are both black.

If, however, a white part of one cord should meet a black on the other, that would correspond to a crest from one source meeting a trough from the other. The one would counteract the other; there would be destructive interference.

It is easy, then, by moving both cords about, to have them cross at different places, and to see where there would be constructive, where destructive, interference. The former, in terms of light, would, of course, correspond to bright fringes, the latter to dark.

Can a spot of light move faster than c?

Question

a. If a beacon spins at a very high speed, will the light far from the beacon cut across the sky with a speed greater than 3×10^8 m/s? What will it look like?

b. What if the light was replaced by a giant rod? Could the end of the rod move faster than 3×10^8 m/s?

c. If my body casts a long shadow, a small motion of my arm will result in a large motion of my shadow. Can my shadow move faster than 3×10^8 m/s?

*These questions were sent to us by **Arthur Eisenkraft**, physics teacher at Briarcliff High School, Briarcliff Manor, NY 10501, and were proposed by student **Becky Hardenbergh** For the answer we are grateful to **Peter G. Bergmann**, professor of physics at Syracuse University, Syracuse, NY 13210. Professor Bergmann worked with Einstein for several years at Princeton, and is the author of, among other works, "Introduction to the Theory of Relativity."*

Answer:

Relativity asserts that c, the speed of light, represents an upper bound for the speed of any material object, regardless of the state of motion of the observer who measures that speed. In other words, no stone, no atoms, not even a photon, can travel faster than c. Additionally, the theory asserts that signals (i.e., the carriers of information) obey the same stricture; there can be no instantaneous transfer of information ("news") from one location to another.

Suppose now that a beacon, or a pulsar, spins at a high rate, and that its beam travels a huge distance before being intercepted. Will relativity limit the rate at which that beam streaks across an object such as our earth? This is not a silly question, considering that the pulsar in the Crab Nebula, for instance, sweeps its beam across our observatory thirty times a second, and that its distance from our solar system amounts to a few thousand light years. If we could replace our solar system by a hollow sphere surrounding the Crab pulsar but at that distance, clearly its beam would sweep across that screen at a speed incomparably greater than the speed of light.

But: It is not the photons composing the beam that travel at that incredible speed, but a quite insubstantial object, the point of intersection of the beam with the intercepting screen. Nor does information travel across the screen; it travels with the beam itself, at the speed c. Hence there is no contradiction between the assertions of relativity and what might be called common sense!

You cannot, however, replace the beam of light by a giant rod. If you were to attempt to spin a rod at an increasing rate, it would suffer disintegration (known as a flywheel explosion) long before its tip would approach the speed of light. If it did not, there would be a violation of relativity, as the atoms of the rod do represent "material objects."

Your shadow could move at a speed exceeding c. This is really just another formulation of the first question, as the shadow represents the absence of a beam of light, so its boundary is also the boundary of a beam.

AN OP ART WAVE DEMONSTRATION

"The quickness of the hand deceives the eye." Coming from the barker of the local fairground, these words were sure to draw kids of all ages to a demonstration of exquisite magic. One of the toys sure to be sold by the same gentleman is shown in Fig. 1. It consists of a sheet of translucent wax paper divided into strips alternately black and clear. (The figure drawn here may be dipped in wax to serve the same purpose.)* This is superposed over a drawing such as that shown in the other part of the same figure. As this drawing is slid beneath the wax paper in the

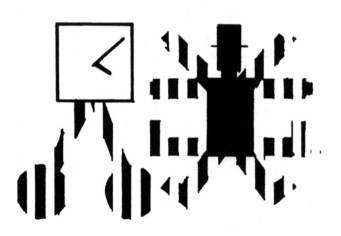
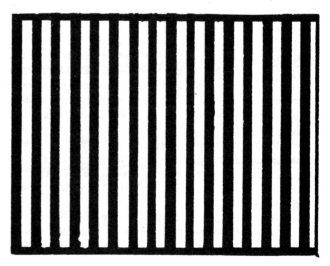

Fig. 1. An old "moving picture" toy.

direction of the arrow, the objects depicted give the illusion of motion in a surprisingly realistic fashion, a precursor of modern op art. A similar effect is seen also in Moiré patterns, where, for example, two gratings ruled with parallel lines are superimposed so that the rulings lie at a small angle to one another. If one grating is displaced slightly in a direction perpendicular to the rulings, it appears that the whole pattern of the crossed gratings moves a large distance in the same direction as the rulings. This "magnification," which is inversely proportional to the angle between the gratings (if the angle is small), has been used to make accurate measurements of displacements—for example, in the location of bubble chamber tracks in particle physics. In addition, many popular articles have been written about the effect.[1]

Recently, the simple toy described above has given way to a more sophisticated model, which may often be seen on dime store pencil sharpeners. Here, the striped wax sheet is replaced by a lenticular grating (Fig. 2), with a drawing similar to that in Fig. 1 underneath. As the eye is moved from side to side above the grating, first one, then the other phase of the drawing is focused upon, again giving the illusion of motion. This technique has also been used to provide three-dimensional pictures, where the left eye focuses on one set of lines which make up the picture, and the right eye the other set of lines, which alternate and are stereoscopic with respect to the first. Instead of two sets of lines, a large number, or even a continuum may be drawn, giving a real illusion of three dimensions from whatever angle the picture is viewed.

One of the more famous brewing companies advertises its wares with a device similar in principle to the first simple toy described. An opaque screen has a transparent slot etched in it which follows the parabolic trajectory of a bouncing ball. A second screen with narrow vertical transparent slots of varigated colors is placed

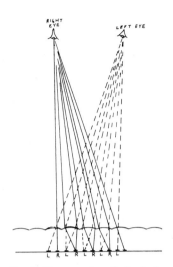

Fig. 2. Diagram of method of obtaining a stereoscopic view using a lenticular grating.

behind the first. The second screen is illuminated from the rear and drawn slowly from one side to the other. Where the two slots cross, a bright spot is seen, which moves as would a real bouncing ball under a reduced gravitational force, since the horizontal component of velocity remains constant.

Crova's disk is another example where the illusion is given of motion in a direction perpendicular to the true motion of the device. This is the well-known and elegant way of demonstrating longitudinal wave motion. Such a disk for traveling waves is shown in Fig. 3. Rings of successively increasing diameter are drawn with centers sequentially 1234567812-- on the figure. A mask having a radial slot is superposed on this disk, which is rotated, the section of lines which are visible moving back and forth as would particles of matter through which a traveling wave passes. A disk for standing waves, which is less common but based on the same principle, is also

---

* Transparent copies of this and the other diagrams in this article may be made on thermofax 3M Type 133 positive projection transparency, or on xerox or other machines giving transparencies. Figures 4, 5 and 6 require negative transparencies.

shown. However, the number of such suitable demonstrations for wave motion is relatively limited. It is not possible to construct a Crova's disk for transverse waves. Nevertheless, a device based on the same principle as the toys described can be made to demonstrate all the most interesting features of traveling waves. In the simplest case, it is merely necessary to draw a sine curve on paper [Fig. 5(a)], then superpose a grid similar to Fig. 1, but with the slits much narrower than the opaque portion (Fig. 4). Then, as the sine wave is moved, the spots seen through the slits go through periodic motion as if a traveling wave were passing over them. A striking demonstration is obtained by making negative transparencies of Fig. 4 and 5 and superposing them on an overhead projector.

It is obvious that this technique cannot be used employing the same sheet of paper for waves traveling in opposite directions simultaneously, or for standing waves. To see how such effects can be demonstrated, suppose we arrange the adjacent slits to be one period of the sine wave apart [Fig. 5(b)]. Each spot seen through the slits moves up and down in phase with all the others, if we move the drawing relative to the grating. If the wavelength is now made a little more than the distance between slits, moving the drawing gives the appearance of waves of long wavelength moving in the opposite sense to the motion of the drawing [Fig. 5(c)]. Similarly, if the wavelength is a little less than the distance between slits, the motion occurs in the same sense both for traveling wave and drawing, but with different velocities [Fig. 5(d)]. Hence, by drawing two sine waves on the same paper, one with a period a little less and the other with a period greater than the slits, we can give the appearance of two waves traveling in opposite directions, as shown in the figure. The addition of these two waves yields something like standing waves [Fig. 5(e)]. In practice, the standing wave pattern moves at the same rate as the paper upon which it is drawn. Physically, this arises because we are combining waves traveling with different velocities in opposite directions. If the drawing is held stationary and the slits moved, a very real impression of standing waves results.

The next two diagrams [Figs. 5(f) and 5(g)] demonstrate group velocity and wave velocity. If we traverse the figure beneath the grid, in one case [Fig. 5(f)] waves arise at the front of the group, travel through it and pass out the back, so the group velocity is larger than the wave velocity. Such waves occur in ripples, and if one examines a patch of ripples, caught up on a smooth pond by a puff of air, it may easily be seen that they appear at the front of the patch, grow in size as they pass backward to the center, then decrease and finally vanish at the rear of the patch. Conversely, deep sea waves do the opposite. The group of waves travels more slowly than the wave itself. This is why an unusually large wave at sea takes twice as long to reach a boat as the waves individually. It appears as though this large wave hands its size back to the waves in the rear, and so on, as in Fig. 3(g). The wave and group velocities are related by the well-known formulas $U = \frac{\partial \nu}{\partial K}$, $v = \frac{\nu}{K}$, where U is the group

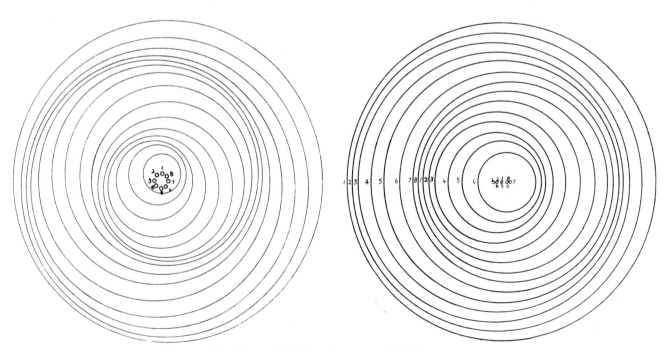

Fig. 3. Crova's disk for traveling waves and standing waves.

velocity, v the wave velocity, ν the frequency, and K the reciprocal of the wavelength.

The steady-state time-dependent solution for wave mechanical problems can be demonstrated using this technique. In Fig. 6(a), a beam of particles is traveling over a region where the potential is slowly changing, as shown at the bottom of the figure. The change in wavelength of the traveling wave may be clearly seen, the wavelength being shorter where the potential energy is smaller.

A demonstration of the time-dependent wave function in wave mechanics for a uniform beam of particles penetrating and being reflected from a potential barrier is shown last on Fig. 6(b). As the grid is drawn over the diagram slowly, it will be seen that a pattern appears which is a standing wave superimposed on a traveling wave impinging on the barrier, within which the exponentially decaying wave function oscillates up and down, and beyond which a traveling wave pattern, which matches the exponential at the

Fig. 4. Slits used for transverse wave demonstration.

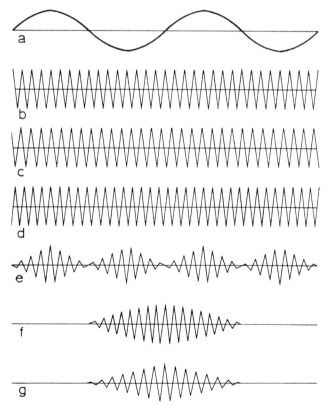

Fig. 5. (a) Simple sine wave; (b) waves having period of the slits, producing a straight line when Fig. 4 is superposed; (c) waves having a period a little more than the slits. If this figure is slid beneath Fig. 4, the waves travel in the opposite direction; (d) waves having a period a little less than the slits. If this figure is slid beneath Fig. 4, the waves seen through the slits travel in the same direction, but faster; (e) the sum of (c) and (d), giving the effect of standing waves; (f) group velocity greater than wave velocity; (g) wave velocity greater than group velocity.

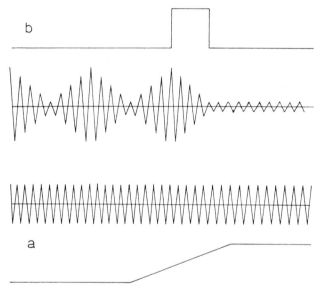

Fig. 6. (a) When Fig. 4 is superposed above the top portion of this figure, then the traveling wave pattern of particles moving in the potential of the lower part of the figure is produced. (b) When Fig. 4 is superposed we have the wave pattern of particles impinging on the barrier shown in the upper part of the figure.

boundary, carries away those particles which have penetrated. The amplitudes of the traveling parts of the wave functions are the same on both sides of the barrier, and the standing part represents that section of the wave which is reflected by the barrier. Such demonstrations are of value, since the steady-state problem is often discussed without mention of exactly how time affects the appearance of the wave solution of the problem.

R. D. EDGE
University of South Carolina
Columbia, South Carolina 29208

References

1. G. Oster and Y. Nishijima, Sci. Amer. **208**, 45 (May 1963).

A sound-level meter: a sound investment

A. David Chandler*
Grissom High School, Huntsville, Alabama 35802

"A sound investment: dB's (see Fig. 1). For more information contact your local physics student." Captions like this on posters around our school this past spring sent a wave of questions and rumors through the student body. The sound-level meter's arrival was evidenced by signs like "70 dB" above a pencil sharpener, "52 dB" in the library, "95 dB" in the band room (see Fig. 2), and a few "What's a dB?" posters in explanation. This marked the beginning of the most exciting and rewarding project I've undertaken as a teacher.

The project started with a grant from the AAPT that enabled us to purchase a sound-level meter. The grant was directed to an innovative high school teaching project with a twofold purpose: to equip physics students for experiments and practical applications of sound measurement; and to arouse interest in our physics course by exposing the student body and community to intriguing applications of physics used in our course.

The meter selected for use in the project is a Scott 450B. It is a rugged, hand-held, survey grade meter with an accuracy of ± 3dB and a range of 20–140 dB. Condensed operating instructions are printed on the back but seldom needed because of the simplicity of operation. The capabilities of the meter met our needs perfectly. It is sensitive and accurate enough for lab work yet durable enough to be sent out with the kids. Each time a student checked it out he was required to read the instruction manual and advised to treat the meter as he would a stack of one-dollar bills the same size. The $275 price tag may seem high, but if I ever transfer to another school a sound-level meter will have a high priority in my equipment budget because the returns on such an investment are well worth the cost.

As soon as the meter arrived it was gone. It's not often that a teacher has to limit involvement in a learning activity, but the interest and motivation in this project required just that. Students signed up on a calendar to schedule the meter in an array of individual projects. The only condition of participants was that they report their results in writing. Noise pollution was a favorite study: motorcycle mufflers, lawnmowers, wood shop, school dances, the rifle range, the airport, and rock concerts. One group test drove several new cars to check interior noise levels. Intensities as high as 110 dB were determined for

*Awarded High School Grant for Innovative Teaching.

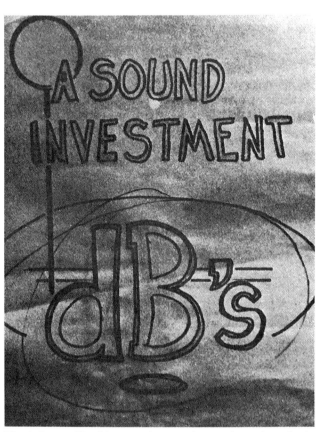

Fig. 1. Posters like this aroused student interest in a unique physics project.

babies' squeeze toys! The reports reflected a high level of interest and motivation as well as originality. Students who usually struggled with regular lab reports turned in top quality work. One pleasant surprise was the response from parents and businesses. Parents were delighted to see their children so interested in school work, and business people were full of questions that gave our project and course effective publicity. The laboratory applications are numerous: interference patterns measurement, sound insulation, human hearing sensitivity and range, and the inverse square law, are examples. The true value of the meter, however, must be measured in terms of the excitement and relevance it brings to students taking their own ideas in their own hands and into their own "lab".

A POTPOURRI OF PHYSICS TEACHING IDEAS—SOUND

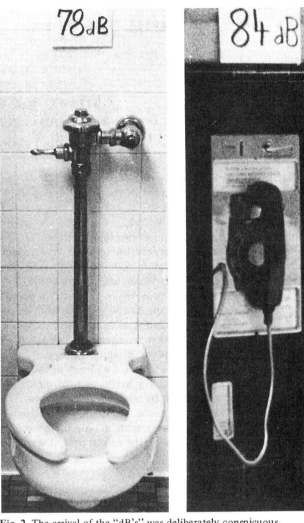

Fig. 2. The arrival of the "dB's" was deliberately conspicuous.

Creating interest in the student body was a lot of fun. When the teaser campaign began with the "dB" posters, it was the talk of the school. Several of my "local physics students" had to review their "dB's" to give satisfactory answers to the curious nonphysics students. The meter was used at pep rallies in order to determine the loudest class in competition cheers. Physics students also conduct an annual loudest-student contest, the Screamathon (see Fig. 3). There is a 10-cent entry fee for four categories: guy, girl, group, and whistler. The school record is 124 dB by a guy shouting at a distance of five feet. The $10-15 profit is used to buy trophies that are presented during a school assembly. It's difficult to make a statement of cause and effect, but eight classes of physics scheduled from an enrollment of 2500 is evidence of the interest in physics at our school.

Physics can be fun. Physics teaching can be fun, too. If you haven't already, let me encourage you to start out on your own innovative project and join the fun. If you think you can't afford it, sound-level meters are now available from Radio Shack for under $40.00. Whether you spend $40.00 or $275.00, I can assure you it will be a sound investment.

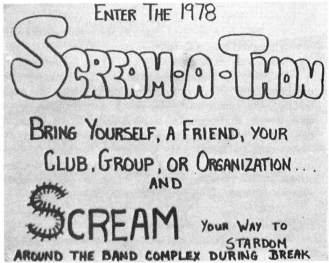 

Fig. 3. The Screamathon drew a crowd and a lot of stares. Nearly 150 students entered the contest.

Demonstrating resonance by shattering glass with sound

Willard C. Walker
Department of Physics, Southern Illinois University, Carbondale, Illinois 62901

A study was done on the feasibility of a classroom demonstration shattering glass with sound. We had in mind the legends describing the way a wine goblet could be shattered by a singer. Our idea was to drive the glass with the amplified output of an audio oscillator and to show that only the natural frequency would break the glass. The method used was developed with the aid of Peter Tappan's suggestion[1] that a small source of sound be used, and a goblet was successfully shattered at 721 Hz. At this frequency four nodes could be seen on the rim of the goblet.

Equipment

The equipment required for the experiment is a powerful audio amplifier and a horn driver (loudspeaker). An audio oscillator is used to drive the amplifier. It is possible to do the experiment with these three pieces of equipment plus, of course, a suitable glass vessel, but an oscilloscope and microphone are very useful and almost necessary to determine the resonant frequency. At first we used a 45-W amplifier and a 30-W horn driver. However, we found it possible to use a 20-W amplifier for shattering laboratory beakers if care was taken to match the driving and natural resonance frequencies precisely. A physically small source of sound is required to excite the glass and can be obtained most efficiently using a high-power driver which has been removed from a horn-type loudspeaker. If a powerful amplifier and a large diameter cone loudspeaker are available, this combination can be used, with the speaker mounted on a baffle which has in it a small hole (2 in. or less). No attempt was made to build a sound absorbing enclosure for the speaker because the horn driver was an ideal source. The sound level in the room could be excessive with this loudspeaker arrangement. The amplifier and driver need to be able to produce at least 135 to 140 dB of sound level at the natural frequency of the glass. The driver was placed behind a baffle to help direct the sound toward the glass and to minimize interference. The baffle used in these experiments was made from plywood (¼ in. thick x 18 in. square) with a hole in the center tapped to match the threads on the driver; thus, it served as both a support and baffle.

Glassware

The selection of the glassware is somewhat critical. The first experiments were done on high quality, thin goblets costing about $6 each. The goblets were nearly cylindrical with a diameter of 3 in. and a wall thickness of 0.04 in. This goblet was selected because of its high-Q resonance. Bell-shaped goblets, martini glasses, and laboratory beakers were also tested. The objects which rang well and had a cylindrical shape in the region of the rim were the only ones shattered in these experiments. The laboratory beakers, if placed on a soft plastic foam pad to reduce damping and movement, and oriented to give maximum response, broke more easily than the high quality, thin wine goblet. The orientation required varied from beaker to beaker and needed to be tested for each one. This is because of the overall cylindrical shape of the beakers and the fact that the thickness of the beakers varies in the vibrating region near the rim. This concentrates the stress at the weaker points. The wine goblet, being symmetric, required no specific orientation.

Studies on the natural frequencies

The natural frequencies of the goblets were formed by striking or rubbing the goblet and measuring the frequency

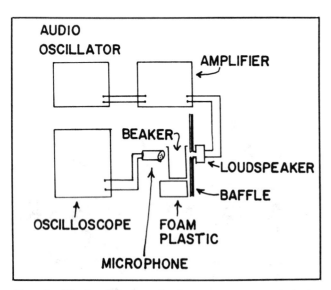

Fig. 1. Block diagram of the apparatus needed for shattering glass vessels.

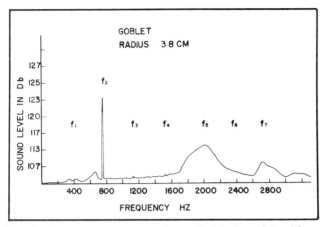

Fig. 2. Frequency response of a cylindrical goblet with a radius of 3.8 cm.

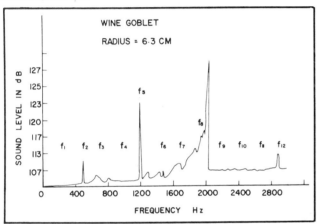

Fig. 3. Frequency response of a flared goblet with a radius of 6.3 cm.

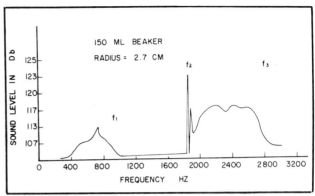

Fig. 4. Frequency response of a 150 ml beaker.

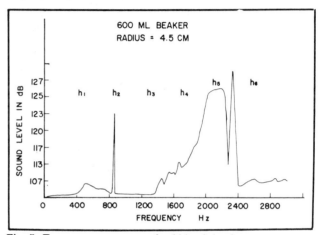

Fig. 5. Frequency response of a 600 ml beaker.

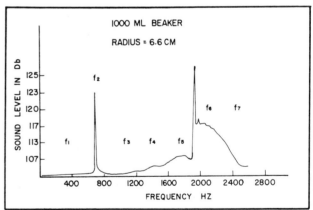

Fig. 6. Frequency response of a 1000 ml beaker.

by matching the tone with an audio oscillator. Striking the glass excited many modes of vibration which always damped out to the lowest natural frequency. The duration of the sound produced by this method was short and varied in amplitude making measurement of these frequencies difficult. The method which proved most successful was to drive the glass with a low volume sound from the horn driver and detect the response of the glass with a microphone connected to an oscilloscope. The glass was set very close to the driver with its rim just above the opening in the driver, and the microphone was placed close to the glass. The glass acted as a resonator and the amplitude picked up by the microphone depended upon the natural frequencies of the glass. At frequencies of least response the presence of the glass had little or slightly negative influence on the amplitude detected. Since the amplitude from the driver was not linear with frequency, the peaks of response were located by comparison with the minimum on either side of the maximum.

The glass was found to respond in many complex modes of vibration, but the modes of a plane circular medium were the most dominant. The outstanding response in all the glassware was noted at the lowest natural frequency which was the main frequency produced by the glass when it was struck. This mode was identified by observing the four nodes around the rim of the glass. Even though strong resonant conditions were noted in certain glassware at other harmonics, this was the only mode in which glass was shattered. A search for other natural frequencies showed that the frequency response curves were similar for vessels of the same size and shape. The tendency for four, six, or eight nodes to form on the rim of the vessel was noted when driven at one of its natural frequencies. The relations between the lowest natural frequency f_ℓ and the others, f_n, was found to satisfy the equation, $f_n = n/2\, f_\ell$, where $n = 2, 3, \ldots$. This is consistent with the expected relationship

$$n\lambda_n = 2\pi r \text{ where } \lambda_n = \frac{v}{f_n} \text{ and } \frac{2\pi r}{v} \text{ is constant.}$$

Large resonant conditions were usually found at $n = 2$ and $n = 5$ with resonance weak or absent at other values of n. The frequency response curves of various vessels show the calculated values of f_n compared to the experimental values. Four symmetrical nodes could always be detected at f_ℓ, but as the frequency was increased the nodal points became more difficult to locate.

The natural frequencies were recognizable by their small damping factor. Thus, when the driver frequency was changed close to one of these modes, beats between the driver and the glass could be observed during the time required for the oscillations in the glass to damp out. The unique difference between the lowest natural frequency and the others was the very narrow frequency response in this region. The band width of response was so very narrow that it could easily be passed over undetected if the frequency sweeping rate was too fast.

When the oscillator was tuned precisely as indicated by the oscilloscope, the goblets and beakers would break at around 130 to 140 dB of sound level from a small source. The source must be small because the wavelength of the sound is large compared to the dimensions of the glass. When vibrating, the opposite side of the glass is out of phase with the driver. If the source of the sound is small, the wave amplitude will decrease nearly in proportion to the inverse of the distance. Thus a small source will drive the side of the glass next to it while allowing the other sides to vibrate freely. With a source large compared to the glass diameter, on the other hand, there would be little drop in amplitude at the far side.

Classroom demonstration

For classroom demonstration the lowest natural frequency of the glass and the orientation in which the glass gives maximum response should be determined before class in order to conserve time and avoid failures. The demonstration should show that loud sound in itself will not break the glass, but sound at the natural frequency of the glass is required. If the audio oscillator is adjusted slightly off resonance and the volume is increased to a maximum the glass will not break. Another method is to increase the volume to the desired level and very slowly change the frequency until the glass is shattered. In the second method the vibration of the glass will aid in determining the resonant frequency. An object like a ball on a string or a pencil can be touched to the glass as a vibration detector. A period of time is required for the oscillations to build up in the glass; therefore, the second method could fail if the driver frequency is changed too rapidly. A smaller amount of power will be required for the experiment if the audio oscillator is precisely tuned as in the first method. When the demonstration was done in a large classroom, the sound level was surprisingly small when considering the power being dissipated. At the levels used, the higher frequencies (1000-2000 Hz) associated with small beakers were more irritating to the ear than the lower ones.

Discussion

The demonstration proved to be dramatic in showing how glass could be driven past its elastic limit, but it should be mentioned that conditions like the small damping factor, resonance, and intensity of the sound are all critical and that, therefore, an opera singer unaided by amplification is not likely to go around breaking glass!

Acknowledgment

My thanks go to Dr. John Cutnell of SIU for his encouragement and to Peter Tappan of Bolt Beranek and Newman Inc., for his helpful suggestions.

References
1. Peter Tappan, "Shattering Goblets with Amplified Singing," published by Audio Engineering Society, 1973.

Seeing the Science of Sound

Brother Shamus Mahoney CFX
Notre Dame High School
Utica, New York 13502

Most physics teachers are familiar with the Bell Telephone record *The Science of Sound* (available on loan from the Bell Telephone Company, or for sale from several science supply companies). An instructive visual aspect can be given to this sound demonstration by connecting the vertical deflecting plate terminals of a cathode ray oscilloscope to the loudspeaker terminals of the record player. Alternatively, most record players have a plug for an external loudspeaker, and the CRO can be connected to this external loudspeaker. The waveforms displayed on the oscilloscope are particularly valuable in illustrating 2 (Frequency), 3 (Pitch), 8 (Fundamentals and Overtones), 9 (Quality), 10 (Filtered Music and Speech) on the record.

The frequency response of the loudspeaker is also demonstrated when very low and very high frequencies are sounded.

A POTPOURRI OF PHYSICS TEACHING IDEAS—SOUND

Transmitting sound by light

J. B. van der Kooi
Groningen, The Netherlands

In some lectures to students from Indiana State University, Terre Haute, Indiana and a number of senior high schools in Indiana and Michigan the demonstrations described herein proved to raise the interest of both students and teachers. Professor L. A. Poorman (I.S.U.) suggested that I submit them to *The Physics Teacher*.

Exp. 1. The output of a tape recorder is connected via a capacitor with a battery-circuit thus influencing the voltage over a bulb, e.g., 6 V. A flashlight produces the best results.

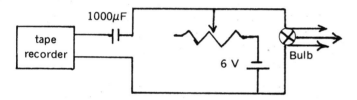

One starts the tape recorder not yet connected with the circuit. The students *hear* the music. Now the output of the recorder is put into the circuit. The sound disappears and the bulb starts flickering. The students *see* the music, but this produces no sensation of music. This phenomenon deserves some discussion.

Hereafter the light beam is directed towards a receiver, for instance, a solar cell or a phototransmitter each in combination with an audioamplifier and speaker. The easiest way, however, is to use a device from Metrologic Instruments Inc. called: Laser-Communicator-Receiver. Surprisingly enough the music is back again, its quality depending on the ratio between the direct voltage from the battery and the alternating voltage produced by the tape recorder as will be shown in the last paragraph.

Exp. 2. Instead of transmission through air a fiber can be used. This demonstration is particularly interesting because of its future application by telephone companies to handle short-distance calls in the eighties. Physics teachers should be delighted to have a demonstration ahead of advanced technology!

Exp. 3. Those among us who prefer hard-core physics above fun can replace the tape recorder by a signal generator and connect the output of the receiver with an oscilloscope. When the direct voltage on the bulb exceeds the pulses from the generator the oscilloscope produces the same graph as if it were connected directly with the generator. However, changing this ratio by decreasing the direct voltage produces a gradually changing graph. This shows twice the frequency of the generator-produced signal at the moment the direct voltage can be neglected compared to the generator's output.

Velocity of Sound Using an Oscilloscope

Measuring the velocity of sound, using permanent magnet radio speakers and an oscilloscope, is especially instructive because it utilizes the phase method of measuring wavelength and also serves as a preliminary lesson on Lissajous pattern interpretation. Any speakers, from 3–6 in. in diameter, are mounted with output transformers on wooden blocks. Cut a groove in the bottom of the blocks or make rails to fit a horizontal meter stick laying on a table, as in Fig. 4.

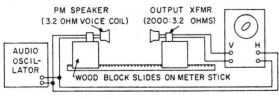

Fig. 4. Block diagram of velocity of sound apparatus.

Connect an audio oscillator to one of the speakers and also to the horizontal input terminals of the oscilloscope; connect the other speaker to the vertical input terminals of the oscilloscope. In operation, the speaker connected to the oscillator produces sound waves, and the other speaker acts as a microphone. Set the oscilloscope for external sweep operation and set the oscillator for about 500 c/s. After a few minutes warm-up, turn up the oscillator gain until an audible tone is heard. Slide the speakers apart, and have the class note the Lissajous patterns that are produced on the oscilloscope display. Each time the speakers are moved a wavelength apart, the Lissajous patterns will indicate a complete 360° cycle, as shown in Fig. 5. Take the average distance per wavelength and calculate the velocity of sound using the relationship $v = f\lambda$.

Fig. 5. Lissajous patterns which appear on oscilloscope.

RESONANCE TUBE FOR MEASUREMENT OF THE SPEED OF SOUND

Having suffered for many years the leaks and squeaks of commercial speed-of-sound resonance tube apparatus we decided to try our own design (Fig. 4). It consists of a 1-m length of clear acrylic plastic tubing, 2-in. i.d. and 1/4-in. wall ($17.50), an aluminum piston 2 in. long, turned to fit the tube snugly, and a 1/4-in. diam push rod of aluminum. The push rod is threaded and screwed into the piston, and has at its other end a knurled knob. At one end of the tube is glued a 3/4 in. thick plastic end-cap with a center guide hole for the push rod. The resonance tube is mounted, along with a meter stick on a wooden base. Twelve units were constructed in an average time of less than 1 hour. Total cost to our department for each unit was approximately $36.00 including labor.

The 1-m length of the tube does limit the frequency, when using the first and second resonance point method, to the D above middle C. In addition, it should be noted that the resonance tube detects harmonics of the tuning fork, especially if the fork is incorrectly struck.

A few problems arose with our prototype: it was constructed from 1/8 in. wall tubing, which turned out to be susceptible to warping, and piston sticking. The original scale, consisting of adhesive tape, proved to be less than student-proof so it was replaced with a trusty old meter stick glued next to the resonance tube. This also was not student-proof, so in the final version the meter stick was screwed down.

The Fisher Scientific Company catalog lists a similar resonance tube made of glass and supported by two wood blocks. The resonance length, however, cannot be made less than 30 cm.

It is a pleasure to acknowledge the able help of our chief technician, Richard Boker, in the design and construction of the resonance tubes.

FRANKLIN R. MUIRHEAD
San Jose State University
San Jose, California 95192

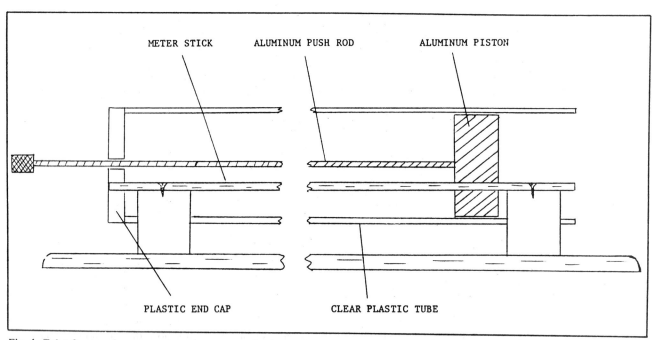

Fig. 4. Tube for sound resonance experiments uses aluminum piston instead of water to change effective length of tube.

Sound on a Light Beam

William M. Zeitz and
Terrence P. Toepker
Xavier University
Cincinnati, Ohio 45207

While amplitude modulation of a light beam is by no means a newly discovered phenomenon, the apparatus described here was designed primarily as an effective but inexpensive lecture demonstration.

Transmitter. The transmitter (Fig. 3) consists of a dc power supply, an ac signal source, and a lamp. The dc power supply should be capable of providing 24 V at about 1 A. The ac source may be any radio or tape recorder, provided that the output impedance is in the range of 2–20 Ω. An ordinary flashlight housing may be used effectively by replacing the 3-V bulb with a 12-V bulb and by mounting the resistor and capacitor in the barrel with appropriate leads for connections to the power and signal sources.

Adjustment of the transmitter is not critical. However, optimum performance will be obtained when the ac voltage delivered to the light source is one-half that of the dc output. Less than half will cause high noise and low volume; while more than half will result in high volume but a great deal of distortion due to over-modulation. The dc source should provide enough power to the lamp to cause bright illumination since the ac signal is superimposed on this beam. The dc should provide enough energy to the lamp so that the detector may be operated in its most sensitive range. Thus small changes are more readily detected.

While ambient light conditions are not critical, a hood at the detector end may be helpful in the elimination of 60-cycle noise when the apparatus is used under fluorescent lighting. Incandescent lights have little effect since the dc is capacitively blocked in the detector. Of course, overhead lights may be turned off.

The range of the transmitter is approximately 6 ft. However, if a lens in front of the detector is used to focus the beam, then distances of up to 15 ft can be obtained.

Receiver. Since most college or high school physics departments have access to a motion picture projector, detection may be accomplished by reflecting the modulated light beam into the photodetection system of the projector. When using this method it is necessary to mask or remove the exciter bulb from the projector. A Bell and Howell Filmosound 285 has worked quite well. Thus a simple, yet impressive, demonstration can be used and the cost is practically zero!

This apparatus can be used not only as a demonstration of modulation but also touches on the uses of lenses, detector response, and mechanical resonance.

Lenses. The distance between the transmitter and detector may be greatly increased by focusing the beam with a converging lens. By changing the position of the lens, one can easily decrease or increase the output of the detector as the lens concentrates or spreads out the energy of the transmitted beam. The use of the lens is essentially the same as the objective lens of a telescope in that it collects enough of the light to reach the threshold of the detection device.

Detector response. Place a red filter (Corning Glass filter, color spec. 2-64, code No. 2030) in the path of the beam and note the decrease in the volume of the sound. Now replace the red filter with a blue one (Corning Gass filter, color spec. 4-71, code No. 4305) and again note the volume change. The red makes very little change while the blue almost decreases the volume to zero.

Now hold both filters up to the light source and judge by means of the eye which filter transmits more light. It appears that the blue transmits much more than the red. Since the response curve of the eye does not match the response curve of the photocell, looks are deceiving.

Mechanical resonance. By lightly rapping the transmitter lamp or housing with a finger, a characteristic ringing sound is produced at the detector. This ringing was measured to be about 833 Hz. It was suspected that this was the resonant frequency of the lamp filament.

As a means of testing, an audio oscillator was used as the ac source. When an 833-Hz signal was applied to the filament, a characteristic ring was produced from the speaker of the detector. Now the audio oscillator was disconnected from the circuit but the mechanical vibration of the filament continued the same 833-Hz output. Thus, this combination video, audio, and mechanical feedback mechanism was found to be resonant at about 833 Hz.

This phenomenon is related to the problem of microphonics in vacuum tubes where forced vibrations of the component parts can produce unwanted noise. This problem ranges from jukeboxes to rocket payloads. The special design of vacuum tubes to reduce microphonics can result in large price differences from one tube to another.

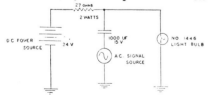

Fig. 3. Schematic diagram of modulated light beam transmitter.

COLORFUL CHLADNI

An element of color can be added to the already beautiful Chladni figures obtained with vibrating plates. If the usual fine white sand[1] is mixed with a bit of fluorescent tempera poster paint[2] and then spread out and allowed to dry, it will produce brilliantly colored Chladni figures when viewed under ultraviolet light. I have found that four or five drops will color 50 cm^3 of sand and that plates made of 1/4-in. aluminum approximately 12-in. across and painted flat black work very well.

R. C. NICKLIN
Appalachian State University
Boone, North Carolina 28607

References
1. Sargent-Welch 3250-C2.
2. Item No. 66-5520 – Six 3/4 fl. oz. jars, a Sargent product of Westab, Dayton, Ohio 45402.

Binaural hearing

Morris G. Hults
Concordia High School, Fort Wayne, Indiana 46805

To demonstrate binaural hearing, I developed an idea given me by my brother, with my own innovations. I used an old department-store mannequin head and bust, which I equipped first with a wig purchased for $.25 at a Goodwill Store. Clipped to the hair at each ear, I positioned a microphone.*

The microphones were fed into a stereo amplifier, placed with the dummy in the physics storeroom, and the output of the amplifier was equipped with a long phone cord which I passed under the door to the stereo phones in the classroom. A block was provided, having multiple phone jacks.

Several students, male and female, stayed with the dummy, and talked to each other as they milled about, on all sides of the dummy, which was mounted at head height of the students (Fig. 1).

The effect is so good that the students with the phones will actually turn around, thinking someone is speaking behind or to one side of them. Frankly, I myself was fooled when an unexpected person walked up to the group with the dummy, and began to talk; I thought someone had stepped up behind me! And I had just set up the demonstration!

*The microphones used were: SONY ECM-50

Sound waves

Morris G. Hults
Concordia High School, Fort Wayne, Indiana 46805

While studying wave behavior, my students found it hard to comprehend that it is not the *medium* that travels, but only the *wave*. So to demonstrate, I dipped the bell of a cornet into a shallow pan (from a laboratory beam balance) full of soap-bubble solution.

Raising the cornet, I then played an entire tune (a bugle call) without breaking the bubble that formed on the bell. The students laughed and cheered at my horrible music (I was a tuba player, not a trumpeter), but were delighted with the demonstration. I had them note that the air with which I played the tune stayed in the quart-sized bubble, but did *not* travel to the back of the room, yet the "music" certainly did.

I commented, too, that if it were necessary for the medium (the air, in this case) to travel to the hearer, there would be quite a windstorm in every classroom, and they might know, even in the back of the room, that I had eaten onions before coming to class.

A POTPOURRI OF PHYSICS TEACHING IDEAS—SOUND

RESONANCE AND FOGHORNS

A rather spectacular demonstration of resonating air columns can be done with the use of very simple, inexpensive, and easily obtainable equipment.

To cause an air column to resonate, we stimulate one of its natural modes of vibration. We can do this by providing a single tone of just the proper frequency, or, we can supply simultaneously a wide band of tones containing the frequency corresponding to one of the air column's natural modes.

The first method is well known and usually employs an audio oscillator, amplifier, and a speaker which is placed near the end of a hollow tube of some kind. When the proper frequency tone is generated, the tube is heard to resonate.

The second method, similar to the "singing tubes" described by Sutton* should be of interest to readers. Light a Fisher burner (high temperature Bunsen-type burner with wire mesh over a large orifice). Turn the flame to a high setting. Lower a *large* cardboard tube vertically over the end of the burner and a *tremendous* booming roar closely resembling a foghorn is heard (Fig. 3). (Be careful to avoid igniting the tube!)

Eddy currents resulting from the gas passing through the wire mesh result in the generation of tones covering a wide band of frequencies. The tube responds (resonates) strongly to one of these frequencies.

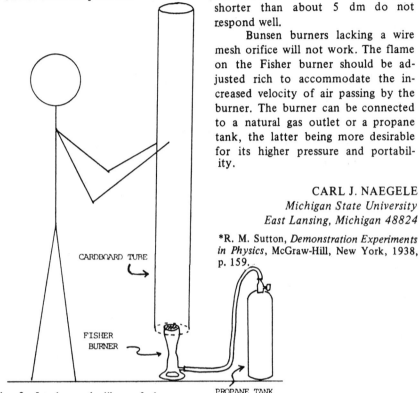

Fig. 3. Loud sounds like a foghorn are emitted from large cardboard cylinder energized by a Fisher burner.

Cardboard tubes of the type carpets are rolled onto are excellent. Try different lengths and diameters. Lengths of about 2 to 3 m with diameters of 1 to 2 dm are exceptionally effective. Tubes with lengths shorter than about 5 dm do not respond well.

Bunsen burners lacking a wire mesh orifice will not work. The flame on the Fisher burner should be adjusted rich to accommodate the increased velocity of air passing by the burner. The burner can be connected to a natural gas outlet or a propane tank, the latter being more desirable for its higher pressure and portability.

CARL J. NAEGELE
Michigan State University
East Lansing, Michigan 48824

*R. M. Sutton, *Demonstration Experiments in Physics*, McGraw-Hill, New York, 1938, p. 159.

Sound in a Vacuum Demonstration

Alfred M. Eich, Jr.
Charles F. Brush H.S., Lyndhurst, Ohio

The usual "Bell in a Vacuum" demonstration (See *Modern Physics*, Dull, Metcalfe, and Williams, p. 301, 1964 edition) can be made more meaningful if one has an amplifier, loudspeaker, and transistor radio earphone available in addition to the bell jar, door bell, vacuum pump, and base plate. Much of the attenuation in sound energy that is noticed when the bell jar is evacuated is due to the mismatch in impedance between the air in the room and that which remains in the bell jar. However, the effect of the reduction in pressure inside the jar can be detected by suspending an earphone (used as a microphone) inside the jar near the sound source and connecting it to an external amplifier and loudspeaker. If an audio-oscillator is available, its output may be connected to another earphone (used as a loudspeaker) inside the bell jar to provide a better source than the door bell.

Demonstrations of Reflection and Diffraction of Sound
Lester Dwyer
St. James High School, Chester, Pennsylvania 19013

Some physics demonstrations that help in the study of wave motion at the high school level are the interference of sound in one and two dimensions. One-dimensional waves are usually demonstrated with a rope, spring, slinky, or wave machine. They illustrate wave motion in a line. A ripple tank extends the study of waves to two dimensions on a surface. A student *sees* the wave properties in these experiments. Sound stimulates *hearing* and is another sense aid to approach a better understanding of wave concepts.

Figure 1 shows schematically the apparatus. The source, an audio-oscillator, produces the tones to a speaker. The speaker is mounted at one end of a box about 1 ft square and 4 ft long. By making the box long, the waves emerge from the end as nearly straight or plane waves. Line the box with a sound absorbent material, such as glass wool. To make the sound source loud use an amplifier or connect an audio-output transformer between the oscillator and the speaker. The transformer matches the high oscillator to the low speaker impedance. The receiver, a microphone connected to an amplifier, detects the sound from the box opening. Across the amplifier's output terminals attach an 8-Ω resistor, with several watts power rating, and an ac voltmeter of 0–10 V range.

The first demonstration **reflects** sound waves off a plane barrier. The purpose of the long box with sound absorbent lining is to produce nearly plane waves as illustrated in Fig. 2.

Tune the audio-oscillator to a 500-Hz frequency. This tone has a wavelength of about 2.3 ft. Face the microphone away from the box opening. Place a large plywood sheet, or any hard smooth surface, some 5 ft from the box opening. Standing waves can be observed between the box and the barrier's reflecting surface. Move the microphone slowly away from the board. A dip in the meter occurs every half wavelength from the reflector. A student can confirm this by using his ear instead of the microphone. A maximum of sound can be seen on the meter or heard by the ear halfway between the nodes. Locate and measure the wavelength. From the wave equation compute the frequency. Check your result against the oscillator's dial setting.

Another simple demonstration uses a curved reflector. Shape a cardboard sheet into an approximate parabola by flexing it. The focus can be located by moving the microphone to where a sound maximum is observed.

The same apparatus can be adapted to **interference** from two sources in phase by adding a board at the end of the box to form two slits. Generate a 5000-Hz frequency. Make the slits 1 ft apart and 1 in. wide. The sound waves from the two apertures form an interference pattern similar to the photograph on the outside back cover of the *PSSC Physics* textbook. Slowly rotate the microphone on a swivel in a 3-ft circular arc about the box front. Nodes can be observed on the meter or heard with the ear placed at the same spot. A quantitative result can be obtained by using the formula in the *PSSC Physics* text for interference

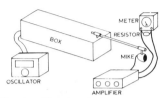

Fig. 1. Sketch of the apparatus.

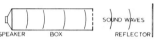

Fig. 2. Nearly plane waves emerge from the box, reflect off the flat barrier, and produce a standing wave pattern.

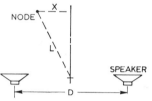

Fig. 3. Two speakers, D meters apart, cause a node, X meters from the center line and L meters from the midpoint between the speakers.

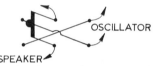

Fig. 4. A schematic of a double pole, double throw switch, which is connected between the oscillator and the speaker to permit phase reversal.

from two point sources. Calculate the wavelength and, again, from the wave equation solve for the frequency. The student should read the dial on the audio-oscillator and compare with the computed result.

An alternate method is to produce interference with two similar speakers spaced a few feet apart. First phase the speakers by touching their leads to a flashlight battery and observing the combination that causes both speakers to move in the same direction. With an audio-oscillator and an amplifier produce a 1000-Hz tone through the speakers. Locate a student across the room midway between the speakers; instruct him to cover one ear and walk slowly from the midpoint till he reaches a minimum of sound. He should be able to pick out the nodes and again using the same equations calculate the frequency. Figure 3 illustrates the needed measurements D, L, and X. With two speakers, you can change the phase of one speaker by switching the leads and quickly shift the interference pattern. Use a double pole, double throw switch connected between one speaker and the audio-oscillator as shown in Fig. 4.

Consider again the demonstrations with the box. Reduce the opening to **one slit** that is slightly larger than the wavelength of the tone used. A typical single slit pattern of interference can be detected. A 5-in. opening with a 3500-Hz frequency provides a convenient choice of parameters. The box front limits the size of the opening and the frequency choice.

Sound is produced by the motion of air molecules vibrating longitudinally. Fortunately, the amplitude of the to-and-fro motion is small compared to the wavelength. These demonstrations are similar to other wave experiments and can aid understanding concepts, generate class interest, and encourage some students toward independent study or projects in sound or wave motion. The *Science Study Series* (published by Doubleday) has several paperbacks that suggest such activities; they are:

Echoes of Bats and Men	by Donald Griffin
Electrons and Waves	by John Pierce
Horns, Strings and Harmony	by Arthur Benade
Waves and the Ear	by W. Bergeijk
Sound Waves and Light Waves	by Winston Kock
Man's World of Sound	by John Pierce and Edward Davis (cloth bound).

Physics on a dinner plate

Jeffrey May

26 Chatham Street, Cambridge, Massachusetts 02139

One evening at dinner I reached carelessly for the narrow neck of the salad-dressing bottle. Instead of grabbing the neck, I hit it with my knuckles and knocked the bottle over. To my wife's horror, the neck of the tumbling bottle struck a sharp blow to the edge of her porcelain dinner plate. A single, narrow, crescent-shaped chunk of porcelain, about 10 cm long, had snapped off. To ameliorate the situation, I suggested that I could easily repair the damage with a little glue. Looking down, my wife replied, "And what about this piece?" I looked in disbelief at the opposite and untouched side of her plate. There, just beyond the mashed potatoes, was another crescent-shaped "bite" in her plate. Somehow, the destructive energy of the salad dressing had been reflected from one side of the plate to the other.

I was surprised to find that this phenomenon is not very unusual. In *The New Science of Strong Materials* (Penguin Books, Toronto, 1968), J. E. Gordon discusses how stress energy is transmitted through materials. He cites some occurrences similar to what I observed at dinner.

Stress produced by an impact on an object travels in waves at about the speed of sound in the material. This speed is a function of the density of the material and its Young's modulus (a measure of the material's ability to resist deformation):

$$\text{Velocity} = (\text{Young's modulus}/\text{density})^{1/2}$$

The speed of sound in ceramics is several thousand meters per second. Since the duration of an impulse from an impact may be on the order of 0.01 sec, stress waves in a small object have the opportunity to undergo many reflections during the actual time of an impact. Even upon repeated reflection, the wave energy is not attenuated to any great extent, and when reflected waves meet incoming

Fig. 1.   Plate before   Plate after

waves, stress concentrations form and may initiate fracture at locations remote from the site of impact.

Gordon points out that in one routine impact strength test of the British Ceramic Research Association, a weight is dropped onto the center of a loosely supported ceramic tile. Usually, the tile is shattered where the weight strikes, but sometimes there is little damage at the point of impact. Instead, all four corners of the tile fall off!

The author further notes that when a bullet penetrates a tank containing liquid, the exit hole is always larger than the entrance hole, this due in part to the impact energy that is transmitted through the liquid to the opposite side. Similarly, impact energy can be transmitted through a human head, which in a sense is also a fluid-containing tank. A sharp blow to the forehead that does not even penetrate the bone can cause severe fracture to the bone at the rear of the skull. The purpose of the interior band in a crash helmet is to prevent this sort of bone injury.

HOW TO HEAR A MOUSE ROAR

James Burgard

Louisiana State University of New Orleans
New Orleans, La. 20122

The intensity of an observed sound depends, among other things, on the surface area of the emitter and its distance from the observer. A tuning fork, for example, when mounted on a sound box, is much louder because of the increased surface area of the box. Similarly, the vibrating strings of a guitar use the resonant quality and much larger surface area of the guitar body to amplify sound. A large surface area is necessary to couple the vibrating source to the air. The efficiency of this coupling is a function of frequency. Efficient emission of low-frequency sounds requires a large surface, hence the large size of hi-fi "woofers."

Intensity is also related to the distance at which sounds are observed. Since loudness is inversely proportional to the square of the distance of an object to an observer, sound energy rapidly disperses unless it is focused or conducted along a narrow channel.

A dramatic example of these sound transmission parameters can be demonstrated with some simple household items. A piece of tin foil, when suspended from a string and struck with some object, will produce an unmelodius but rather loud, crinkling sound. Contrast this example to an object with a much smaller surface area—for instance, a metal coat hanger. We find that decreasing the surface area decreases the amount of energy emitted into the air as sound.

The channeling effect of sound transmission can be studied by coupling a small surface area object directly to our ears. Cut and shape a metal coat hanger

into the form of a "U." To the ends of this rod attach a piece of thread or string approximately three feet long. Drape the thread over your head so that it hangs across your ears; using both index fingers, hold the thread snugly in each ear. It is important that the rod swing freely from the point where the index finger touches the ear. Now strike the rod. Alas! The faintly audible tinkling sound of a coat hanger looms into the awesome chiming of Big Ben.

A Classroom Apparatus for Demonstrating the Doppler Effect

Ernest Hammond
Morgan State College, Baltimore, Maryland

An effective classroom apparatus to demonstrate the doppler effect consists of a tuning fork mounted on a resonant wooden case attached to a 2-m wire spring stretched approximately 13 m along the laboratory floor. When the mounted tuning fork is at rest the students hear the original frequency; however, the motion of the vibrating tuning fork, when the spring is released, results in a distinct lowering of the frequency or pitch for the students nearest the fork's original position. Those students at the other end of the laboratory can detect an increase in the frequency.

The frictional force between the mounted tuning fork and the floor should be minimal for this experiment. A tiled or linoleum-covered waxed floor and a cork mounting at the base of the tuning fork will ensure ease of operation. Moreover, it is essential to have the prong plane of the fork vibrating in the same direction as the motion of the system for maximum effectiveness.

Vibration in pipes

The vibration of air in tubes is the mechanism whereby virtually all wind instruments sound. It also provides an excellent example of standing waves. All you need for these experiments are a few drinking straws, a sheet of paper, and a pair of scissors. We shall study tubes open at both ends, and closed at one end. Closed at both ends does not work, because the sound cannot get out! The frequency of the note produced by blowing gently across the end of a drinking straw is that of the first normal mode of this tube. It takes a little practice to draw the note out of this instrument, and you may have to vary the angle at which you blow, and the distance from your lips in order to hear it, but with a little practice you should have no difficulty. The wavelength λ of the first standing wave in an open pipe is twice the length of the pipe. Now, close one end of the straw while blowing across the other. The pitch falls an octave, since the first normal mode of a pipe closed at one end has a wavelength four times the length of the pipe. In this mode, air rushes in and out of the open end, the motion diminishing towards the closed end, where it is zero, as shown in Fig. 1. Listen to the timbre of an open straw, and one which is half as long closed at one end. Although they have the same pitch, they sound quite different. The open straw can vibrate with all the harmonics of the fundamental, but the closed straw possesses only the odd harmonics (i.e., three, five, seven times the fundamental etc.).

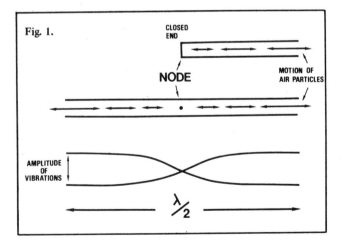

One can make a pitch pipe from a straw, since its frequency depends only on its length, and the velocity of sound, which is 344 m/sec (1126.6 ft/sec) at sea level and room temperature. The top string of a guitar is tuned to E_4 (329.6 Hz) which would be given by an open ended straw 52.5 cm long. It is more convenient to use a straw 12.5 cm long, which, with one end closed, gives the octave of this note. A correction should be made for the finite width of the straw. One should add 0.82 of the radius for each open end, so this amount should be subtracted from the length of the straw given above.

Vibrations in pipes open at both ends are represented musically by such instruments as the flute family, and the open flue pipe in an organ, often called the open diapason. However, most wind instruments employ reeds, which includes the lips, to provide vibrations, and these act effectively to close the end of the pipe. We can make simple reed devices with paper and straws with which to experiment.

Take a 6-9 in. square of paper and fold along one diagonal. Then open it out and proceed to roll the paper tightly around a pencil from one end of the diagonal crease to the other end, so that the diagonal rolls along itself as shown in Fig. 2. If a six-sided pencil is used, do not wrap it too tightly, or the pencil will not come out. When completely rolled it should look like Fig. 2A. Push the pencil out and glue the last fold at A., or hold it in place with a rubber band or strip of sticky tape.

Now from the point marked B in Fig. 3, at one end cut away on each side, in the direction indicated by the small arrows, until the end piece may be opened out into a triangle shape C. The cuts must be at right angles to the main roll and are each a trifle over one-third of the circumference of the rolled tube. Fold the triangular piece at right angles to the tube so that it forms a little cover over the end. Now place the other end of the tube in your mouth, and, instead of blowing, draw in your breath. This action will cause the little triangular paper lid to vibrate and the instrument will give a bleating sound. The noise can be made louder by poking the tube through a hole in the

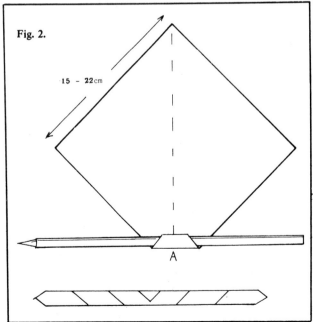

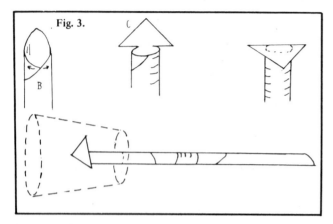

Fig. 3.

bottom of a Styrofoam cup, as shown in the figure, or through a paper cone which you can roll. This type of reed, in which the flap closes off the aperture completely, is known as a "beating reed," and is used in the clarinet, oboe, and bassoon. The clarinet has one reed, and the oboe and bassoon two reeds beating together.

One can make another type of beating reed from a drinking straw. Fold over the end of the straw, as shown in Fig. 4a, and tape it closed. Take a razor blade, or sharp knife, and make a rectangular cut about 2 to 3 mm wide, and 2 cm long. Put the whole of the reed in the mouth and blow, as shown in the figure. It can generally be made to sound by very slightly lifting the reed, and bending it up, so that when at rest the hole in the side is a little open. Such a reed is very similar to that used in the drone of a bagpipe, or antique instruments such as the hornpipe.

So far, we have provided examples of single reeds, but a double reed can be made from a plastic soda straw by cutting slits on either side at the end, as shown in Fig. 4b.

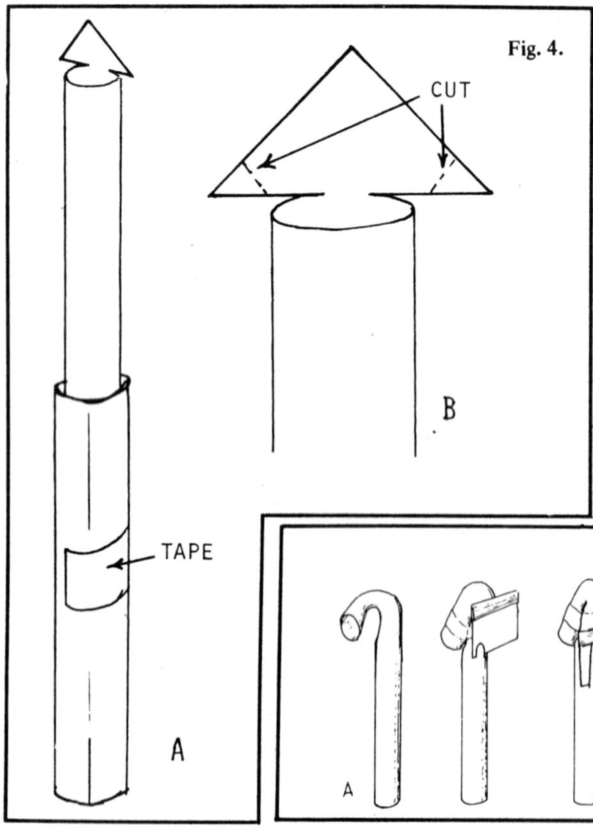

Some practice is required in blowing this device. Dick Minnix at V.M.I. says that chewing the cut end between one's molars so that it is flat greatly improves the response. Press the reed between the lips while blowing — the range of pressure under which you can make the reed sound is rather restricted, and you need to blow rather hard. This is similar to the double reed in an oboe. As before, you can insert it into a paper cone or Styrofoam cup, to demonstrate how the cone matches the impedance of the instrument to the air, and hence greatly increases the volume of sound. Now try cutting the tube of the reed pipes shorter, and shorter. Note how the pitch of the reed goes up and at some point the reed stops functioning. Clearly, the air in the tube is necessary to the functioning of the reed, even though the length does not determine the pitch in the same way as it does for an open pipe. After cutting the pipe shorter, roll a tube of paper around the outside and tape it as shown in Fig. 5a. Slide the outer tube up and down. The pitch goes up and down as you slide the tube in and out, and you can measure the frequency as a function of the length. A puff of air is allowed into the pipe each time the flap opens and closes, and such sharp puffs have a lot of high frequencies in them, in addition to the lowest frequency, or fundamental. Now, cut the corners off the paper reed, as shown in Fig. 5b. (Take care that the reed covers the tube — otherwise it will not work.) Observe that the pitch goes up. The natural frequency of vibration of the reed is higher the smaller the mass oscillating.

In a wind instrument the reed is not free to vibrate at its natural frequency, as here, but is forced to oscillate at the resonant frequency of the tube of the instrument. The paper reed can only vibrate at a low frequency, so you would need a tube about three feet long to be able to bring this reed into resonance with the fundamental mode. If you have such a tube you could try it out.

We can examine the way the voice works using this device. The paper or Styrofoam cup, which we placed over the end of the tube, has its own resonant frequency. The vocal tract (larynx, mouth) behaves similarly in the case of the voice. The resonance is at a high frequency, and tends to emphasize frequencies produced by the reed or vocal cords in this vicinity — these resonant frequencies are called formants in the case of the voice, and determine whether you are saying "oo" or "ah," even if your voice holds the same basic fundamental pitch.

While sucking on the reed, close the cup partially with one hand, then open it again. Doing so alters the formants, and it is quite easy to get the device to say "ma ma" or even more difficult vowel sounds with a little practice. (This is similar to the "wa wa" sound obtained by opening and closing the mute on a trumpet.)

Fig. 4.

Fig. 5.

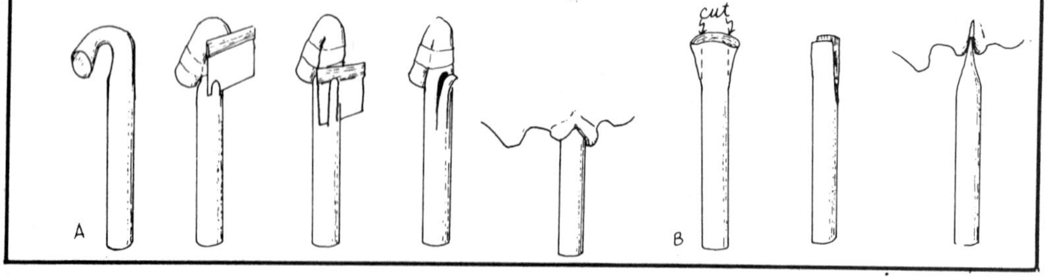

Demonstration Experiment to Show the Effect of Pressure Variation on the Velocity of Sound in a Gas

Kenneth Geo. Keith
Teacher of Physics
Holland, New York

THEORY

It is a well known fact that since the pressure-density ratio in a gas is a constant at constant temperatures, the velocity of a sound wave through that gas should be a constant regardless of the pressure. Devising an empirical proof of this, however, has been difficult since the more common experiments for determining the velocity of sound depend on resonance within containers which are open at one or both ends and are therefore not able to be evacuated nor to contain a compressed gas.

It is, therefore, a prime requisite that this experiment be conducted in a closed container. The container diagrammed in Fig. 1 was constructed and functions in the following manner.

When the aluminum rod is set into longitudinal vibration the disc at the end of the rod causes a sound wave to travel through the air to the opposite end of the tube where it is reflected. If the distance between the disc and the rubber stopper at the opposite end of the tube is $n\lambda/2$ cm, then the waves will be reinforced by the reflected waves and the standing waves will cause the cork dust to collect at various places in the tube called nodes and not at others, called antinodes. (A phenomenon well known from the Kundt's tube experiment used to determine the velocity of sound in a metal.)

The frequency of the wave will depend only on the frequency of the source (i.e., the aluminum rod) and will therefore be constant throughout the experiment. The velocity of the sound through the gas will then be directly related to the wave length which can be found by determining the distance between any two adjacent antinodes and multiplying by two. Any change in velocity as the pressure changes will also cause a deterioration of the nodes and antinodes since the waves will not be reinforced by the reflected waves.

EQUIPMENT

1 glass tube 91.5 cm long and 3 cm in diameter
1 aluminum rod 81 cm long and 0.4 cm in diameter with an aluminum disc of 2.8 cm diameter securely attached to one end.
2—one hole rubber stoppers, size 6½ or 7
1-10 cm length of glass tubing
cork dust
powdered rosin and small piece of leather
vacuum-compression pump and thick walled rubber tubing

The equipment is assembled as in Fig. 1. The rubber stopper should be centered on the aluminum rod.

PROCEDURE

PART 1—With normal air pressure in the tube the aluminum rod is stroked with a small piece of rosin-coated leather. This sets the rod into vibration and standing waves will be produced in the air column. If standing waves are not evidenced by the cork dust the position of the rod may be altered slightly until the distance from the disc to the stopper is $n\lambda/2$ cm where resonance will be observed.

The number of nodes and antinodes is determined and the distance between the disc and the stopper is measured.

PART 2—The pump is now activated briefly and the air in the tube is compressed. After jarring the tube slightly to distribute the cork dust evenly and allowing a minute or so for the temperature to stabilize, the rod is again stroked. The number of nodes and antinodes is determined and the distance from the disc to the stopper is measured.

PART 3—The pump is reversed and the tube partially evacuated. The tube is again jarred slightly and the temperature is allowed to stabilize. Stroke the rod again and determine the number of nodes and antinodes and measure the distance between the disc and the rubber stopper.

RESULTS

When Part 1 of the procedure was conducted, 9 well formed nodes and 10 antinodes were observed and the distance between the disc and the rubber stopper was 51 cm. The wave length of the sound in air was calculated to be 11.3 cm by dividing 51 cm by 4.5 waves.

When the air in the tube was compressed in Part 2, exactly the same number of nodes and antinodes were formed with no measurable change in the distance. Again the wave length was 11.3 cm.

After partially evacuating the tube and performing Part 3 of the procedure the same results were obtained.

CONCLUSIONS

Since no change in the wave length of the sound waves was observed when the air was compressed or when it was rarefied, and since the frequency is constant throughout the experiment, it is reasonable to conclude that the velocity of the sound was not affected by altering the pressure of the gas within the tube.

NOTE

In order to make a more quantitative study it would be advisable to include a pressure gauge and a thermometer in the apparatus although the result is strikingly apparent with the modest equipment described.

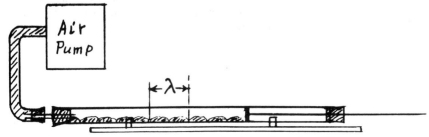

Figure 1. Arrangement to show constancy of velocity of sound at different pressures.

An acoustics demonstration for students interested in music

J.S. Faughn and S.W. Slade
Eastern Kentucky University, Richmond, Kentucky 40475

The average high school or introductory college physics class usually contains a number of students who either play a musical instrument or have an interest in music. This note describes an experiment or demonstration that shows how a musician can visually tune an instrument to a standard source. The desired result is, of course, to capture the attention of a particular group of students. The system described here has a value as a general teaching tool because it *demonstrates* visually such properties of sound waves as frequency, amplitude, and interference.

The basic method discussed here is used in a commercial device called the "Stroboconn," manufactured by G.C. Conn, Ltd.

The necessary equipment consists of an amplifier, a microphone, a turntable, strobe discs, and a strobe light with an external input. Also required is a sound source; this can be a sine-wave generator connected to a loudspeaker or a tuning fork and an instrument to be tuned such as a piano or a trumpet. The apparatus is connected as shown in the accompanying diagram.

In order to tune an instrument, a known frequency note (say 440 Hz) is sounded into a microphone. The output is then fed to the external input of a strobelight. The strobe used was a General Radio type 1531 Strobotac which requires a 2-V rms electrical signal to trigger it. This triggering requirement necessitates the amplification of the microphone output before it is applied to the strobe. The amplifier in most tape recorders or record players should be satisfactory. This input causes the strobelight to blink at the frequency of the sound source.

This light is directed at a stroboscopic disc rotating on a turntable. The disc consists of a number of concentric circles, each of which is broken up into a number of equal-sized divisions by radial lines (e.g., Cenco Scientific Catalog number 74695). When illuminated in this manner, one or more bands on the disc appears to stop rotating. If none of the bands on the rotating disc stands exactly still, the turntable speed can be slightly changed to achieve this effect. A precalibrated turntable speed is not necessary for this demonstration. After noting exactly which band is

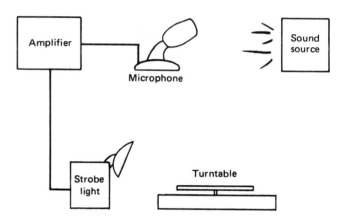

Fig. 1. *Block diagram of experimental set-up.*

stopped, the standard source is removed and the instrument to be tuned (say a trumpet) is sounded into the microphone at what is presumably the frequency of the standard. If the note is flat, the originally stopped band will be observed to slowly rotate clockwise, and if sharp, to rotate counterclockwise. By suitable adjustment of his instrument, the observer can bring himself into tune with the standard source.

This same experimental arrangement can also be used to visually demonstrate beat frequencies. If two tones of about the same frequency are played simultaneously into the microphone, a complex wave is formed which causes the strobe to vary its output intensity at the difference frequency.

With just a little more difficulty, this system can be used to demonstrate the Doppler effect. With a known frequency being emitted by the sound source, note which band of the strobe disc is stopped. When the microphone is hand held and rapidly moved toward the sound source, rotation of this band indicates that the note is not sharp; and when moved rapidly away, the note appears flat.

A slight disadvantage to the demonstrations described is that students must gather around the strobe disc in small groups to observe the effects.

Sound intensity and good health

Hugh Haggerty
Cardinal Mooney High School, Rochester, N.Y. 14615

With the awareness today of ecology, I thought it would be interesting to give our students a method of measuring sound pollution in school. Some of the commercially available decibel level meters are good, but the data the student obtains becomes a little more dramatic if a recording of the sound levels is made.

To this end we enlisted the help of our Health Science Program, and Introductory Electronics course, to set about designing a system that could be used to make a graphic representation of the numerical data that can be obtained from a sound level meter. The results were not only spectacular, but we hope, will lead our student body to take serious action to correct what might be considered an "unhealthy" situation in our school.

A Realistic Music/Sound Level Meter #33-1028, sold for $49.95 by Radio Shack was selected because of its price and more importantly, its circuit design. The meter is essentially a null indicating meter, employing a bridge circuit. The variable scales from 70 dB to 110 dB in 10 dB ranges make it convenient to switch from a less sensitive scale to a more sensitive scale if such is desired, without re-calibrating the recording equipment.

A Heath Co. servo recorder, Model No. EUW-20A was employed as the recorder in the system. The chart speeds of this recorder are variable, and it was found that the 2 min/in speed was best suited for studies of prolonged programs, since the servo mechanism gave a nice profile at this speed. The output of the sound level meter was fed into a simple full wave bridge circuit to insure the servo recorder's response to the full output of the meter. The output jack of the sound level meter was another feature which prompted our purchase of it. Since the servo recorder's sensitivity to the DC output of the sound level meter had to be contained on the 250 mV level, a voltage divider was used to insure that a maximum of 250 mV would be fed into the servo recorder. The actual output of the sound level meter varied from a few millivolts to 2 volts, depending on the sensitivity of the scale selected. Overloading the servo recorder would have caused damage to the equipment, hence the necessity of the voltage divider.

Preliminary tests with an audio frequency generator attached to a speaker, and held near the sound level meter, gave us a method of calibrating the servo recorder. When the level of the A.F. generator output was great enough to cause the sound level meter to indicate 0 on the 70 dB scale, the position of the pen on the servo recorder was adjusted to the 50 mark on the chart scale. This reading of 0 indicated a 70 dB level of sound on the recorder. If the meter is then switched to the 80 dB scale, the 50 mark on the chart now becomes the 80 dB level. Reading the meter for 82 dB or 78 dB, and marking those positions on the chart will show the variations of the servo recorder pen with input variations from the sound level meter. In switching to higher scales on the sound level meter, the 50 mark on the chart indicates the position of the scale selected, and variations from the 50 mark are used to measure the decibel level of the sound recorded.

Having obtained a method of calibrating the servo recorder with a fair degree of accuracy, it was discovered that the response of the servo recorder was not linear with respect to the output of the sound level meter. Since we were only searching for a method to illustrate dramatically sound levels which were injurious to health, and we wanted the ordinary person to be able to interpret our readings, it was at this point that we modified our calibrating technique.

The system could be set up rapidly on a movable table, and transported to any part of the school. The calibration technique consisted of looking at the output of the sound level meter until, on any scale, a definite known value of dB output could be read, then the pen position for that output was adjusted. All other outputs were relative to this position, on all other scales. This eliminated the necessity to calibrate with the AF generator. The new calibration technique was essentially the same as the old technique, but used existing sound, rather than the AF generator.

The block diagram in Fig. 1 shows how the sound level meter, recorder and associated circuits should be connected. Any diodes may be used for the bridge circuit.

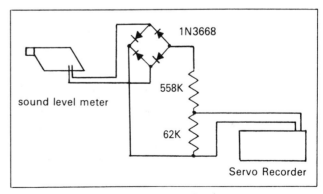

Fig. 1. Sound intensity measuring circuit.

The bridge circuit and voltage divider may be connected on a circuit board, and assembled as a complete unit with an appropriate input jack and output jack, so that the novice electronic, health, or biology student will have little difficulty in connecting the sound level meter to the recorder.

In the event you do not have a servo recorder, an oscilloscope may be used to show the output of the sound level meter. The bridge circuit using the crystal diodes should be used. You can calibrate the oscilloscope in much the same manner as the servo recorder was calibrated, and a visual demonstration of sound may be made to your physics, health, biology, or electronics class. You can use a radio, or some other method of producing sound of various levels, and show the output level on the CRO tube.

Fig. 2 shows the calibration technique using the AF generator, with the sound level meter set on the 70 dB scale, changing to the 80 dB scale, and then the 90 dB scale. There is a corresponding movement of the pen position of the recorder. Here a 1 000 Hz signal was used.

A recording made during a student dance is seen in Fig. 3.

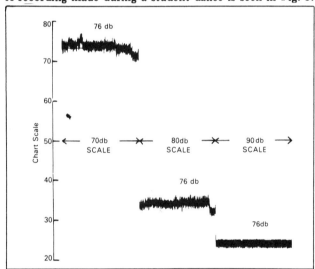

Fig. 2. Changing scales using a calibrated signal.

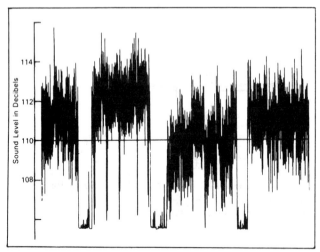

Fig. 3. A dance recording —— sound level in decibels versus time.

This recording was made some 90 feet from the band. At this point we get into the health aspects of what is being recorded. A 103 dB level of sound can cause injury[1] to the auditory system of a human. You will note that the students at this dance were exposed to levels of sound which were well above the 103 dB level. The sound level meter had to be on the 100 dB scale to prevent the servo recorder from going off the scale. Needless to say, even between songs, indicated by the low sound levels (which were measured at 100 dB), the sound was too much to bear. This writer uses wax ear plugs on such occasions, to retain his sanity.

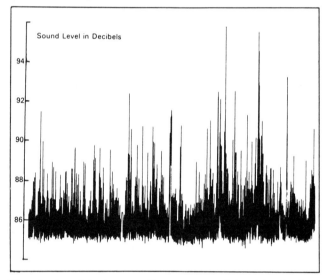

Fig. 4. A cafeteria recording —— sound level in decibels versus time.

Another interesting recording is pictured in Fig. 4. It was made during a Cafe period. The 90 dB scale was used. Our school has three lunch periods with about 500 students in each lunch period. The level of the sound as indicated in the recording was not too bad. It fell into the realm of tolerable.

Student interest in this project has lead some of them to start proceedings with the Student Council to limit the sound level which a given band may use when playing for a student dance. In the past, some students have complained of how it "hurt" them to go to a student dance. Now they

Table I: Sound Pressure Levels[2]

Sound Level in Decibels	Loudness	Noise Source
0 - 20	very faint	whisper, rustle of leaves
20 - 40	faint	quiet home, quiet conversation
40 - 60	moderate	noisy home, average conversation
60 - 80	loud	noisy office, average TV
80 - 100	very loud	noisy factory, loud Hi-Fi
100 - 120	deafening	thunder, boiler factory, jet engine

can "see" why it was "hurting" them.

We hope that our project will start your students on the road to realizing that they are going to damage, or possibly lose their hearing, if they continue to expose themselves to such high levels of sound pressure.

References

1. Realistic Music/Sound Level Meter Operating Instruction, and Service Data (Radio Shack Cat. No. 33-1028)
2. P. Sarnoff, *Dictionary of the Environment* (Quadrangle Books, New York, 1971)

MAGNETICALLY DRIVEN SONOMETER

The phenomenon of resonance, as well as all of the classical experiments performed with a sonometer or monochord, can be dramatically demonstrated with a magnetically driven vibrating string. Although similar demonstrations have been described in the literature,[1,2] this apparatus is apparently not well known to many physics teachers.

One attaches a metal wire to a sonometer box in the usual manner. The ends of the vibrating string are connected in series with a power resistor (5Ω, 10W) to the output of a monaural hi-fi amplifier. (Fig. 7.) A sine-wave generator is connected to the input of the amplifier. A large permanent magnet is placed near the vibrating string so that part of the string is in a strong, constant magnetic field. When the signal generator is set to one of the normal mode frequencies of the vibrating string, the string sings out quite loudly.

This is a vivid example of resonance frequencies and the harmonic series because the students can easily hear the manifestations of these physical concepts.

A beautiful, harp-like arpeggio is produced by smoothly turning the tuning dial on the signal generator through the higher harmonic frequencies. This never fails to produce a favorable audience reaction; I cannot imagine a better presentation of the beauty in physics.

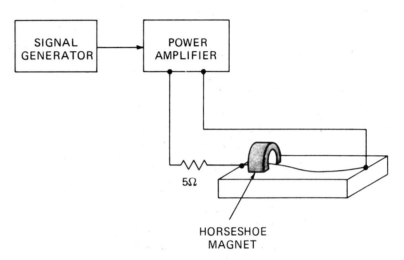

Fig. 7. *Magnetically driven sonometer, connection diagram.*

By using a flexible wire under low tension (e.g. Nr.20 copper wire under about 25 newtons tension) and a stroboscope, one can inspect the shape of the vibrating string for the first several normal modes.

An electronic counter can be employed to measure the driving frequency. If desired, an inexpensive crystal phonograph cartridge, with stylus removed, may be placed on the string. The output of the cartridge can be displayed on an oscilloscope.

One point of caution: the vibrating string may snap during the experiment. One should not stand at the weight end of the sonometer. It would be prudent to insist that people using the apparatus wear safety goggles.

Acknowledgment. I thank Professor Herschel Neumann of the University of Denver Physics Department for guiding my investigation of vibrating strings.

RONALD B. STANDLER
*New Mexico Institute of
Mining and Technology
Socorro, New Mexico 87801*

References:
1. Marcus O'Day, Amer. J. Phys., **8**, 326 (1940)
2. W. David Bemmels, Amer. J. Phys., **17**, 515 (1949)

OSCILLOSCOPE MEASUREMENT OF THE VELOCITY OF SOUND

Almost any oscilloscope can be used to measure the time a sound pulse requires to travel a known distance. This is an easy task for a scope with a triggered sweep and a calibrated time base, but this note describes how the measurement can be made on less expensive oscilloscopes without these features.

The oscilloscope is set to its slowest sweep rate, typically about 10 per sec. Pulses are taken from the scope sweep, fed to the amplifier, and then sent to the speaker across the room (Fig. 4). The loud thumps or clicks from the loudspeaker obviously coincide with the visible sweeps of the CRT beam, so that no initial "pip" on the screen is needed to mark the start of the sound pulse from the speaker. The microphone is connected to the vertical input of the oscilloscope which shows the arrival of the pulse several milliseconds later. The display usually looks something like Fig. 4, where the microphone signal is not a single pulse but many, probably due to room echoes. It is convincing to bring the speaker right up to the microphone and watch the time delay decrease and vanish while the height of the received pulses gets very large.

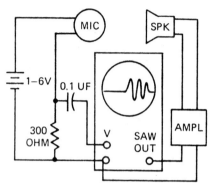

Fig. 4. *Sawtooth output from an ordinary oscilloscope generates sound pulses which are picked up by a carbon microphone.*

The time base can be calibrated with the aid of a tuning fork held close to the microphone.

The accuracy of the measurement will only be within 10%. The time delay before the arrival of the sound pulse is only a few centimeters on the scope face.

A carbon microphone (79¢, Radio Shack) works well in this application. Its advantage is a relatively large output, several tenths of a volt. The disadvantage of a carbon microphone is that a battery and resistor must be connected in series with it.

Some oscilloscopes will furnish sweep pulses more readily than others. The EICO 427 (Macalaster, Frey) has a sweep sawtooth available on the front panel. The Heathkit IO-10 has a binding post connection for an external timing capacitor which provides a small sawtooth voltage. Some scopes have rear terminals that connect to the deflection plates. In all cases the sweep voltage should be attenuated to a low enough level so as not to damage the audio amplifier.

If no sweep or blanking voltage is available, a simple modification is possible that does not require connecting to the scope circuitry. Wrap several turns of insulated wire around one of the leads to the horizontal deflection plates of the CRT. This will couple enough voltage from the flyback part of the sawtooth to give pulses to the amplifier. It would be best to bring this signal to a binding post on the front panel.

This makes an interesting supplement to the treatment of the time-of-flight mass spectrometer in the PSSC course. The use of the oscilloscope as a fast timer and the similarity of the demonstration to sonar and radar make it valuable to any physics course.

ROBERT H. JOHNS
The Academy of the New Church
Bryn Athyn, Pa. 19009

THE HARMONICA, AN AUDIO-FREQUENCY GENERATOR

A chromatic harmonica offers about 40 fairly standard frequencies in the range of 200 to 2000 Hz. If these frequencies are written on the side of the harmonica it becomes a handy source of known audio frequencies in the physics laboratory, and an aid in estimating unknown ones by comparison. Figure 2 shows two scales that can be cut out and glued to the sides of a harmonica. They were prepared for a Hohner Chrometta 12, key of G, which is available in music stores for about $17. The letters are the musical notes, to be put on one side, and the numbers are the frequencies of those notes, to be attached to the other side of the harmonica. The lines should line up with playing holes. The frequency scale is mounted upside down so that it can be seen and read without moving the harmonica very far from the mouth. The scales will last longer if they are coated with transparent adhesive plastic sheets that are used to protect wallet cards.

The underlined frequencies are played with the slide in. The top two rows of notes are produced by blowing, the bottom rows by inhaling. In general the notes progress up in half steps or semitones, but there are many exceptions. The reasons for the redundancies will become apparent if you learn to play songs on the instrument.

The two pairs of G's on my harmonica produce beats when blown together. Beats can also be heard when one sings and blows the same note. Doppler shifts can be estimated with a harmonica. A frequency shift of a little more than two cents per mph is produced by the Doppler effect, so that a car horn passing at highway speeds will drop several semitones in pitch. (There are 100 cents to a tempered semitone.) Car horns can be easily taped with portable cassette recorders.

This has been a valuable laboratory instrument, and still a lot of fun for playing by the campfire!

ROBERT H. JOHNS
Academy of the New Church
Bryn Athyn, Pennsylvania 19009

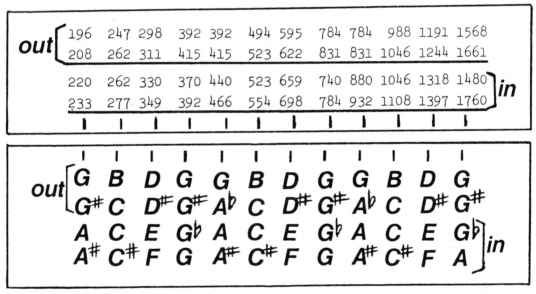

Fig. 2. To help identify the notes and frequencies produced by a Hohner Chrometta harmonica, key of G, make a photocopy of the above and paste it diretly on the harmonica.

A "SOUND" CROSSWORD

Veralee Falkenberg Wisconsin State University, River Falls, Wisconsin 54022

Much concern has been shown in the development of techniques that make physics a subject more appealing to high school students. Choosing games which are both familiar to the student and able to enhance his understanding of physics is an important step in this development. A crossword puzzle, as will be seen, can do both of the above. The following crossword puzzle has been revised and is ready for use with a unit on sound.

The original puzzle was designed as an activity for a teaching presentation, required in the course "Curriculum Physics" offered at Wisconsin State University—River Falls. This original contained names of classmates. Spaces containing names are Across: 10, 50, Down: 9, 37. Those using the puzzle may wish to alter other parts of the puzzle to accommodate other names, especially when small classes are taught. Examples for names to be used are those of the principal, superintendent, physics instructor, prom queen, basketball star, or custodian. The choice is unlimited. After reading through the "Across" and "Down" selections, it is evident that the puzzle is not entirely serious physics. There are many questions needing thought and imagination in order to be answered. As a result, this crossword puzzle should prove to be an enjoyable learning experience for the student. Good luck in its use!

SOUND—A POTPOURRI OF PHYSICS TEACHING IDEAS

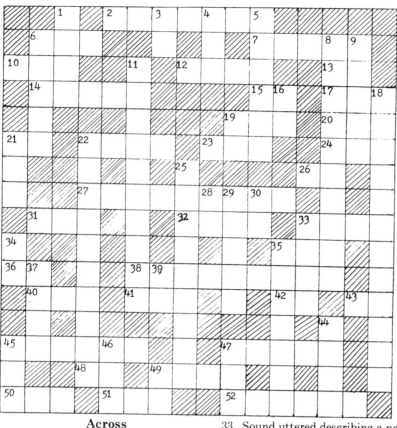

DOWN

1. Do you _____ the table vibrating?
3. Sound signifying the affirmative.
4. _____ should transmit sound more efficiently than its liquid form, water.
5. In the equation $v = f\lambda$, v stands for _____, spelled differently because we stutter!
6. Disturbance in matter to which the ear is sensitive.
8. Another word for superposed.
9. A famous man of 1492 would call his first son this if he were living today.
11. λ represents _____.
16. "Do, a deer, a female deer, _____, a drop of golden sun, ..."
18. Waves producing sound are _____.
19. "_____, a note to follow so, ..."
21. _____ $= \log I/I_0$.
22. These vibrations are so low that they are not sensitive to the human ear.
25. Two waves traveling in opposite directions on a rope always _____.
28. Every other vowel beginning with the second vowel.
29. Sound produced by irregular vibrations in matter.
30. Third and twelfth letters of the alphabet.
33. This object beckons us to classes.
34. "_____ a long, long way to run ...."
35. The _____ Effect.
37. This man said "I do" to Julie Nixon.
39. A short greeting.
44. We usually begin determination of a wave period at time $t = $ _____.
45. A graduate student in science education works for his _____ degree.
46. Sound uttered when you are disgusted.
47. A young animal that makes a barking sound. You probably have one at your house.
49. Sound issued forth when you look at something good to eat.

Across

2. Modern _____.
6. Unit of time (abbr.)
7. Having to do with sound.
10. Deceased senator of Wisconsin. He liked to "toot his own horn."
12. The speed of sound is greater with increase in _____. (abbr.)
13. Two words for discordant frequencies. (init.)
14. Those vibrations in matter above 20 000 cps are _____sonic.
15. Sound uttered when you have said the wrong thing.
17. Another word for path without the vowel sounds.
19. Less disturbance of sound waves results on this side of a mountain.
20. The middle vowels.
21. Two letters that sound a description of eyes.
22. A tuning fork in an _____ of apparatus used when studying sound.
23. &*$%¢=*%&¢ are "sounds" uttered when Sarge is _____ at Beetle.
24. Risen is not much without its vowel sounds: _____.
26. Sound made when giving a condition whereby something must be done before something else is allowed.
27. f stands for _____ in $v = f\lambda$.
31. We hear by means of one _____ when the other is covered.
32. A stone dropped into a _____ of water creates circular waves.
33. Sound uttered describing a naughty child.
35. The number 455 represents these letters of the alphabet when letters are numbered $A = 1$, $B = 2$, $C = 3$, etc.
36. Rather than yelling in the streets "for sale," we place an _____ in the newspaper.
38. There is a _____ of hearing.
40. The bel is named after a famous man. Suppose that he wished to change his last name to Olson; what would his initials be?
41. Another sound for "mod."
42. Abbreviation for please.
43. "_____, a drink with jam and bread."
45. Sound travels through a _____.
47. The doctor hears your heart beat, but feels your _____ at the wrist.
48. The first and last letters of the word meaning summoning.
49. Sound issued when you wish to look at the selection of food at a cafe.
50. Mr. Hanson's first name (shortened version). This man took a grass seed from New York to the Carolinas in the 18th century.
51. Sound made when you carry a tune, but do not pronounce the words.
52. The time it takes for a wave to complete one cycle of its vibrations. Also, the inverse of the frequency.

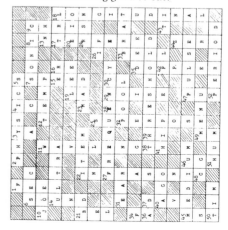

Children's toys

D. P. Jax Mulder
*Laboratorium voor Algemene Natuurkunde
Rijksuniversiteit, Westersingel 34, 9718 CM Groningen, The Netherlands*

Looking at old volumes of *The Physics Teacher* I noticed the article by Susan S. Welch: "What makes it turn?"[1] Perhaps this is still a question? Or has the problem been already solved? In any case, I thought I might convey to you my suggestion for Dutch physics teachers to use the device to demonstrate circularly polarized waves.

Circularly polarized waves can only be completely explained theoretically. You can try to demonstrate two

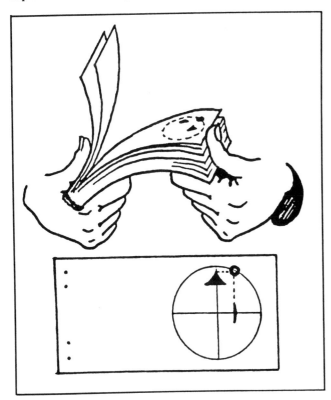

vibrations, perpendicular to each other, with a 90° difference in phase, and its resultant: a circular movement! However, that requires swinging both arms and legs, producing a knotted demonstrator. It is virtually impossible!

With a small moviebook it can be shown. It's a little booklet, about 5 x 10 cm with 30 to 40 pages of drawing paper, stapled together on the narrow side. On every page is drawn a circle with two vibrations, perpendicular to each other, in several phases, with their resultant on the circle.

Take 12 phases for one rotation, and when the pupils quickly run over the leaves of the booklet with their thumb, they will see the circular wave movement as a kind of cartoon film.

Once, at an agricultural school, a colleague showed me a child's toy of the sort described in reference 1; a wooden stick ribbed on one side, which turned a propeller arbitrarily to the right or left when the stick was stroked. But how does it work? Or, as Susan Welch would say: "What makes it turn?"

This mill is a very amusing demonstration of circularly polarized waves. The cost of this experiment is about 50 cents. The time needed to make it yourself about 30 minutes.

Take a small stick, of square cross section, about 2 cm and a length of about 40 cm. Use redwood, pine, or a similar wood, but do not use wood that is too strong or too heavy.

On one of the edges, cut notches every centimeter for 2/3 of the length of the stick. The sharp edge can be smoothed with sandpaper.

Rubbing the stick with a back and forth motion on the notches using a pencil or your housekey, will cause the stick to vibrate. The part without notches is the handle. On the end of the stick away from the handle we mount a small propeller: a flat piece of thin wood, about 2 x 8 cm with a hole in the center of 3 mm diam. The propeller is mounted freely on the end of the stick using a thin nail through the hole. Free and easy movement of the propeller

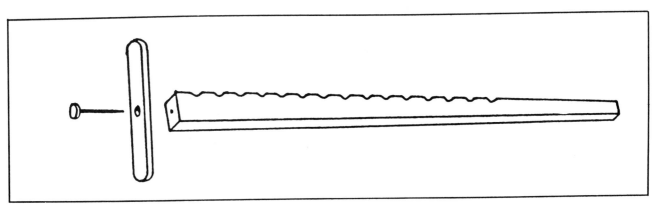

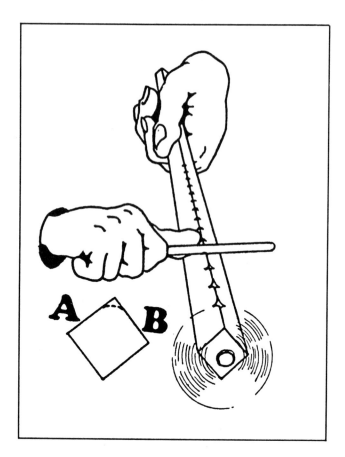

is required and the nail needs room for circular vibration in the hole. What makes it turn?

Well, when you rub the stick with your pencil or housekey, you produce vibrations in the stick along the diagonal of the cross section. This diagonal vibration is resolved into two vibrations perpendicular to each other in the direction of the two small dimensions of the stick. For here the flexibility is greater than in the diagonal direction.

The propeller then only rattles, for both vibrations are equal and coherent, assuming the stick is made of tolerably homogeneous wood.

If you hold your thumb along side A while stroking the notches, the mass of your thumb will reduce the vibration perpendicular to plane A and will also delay the phase of these vibrations (the extra mass makes the oscillation lag).

The result is that the nail is moved by two vibrations, perpendicular to each other, of which one is retarded in phase by the mass of your thumb.

The nail vibrates with a circular motion and the propeller starts to turn.

When you rub with your finger on side B while stroking the stick, the situation reverses and so does the propeller.

For schoolchildren it's a rather nice experiment. You can also keep your mouth shut about circularly polarized waves, but with a little dexterity you can turn the propeller at will to the right or left. It is an intriguing matter and interested pupils will often make a mill for themselves.

References

1. Phys. Teach. 11, 303 (1973).

A POTPOURRI OF PHYSICS TEACHING IDEAS—TOYS

What makes it turn?

Susan S. Welch

University of Massachusetts, Boston, Massachusetts 02116

We recently received the following letter and drawing from a student of Prof. Kenneth Ford at the University of Massachusetts, Boston. Readers are invited to send explanations of the effect.
—the Editors

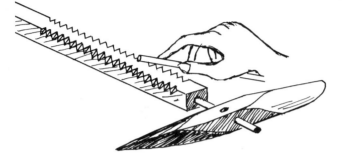

Fig. 1. *The perplexing stick-propeller device.*

The stick-propeller device shown in Fig. 1 appears to produce angular momentum from nowhere, as the stick is stroked with a back-and-forth motion, the propeller, which is free to rotate on its shaft, will turn as follows:

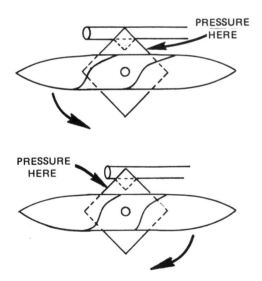

Fig. 2. *Motion induced with a stick of square cross section.*

If the stick has a *square* cross-section and is simply stroked, the propeller will not turn. If the stick is rubbed with a finger along one side while the notches are stroked, the propeller will turn as shown in Fig. 2. A stick of *rectangular* cross-section will rotate the propeller when only the notches are rubbed (Fig. 3).

The propeller on my stick was cross-drilled to see if the slant of the blades had any effect—none was evident. (A flat piece of wood will also behave in the same manner, provided the hole is not too loose on the shaft.) I have thought that per-

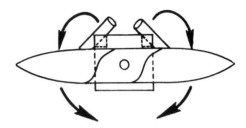

Fig. 3. *Motion induced with a stick of rectangular cross section.*

haps the stick responds to this stroking of the notches by vibrating with some angular momentum, as shown in Fig. 4. An evenly stroked stick would have no net angular momentum, while a stick with pressure applied along one side, or of rectangular cross-section, might have one mode of vibration preferred over another, and therefore impart net angular momentum to the propeller.

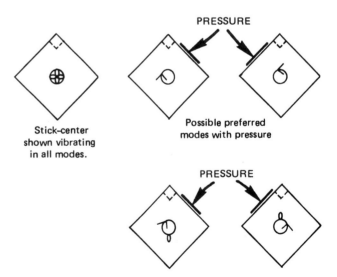

Fig. 4. *Possible vibration induced by stroking the stick.*

I was going to work with an advanced physics student at the University of Massachusetts and check up on my idea using a stereo phonograph needle hooked up to an oscilloscope to try to get a magnified trace of the stick's motion, but time considerations forced me to lay the project aside. The stick, propeller, and stroking stick I use are whittled out of walnut, with the shaft fitted fairly carefully to the hole in the propeller. For such an intriguing device it is very easily made!

A mechanical toy: The gee-haw whammy-diddle

Gordon J. Aubrecht, II
Department of Physics, The Ohio State University, Marion, Ohio 43302

The physics teacher at a small college or regional campus, such as The Ohio State University at Marion, has the opportunity to learn about interesting phenomena from other, nonscience, faculty and then to attempt an explanation. One such discovery, which I have since used in class, arose from the teaching of a course on Ohio folklore. A Project-60 student (someone 60 years old or over who may attend classes free) brought one of his childhood toys to the class. His teacher and he came to me to ask me to explain the action of this folk toy, which he called the gee-haw whammy-diddle. It is (or was) used in Ohio and Kentucky, and consists of a notched stick with a loosely mounted propeller on the end. When another stick is moved back and forth across the notches, the propeller spins (Fig. 1).

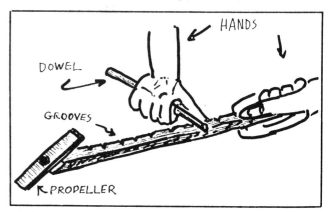

Fig. 1.

The toy can be made to work in ways that seem mysterious. It may even seem that it is "supernatural": "When my eyes are closed, the propeller goes clockwise; when they're open, it goes counterclockwise."

Does the propeller reverse itself spontaneously? How is it that a back-and-forth motion of a dowel on the notches cut into a stick can cause rotary motion?[1]

Of course, it is not mysterious at all.[2-6] It did provoke spirited student discussion in my physics class when I asked them for an explanation. In the end, the class did see how it worked (and the applicability of physics to everyday phenomena).

One apparent mystery is how the motion of the little dowel over the grooves can produce rotary motion. Eventually students realize that an intermittent force is being transmitted from the jiggling end of the notched wood shaft, which must be hand-held, to the propeller by contact with either the end of the shaft or the nail head. Because the propeller is a very loose fit on the nail, the force is applied at some distance from the axis of rotation. Scott[5] suggests that the propeller is driven in the same way a child keeps a hula hoop turning.

The rotary motion is in a direction determined by the direction in which the stick is free to vibrate. In Fig. 1 we see the finger position leading to clockwise motion. The thumb of the hand holding the dowel slides against one side of the stick, damping out the vibration in that direction. Thus the shaft is free to transmit only one component of the force (the other being absorbed by the hand holding the dowel). To obtain counter-clockwise motion the forefinger is extended past the dowel so that it bears on the other side of the stick, thus absorbing the other component of the vibration.

It is also possible to use the hand holding the whammy-diddle itself to absorb one of the degrees of freedom, leading to an equivalent result. This takes a lot of practice.

The spacing of the notches is about a centimeter on my whammy-diddle. The toy works even if the spacing is not uniform (I built one that way and tried it). The mounting hole in the propeller has a diameter about twice that of the nail. The size of the propeller does not seem to matter, as long as the mass is not too great. I made test propellers from 20 to 40 cm long, using various materials, including metal, and all worked.

The greatest advantage of the piece of apparatus is that it provided me with the opportunity to have fun by amazing and mystifying my colleagues and students. Then came the unadulterated pleasure of having nonphysics colleagues sit still for an explanation of physics. (It was the first time for some that I had ever been able to capture their attention for physics.) It also goes to show, as Miller pointed out,[3] that "not all physics is in books." I hope you will have as much fun with it as I did!

References

1. S. S. Welch, Phys. Teach. **11**, 303 (1973), asked this question. She received no answer in *The Physics Teacher*, though it had been discussed in the literature.
2. R. W. Leonard, *The American Physics Teacher* (now *American Journal of Physics*) **5**, 175 (1937).
3. J. S. Miller, Am. J. Phys. **23**, 176 (1955).
4. E. R. Laird, Am. J. Phys. **23**, 472 (1955), calls it an "Indian Mystery Stick."
5. G. D. Scott, Am. J. Phys. **24**, 464 (1956).
6. J. Walker, *The Flying Circus of Physics with Answers* (Wiley, New York, 1977), discusses this in Question 2.68, pages 41, 237, 238 based on Refs. 2-5.

A "perpetual motion" toy?

Donna A. Berry
Shaker Heights High School, Shaker Heights, Ohio 44120

Many toys are designed to educate as well as to entertain. It is interesting to note the number that can easily illustrate basic concepts of physics.

A toy often creates enthusiasm in a physics class that has little interest in conventional demonstration equipment. Toys are often an inexpensive way to demonstrate physics principles. These unique demonstrations are useful to vary every teacher's standard classroom procedure, and to illustrate everyday applications in an amusing manner.

An interesting example is the "kinetic ballerina." "She is poetry in perpetual motion as she pirouettes in dainty whirls and twirls and twists above the silvery metal stage. Dancing by kinetic action, she'll entertain you for hours."[1,2]

This toy is an ideal way to use the inquiry approach. The class was intrigued by the skater and thought-provoking questions of how and why it worked led to a discussion of perpetual motion and kinetic energy. The students were given time to formulate their hypotheses on the underlying source of energy of the kinetic skater and to pose a way of testing their predictions.

As hypotheses were tested and eliminated, one group suggested iron filings be placed on the platform (Fig. 1) revealing a circular magnet. The skater's foot also contained a small magnet (Fig. 2). The toy is nothing more than a simple pendulum. The initial release of the skater sets the oscillations in motion and the magnetic fields do the rest. The random oscillations continue, but they eventually die down. These observations led to a more complete discussion of magnetic fields and further physics concepts.

"Perpetual" she is not, but "practical" she is.

References

1. "Kinetic ballerina," *Taylor Gifts*, 355 E. Conestoga Road, P.O. Box 206, Wayne, PA 19087, $10.98.
2. I purchased the "kinetic skater" at a gift shop for $9.95. (Magnetic Kinetic, kinetic art by *Otagiri* (#15/53 Figure skater), c. 1979, made in Japan)

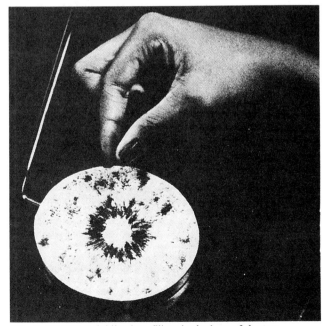
Fig. 1. Student sprinkling iron filings in the base of the toy.

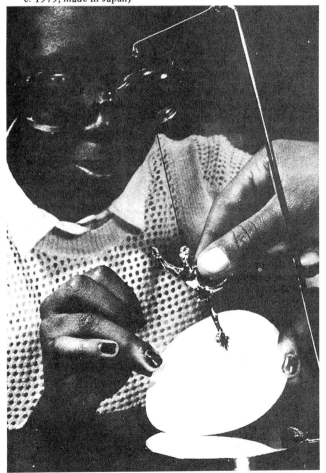
Fig. 2. Students showing the magnet that is in the skater's foot.

More toy store physics

David P. Koch
Hoover High School, North Canton, Ohio 44720

From time to time *The Physics Teacher* has published notes calling attention to toys that can be used to demonstrate physical concepts. Here is another example of a toy that is useful in introductory physics classes.

An inexpensive version of a physical model of the inverse square law for gravitational attraction can be found in the children's toy department of your local discount store. It is manufactured by Playskool, and can be purchased for about $4.00. The toy is the Baby Drum Drop.

The drum has a flat, 14-cm diam surface on one side, but the other side curves down toward the center. Three plastic balls, 4 cm in diameter, come with the drum, and work well for demonstrating elliptical orbits. They can be retrieved through an opening in the side of the plastic drum.

Although it would seem that a smaller projectile, such as a marble, would remain in "orbit" longer, best results were obtained with the hollow balls supplied. The orbit of the ball in the photograph lasted long enough for the ball to be placed in motion, the camera and flash unit to be aimed, and the photograph to be taken.

Had this drum not belonged to my 20-month-old son, the next project would have been to paint the drum black, paint the ball white, and try using stroboscopic methods for photographs. That could be a project for some of your students.

Fig. 1. Baby Drum Drop with ball in "orbit" under a force law that is approximately an inverse square law of attraction toward the center.

A POTPOURRI OF PHYSICS TEACHING IDEAS—TOYS

The physics of toys

HENRY LEVINSTEIN

In teaching various physics courses we all have over the years accumulated a number of toys which are not part of our conventional demonstration equipment. The collection which varies from person to person and which we accumulate in drawers, cabinets, and pants pockets include such things as the dunking bird, gyroscopes, yo-yo's, a tippy top, a propeller on a notched stick, Newton's cradle, Slinky, coupled pendulums, and more recently "rattlebacks."[1] About 10 years ago I decided to gather my various "toys" and teach a one-hour-per-week mini course devoted to the physics of these devices. While the course can be taught at any level,[2] I chose to teach it at the lowest possible level to interest those students who never had a physics course, never intended to take one but were intrigued by the course title. Quite a few found that physics wasn't so bad after all and have followed this course with a more conventional one. Those who do not continue (the larger fraction) obtain some feeling for science which they could otherwise not have had. The course consists of 13 or 14 one-hour lectures. Enrollment has varied between 100 and 300, depending on the publicity the course has received in the school paper. I started the course with about 50 toys, but it soon became evident that since I did not want to go deeply into the physics of each toy, I did not have enough material. I began searching at toy stores, in mail-order catalogs, at flea markets and antique shows, for children's and adult toys and appropriate gadgets. My collection has grown by about 50 a

Henry Levinstein *obtained his Ph.D. in 1947 from the University of Michigan. Since then he has been teaching physics at Syracuse University. His research is on infrared detectors. He has been a consultant for industry and government in connection with this work. He has been chairman of the New York Section of the American Physical Society. The toy course, an outgrowth of several lectures, was designed to reduce scientific and technological illiteracy. (Department of Physics, Syracuse University, Syracuse, New York 13210)*

The author surrounded by toys from his collection.

year to the point where I have about 600 now. Most have an advantage over conventional demonstration equipment in their relatively low cost and in the fact that students relate well to them. Their disadvantage is that they are not constructed to be used over and over again, that they are often no longer available when one looks for a replacement. Furthermore, it is not always obvious which toys are suitable for teaching physics, especially when they are in sealed boxes which can only be opened in the store under some risk.

The course has had a variety of side effects which I had not anticipated. Collecting the right kind of toy takes considerable time. The search for new and antique toys and gadgets never ends. As the collection grows, considerable time is spent in organizing the toys and repairing them. As stories about my collection have spread, there have been a large number of requests for "Toy Talks." These, just as the course lectures, can be given at any level to virtually any audience for any occasion from physics colloquia to after-dinner talks merely by adjusting the level of presentation and the type of toys to be used. The main problem is the transportation of the toys. A carry-on suitcase usually holds sufficient toys if the trip requires travel by plane; several shopping bags are more appropriate when I travel by car.

Because of the many talks I have given, I am often considered a toy expert by the local media. It is usually forgotten that I collect toys which can be used to teach and explain physics and that these are not necessarily "good" toys for a 10-year old. As a matter of fact, some of "my" best toys are those that don't work well because the designer has neglected some physical principle, or toys which are not appealing to children for one reason or another. I must admit, however, that as I look at more and more toys it becomes fairly easy to predict the "flops."

I use about 30 or 40 toys in each lecture. These lectures may take various approaches. Some are devoted to concepts such as centripetal force and conservation of momentum. The concept is discussed first, then illustrated by a variety of toys. In other lectures, a historical approach is used such as "From Edison to HiFi." Other lectures are devoted to technological changes without too much emphasis on physical principles but rather on the physical basis of the breakthroughs which have led to the changes. The lecture on programming, both mechanical and electronic, is such an example. One fact becomes clear to the students at the outset. In real life, it is impossible to compartmentalize science into narrow categories as is usually done in texts (statics, dynamics). A toy may illustrate a principle in mechanics and at the same time a concept in electricity. A phonograph with a crystal cartridge is such an example.

I have pondered a great deal on how a paper such as this should be written. It is quite obvious that I cannot include all 600 toys. To do so would make this a book. Perhaps I should describe only those toys which are readily available. In trying to do this, I found that it would really not give the flavor of the course and furthermore I have noticed that some toys appear on the shelves, stay there for a few years then disappear only to reappear in a slightly different form a few years later. I will therefore outline the course stressing mainly toys which are available and those which may be easily fabricated or will probably be available soon again.

Fig. 1. Clown on a monocycle.

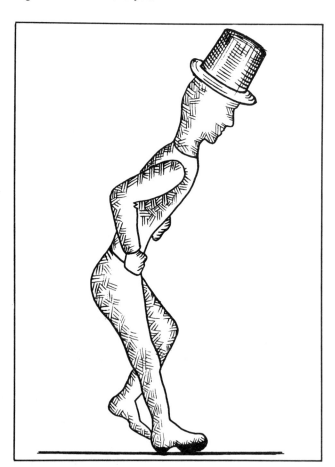

Fig. 2. Drunk falling backwards as center of gravity shifts.

The course follows as much as possible along classical lines. The lectures which usually cover one or two topics and may extend over two periods vary from year to year, depending to a large extent on the type of toys which become available and which provide a nucleus around

A POTPOURRI OF PHYSICS TEACHING IDEAS—TOYS

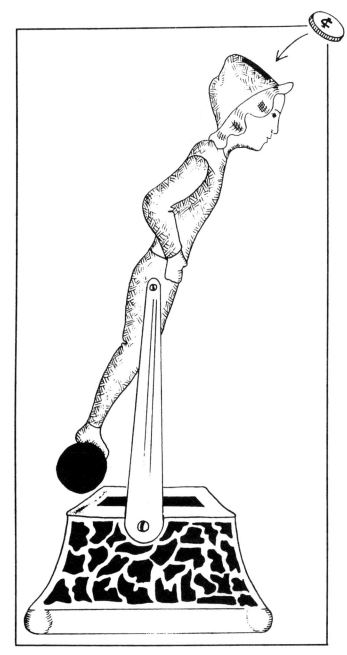

Fig. 3. "Boy on a Trapeze" Bank

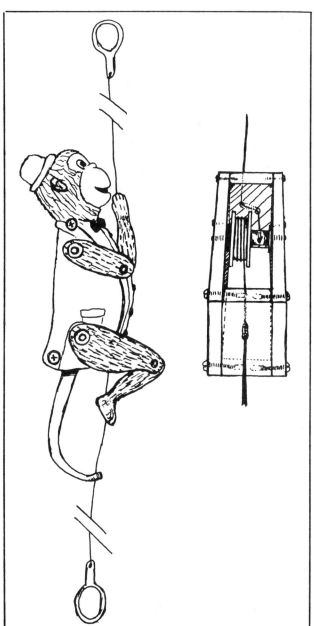

Fig. 4. "Climbing Monkey" and viewgraph model of mechanism.

which a lecture can be built. The most recent sequence of lectures is (1) Equilibrium (2) Force and Torque (3) Conservation of Linear and Angular Momentum (4) Conservation of Energy, its storage and conversion (5) Flying — a historical approach (6) Vibrations including resonance, coupled vibrators, art from vibrating systems and keeping time (7) Vibrations and Music including musical instruments and electronic music (8) Programmed Music, both mechanical and electronic (9) From Edison to Hi-Fi, the story of recording (10) Vibrations and Light including reflection, polarization, light shows, holography (11) Optical Instruments — the development of photography (12) Optical Illusions — the development of motion pictures (13) Magnetism, Electrostatics, Generation of Electricity (14) Electromagnetic Waves (UV, IR, Radio, TV, radio-controlled devices) sometimes called "seeing the invisible" (15) The Story of Communications — From Bongo Drums to Fiber Optics.

Balancing toys for the lecture on equilibrium are available in a never-ending variety. Various kinds of mobiles are seen in every store. A clown on a monocycle or bicycle (Fig. 1), cycling down a string, appears in various modifications. A drunk (Fig. 2) who falls over as the center of gravity changes because a viscous fluid drains from his head to lower portions of his body, is another example. Then there are reproductions of an old mechanical bank, The Boy on a Trapeze (Fig. 3), where placing a coin into the boy's head causes him to revolve, depositing the coin into the box below. While a great many toys specifically on torque are available, various forms of jumping jacks can either be put together or may be purchased from gift shops. Various types of Nut Crackers fall into the gadget variety. A somersaulting dog,[3] whose rear legs apply sufficient torque to cause the dog to stand again on his feet after a complete rotation, is a copy of an old toy and is usually available in specialty shops. A climbing monkey, consisting

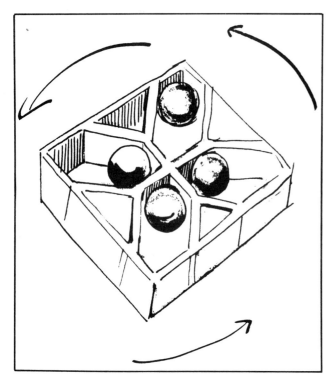

Fig. 5. Puzzle.

Fig. 6. Clown Top.

of a pulley system with a small and large radius pulley, is a copy of an antique toy (Fig. 4). Its operation can best be explained by a home-made transparent model which can be placed on an overhead projector. Probably very few concepts have intrigued toy makers more than centripetal force. Among the best are "Hot Wheels" or "Darda" cars with various types of tracks. It is interesting to note that there are at least three mechanical methods for giving the car sufficient kinetic energy to go around the loop: 1) simply elevating the track where the car starts, 2) catapulting the car and 3) spring winding a special car. Some tracks have banked curves. It is unfortunate that the banking of the track cannot be changed but it is relatively simple to make an unbanked curve by piecing small sections of track together. Then there are various puzzles or skill games. In one, 4 marbles are supposed to be brought to the 4 corners of a square enclosure (Fig. 5). This can only be accomplished by rotating the configuration rapidly. Among other toys are Yo-Yo's or tops which light up as they spin fast enough for a centrifugal switch to close, or a clown whose collar expands as he spins (Fig. 6).

Conservation of linear momentum is illustrated by such toys as rockets, balloon cars, or boats where the momentum of air from the inflated balloon causes the car or boat to rush forward.

Newton's Cradle is available in many forms. Five balls of equal mass are suspended or placed on a track. When one is caused to collide with the others, one at the other end shoots out with approximately the same velocity. An interesting modification was available some time ago. About twenty steel balls of decreasing size were suspended. When the most massive ball was caused to collide with the others, the smallest ball shot away at a very much greater velocity.

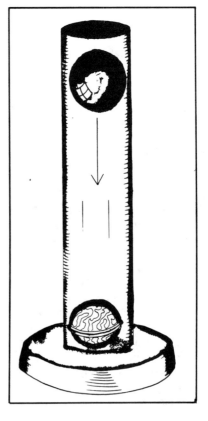

Fig. 7. Newtonian Nutcracker.

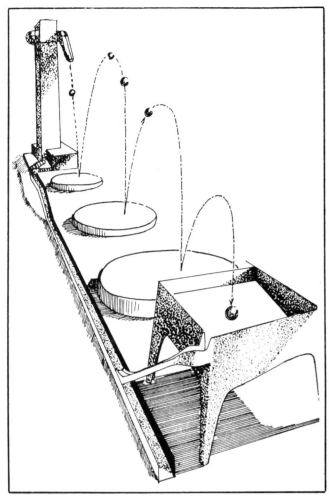

Fig. 8. Cascade Toy demonstrating elasticity and parabolic trajectories.

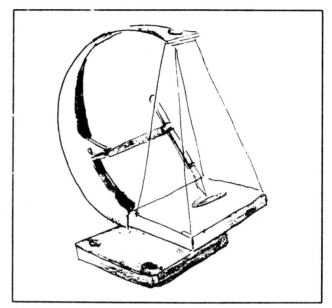

Fig. 9. Pendoodler.

The fact that force represents time rate of change in momentum may be illustrated with a Newtonian Nutcracker (Fig. 7), which may no longer be available but is easily reproduced. A fairly large steel ball is dropped in a plastic tube onto a nut which rests on a brass base. The nut is shattered. When the brass base is replaced by a cushioned base, the nut does not break.

A toy which fits into several categories and is no longer available but may be reproduced is "Cascade" (Fig. 8). In the original version, ball bearings drop from a tower at an angle of about 60° onto the first of three drums containing a flexible surface. From there they bounce in a parabolic orbit onto a second and third drum and finally to a container from where they are returned to the top of the tower. In the homemade version the flexible drums may be replaced by a heavy steel or glass plate; the bearings, by glass marbles. The return may be eliminated.

Toys illustrating rotation and angular momentum are plentiful. There are all kinds of gyroscopes and tops, some of them battery operated. They frequently show other phenomena as well, such as the color-changing top where geared rotating colored disks show color mixing. Gyroscopes in toy cars and motorcycles produce the expected results. I have generally found that motor-driven gyroscopic toys do not work as well as simple hand-operated ones.

The physics of flying can be represented by a whole series of toys thus making it a unified lecture with a historical approach. There is, of course, the bird[4] with flapping wings whose flight path is adjusted by its tail. Then there are a number of helicopter toys from the free-flying helicopters usually wound by a spring or rubber band to helicopters attached to rods and operated by batteries and a motor where elevation and direction of rotation about a center post may be controlled by the blade or rather the helicopter inclination. The Bernoulli principle may be illustrated by keeping a light ball afloat in an airjet or more dramatically not by a toy but by blowing air from a compressor over the end of a screwdriver making it "float." Of course, there are helium-filled balloons and more recently balloons which expand and rise when exposed to sunlight, heating the air inside. Then, there are various forms of rockets and simple gliders, frisbees, etc. After balloons are discussed in connection with flying, buoyancy can also be discussed in general in connection with water toys. Various kinds of Cartesian-diver toys are available. The best and most recent is "Something Fishy" where a fish "swims" up, down, left, or right in a plastic container.

A large part of the course is devoted to "vibrations" because of the many toys available and because of the importance of the subject. There are simple vibrating toys such as a bird on a spring, figures sitting on a swing, and balls attached to bars suspended on a string which may be twisted, illustrating a torsion pendulum. These devices can be used to explain period, frequency, and amplitude for various vibrating systems. They also form a basis for the lecture on keeping time. Then there are coupled pendulums which are often considered "adult" toys. These include two pendulums of equal period coupled by a string, or two vibrating objects of equal period coupled by a rubber band, or a toy variation of the Wilberforce pendulum. Lissajous figures form the basis for many toys on "vibrations and art." In the Harmonograph,[5] two weights vibrating at right angles are coupled together and to a pen. Various types of Lissajous figures, depending on the ratio

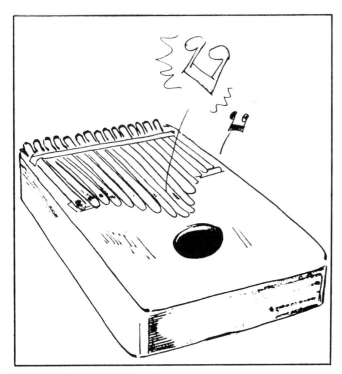

Fig. 10. Kalimba.

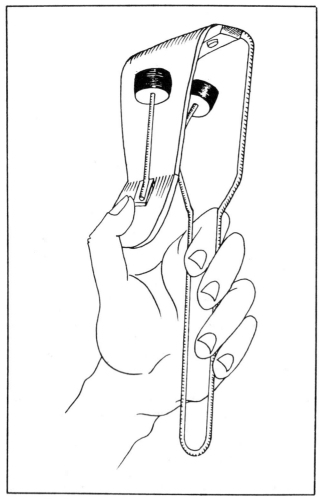

Fig. 11. Flexicon.

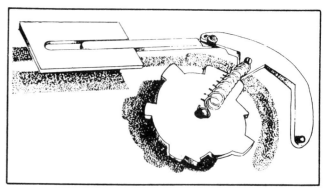

Fig. 12. Viewgraph Model of Programmed Slide Whistle.

of the pendulum lengths and on the amount of friction between pen and paper, are obtained. Commercial versions in a somewhat different form are available as a sand pendulum, where a swinging pendulum traces patterns in a sand-like material, and as "Pendoodler" (Fig. 9) where a pattern is drawn on paper resting on a rectangular board attached by four strings to a support. The pen is stationary while the board with paper on it vibrates. A stimulating lecture can be prepared on the subject of vibrations and time-keeping toys — there are clocks with swinging pendulums, vibrating pendulums, torsion pendulums and such sophisticated clocks as Ignatz where time is kept by a pendulum wrapping itself around one post, then unwrapping itself and winding itself around another post. A variety of adult toys based on this principle are available,[6] while the actual reproduction of this clock is becoming rare. Of course, at this point one should not forget the quartz time pieces where vibrations at a much higher frequency are reduced to the required frequencies by "chips." Vibrations and sound introduce perhaps the largest variety of toys. I usually begin this by showing how the length of the vibrating device determines pitch as with a "Kalimba" (Fig. 10) or thumb organ consisting of a series of reeds of different lengths and with a simple slide whistle. Since pitch can also be varied by a change in velocity in a medium, I use the "Flexicon" (Fig. 11) available in music stores. It consists of a piece of thin steel, which can be strained. Hammers hitting the metal produce an eerie variable pitch sound. A siren shows another technique for changing pitch. In the mechanical version, air is blown through a disk with holes rotating at a variable speed. Of course, there is now the whole area of electronic music where the oscillations, anywhere from sinusoidal to square wave to ramps to random pulses, try to reproduce various instruments as well as white noise by changing the harmonic content. Two devices may be specially recommended: "Sound Gizmo" which demonstrates various sounds including white noise and sirens and VL Tone by Casio, which demonstrates various voices as well as electronic programming. The output of these devices may be displayed on an oscilloscope and the harmonic content can be seen on a spectrum analyzer if one is available. From simple musical instruments we can immediately go into mechanical music and programming. The Kalimba becomes a music box when a cylinder is "programmed" by having small protrusions which cause reeds of various lengths to vibrate in a predetermined manner. The slide whistle becomes the locomotive "Toot-A-Loo" or a whistling bird music box where a programmed rotating wheel controls the length of the slide whistle and plays a tune. A viewgraph model usually helps to illustrate the principle (Fig. 12). In another version, a programmed rotating wheel simply controls the airflow to pipes of different fixed lengths.

Programmed electronic music is now available in many toys much less sophisticated than the VL Tone. Some have tunes in a memory, others make it possible to play a tune and enter it in the memory. One good toy for my purpose but inadequate for a child (it won't be available much longer) is "Colortone," a musical pencil. Light from the "pencil" is reflected from a surface with varying reflectivity onto a photoconductive cell which is part of an oscillator circuit. As the pen moves along the multicolor path, a tune is played, depending on the color variations of the path and therefore the amount of reflected light. One form of programmed sound is the phonograph. This is best illustrated by "Talking Cards,"[7] where the thumbnail sliding over a ribbon with ridges produces sounds such as "Happy Birthday" etc. This can be followed by an old Edison phonograph or its modern nonelectric toy version (dolls with pull strings that talk or bags that laugh, etc.) and phonographs that use various types of cartridges and amplifiers, tape recorders (this requires some discussion of magnetism which is again covered later) and finally electronic voice synthesizers such as "Speak and Spell" and "Talking Clocks" and "Talking Calculators."

The variety of toys for vibrations and light, just as for sound and music, is limitless. I cover such topics as photography and cameras and how one can produce the illusion of motion. I use old slides where motion is shown by actually moving objects on the slide, then flip books and mutoscopes (mechanized flip books) and finally a modern toy film projector.[8] I frequently add the illusion of motion as demonstrated by multiplex holograms, Moiré patterns and polarized light. Toys using reflections and mirrors include kaleidoscopes, anomorphic pictures,[9] rangefinders, periscopes, and, of course, "Mirage"[10] where an image of a coin or similar object is formed by two spherical mirrors and appears above the mirrors. Polarized light and color are beautifully illustrated by the rather expensive "Aurora" clock but more simply by placing pieces of cellophane between crossed sheets of polaroid. The effect is also demonstrated in kaleidoscopes and the Edmund Scientific Polarized Color Wheel when used in conjunction with the visual effects projector or a conventional overhead projector.

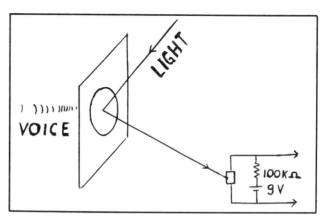

Fig. 13. Modified version of Astrophone. Voltage across the resistance goes to amplifier and speaker, such as "Miniamplifier" from Radio Shack Cat. No. 277 1008A.

It is obvious that the discussions of light lead to various light sources found in toys from the antique oil lamps to early incandescents to current flash guns, fluorescent lights, and finally to lasers. This discussion includes UV demonstrations of a psychedelic nature and IR demonstrations using a snooper-scope (rather expensive toy). Since I have covered the visible, UV and IR, it is obvious to follow this with toys using longer waves. Toy radios and TV's and radio-controlled cars are best. It is worthwhile to compare radio-controlled cars with sound-controlled cars where the controls are much more primitive. This comparison between sound and electromagnetic waves can be carried a bit farther by the lecture on communication "From Bongo Drums to Fiber Optics." The lecture begins with communication by sound, the string telephone, the transmissions of sound through tubes as is found in airplane sound systems to conventional toy telephones. Unfortunately, a reproduction of Bell's "Photophone," the "Astrophone" where light is modulated by a vibrating mirror and reflected onto a photocell and amplified, is no longer available but can easily be reproduced (Fig. 13). Light from a flashlight is reflected onto a 1/4-mil reflecting milar film mounted on cardboard with a 2-in. hole cut into the center. As sound waves strike the mirror, light is modulated by the vibrating mirror and reflected onto a CdS or PbS photoresistor illustrated in the circuit shown. The voltage across the resistance is amplified. This experiment may be modified by focusing light from the mirror onto a fiber optic bundle and then onto the photocell, thus giving an introduction to modern technology. The action of glass

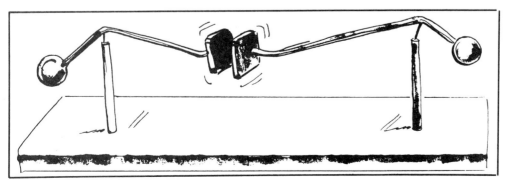

Fig. 14. "Courtship" or "Flirtation" adult toy.

fibers may also be illustrated by fiber optics lamps, available in most gift stores.

Magnetism, electrostatics, and the "generation" of electricity form the basis of one or more lectures, since there are so many toys. Magnetic toys may fall into the building-block variety where blocks of iron are placed on magnets and structures or balancing figures may be formed. There are also toys in which iron filings and magnets are used for "drawing" as in the toy "Superdraw."

One of my favorites is an adult toy "Courtship" or "Flirtation" shown in Fig. 14. Two magnets are attached to balanced rods such that unlike poles face each other. As one rod is spun around, the other spins also until, after much rotation of each — often accompanied by several energy transfers they are slowed down sufficiently so that they come at rest at their starting position.[11]

In a new electrostatic toy "Mystic" a plastic tube is charged by rubbing it with a thin piece of fine plastic sponge that is supplied, then a film insulator which may have various shapes is charged by the same material. The very light film thus repelled by the tube is caused to "fly." Toys which illustrate generation of electricity include photovoltaic cells operating motors which drive "wind mills," bicycles and helicopters as well as generator flashlights and piezoelectric gas lighters. In discussing motors, students are usually intrigued by two "perpetual motion" machines: "Top secret" or "Mystery Top,"[12] a spinning top and "Space Wheel,"[12] a device which rolls endlessly back and forth on a track, slightly depressed at the center. Since the mechanism is cleverly disguised, they make an excellent topic for discussions.

Because of the great interest in programming and computers, additional time is often devoted to programmable cars, both mechanical and electronic and this year to programmable light-emitting diodes as used in "Sky Writer" where a row of seven light-emitting diodes, mounted on a wand light up in a programmed manner as the wand is rapidly waved back and forth. Words or sentences appear because of the persistence of vision of the eye.

As I have pointed out earlier, this paper is not all inclusive since it covers only a small fraction of the devices I have accumulated and use. I have, from time to time, grouped toys differently such as: How do we imitate whistling of birds, comparing electronic with mechanical methods, or how we propel toy cars (steam, air pressure, fly wheels, spring and battery motors, etc.) and how we control their motion. I have combined toys which are really relaxation oscillators, either mechanical such as the "Dunking Bird" or "Lava Lamps,"[13] or electrical such as the "Nothing Box"[14] where neon bulbs flash in a random fashion.

Many other approaches are possible. Occasionally I have included lectures on how good scientific principles are used to perform medical quackery, or on the physics in magic (Chinese sticks, for example) and how the materials used in toys — paper, sheet metal, plastics have reflected the use of materials in everyday life, actually how the type of toys we use reflect our civilization.

It is my hope that this paper will whet the appetite of some of the readers to look in toy stores for their demonstration equipment and that it will be of some help to the many who have requested information about my course. There is even a faint hope that this paper will fall into the hands of a toy manufacturer who might realize that some of the products his company turns out could have a positive effect on our scientific future.

I would like to express my appreciation to the many students who have shown unequalled enthusiasm, to my wife who has learned to live in a museum, and to my granddaughter who has once again shown me how to play.

References

1. Rattlebacks have been described by J. Walker in *Scientific American*, October, 1979. They are boat-shaped objects which first spin in one direction then reverse themselves. They may be obtained commercially under the name of "Space Pet" from Toltoy, Inc., 5439 Schultz Drive, Sylvania, Ohio 43560.
2. A course at a much higher level on mechanical toys is given by W. Bürger at the University of Karlsruhe and is described in "Gesellschaft für Angewandte Matematic und Mechanic," July 1980.
3. Available in gift shops and currently available as "Acrobat Dog" from Walter Drake, Drake Bldg., Colorado Springs, Colorado 80940.
4. Available from Edmund Scientific, 101 E. Gloucester Pike, Barrington, New Jersey 08007.
5. A homemade harmonograph "A Double Pendulum Art Machine" is described by R. H. Romer, Am. J. Phys. **38**, 1116 (1970).
6. See for example "Perpetuo — The Executive Toy" by Edmund Scientific (Ref. 4).
7. Available from The Greeting Card Co., 3749 Costa Del Ray, Oceanside, California 92054.
8. Two books, *Photo and Scene Machines* and *Paper Movie Machines*, Troubador Press, 126 Folsom St., San Francisco, California, show how some of these old devices may be constructed.
9. See a *Moving Picture Book* and *The Magic Mirror, An Antique Optical Toy* — Dover, New York.
10. "Mirage" is available in many gift shops and from Edmund Scientific (Ref. 4).
11. This toy can either be constructed or is available from Don Conrad Mobiles, Ghiradelli Square, San Francisco, California.
12. Both are available from Edmund Scientific (Ref. 4).
13. Lava Lamps are devices where a low melting point solid immersed in a liquid is heated at the bottom by a light bulb. It melts, expands, rises, thereby cooling and sinking only to expand and rise again.
14. The flashing of a neon bulb is produced by a battery charging a capacitor through a resistance. A neon bulb connected across the capacitor discharges it when the voltage reached is about 90 V.

# Ideas for the amateur scientist

Jearl Walker

Here is a list of ideas for experiments for the amateur scientist, some of which are based on my articles already published in "The Amateur Scientist" department of *Scientific American* and others of which are planned for future articles there. The experiments could be used as class projects, science fair experiments, special work for individual students, or done just for the fun of it.

If you or your students would like to work on a project, I would like to consider your work for possible publication in my articles. I do not need a polished paper, only an outlined description of the experimental design and results from which I could write the article. What you will have to do is develop an experiment from one of the following ideas, keeping it reasonably within the grasp of an amateur scientist or a typical high school or college student. Any contribution toward one of my articles would be properly credited to you. Since an estimated two million people read the magazine, you would get immediate and widespread exposure.

Many of the projects could also be submitted for possible publication in *The Physics Teacher* or the *American Journal of Physics*. (Indeed, many of these projects have never been treated except in fairly sophisticated research journals. A paper in *The Physics Teacher* or the *American Journal of Physics* would put the physics within the grasp of teachers.) If you like, I may be interested in coauthoring such papers. In that case I could supply my ideas for the design of the experiment, whatever experimental results I already have, and my store of references already published on the experiment. I could also write the paper itself. However, my primary goal here is to get people working on these projects for the fun of it, regardless of whether anything is ever published or whether I am associated with the publication.

Much of the literature available about the projects through 1974 can be found listed in my book, *The Flying Circus of Physics with Answers*.[1] Where this is true, I have added the Flying Circus problem number to the project description. For example, in the rainbow project FC 5.32-5.40 refers you to problems 5.32 through 5.40 in my book. For each of those problems there are bibliography references that may aid you in understanding the physics or designing the experiment.

Articles already written

The following are experiments already described in "The Amateur Scientist." You might like to repeat the experiments, or you may prefer to follow the suggestions here and in the articles for extended experimentation. It is the latter that I could consider for future "Amateur Scientist" articles. However, with nearly all the projects, papers could be prepared for submission to *The Physics Teacher* or the *American Journal of Physics* because little has ever been published about adapting the material for classroom use. In addition to the references in my book, other references are listed in the bibliography on the last page of that issue of *Scientific American*.

1. **Rainbows in a single drop** (July 1977)

   About a dozen rainbows (the common first two orders plus about 10 higher orders) can be observed in a drop of water suspended from a wire and illuminated with light from a projector. Other fluids with different indices of refraction could be substituted; the rainbow colors will appear at different angles of scattering. I now have a better method of suspending the drop than was described in the article. I would especially like to see more work done on the interference that creates the supernumerary bows (the brightest interference peaks are the colors we normally see in the sky but the interference pattern contains hundreds of less intense peaks that we only occasionally see in the sky). How do the supernumerary bows move in angle as the drop evaporates and becomes smaller? (FC 5.32-5.40)

2. **Leidenfrost phenomena** (August 1977)

   Water drops sprinkled onto a skillet whose temperature is between 100° and 200° C will evaporate in about 2 sec, but when they are placed on a skillet hotter than 200° C they may last up to 1.5 min. With the hotter skillet there is a thin vapor layer beneath the drop, protecting and supporting the drop. The temperature at which the transition from a 2-sec drop to a 1.5 min drop occurs is called the Leidenfrost point. What are the Leidenfrost points for some common fluids besides water? How does a drop of a fluid mixture behave? How does the Leiden-frost point for a particular fluid depend on the surface roughness and composition of the skillet? You might also want to examine the oscillations of the drops by illuminating them with a strobe: not only do they sometimes hop up and down, but they oscillate radially in normal mode patterns. (FC 3.65 and 3.69)

3. **Cooling and freezing initially hot water** (September 1977)

   In some cases initially hot water will reach the freezing point sooner than an equal amount of initially cooler water. Although I presented data to support this result, the result is nevertheless still controversial. A huge number of parameters can be varied in the experiment to try to isolate the cause of the result, if in fact I am right about the result. My best guess is that the larger evaporation from the initially hotter water removes significant heat and some of the water mass, cooling the water below its competitor's temperature within the first few minutes. (FC 3.40)

4. **Density oscillators** (October 1977)

   In the experimental setup a cup of (dyed) salty water is lowered into a glass of fresh water. The cup has a small hole in its bottom. Because of the density differences, the salty water begins to stream downward through the hole, but soon it stops and then the fresh water begins to stream upward. The streams may oscillate like this for hours or even days. How does the oscillation period depend on the hole size, the density difference, and the running time? One 1978 Science and Engineering Fair finalist did a very nice study of this project. (FC 4.16-4.18)

5. **Physics in a cup of coffee or tea** (November 1977)

   In a cup of coffee or tea you can demonstrate secondary flow and the development of vortexes. Over the hot fluid tiny water drops are suspended and mark the circulation systems commonly called Bénard cells. I try to examine the age-old questions about what parameters affect the cooling rate of hot coffee, the color of the mug, whether you should add the cream immediately or later on, if you should stir the coffee, and if you should leave a metal spoon in it. The property I would like investigated further is the change in pitch of a spoon clinking against the mug when a powder is added to the hot coffee. I describe the pitch change and give a published explanation of it, but I may be wrong because several people have now described how the pitch actually varies several times after the powder has been added. (FC 1.22 and 3.91)

6. **Polarization, wave plates, and vision in polarized light** (December 1977 and January 1978, correction in August 1978)

   I describe how you can make your own quarter- and half-wave plates from common food wrapping and cellophane tape. A dodecagon of such half wave plates can be used to examine the polarization of the sky. Common polarizing sunglasses can be used to measure the size of smog particles and to demonstrate the polarization characteristics of light scattered from a suspension of milk in water. The double refraction properties of your eyes can be demonstrated with the homemade wave plates. More work could be done on determining what you see when you view elliptically polarized light. A Lyot filter can be constructed from the cellophane tape and then used in some of the standard spectroscopy experiments. (FC 5.50-5.52 and 5.55-5.57)

7. **Kites** (February 1978)

   I explain the aerodynamics of several of the basic flat and bowed kites, but there is a large number of other kites that could be investigated, especially the curious designs such as tetrahedrals, boxes, circular frames, rotating kites, and parafoils. Many good kite books have been published recently from which you can obtain the kite designs, but explaining the aerodynamics will not be so easy.

8. **Visual latency and the Pulfrich illusion** (March 1978)

   The response time of your eye decreases as the illumination on it decreases. This results in several visual illusions, notably the Pulfrich pendulum illusion in which a pendulum swinging in a plane appears to swing about in a circle. I describe a simple apparatus that can be used to measure your eye's response time as a function of the illumination level. I am very interested in determining how this effect varies as the illumination stimulus is moved away from the fovea, the pitlike structure on the retina where vision is most acute. (FC 5.117)

9. **Candle flames** (April 1978)

   The article describes how a candle burns and also points out that a clear picture of the burning is still not available. I would like the simple spectroscopy of the article developed into a student lab. I would also like the flickering of narrow (Christmas) candles investigated. The flicker frequency can be correlated with the wick length, the

A POTPOURRI OF PHYSICS TEACHING IDEAS — ODDS AND ENDS

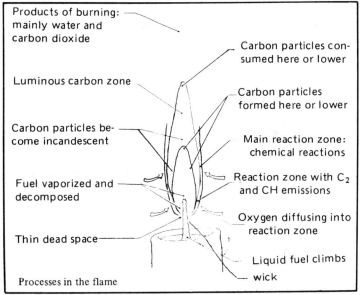

Processes in the flame

system acting as a nonlinear oscillator. The experiment can employ a (cheap) phototransistor as a detector and a stripchart recorder as the recorder.

10. Smokestack plumes (May 1978)

The smoke plumes (streams) rising from chimneys can take on a variety of shapes. An interesting experiment would be to measure the vertical temperature distribution near a chimney, using a helium balloon to lift the instruments, and then to see if the plume takes on the theoretical shape for that distribution. The results would be especially interesting for the curious shape in which a plume first narrows and later expands in its ascent. You might also want to take hourly pictures of the plumes and correlate their shape with the change in the air temperature distribution during the day. (FC 3.37)

11. Floating water drops (June 1978)

Water drops squirted on a body of water disappear into that body in about a second. With soapy water they may last for 10 sec. But if you vibrate the body of water, the drops may last for 18 min or longer. I give an explanation for the effect that may be wrong and which should be tested by further work. Other fluids could be substituted for the water. (FC 4.121)

12. Chemical oscillators (July 1978, slight correction December 1978)

When chemicals are mixed they are expected to give one particular color, but when the chemicals listed in this article are mixed, they give a solution oscillating in color. For example, one mixture will oscillate between colorless, gold, and blue about once a minute. In some cases these solutions will also display color "waves" spiraling and spreading through the solution. (Inexpensive samples of filters having these waves can be purchased from Art Winfree whose address is included in the article.) There are more oscillating reactions to be investigated.

13. Scratches in free abrasive grinding (August 1978)

How can rubbing a piece of glass over a fine grit result in a smooth, optically clear surface, especially since the grit has oversize particles that can scratch the surface? An explanation is given but needs much more work to be accepted.

14. Heat radiated by a fireplace (August 1978)

Most of the heat available from a fire is in its radiation, but much of that is radiated up into the chimney. A new log grate, the "Texas fireframe" slot design, is checked with a very simple experiment. The slot design apparently beams the radiation into the room and is therefore superior to the conventional stacking of logs. I would

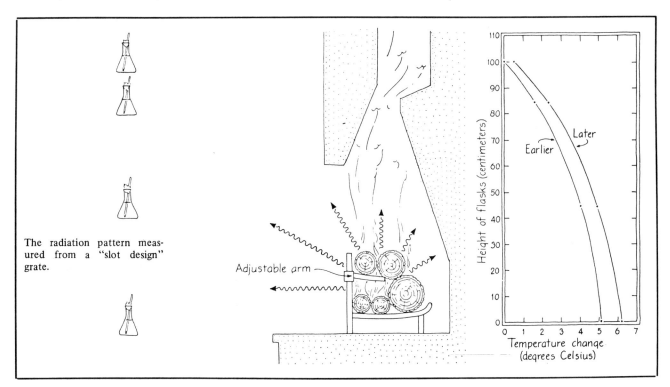

The radiation pattern measured from a "slot design" grate.

be interested in such radiation patterns being monitored in time and in angular distribution. How do the common glass and wire fireplace covers alter the heat radiated into the room?

15. Colors of a soap film (September 1978)

The colors you see in a thin soap film depend on the temperature of the emitting surface of the "white" light source. For example, sunlight (which corresponds to a surface temperature of 5500° K) gives more blues than does a projector lamp (whose filament temperatures are lower). With a series of photographs you could correlate the colors available in a soap film with the filament temperatures of several lamps. Care must be taken that the film and photographic filters do not distort the results.

16. Whispering galleries (October 1978)

Along some curved walls a person can whisper along the wall and a listener can hear that whisper even at a considerable distance. For example, in the whispering gallery of St. Paul's Cathedral a whisper can be heard clearly when the two people are at opposite sides of the circular gallery whose diameter is 32 m. I describe an experimental setup similar to that employed by Lord Rayleigh to demonstrate the effect. Many parameters in the setup can be varied. You might also want to investigate the acoustical focusing found beneath ceilings that are portions of spheres or ellipsoids. (FC 1.31)

17. Non-Newtonian fluids (November 1978)

The viscosity of most fluids change only if the temperature changes, but some fluids (called non-Newtonian fluids) change their viscosity when the fluid is put under stress and shearing. Common examples include ketchup, Silly Putty, quicksand, one-coat paints, and margarine. A mixture of corn starch and water provides an interesting demonstration: the mixture flows with a low viscosity when poured but if you slam your fist into it the viscosity increases so rapidly that the mixture will not splash. The new product called Slime is a good example of an elastic fluid, one that will bounce and will even siphon itself out of a container. A lot more work can be done with what is called the Weisenberg effect. In this experiment a rotating rod stirs an elastic fluid. Instead of the fluid moving away from the rod to form a concave surface (as water would do), the fluid climbs the rod. (FC 4.122-4.130)

18. Visual toys (December 1978)

Many of the visual toys that were popular in the Victorian age are investigated: Moiré patterns, kaleidoscopes, etc. I am especially interested in having more work done on the stereoscope. A basic explanation is given in the article, but the several visual clues which give rise to the illusion of depth in the stereoscope are still not clear.

Ideas for future articles
Sound

1. Homemade and primitive musical instruments

Instead of examining the acoustical properties of complicated or sophisticated musical instruments, I want to study either very simple homemade instruments or instruments that have been around since primitive times.

The whispering gallery of St. Paul's Cathedral

Examples include the nose harp and the vibrating saw. How is such an instrument excited? What is the resonant body? How can the player control the frequency produced? How might a reader build one of the instruments?

2. Sounds of bubble formations

A brook babbles largely because bubbles are formed beneath the water surface. This effect can be demonstrated in the lab with a simple arrangement once devised by Minnaert. I would also like an investigation of the sounds of water set to boil: at first there is a hissing, then a harsher sound, and finally when full boil is reached there is a softer sound. These sounds should be correlated with the bubble formation on the water kettle's bottom, where the bubbles burst (which is probably higher up), and if there is splashing on the surface of the water. (FC 1.12 and 1.13)

3. Tin can telephones

What is the frequency response of a tin can telephone? Does it depend on the can size, string length, string density, or string tautness?

A POTPOURRI OF PHYSICS TEACHING IDEAS — ODDS AND ENDS

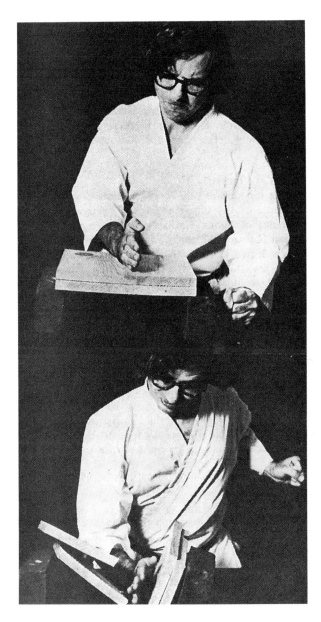

Breaking boards

Mechanics

1. The physics of karate and judo

Karate has recently been thoroughly investigated by Michael Feld of MIT. You might want to study the impact forces commonly found in karate strikes with the hands and feet. Judo, on the other hand, has never been investigated for its physics. What are the physics principles behind the common throws? How is leverage used in the holds one employs once the opponent is on the ground? Why does one slap the mat when thrown down?

2. Earth vibrations near waterfalls

John S. Rinehart once noticed that the earth vibrations created near a natural waterfall have a frequency distribution that is inversely proportional to the height of the waterfall. His suggestion was that the vibrations may be dominated by a standing wave of sound set up in the falling stream of water. I would like several waterfalls investigated in a fashion similar to what Rinehart did.

You would need to build an oscillator whose resonant frequency could be varied across the frequency spectrum of the earth vibrations. (FC 2.65)

3. Coupled oscillators

Some types of oscillators will pass energy from one type of oscillation to another. For example, a certain type of spring pendulum will periodically switch from oscillating purely as a pendulum to oscillating purely as a spring. I would like several types of these oscillators built and studied. (FC 2.62)

4. Traffic waves

When a car becomes disabled on the freeway during rush-hour traffic, the effects of that car can be seen long after it has been removed. For as long as there is dense traffic on the freeway, cars will be forced to slow around the place where the car was disabled even if the disabled car has been removed. The slowing effect will also occur behind that place, propagating down the road as a wave. Similar wave propagation occurs when the freeway density passes a certain critical value, even if there are no disabled cars. I would like a photographic study made of the "slowing wave" and its occurrence correlated with the freeway density, time after a disabled car has been removed, and other paramenters.

5. Amusement park physics and physiology

Experiments could be done on some of the common amusement park rides to measure accelerations, rotational speeds, frictional forces, etc. Prime candidates are roller coasters (the usual type, the twister type, the new world's highest at Cedar Point, Ohio), bumper cars, and the rotating cylinder whose floor is removed once sufficient rotational speed is gained.

Thermodynamics

1. Greenhouse effect

You could build a small house of glass and compare the air temperature inside it with the temperature that would be there without the glass. Although some researchers discounted the so-called greenhouse effect (where the glass somehow traps infrared radiation inside the house and thereby increases the air temperature), others still argue for its existence. (FC 3.83)

2. Heat islands

The groundlevel air temperature increases as you travel into a city, creating a hot air "island" in the city. I would like temperatures measured across a city for a number of days and for at least two seasons. How does the "island" change from season to season, from early morning to late afternoon, from a normal workday to a calm Sunday? (FC 3.93)

3. Ice skating

Measure the friction from the ice as a function of temperature, weight on the blades, width of a blade and type of metal. Is there really melting beneath a normal blade due to the pressure on the ice? What is the lowest temperature at which one can ice skate? (FC 2.54)

4. Patterns in drying soil and in ice fields

When soil dries and cracks, the surface develops a geometric design with the cracks. Something similar occurs in ice layers. At what angle do cracks typically intersect each other? If it is random, then the pattern will be random. If it is typically some value, then the surface will develop polygons. Why a particular angle is preferred would then have to be explained. (FC 3.113-3.115)

5. Drying out a masonry wall

One way to prevent a masonry wall from becoming damp with water drawn up from the ground is to connect a wire from the wall to a metal stake in the ground, thereby electrically shorting out the wall. If this prevention really does work (and it will first have to be proven), why does it work? (FC 3.107)

Fluids, including aerodynamics

1. Vortex direction in draining bathtubs

Bathtubs in the Northern Hemisphere are commonly believed to drain counterclockwise because of the Coriolis force, although years ago Shapiro showed this force did not play a significant role. Two things could be done here. Survey a large number of bathtubs to see which way they drain. Is the sense of circulation due merely to the fact that more cold than hot water enters the tub and that in the United States the cold water enters from the right as you face the faucet? The second thing that could be done is to repeat Shapiro's experiment in which he built a carefully designed tub where the tiny Coriolis force could actually be demonstrated. (FC 4.67)

2. Shapes of rising bubbles

Depending on the viscosity of the fluid and the rate of ascent of the bubbles, air bubbles rising in a fluid can take on a variety of shapes from spherical to umbrella-like. Their ascent can also vary from being straight to being zig-zag. All of these behaviors could be correlated with the ascent rate, bubble size, and fluid viscosity. (FC 4.81)

3. Splashes

I would like a photographic study of liquid drops splashing on a flat surface, shallow fluids and deep fluids. Under some conditions the splash can leave a doughnut area of wetness. Under other conditions a drop can throw up a crown of droplets and then send up a central jet of fluid. (FC 4.108 and 4.113)

4. Aerodynamics of seeds

A variety of seeds have curious aerodynamical behaviors. One type, what I call a helicopter seed, twirls as it falls. I would like these behaviors studied.

5. Water bells and sheets

When a narrow stream of water falls upon a small flat surface, the water spreads out into a sheet which can then close back on itself as it falls, forming a bell shape. What parameters determine the shape of the bell? The shape could be correlated with water speed, water surface tension, the diameter of the flat surface, and other parameters. The reduced pressure of the air trapped inside the bell could also be measured. If the sheet does not close back to form a bell, it eventually breaks up into droplets. What determines where the breakup occurs? (FC 4.114 and 4.115)

Optics

1. Suntan lotions

The wavelength range for tanning is somewhat different than the range for burning. Do any of the commercial lotions, creams, and oils filter out the burning range but leave a portion of the tanning range? Can you rank the products according to their protection (some of the products already have this ranking on the package)? (FC 5.109)

2. Colors of a butterfly

The top surface of a Morpho butterfly's wing is blue not because of any blue pigment but because of the wave interference of light. (Similar interference gives colors in thin soap films and oil slicks.) I want to measure the wavelengths reflected as a function of tilt of the wing and then calculate the thickness of the material causing the interference. I would also like to investigate the scattering properties of some beetle backs that produce circularly polarized light. (FC 5.94)

3. Kinetic art and art from special optics

Some forms of modern art depend on the polarizing properties of polarizing filters and stressed plastic. Optical fibers and lasers have also attracted some attention. I saw one huge display at the Smithsonian in which a viscous fluid slowly circulated (I suppose because of heating) inside a glass container. The fluid contained some type of suspension that gave a sparkling look to the fluid. These and other visual effects in art could be investigated.

Electrostatics

1. Measure the charge on a moving car

When a car moves along the road, there is a charge transfer between the rubber tire and the road surface, leaving the car charged. What is the voltage difference between the car and ground? Years ago trucks (especially gasoline trucks) would drag a chain on the ground to drain off the charge in order to lessen the chance of an explosion sparked by the charge. Did the chain do any good? Some people now say that a chain actually made the situation more dangerous. (FC 6.13)

2. Electric fields from splashing water

The drops splashed into the air by falling water (such as from the shower in your bathroom) can be highly charged, setting up large electric fields in the air. You could measure these fields in your own bathroom or near a natural waterfall. (FC 6.14)

References
1. Jearl Walker, *The Flying Circus of Physics with Answers*, (John Wiley & Sons, New York, 1977).

Humor in the physics classroom

IVARS PETERSON

Peter Ustinov echoed the memories of many when he wrote, "Of physics I could understand nothing at all. Why imaginary wheels should gather speed running down hypothetical slopes, and create friction, I could neither understand in the terms in which it was taught, nor care about."[1] A physics teacher should not leave a student with this impression of physics.

Doing physics can be a stimulating, exciting experience. It can also be fun. If the concepts and applications of physics are approached with a sense of humor and with imagination, then all the hard work involved will seem lighter.

Writers such as Arthur Koestler[2] have argued that scientific discovery, artistic originality, and comic inspiration are closely related. The same kind of creativity that is required to find the humor in a situation is also needed to solve a problem or design an experiment. Laughter, in Koestler's words, arises out of seeing an idea in two self-consistent but usually incompatible frames of reference. It involves a spontaneous flash of insight that shows a familiar situation in a new light. This leads to learning. There is an element of riddle in every good joke.

A similar flash of insight is involved in many scientific discoveries. An early biographer of Newton wrote, "We went into the garden and drank tea under the shade of some apple trees... he told me he was just in the same situation as when formerly the notion of gravitation came into his mind. It was occasion'd by the fall of an apple, as he sat in a contemplative mood."[3] There is an element of surprise in good humor and in great scientific discoveries. Remember Archimedes' shout of "Eureka!"

A good place to start, if you are interested in adding a little humor to your courses, is with cartoons. In less than five years I was able to collect several hundred cartoons with subjects related to science. I made it my practice to use cartoons in my lessons and to post cartoons on the bulletin board.

The "cartoon of the day" was changed daily. The most effective location was a notice board outside the classroom. Very quickly students, including many not enrolled in physics, began to go out of their way to check the latest offering (and there were strong complaints when I was late or failed to change the cartoon). One bonus was that some students would ask about a cartoon which they could not understand or did not find amusing. Sometimes I was able to give a quick physics lesson to an interested boy or girl.

Cartoons can be used effectively to illustrate important points in a lesson, to introduce a topic and to indicate common misconceptions in physics. The following dialogue appeared in a cartoon.

Archie: How's your roller skating job, Betty?
Betty: How *was* it, is more like it!
Archie: What happened?
Betty: Well, I had this great big platter of food. I made a sharp *right* turn and the platter made a sharp *left*!

When used in a discussion of centripetal force, this cartoon highlights a common error that many people make.

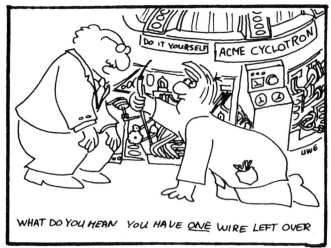

Cartoons sometimes contain good physics problems. One well-known cartoon, discussed in *The Physics Teacher*,[4,5] showed the determination of the depth of a well in terms of the time it takes one to hear the splash of a stone dropped into the well. A 16-s well works out to be 868 m deep!

I have also used cartoons as the basis for a student assignment. "Select a cartoon and make up a physics problem for the situation shown. Evaluation will be based on the ingenuity and originality of the problem and on its correct use of physical principles. A complete solution must be included. State all assumptions you make." This exercise forces students to look for physics in "everyday" situations and to learn about the assumptions and simplifications that must be made in order to solve a problem. Remember that "a physicist is a person who can't solve a problem until it is simplified."

The best cartoon problem submitted by a student showed an ant lifting a snowball. The question was how many ants would be required to throw the snowball so that it would hit a man in the face. The data for this problem in projectile motion came from measurements made on the drawing itself, scaled appropriately.

A pun has been described as "two strings of thought tied together by an acoustic knot."[6] Puns, no matter how terrible, do give listeners the satisfaction of being clever enough to see the joke. Learning words or definitions can be made easier and less tedious by using puns. Metric prefixes, in particular, lend themselves to this approach. There are groans in the classroom but it is amazing to see how diligently and quickly students work through long lists which may contain the following: 10^{12} bull; 10^9 le; 10^{-9} goat; tro 10^{-12} 1; 10^{-2} fic; and 10^{-9}, 10^{-9}.[7]

A "visual pun" is a picture in which the letters of words are modified to suggest an aspect of the concept involved. Students enjoy creating these and at the same time begin to learn important definitions and symbols. Here are some examples.

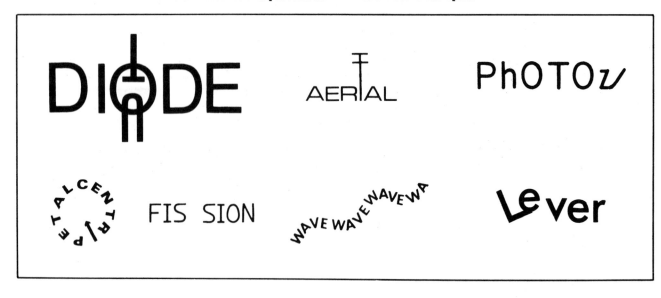

Physics requires the frequent use of symbols, including Greek letters. In order to familiarize students with the symbols I use the following.

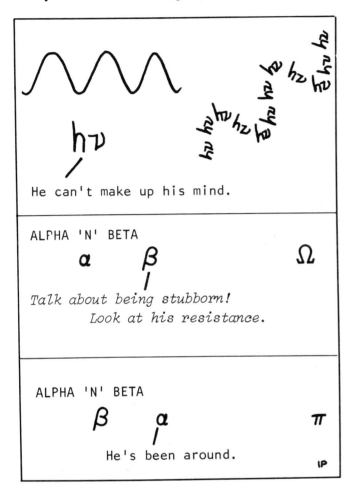

While puzzling through these, students learn some physics, too.

Central to almost any physics course is the solving of problems. Every problem does not have to involve mass A connected by a massless rope to object B sliding down a frictionless plane C. It is worthwhile to use problems that relate physics to real-life situations or to put problems in an amusing setting, as long as the basic principles involved are not obscured by too many irrelevant details.

One simple way of achieving this is to use unusual units of measurement. This emphasizes the fact that the name of the unit is not relevant to the concept involved. Pressure is pressure whether it is measured in atmospheres, pascals or ghlerfs. A ghlerf sounds like an appropriate unit in a problem related to submarines.

It is worth the effort, just to see the surprise on a student's face, to create and insert a problem like this one into an assignment or even a test or exam.

In the far corner of our galaxy (if galaxies have corners) hangs the planet of ID, with one moon slowly orbiting it. The inhabitants of ID (the IDiots?) wish to send a rocket to their moon. If their moon has a mass of 1000 finks, and their planet has a mass of 81 000 finks, at what shortest distance from the centre of ID will the rocket experience no net gravitational force? The distance between centres of ID and its moon is 100 000 wizards.[8]

It isn't the "same old problem" and the unusual setting may trigger new ideas in the mind of the student.

You can't always trust the physics that you see on the screen or read in print. Flash Gordon, for example, lives in a very curious universe.

"Velocity means speed multiplied by the weight of the object, of course, to quote a bit of elementary physics." Zarkov grinned. "You would have to regulate the speed with the weight of the object transmitted. Too much speed would crush a large object; too little speed would cause a smaller object to explode."

"Oh," Dale said, impressed."[9]

There are numerous examples of scientific errors, sometimes with amusing consequences. Once the idea is introduced students will discover many more examples, particularly in television programs.

In the novel *Lord of the Flies* a group of schoolboys find themselves on a deserted island. One of their first necessities, in order to survive, is a fire. The fire is started using the lenses from the spectacles of one of the boys, someone who is extremely short-sighted.

Ralph moved the lenses back and forth, this way and that, till a glossy white image of the declining sun lay on a piece of rotten wood. Almost at once a thin trickle of smoke rose up and made him cough.[10]

The situation described can lead to an interesting discussion in the physics classroom, particularly if the novel is on the course of study in English. Since the boy's lenses must be concave, a fire could not be started unless they were concave meniscus lenses and were arranged as shown below.

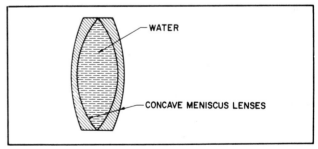

However this does spoil a good story. Another excellent book to use is *The Invisible Man* by H. G. Wells.[11]

It is sometimes possible to illuminate a concept by turning the physics inside out. An article that appeared in the *Journal of Irreproducible Results* begins,

There is no such thing as light. What there is in the universe is dark. It is obvious from simple observations that this is so.

What we call light is merely the absence of dark. Dark is continually created. As fast as it is whisked away, more fills up the space.[12]

The article continues with a discussion of dark-sinks, the speed of dark and darktons. Using this model, how would you explain how a "dark bulb" works?

In a similar vein, "what if" questions can lead to deeper insights even while providing amusement. What if gravity were "turned off" under a small area? What if gravity fluctuated from day to day in the same way as the weather — today's gravity report? What if you tried to bake a cake in Skylab? What would you wear if there were no friction? The possibilities are endless and reminiscent of Einstein's question, "What would a light wave look like to someone keeping pace with it?".

Decorating the classroom with posters also helps to enliven the atmosphere. Sometimes a quotation or an aphorism captures the essence of a physical concept better than a long mathematical derivation or a detailed definition. These are very suitable subjects for poster work.

In questions of science the authority of a thousand is not worth the humble reasoning of a single individual.
Galileo

An echo is like hearing yourself in the mirror.
ABSOLUTE ZERO IS O K
IMPULSE CHANGES MOMENTUM
FRICTION IS A DRAG
Pornography in physics is a bare naked number. Clothe it in units.
AXIS: SOMETHING TO TORQUE ABOUT

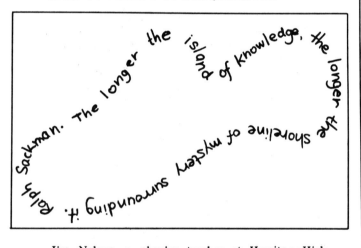

Jim Nelson, a physics teacher at Harriton High School in Rosemont, Pennsylvania, has an interesting way of reviewing work at the end of a term. The classroom becomes a game show for one period. He uses a wild mixture of questions, requiring answers that may be outrageous puns or historical facts. These are a few questions selected from his collection.

1. What is the center of gravity?
2. What element is named after an Italian physicist?
3. Name a unit of measurement named for an American scientist?
4. Who invented the radiometer?
5. What is the definition of a "law of nature"?
6. What is a little joule?
7. What is the mass of a 90-kg man on the moon?
8. What do you call people who vote "yes" for better tractors?
9. Who said, "It is by logic that we prove, but by intuition we discover"?
10. What is the beginning of eternity and the end of space and time?[1][3]

I have a special problem set that I hand out as a "Christmas treat."

PROBLEM SET

TO BE HANDED IN BY 12:00 PM (Midnight), December 24 (or else Santa won't come!)

If you feel essential data is missing, invent your own.

1. Calculate the force Rudolph exerts when he stumbles against a cloud and brings Santa's sleigh and its eight tiny reindeer crashing to the ground. Comment on the expression on Santa's face as Rudolph tries to explain the laws of physics to Santa.

2. Prove by direct experiment that the total work done in carrying a 25-kg Christmas tree upstairs to the third floor and then down again to the starting point is zero.

3. Plot a displacement-time graph for a chicken crossing the highway. Comment on the points when the acceleration is greatest, and how the chicken selected the direction of his initial velocity. Determine the minimum speed of a car which can make an impact with coefficient of restitution equal to zero with the chicken crossing the highway. Describe the subsequent motion of the car.

4. Calculate the initial velocity and trajectory a fox must have in order to reach a bunch of grapes. Prove to your and the fox's satisfaction that the grapes were sour anyway.

5. Junior delivers a horizontal blow with a blunt instrument to the Christmas tree at a point 0.5 m up from the floor. Determine the motion of the star at the top of the tree and the reaction of the family.

6. Measure the minimum force necessary to crack each Christmas nut, and draw a histogram showing the number of nuts in each suitably chosen force range. Determine the median, mean and root mean square force and the quartiles. Compose an appropriate letter of complaint to the nut-growers.

BONUS

Two identical dishes, loaded with cranberry sauce, and each of mass M, rest on a frictionless Christmas table. A perfectly rough insect of mass m jumps from the center of one dish to the center of the other and back to the center of the first. Show that the ratio of the velocities of the dishes is M + m : m and determine the fate of the insect.

So far, I have received three very amusing solutions to the problem set, one with a grape (for testing and tasting) attached. Parents have also remarked on the curious things that seem to be going on in the physics classroom, often referring to this particular "assignment."

The physics teacher does not need to be a comedian,

continually cracking jokes, inventing puns and tossing off one-liners. But the judicious use of humor at just the right times can lead to a more happy and creative atmosphere in the classroom. Besides, your students want a chance to show you how clever they can be.

Notes and References

1. Peter Ustinov, *Dear Me*, Penguin Books, Baltimore, 1977.
2. Arthur Koestler, *The Act of Creation*, Pan Books Limited, 1964.
3. Dr. William Stukeley, *Memoirs of Sir Isaac Newton's Life*, 1752, pp. 19-20.
4. Bradlee Chang, "The depth of a well," Phys. Teach. 16, 489 (1978).
5. "More on the depth of a well," Phys. Teach. 17, 327 (1979).
6. Arthur Koestler, *The Act of Creation*, Pan Books Limited, 1964, p. 65.
7. Answers: terrible; giggle; nanny goat; tropical; scientific; nano, nano (Mork's greeting on the TV program *Mork and Mindy*).
8. Phil Eastman, *A Decade of SIN*, Physics Department, University of Waterloo, Waterloo, Ontario, 1979.
9. Alex Raymond, *Flash Gordon: The Time Trap of Ming XIII*, Avon Books.
10. William Golding, *Lord of the Flies*, Faber and Faber Limited, London, 1954, p. 45.
11. Ivars Peterson, "Blind Sight," *Photon*, No. 27, March, 1980.
12. Kirk R. Smith, "The Age of Enlightenment Ends," *The Journal of Irreproducible Results*, Vol. 20, No. 1, June 1973.
13. Answers: 1. the letter "v," 2. Fermium, 3. henry, bel, or sabin, 4. Crookes, 5. "A guessed generalization based on imperfect observation of nature," 6. erg, 7. 90 kg, 8. protractors, 9. Poincaré, 10. the letter "e."

A TRICK DEMONSTRATION

I really enjoyed reading Ivar Peterson's article "Humor in the physics classroom" [Phys. Teach. 18, 646 (1980)] and encourage physics teachers to try some of his ideas. I have heard him speak at several professional meetings and have started my own collection of "physics humor."

One demonstration I learned from a fellow teacher is fun to do when studying circular motion. Make a pail out of a 48-oz. juice can using a coat hanger to make a handle. Get a large absorbent sponge to fit in the bottom. Talk about how water stays in a pail when you swing it around in a circle. Pour water into the can so it is absorbed by the sponge. Then swing it around in a circle and ask the students what will happen if you stop it at the top. You will love their expressions when you stop it at the top and the water does not come out! (Warning: practice it ahead and make sure you stand back far enough so students can't see the sponge.)

Judith L. Doyle, *14 Sunset Hill Drive, Granville, Ohio 43023*

Proclamation

WHEREAS: Isaac Newton is the founder of classical Physics, and,

WHEREAS: the birth of such an intellectual giant should be honored,

BE IT Therefore Resolved that the Physics classes will be dismissed from school on the anniversary of his birth...

_____ Principal
signature

This is a bit of trivia I sometimes use to make all students aware there is a physics department. I put this in the school paper with the signature (and approval) of the principal. Perhaps it may be of some use to others.

MICHAEL SCOTT
Stephen Decatur High School
Decatur, Illinois 62525

Ed. note: Isaac Newton was born on Christmas Day, 1642.

Pocketbook science

James Watson, Jr.
Department of Physics and Astronomy
Nancy T. Watson
Department of Laboratory School
Ball State University, Muncie, Indiana 47306

At Ball State we teach several levels of physical science, ranging from elementary school and middle school to the physical science content course for future elementary teachers. Elementary schools usually are poorly equipped for science experiments. The teachers and their students do not even have the simplest equipment. Even on the university level, simple equipment is not readily available for students to try demonstrations away from class and laboratory.

Our solution is Pocketbook Science. The following is a list of simple demonstrations that can be performed using materials customarily carried in a woman's pocketbook. Most of the experiments have been described elsewhere in connection with more formal classroom activities. We bet that once you've been reminded of these possibilities you'll think of many more demonstrations that can easily be done with the wealth of "apparatus" usually found in a pocketbook.

Mechanics

1. The classic tablecloth demonstration can be executed with a handkerchief and almost anything else found in the pocketbook such as a lipstick or pill bottle. The handkerchief is spread out on a table with about one-half of it hanging over the side of the table. The lipstick and pill bottle are placed on the handkerchief. Holding the end hanging off the table in one hand, karate chop between there and the table. The handkerchief will be quickly jerked from under the objects, leaving them still standing.

2. A variation of the tablecloth demo can be done with the use of a dollar bill and a lipstick. Have one-half of the dollar bill hanging over the end of a table. Place a lipstick on the other end. Hold the loose end with one hand and chop down with one finger between your hand and the table.

3. A similar demonstration is easily accomplished with a stack of coins. Place a stack of coins on a flat surface. Take one coin and quickly slide it along the surface so that it hits the bottom coin of the stack. The momentum transfer will cause the bottom coin to move from under the stack. The stack of coins will remain at rest and still be standing after dropping down, once the bottom coin is gone.

4. A nice trick can be demonstrated with a pill bottle, a pill, and two credit cards. Place a pill on a credit card or piece of stiff paper. This is placed on top of an open pill bottle with the pill centered on the opening. Hold another credit card close to the bottom of the pill bottle so that it extends up past the other card with the pill. Hold this card in place with one hand and slightly bend it away from the bottle with the other card. Let go and the pill should remain at rest as the card flies off. The pill will then drop into the bottle.

5. The old "shoot the monkey" demonstration can be performed for two cents! Place two coins on the table one on the table edge in front of the other, but off to one side. The back coin is flipped with your finger so that it glances off the other and goes flying off the table causing the front coin to drop off. Both coins will hit the floor at the same time since each only has a vertical force on it due to gravity.

6. Most people now carry inexpensive calculators with them so an experiment requiring a quick calculation can be made. This requires two people. Fold a dollar bill lengthwise, then hold it so that it hangs vertically between another person's thumb and forefinger at George's head. Drop the bill. The other person cannot catch it since his reaction time is too long. He has to see the bill drop and then tell his fingers to close. Let's calculate your subject's reaction time using the standard formula $d = gt^2/2$ where d is the distance of fall, g is acceleration due to gravity which has a value of 980 cm s^{-2}, and t the reaction time. Solve for the time and one obtains $t = \sqrt{2d/g}$. You will need a ruler. If you don't have one, make one right quick with a sheet of paper. Make it 30 or 40 cm long. A thumbnail is about 2 cm wide or a dollar bill is 15.5 cm long. Hold the ruler so the other person's fingers are at the zero mark. Drop the ruler several times and read where the person grabs it. This is d. Take an average.

Using the calculator, calculate 2d/g. Now you need the square root of this number. If you have that key on your calculator, push it and up comes the reaction time. If you have a "cheapie," look at the number and guess what time squared will give 2d/g. Try it on the calculator. Keep guessing until you narrow to within two decimal points.

7. The idea of center of gravity can easily be shown with a pencil and a paper clip. Straighten out the paper clip. Stick one end into the eraser end of the pencil. Bend the paper clip into a large curve. Place the end of the paper clip on your finger allowing the pencil to hang below it. Bend the paper clip until you can support the whole thing with your finger.

8. A lipstick or lighter and a table are all that is needed to show that males are different from females. Have a person kneel on a table or the floor and then bend down toward the table with elbows snugly against knees. Extend an arm and hand flat against the table and place the lipstick at the tip of the fingers. Then placing the hands behind the back, have your experimenter bend over and try to knock over the lipstick with his or her nose. Careful, could result in a flat nose! Females can do it, males can't.

9. Center of gravity can also be demonstrated by supporting a horizontal pencil with an index finger under each end. Move the fingers toward each other. They will always meet

A POTPOURRI OF PHYSICS TEACHING IDEAS—ODDS AND ENDS

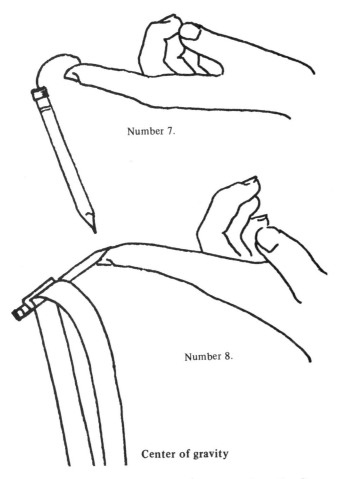

Number 7.

Number 8.

Center of gravity

under the center of gravity no matter where the fingers were originally placed.

10. If a belt and a mechanical pencil or pen with a pocket clip are available, the idea of change of center of gravity can be shown. Try to support the pencil just at the point. Move the center of gravity by draping a belt over the pencil with the clip holding it in place.

Sound

11. The idea of frequency of a vibrating object and its relationship to length can easily be seen with the use of a credit card. Hold a credit card so that only a small part hangs over the edge of a table. Pluck it and note the frequency of vibration. Allow more of the card to extend over the edge. How is the length of the vibrating object related to the frequency?

12. Rub your finger nail along the teeth of the comb and listen to the sound. Hold the end of the comb against a door or table top. Rub the teeth and note the louder sound. The door acts as a sounding board.

Momentum

13. Transfer of momentum can easily be demonstrated by lining up coins. Take one coin and slide it so it will hit the line of coins. As the coin hits the line of coins, the momentum is transferred and the last coin in line will move away from the other coins.

Air pressure

14. The effect of air pressure is demonstrated with an empty pill bottle and a credit card or piece of paper. Fill the pill bottle with water, cover with a piece of paper and invert. The water in the bottle does not come out. The air pressure pushing up on the paper is greater than the pressure of the water pushing down.

15. The Bernoulli effect can be shown by using a note card or a piece of paper. Fold the ends to make a U shape and stand the paper on the ends. Blow directly under the standing card. The card will not move.

16. The Bernoulli effect can also be demonstrated by placing a dime on a table and giving a quick puff over the coin. The coin will jump into the air with the lower pressure on top.

Light

17. If two mirrors are available, place them parallel and facing each other with the lipstick between them. If you look into one of the mirrors you will see many lipsticks. This is an example of multiple reflections.

18. Refraction can be shown by filling a clear pill bottle with water and putting a pencil in the bottle. Look at the pencil – it appears broken.

19. Print your name while looking at the paper and pencil in a mirror. It is extremely difficult to do, and the result will be a mirror image.

20. A scarf made of a mesh material will serve as a diffraction grating. Hold the scarf up to your eye and look at a small light source far away. You will see a spectrum.

Electricity

21. Static electricity is demonstrated with a comb and a piece of paper that has been torn into small bits. Tear up a piece of paper and put the pieces in a pile on the table. Take the comb and either rub it through your hair or rub it on some wool clothing. This charges the comb. Bring it near the pieces of paper. The comb will induce an opposite charge on the paper and the paper will jump to the comb. The paper may then get a like charge and be repelled and jump off.

22. If two bandaids or two pieces of sticky tape are available, press on them and stick them to the table. Rip them off and bring the sticky sides together. They should repel due to the charges picked up by ripping them off the table.

If all else fails

23. Circular motion can be shown by using the whole pocketbook. Anything going in a circle is forced to do so by a centripetal force. Open the pocketbook and hold it by its straps. Swing it quickly in a complete vertical circle. Why doesn't everything fall out when it is at the top of the circle?

References

1. Buckwash, Phys. Teach. **14**, 39 (1976).
2. Toepker, Phys. Teach. **15**, 241 (1977).

The physics "flub-stub"

Wayne Williams
McClintock High School, Tempe, Arizona 85282

An important part of science education is teaching the concept of a questioning attitude. Students should learn not to accept blindly everything they read in science books or are told by a science teacher.[1] Many students have never known a teacher to confess he was wrong. A teacher should be willing to admit mistakes. How often as a teacher have you drilled into your classes the importance of significant digits and units, and then completely ignored them when you worked problems on the chalk board?

A good technique for encouraging a questioning attitude is the physics "flub stub." This is a small slip of paper entitling the holder to certain privileges. In our classes a specified number of "flub stubs" can be exchanged for an assignment. The number required depends on the particular assignment. This allows a student an option concerning an assignment. If he feels it is a review he does not need, he can turn in a "flub stub" instead.

To earn a "flub stub" the student must catch the teacher making a mistake, raise his hand, and point it out. If a student finds a mistake in physics in magazines, textbooks, or other publications either in or out of the classroom, he may bring it in and receive a "flub stub."

This technique helps to keep the attention of the students. Occasionally the teacher can purposely make a mistake to see if students are really paying attention. You will be surprised how alert your students will become and how well they learn to question and put forth evidence for their point of view. This has helped me also as I have learned new things from some of my students who are more

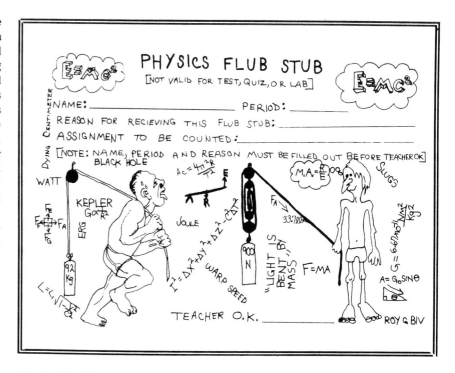

Fig. 1. The Physics Flub Stub awarded to alert students who catch their teacher in an error.

knowledgable in certain areas than I am.

The "flub stub" I use (Fig. 1) was designed by one of my physics students, and I accepted it without quibbling over one or two slight defects which readers and their students may enjoy finding for themselves. We run off copies on a ditto machine.

Reference

1. A similar view was expressed in the editorial of the October 1975 issue [Phys. Teach. **13**, 390 (1975)].

HOT DOG CONDUCTION.
AN EDIBLE EXPERIMENT

Roasting a hot dog by applying standard line voltage across it is not a new idea. However, as familiar as it is, the demonstration provides an interesting background for learning good, solid physics.

The demonstration requires a variac, two nails about 6 cm long, a hot dog (plus bun and mustard), a timer, two clip leads, and an ac ammeter. The ammeter should be easily visible to the whole class because it is interesting to watch the current rise to a maximum, decline slowly and then very sharply drop to zero. (Fig. 1.)

Because of the exposed line voltage, suitable safety precautions should be observed. Also be sure the hot dog is raw and unfrozen since conduction depends on the hot dog's juices.

Typical data ranges are: current = 0 - 1.5 A, resistance = 80 - 110Ω and cooking time = 50-100 sec. Taking current readings every ten seconds provides a manageable amount of data. Student activities could include computing the hot dog's resistance, cost of cooking and amount of heat energy used.

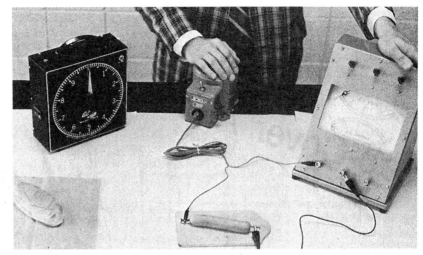

Fig. 1. Current supplied by Variac is monitored by an ac ammeter as it cooks a hot dog. The bun is kept ready for the end of the demonstration.

All of the above investigations require values for current. Because the current does not change linearly, a simple average is not the best value. I have my students plot a graph of current vs. time. They then find the area under the graph line by counting squares, which is equal to total charge ($\int I dt = Q$). Dividing the total charge by the total time yields a better average current value.

For completeness you may want to monitor the voltage across the hot dog, pointing out that the potential difference provided by the power company is essentially constant.

A nice touch at the demonstration's end is to offer the cooked frank to a class member complete with bun and mustard!

I wish to thank Peter Galindez for taking the photos of the apparatus.

ROBERT F. NEFF
Suffern High School
Suffern, New York 10901

Hot dog physics

George B. Barnes
Washington State University, Pullman, Washington 99163

Anyone who has ever tried to fry hot dogs knows that they were not designed to be fried. They curl away from the pan as soon as they are heated up and once they have curled it is extremely difficult to turn them over to cook on the other side. Hot dogs must have been designed to be boiled. When hot dogs cook at a low boil, they cook for about thirty seconds on one side, roll over cook on the other side, roll over again and so forth.

Cooking hot dogs and figuring out the physics involved can make for an interesting laboratory period. First, have students fry some hot dogs and ask them why the hot dogs curl. After they have figured out that hot dogs curl because of their high coefficient of thermal expansion, students will be ready to attack the question of why hot dogs roll over when they are boiled. Without much trouble students will be able to figure out that when the submerged portion of a hot dog heats up and expands enough it will float up turning the hot hot dog over. Once the hot dog has turned over the top portion cools off and the bottom portion warms up so that the hot dog soon turns over again. *

From this understanding some students can be led to understand that simultaneously cooling the top of the hot dog and heating the bottom of it raises its center of mass and lowers its center of buoyancy. When the center of buoyancy gets below the center of gravity the equilibrium becomes unstable and the center of gravity falls below the center of buoyancy, turning the hot dog over.

After students have gained some understanding of how hot dogs roll over, they will be ready for hot dog races. The object of these races is to get the hot dog to turn over as frequently as possible. Students should be encouraged to experiment with such things as fanning the water to cool the top of the hot dog faster, adding salt to the water to make it float higher, cutting the center out by running a plastic straw through it and cutting it in two. A hot dog laboratory period is a good way to promote creative problem solving, interest, and involvement in physics. It is also a good way to make physics fun.

---

* The less expensive brands generally work best.

Lab-Cover Art

by Craig Fletcher,
Polytechnic School, Pasadena, CA 91106

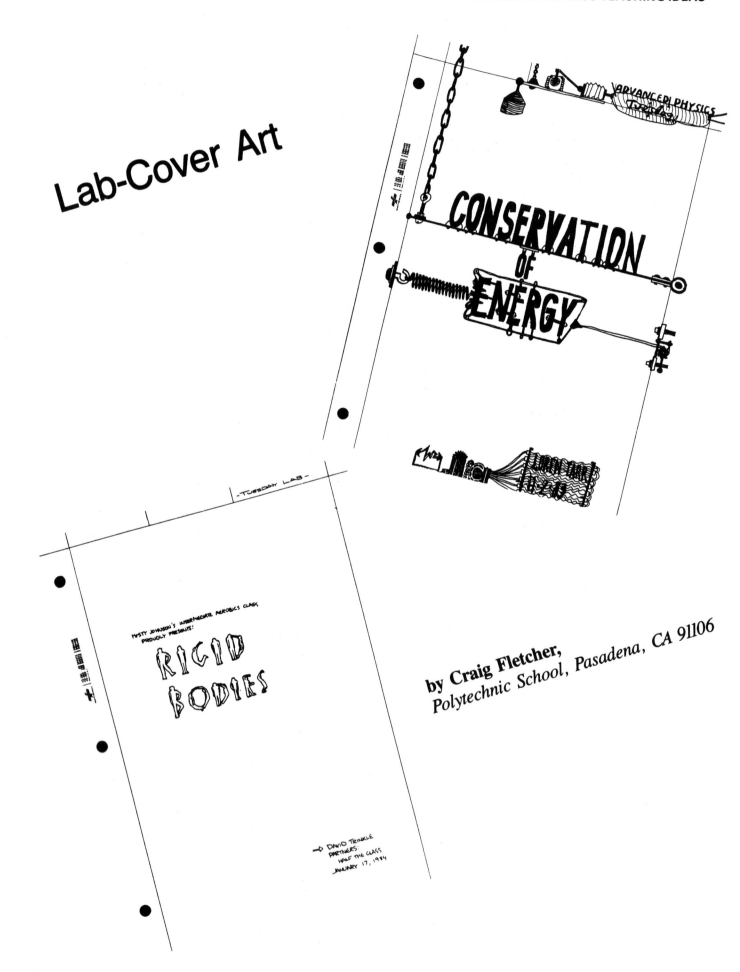

A POTPOURRI OF PHYSICS TEACHING IDEAS — ODDS AND ENDS

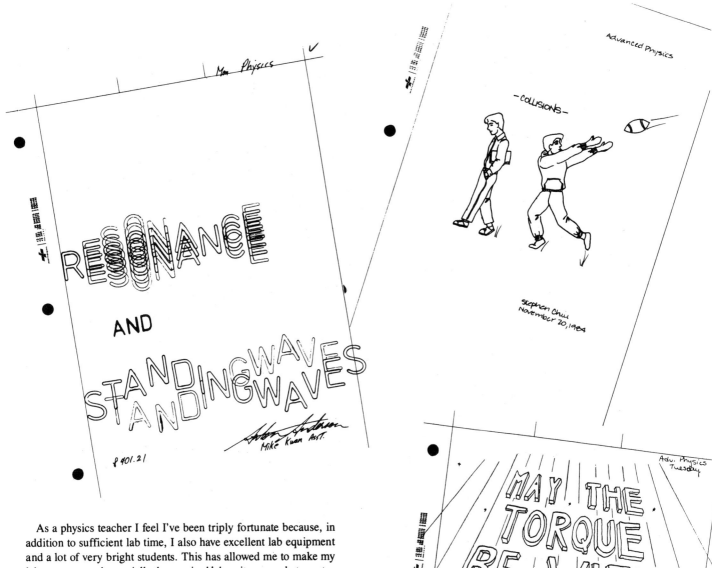

As a physics teacher I feel I've been triply fortunate because, in addition to sufficient lab time, I also have excellent lab equipment and a lot of very bright students. This has allowed me to make my lab program, and especially the required lab write-ups, what my students affectionately call "challenging." The labs take time, and in most instances, they are rigorous and difficult. Thus, I am constantly being amazed by my students and the artistry they display with their lab covers (I require a write-up cover that includes the student's name, the experiment's title, etc., with every lab done).

I say "amazed" because if I had just finished spending three to six hours crashing around with theoretical calculations, I'd probably be in no mood to spend an additional hour or two being creative with my lab cover (though I must admit, in all seriousness, that such an effort would undoubtedly be pleasant and therapeutic). But to my delight, that's exactly what many of my kids do whenever they aren't buried in homework from other classes.

At the beginning of each school year, I show my new students what previous students have done. I then invite them, if they have the time and inclination, to follow along in the tradition. I make it clear that it has absolutely nothing to do with their grades—that it's purely for their own amusement and entertainment. And I let it be known that if they do a particularly nice cover, they shouldn't be surprised if, when they get the lab back, they find that their cover has mysteriously disappeared. Mr. Fletcher filches good covers.

Although I've been saving covers for only the last few years, I thought I'd send along a sampling in the hopes that other physics teachers across the country might start up a tradition of their own, assuming they don't already have one. □

SCALING AND PAPER AIRPLANES

I have long wanted some experimental work to augment a discussion of scaling things up or down in size. The paper airplanes shown in Fig. 4 illustrate several properties of scaling. The middle-sized one is standard, made from a full sheet of notebook paper. It sails well. The large one is about 2.5 times larger. With some selection at the stationery store, poster paper about 2.5 times the thickness of the notebook paper (0.1 mm) could probably be found. The one shown is too thick. When students have compared the wing areas and weights by squaring and cubing the linear ratios, they correctly predict that the large one will fly "like a rock," or have to go a good deal faster to sustain flight. It does.

The smaller one, about ¼ the scale of the standard model, obviously has a much lower weight-to-surface area, and flies very slowly. It is made from toilet paper.

If these are prepared as a demonstration, some attention must be given to the "trim" of each glider so that they perform well. There are several ways the paper airplanes could be developed into a lab experiment, but it would be a wild one!

An interesting assignment on scaling is to ask students to design things for either the Brobdingnagians or the Lilliputians of *Gulliver's Travels* (houses, chairs, clothing, cooking utensils, automobiles, etc.). They must sketch their designs and support them with mathematical arguments. They come up with many interesting conclusions, e.g., a Brobdingnag roast of beef would take quite a while to cook, and a Lilliputian shirt is bound to be relatively coarse and uncomfortable.

Does anyone have other experiments on scaling?

ROBERT H. JOHNS
Academy of the New Church
Bryn Athyn, Pennsylvania 19009

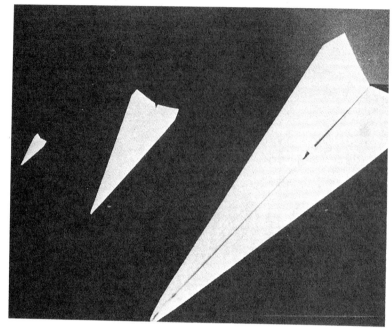

Fig. 4. Paper airplanes illustrate properties of scaling by flying them in the physics class.

Enchanting things to think about

Julius Sumner Miller
16711 Cranbrook Avenue, Torrance, California 90504

INTRODUCTION: In a recent course on my Demonstrations in Physics for a heterogeneous class — students from biology — chemistry — history — psychology and such — *and* from physics — from freshmen onward — I gave as the final examination an array of questions which I call Enchanting Things to Think About. In every course I have ever taught — over some 40 years in the classroom — my principle endeavor has been to cultivate THE THINKING OF PHYSICS. This is best done — as I see it — by extending the physics of the classroom and the lecture and the textbook to THE THINGS OF NATURE. The seeing and contemplation of these things which so abound must be inculcated. It makes for a fuller life. As Faraday (following Shakespeare) put it:

"...and makes the observing student find tongues in trees, books in the running brooks, sermons in stones, and good in everything."

Further to my philosophy I have tried these questions on some good physicists both of the younger and the older generations. The older do pretty well; the younger are often helpless. I do not believe that the THINKING OF PHYSICS is now being taught!

Here now are 10 of the 49 I set in the examination. With the accord of the editors of *The Physics Teacher* I will be pleased to give you *my* answers provided you submit your own to me first. On your request I will also send you a copy of the whole examination if your interest dictates it.

1. PASSING UNDER A FLOWERING TREE: In the morning I pass *under* a flowering tree and I smell it *much*. In the afternoon — on a sunny day — I pass again under this tree but now I smell it *less*. What accounts for this and what seeming PARADOX arises?
2. THE FREEZING OF A LAKE: Talk about the FREEZING of a quiet pond or lake.
3. TOASTED BREAD ON A PLATE: At breakfast you toast a slice of bread in a toaster. You now quickly place it on your bread-plate. In a few seconds you lift it up and lo and behold! the plate under the bread is WET! What goes on here?
4. SEA BREEZE: We stand by the sea — and at certain times of the day we feel a breeze FROM the sea — and at other times a breeze TOWARD the sea. What do you make of this?
5. WHY I DON'T FEEL COLD: My body temperature is about $98.6°F$ — which is about $37°C$. The room temperature is about $23°C$ — say. Why don't I feel cold?
6. A MIRAGE: Can a MIRAGE be photographed? Don't just say YES or NO. I want more than that!
7. MERCURY DROPLETS: A tiny mercury droplet on a clean glass plate can be spherical. Add mercury to the droplet. It coalesces. The drop can be made bigger and bigger. It now flattens out. Give an ENERGY argument on HOW MUCH DOES IT FLATTEN OUT?
8. THE SNOW CREAKS: Have you ever walked on *dry cold* snow? It "creaks" as you may know. (This word "creak" is nearly like "splash" or "bang" — being onomatopoeia — or onomatopoetic). What do you make of the snow creaking?
9. ICE CUBE IN A GLASS OF WATER: At random you drop a tiny chunk of ice into a glass of water — the glass not quite full. If you are alert you will see it drift toward the wall of the glass. Why does it do this? Now we wish to have this chunk of ice STAY IN THE MIDDLE of the water surface. What action can you take to accomplish this?
10. HOUR-GLASS ON A SCALE: An hourglass — you know what this is — rests on a scale or on a platform balance. The sand is in the UPPER chamber at time zero. The system is in equilibrium — having weights on the other pan. We release the sand to dribble down into the *lower* chamber. Explore the several conditions which arise. HINT: In the beginning we have a stream of falling sand NOT COMPLETE. Soon thereafter we have the first grains hitting the bottom. There is *now* a *full* stream. Follow our thinking? And now how does the reading of the scale go?

Further enchanting things to think about

Julius Sumner Miller
16711 Cranbrook Avenue, Torrance, California 90504

INTRODUCTION: In the May 1977 issue of this journal[1] I posed an array of TEN ENCHANTING THINGS TO THINK ABOUT and in the Introduction there I declared — rather boldly, I suppose — that "I do not believe that the THINKING OF PHYSICS is now being taught!" I WAS RIGHT! Nor is the THINKING OF PHYSICS being cultivated at any level! The instruction is textbook bound — classroom bound — computer-bound — and THE SPIRIT OF PHYSICS is rather silent. I have found it so nearly everywhere in the WORLD! The old-fashioned classroom where DEMONSTRATIONS abounded is pretty well dead! Ask the graduate students what they think about their earlier courses! These are often taught by the *high-level* staff whose interests are pretty remote from Galileo and Pascal — from Hooke and Archimedes. These lower courses are so filled with mathematical manipulation and so scant in the description of NATURE that the FEELING FOR PHYSICS is altogether lost.

Returning to my phrase above: I WAS RIGHT! This is based on the replies I received to my TEN original THINGS TO THINK ABOUT. Since it may well be claimed that *one* such exploration of my charge has little reliability I am trying it *again*. So here is another 10 of the 49 I posed in that examination I referred to. As before — I will be pleased to give you MY answers provided you submit your own to me *first*. And — on your request I will also send you a copy of the whole examination if your interest dictates it. Parenthetically — many wrote to me saying that they enjoyed the first 10 and asking for the whole set of 49 — *but they never did send me their answers!* I *still* sent to them the whole set but I heard no more!

Here then another set of TEN ENCHANTING THINGS TO THINK ABOUT —

1. SLICING AN ONION: Some people cry when slicing and dicing an onion. This can be averted by cooling the onion in the refrigerator. What is the PHYSICS of this?

2. ICE CUBES IN A BOWL: We place some well-formed ice-cubes in a deep glass bowl — say — a heap of them. We soon find them frozen together *at their edges.* What goes on here? Or — as Maxwell would say — "What's the go of it?"

3. COPPER BALL FLOATS IN WATER: A so-called "empty" copper ball — some 3 in. in diameter — of thin wall — is so designed as to *just float* in a beaker of water at room temperature. It floats — say — just half submerged. What do you predict if we A) cool the water — B) warm the water? Why do I put "empty" in quotations?

4. HOT AIR: The day is hot — the air is hot — you are hot — the tire on your car is hot — the air IN the tire is hot. You depress the valve stem. Out comes some of the HOT air. You do this on the road in the desert valley and again on a mountain top. Talk about this.

5. FILLING A NARROW-NECKED BOTTLE: You have a narrow-necked bottle — like a pop-bottle — say. You wish to fill it at the kitchen sink with water. Where — under the flowing water — is it best to hold it? And what *acoustic* phenomena can be witnessed? (I add this NOTE: What is the *shape* — the *geometry* — of the falling stream of water between tap and sink and *where* is the center of mass of this stream?)

6. ON THE LOAD ON A TRUCK AND TRAILER RIG: We see an enormous truck and trailer rig with an enormous load thereon. *Knowing* a thing or two *and* some PHYSICS how could you calculate the load on the vehicle?

7. A TEA KETTLE IS BOILING: So we say. Of course — it is the water that is boiling! Now I ask: The water is boiling. *Something* emerges from the spout. If you look sharply — closely — at the spot NEAR the orifice of the spout — you will SEE NOTHING. A few millimeters away you *do* SEE something. Tell us about this. The language in the answer must be PRECISE!

8. ON WHIP-LASH: Your car is at rest. You are at the wheel. You are at rest. Your car is now struck sharply from behind. Your head is jerked backwards. *This is what is said.* This is the claim that is made. It is called whip-lash. The legal aspects ensue and this is the language that prevails. What is the REAL PHYSICS in this engagement?

9. THE ARM AND THE ELECTRIC LIGHT: We have a 100-W lamp — say, clear-glass — unlighted. We stand near it with the under-forearm bare — say 6 in. from the "bulb." We now energize the lamp — we turn it on, as we say. NEARLY INSTANTLY we FEEL the "HEAT" on the exposed flesh of the sensitive arm. Now we *quickly* FEEL the glass or the lamp with the hand. The glass IS NOT HOT! What goes on here?

10. AN EXTRAORDINARY DEVICE: The device shown is a tightly-fitted chamber with mercury in the one side and water in the other. A cylinder or disk is mounted in the center — on a horizontal axis. Clearly — the *óne* side — that in the mercury — has a larger buoyant force on it than does the other side in the water. Accordingly — the disk turns clockwise. No question about it! So a central shaft in the disk can be belted on the outside to drive a device outside this engine. What do *you* say about this?

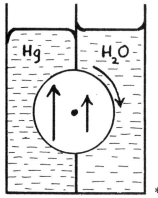
*

EPILOGOS: As I reported in my first note the older men in the business do pretty well with these little investigations of NATURE. The younger ones show little talent for THINKING PHYSICS. So for them a few words: The requirement you have before you in each exercise is *not* to find a formula that fits nor a Law of Physics that you can quote. You wish to *un*cover the PHYSICAL MECHANISM of WHAT IS GOING ON. To jump quickly and impulsively into an answer — the final answer — is utter folly. What you must do is this: YOU MUST THINK! And what does THINKING *mean*? It means ASKING YOURSELF QUESTIONS! The *first* questions become the foundation for subsequent questions and therefore these first questions must be of a simple sort. (Of course — these "simple" ones may not be trivial!). By this sequence of questions you add to your body-of-knowledge of the matter at hand — by infinitesimal accretion, as it were, layer upon layer — until AN UNDERSTANDING EMERGES. This is a far cry from just KNOWING something!

Consider NUMBER 2 — ICE CUBES IN A BOWL: *Why* do I say *well-formed* ice cubes? Why in a *deep* bowl? Why a *heap* of them? And what is the significance of *edges*? Each of these — in turn — brings light into the darkness!

Reference

1. J. S. Miller, Phys. Teach. 15, 296 (1977).

* NOTA BENE: Often a man claims an idea as his own coin when in Truth he learned of it on a far-gone day. So it is with *this* device. But I have used it in class for 40 years. If it is yours I am beholden to you and I say my THANKS. JSM

A game of gravity

Gus A. Sayer
Franklin High School, Franklin, Massachusetts 02038

Physics can be entertaining, as well as challenging, to students of a wide range of abilities! As part of our course we have employed a number of activities which raise interesting questions not usually discussed in physics textbooks. One successful activity has been the "gravity game," a quiz contest which poses paradoxical problems for which the solutions are not immediately obvious.

The rules are simple. The class is divided into six teams according to their seating locations. The first question goes to the first team who may either attempt to answer or pass. The questions go around the room in rotation until the correct answer is provided; then the next question goes to the next team. Correct answers receive five points; incorrect ones, and answers out of turn, lose two points; partial answers receive two points.

Gravitation is not an easy concept for most high school students to grasp. In particular, the student's everyday experience in a constant gravitational field makes it difficult to comprehend the significance of the inverse dependence upon the square of distance in Newton's gravitational force law. Furthermore, students don't easily accept the implication of the universality of gravitation that *all* matter, even small pieces, exert forces on each other in proportion to their masses. The problems in the "gravity game" draw attention to these important relationships and provide examples to students to help them better understand the phenomenon of gravitation.

Most of the problems have been culled from two sources: Martin Gardner's *Space Puzzles*[1] and Jearl Walker's *The Flying Circus of Physics*[2]. A sampling of some of the most instructive and curiosity-arousing problems follows.

(1) The moon's gravitational pull on the waters of the earth creates a tidal bulge facing the moon. Why, then, do we have in most places not one but two high tides a day?

(2) By direct calculation the sun exerts 169 times greater force on the earth's oceans than the moon does. Why is the sun's effect on the tides only about half as great as the moon's effect?

(3) If a person could stand at the center of the earth he would feel no force due to gravity. Why not?

(4) If a hole were drilled through the center of the earth and a ball dropped down the hole, what would happen to the ball?

(5) Each team is asked to estimate the time for the ball (in question No. 4) to reach the opposite side of the earth. The team closest to the theoretical value of 42 minutes receives five points. (An exact calculation requires not only calculus but knowledge of the earth's internal structure, but an approximate answer can be quickly found assuming that the average acceleration due to gravity is halfway between its value at the center and its value at the earth's surface.)

(6) Lewis Carroll proposed a system of gravity trains which would link major cities by tunnels drilled through the earth. *Ignoring friction*, would such a system work? What would be the fuel cost for a trip between two cities? How fast would a train be travelling when it reached its destination? How long would a trip between Boston and Los Angeles take? (Surprisingly it takes 42 minutes for any tunnel going through the earth!)

(7) The sun pulls on our moon with a force twice as great as the earth does. Why then does the moon stay in orbit around the earth?

(8) Why can't the United States maintain a permanent "spy satellite" directly over Moscow 24 hours a day?

At the completion of the game, prizes are awarded. Students on the winning team receive large, juicy apples to remind them of Newton's original inspiration for the law of gravitation.

References

1. Martin Gardner, *Space Puzzles* (Archway Paperbacks, New York, 1972).
2. Jearl Walker, *The Flying Circus of Physics* (Wiley, New York, 1975).

AN EXPERIMENT ON POPULATION GROWTH AND POLLUTION

M. Jeffries

Macalester College
St. Paul, Minnesota 55101

The Foundations of Physics course at Macalester College is a one semester introductory course for nonscience majors with no mathematics prerequisites. Most of the students have a very limited mathematics background and prefer to avoid mathematical analysis if there is any other way of solving a problem. An attempt has been made in the laboratory to provide students with insights that will be useful to them in their approach to other fields of study, and to demonstrate the relevance and necessity of physics in understanding some of the problems affecting us today. One of these problems is over population, closely linked with pollution. The full implication of world-wide over population and pollution depends upon a realization of rapid increase of the exponential function, a concept which many students outside of the sciences do not understand.

This experiment was designed to provide students with an insight into the mathematics of population growth, pollution increase, and some of the problems affecting our environment.

The Experiment. Fifteen jars or flasks were set up in a row, beginning with small jars and increasing to larger ones and a large dish of aluminum or copper pellets was placed nearby. After the students entered the laboratory, they were told that all that was required for this experiment was to place one pellet in the first jar, two in the second, four in the third, etc., until they filled the 15th jar. Usually there were a few students who quickly realized what this entailed, but most of the group would begin filling the the jars. After five or 10 minutes it became apparent that it would have been an impossible task to fill the 15th jar by the end of the two-hour period. When everyone had realized the magnitude of the number of pellets necessary, the experimental write-up was distributed. An appendix was attached giving a brief explanation of exponentiation in terms of powers of two. This was explained by the instructor and then the students went ahead with the experiment, setting up the problems and solving them with the aid of a canned program on an IBM 1130 computer.

Experimental Write-up. (Discussion of population growth has been eliminated for brevity.)

Real Life Problem No. 1. There are now three billion (3×10^9) people on the earth. Assuming that the world's population has been growing at its present rate (doubling every 30 years), calculate how many years ago Adam and Eve existed. If this growth continued, this many years in the future will see three billion human beings on earth for every couple today. Obviously some of our assumptions are going to have to change.

Real Life Problem No. 2. The cities of St. Paul and Minneapolis have a population of 1.2 million and an area of 120 square miles, giving a population density of about ten thousand people per square mile. The population of the United States is 200 million and its land area is three and one-half million square miles. With the U.S. population doubling every 40 years, how long will it be before the entire country has the same population density as the Minneapolis-St. Paul area and there are no longer any farms, forests, or wilderness?

Pollution

As fast as our population is increasing, pollution is increasing even more rapidly. Assuming that social and political factors have no effect, as they have not until very recently, and that industrialists are unwilling to spend part of their profits in controlling the problem, then pollution depends closely on the gross national product. Everything we use contributes to pollution. The great majority of all our energy comes from burning hydrocarbons and contributes to air pollution. The pencil I wrote this with has contributed to pollution in several ways: first, the electrical energy necessary to run the machinery in making the pencil was probably produced by a steam generator burning coal; second, the metal needed to hold the eraser had to be smelted from its ore and produced chemical wastes, and the eraser itself required similar chemical wastes to be produced as a by-product of its manufacture. Anyone who has driven by a paper plant should be aware of the air and water pollution caused by the manufacture of this paper. Thus, almost anything we use in our daily lives has contributed to pollution.

The economy has been growing at an even faster rate than the population, averaging about 6 or 7% a year, thus doubling every 10 years. This means that the amount of pollution in the air and water and the amount of garbage and waste is doubling every 10 years.

Real Life Problem No. 3. Suppose a steel mill is pouring smoke and fumes into the air at a rate that the people in the neighborhood consider intolerable. They bring a suit through the State Pollution Control Board against the company. The home owners demand an end to the pollution. This would involve an initial expenditure of several million dollars by the company and one hundred thousand dollars per year maintenance. The company offers a compromise which would cut down pollution by 75% and cost the company only one-quarter million dollars initially and thirty thousand dollars a year for maintenance. If the home owners do not want to accept the company's proposal, the company threatens to tie the case up in court for up to five or six years, thereby saving themselves the maintenance costs for that period. In addition, there is no guarantee that the State and home owners would win the suit.

In the meantime, this company is fairly successful and is increasing its sales and production by about 10% a year. Assuming the pollution increases with production, doubling every seven years, how much will the pollution increase in the five years of litigation with nothing being done?

If the home owners accept the proposal, how long will it be before the pollution returns to its original level?

Real Life Problem No. 4. Detroit auto manufacturers are working on a method to cut down automobile pollution by 90 percent. If the number of cars increases at the same rate as the economy, how long will it be before the pollution is just as bad as before?

Discussion and Conclusions. The first two problems on population growth seem, in retrospect, to be less impressive to the students than the third and fourth. One reason seems to be that the answers, although quite unexpected, were so far into the future that it was difficult for them to see a direct effect on their lives. In performing this experiment again, it might be more advantageous to use times and numbers closer to contemporary situations. One such problem is:

If we assume that the population of Mexico is 40 million and doubling every 20 years, and that the population of China is 600 million and doubling every 50 years, how long will it be before there are as many Mexicans as Chinese? What will the population of either country be then? When will there be as many Mexicans as Americans?

Answers: (a) about 130 years, (b) 3.65 billion, (c) about 90 years.

The problems on pollution were eye opening to many students and demonstrated how quantitative knowledge is essential in understanding environmental problems.

This experiment was done by four different groups of *Foundations of Physics* students and once by a visiting group of eighth grade honors students. The response was generally enthusiastic and the laboratory periods ended with interested discussions or students working out additional problems using the canned computer program. As written, this experiment would be suitable only for nonscience majors in college or for high school physics courses. With a somewhat more intensive mathematical approach, a similar experiment might prove useful in introductory laboratories or as part of an ecology course.

Acknowledgments. The author is indebted to Dr. R. Mikkelson and Dr. Sung Kyu Kim for their assistance and suggestions in the preparation of this article.

The trigonometry "laboratory"

Chandler M. Dennis Jr.

Natural Sciences and Mathematics, Stockton State College, Pomona, New Jersey 08240

An important recent study[1] of introductory physics students revealed that while all persons surveyed had completed a standard algebra and trigonometry course, most of them could not successfully apply the basic trig concepts. Without a firm trig foundation, students find vector components very, very formidable.

This note describes a teaching attack that has been successful in conveying the elements of trigonometry to the author's pupils. Some readers will object to sacrificing any of their precious contact hours to such a decidedly nonphysics topic; however, as food for thought for those persons I restate the conclusion of the study done by Hudson and Rottman:

It seems clear that the teacher of introductory physics should seriously consider formal remediation of mathematical skills necessary to the study of physics at the beginning of the semester.[2]

As a necessary background, we will review the apparent psychological state of students entering general physics. Everyone reading this article has overheard comments like, "I've heard physics is very heavy; it doesn't relate to anything practical; avoid it if you can." We've also seen in the eyes of an otherwise bright person a strange look which reveals that at the first little reversal this student will panic and lose the *will* to *struggle* with physics for the rest of the year. Vectors, appearing early in many course sequences,[3-4] confirm the worst predictions about physics and possibly may permanently psyche out our normally intelligent scholars.

It should be pointed out that a student's lack of understanding of vector methods does not necessarily surface clearly during your planned vector discussions and exercises. A good student with only a glimmering of comprehension might not display this ignorance until motion on an inclined plane is discussed.[5-6] Since in most vector work prior to that of the inclined plane the horizontal component has been proportional to the cosine and vertical directly related to the sine, the student has come to associate horizontal with cosine (and vertical with sine). He is not applying trig: he is regurgitating problem solutions. However, on the inclined plane, one finds that the "horizontal" components are related to the sine, while the "vertical" are proportional to the cosine. The student finds himself in difficulty (again).

By pointing out that the trig functions actually concern the relationship between sides of right triangles, you can help your students form a solid trigonometric foundation. The writer emphasizes that the cosine, for example, is the name for a *ratio*. Students are taught that only certain combinations of lengths actually result in a *right* triangle (Pythagorean relation). Trig, as it applies to physics, can and should be more practical than the trigonometry discussed in a math class. Scouting manuals provide supplementary applications for instructors and students.[7]

In moving from the trig identities to the inverse trigonometric functions a teacher finds himself going "out of the frying pan and into the fire." The ability to determine an angle given its tangent, for example, is a critical skill for a first-year pupil. Yet, it has been the writer's experience that students almost unanimously interpret $\tan^{-1} X$ as the reciprocal of $\tan X$!!

If one makes suitable restrictions on the angles considered[8] there exists a one-to-one relationship between the angle and the value of a given trig function. One can attempt to drive this "monogamous" relationship home by use of the trigonometry tables found in the appendixes of most texts.[9] By consulting these tables you can demonstrate, using one finger, that each angle has a unique value of the cosine (or sine or tangent) and conversely each cosine value is associated with a definite angle. Hence, the inverse trigonometry function has been reduced to its essentials. Scientific hand-held calculators can then be examined and their inverse trigonometric capabilities displayed.

While students are becoming familiar with the one-to-one relationship between an angle and the value of any trig function, they are also drilled to translate a *mathematical* question such as:

$\sin^{-1} 0.5000 = ?$

into its *English* equivalent, possibly:

"What is the angle whose sine is 0.5000?"

After two class periods on the definitions, the author supervises an activity period, in which groups of three or four persons determine the height of individual group members by trigonometric means. Each group is supplied with a meter stick, protractor and about 10 ft of string. A group member stands and the string is stretched past the top of the subject's head to the floor (Fig. 1). One can identify a right triangle: the string forms the hypotenuse, the person is one side and the floor, from the student's feet to the point at which the string touches the floor, forms the third side. The group measures the string's length, the distance along the floor and the included angle. They can now use either the tangent or sine relationship to find the student's height. (Use of the Pythagorean relationship is discouraged).

Most groups quickly eschew use of the protractor — this is a sure sign that understanding is growing. The protractor gives only a rough approximation of θ and the students realize that the angle can be more accurately appraised by use of an inverse function.

In conclusion, vectors present no hardship to a student competent in trigonometry. If our task is to teach physics, it seems clear that we must delve into some lecture topics which are not traditionally physics, in order to communicate effectively. One of the amendments to Murphy's law puts it succinctly: whatever you want to do, you have to do something else first.

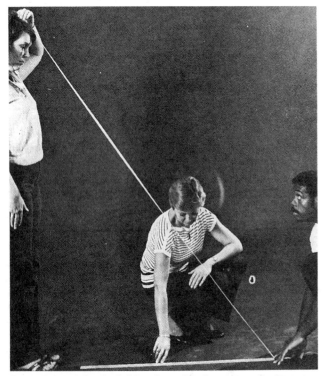

Fig. 1. Determining the height of a student by trigometric means.

References

1. H. T. Hudson and R. M. Rottman, "Let the machine do the bookkeeping!" Phys. Teach. 18, 305 (1980).
2. *ibid.*
3. F. W. Sears, M. W. Zemansky and H. D. Young, *University Physics* 5th ed. (Addison-Wesley, Reading, MA, 1976), pp. 9-11.
4. F. Bueche, *Principles of Physics* 3rd ed. (McGraw-Hill, New York, 1977), pp. 6-9.
5. F. Miller, *College Physics* 4th ed. (Harcourt Brace Jovanovich, New York, 1977), p. 86.
6. J. W. Kane and M. M. Sternheim, *Physics* (John Wiley, New York, 1978), pp. 50-51.
7. E. Hillcourt, *Official Boy Scout Handbook* 9th ed. (Simon & Schuster, New York, 1977), pp. 200-201.
8. E. F. Beckenbach and I. Drooyan, *Modern College Algebra and Trigonometry* 3rd ed. (Wadsworth, Belmont, CA, 1977), p. 302.
9. S. MacDonald and D. Burns, *Physics for the Life and Health Sciences* (Addison-Wesley, Reading, MA, 1975), pp. 707-711.

Lecture Demonstrations for the High School Science Teacher

Thomas J. Parmley, J. Irvin Swigart and Ray L. Doran
University of Utah
Salt Lake City, Utah

THE purpose of this paper is to describe a number of lecture demonstrations which, it is hoped, will be exciting to science and physics teachers in both our junior and senior high schools. As will be seen, each experiment requires the building of some pieces of equipment. This would seem to have some advantages because many high schools have wood and metal shops in which the student receives training. Thus students and teachers become involved in a stimulating and motivating project.

Figure 1 is a double exposure of a pirate in the two indicated positions as he walks "the plank." He toddles forward on the rough, bakelite board and stops on reaching the end of the plank. The freely hanging weight gives a constant tension to the string in slowly pulling the pirate forward. One can see from the photograph that the forward component of the tension decreases very considerably near the end of the path. Thus the pirate automatically stops and does not make the final leap. The counterpart of this equipment is to replace the pirate with a block of wood and the rough surface with a glass plate. The block will slide forward until it stops near the end of the plank.

Figures 2 and 3 are fine examples of replacing the supports that hold bodies in equilibrium by strings, which assume the direction of the resultant forces at the points of support. In each of these cases, the problem is simplified by having one of the forces in a horizontal direction. The tension in each string can be determined by measuring the stretch in a calibrated coiled spring (not shown in the photographs). Thus the student learns the significance of the basic step of making a "free body" in preparation for the solution of an equilibrium problem in mechanics.

Figure 4 illustrates a loaded truck on a bridge. The two spring balances, "c" and "d," and truck would be the main items which could not be built in the shop. In the photograph, arbitrary spacings "a" are indicated along the upper edge of the bridge for easy calculation of moments of force. In this case the bridge has its center of

Figure 2. Ladder as a "free" body in which the supports are replaced by the pairs of strings "a" and "b."

gravity at its midpoint. Inside the truck is a fairly heavy, solid, iron cylinder "e" with a small hole drilled along its central axis. This cylinder will fit over any of three short upright rods (not visible in the photograph) built into the body of the truck. Thus the center of gravity of the truck can be changed by merely moving the cylinder to any one of three positions. The corresponding positions of the center of gravity are indicated by three, black vertical lines ("b" indicates one of them) on the body of the truck. This demonstration lends itself quite readily to mathematical treatment before a lecture class.

The spool and bicycle in Figures 5 and 6 are fine demonstrations by which one can illustrate moments of force, especially inasmuch as their performances seem contrary to "one's common sense." The wheels of the spool are made from clear plastic while the central section is made from a colored plastic. For a high school demonstration a large, wooden spool would serve admirably well. As a further refinement, grooves can be cut

Figure 1. Double exposure of a pirate "walking the plank."

302 ODDS AND ENDS—A POTPOURRI OF PHYSICS TEACHING IDEAS

Figure 3. Gate as a "free" body in which the supports are replaced by the pairs of strings "c" and "d."

Figure 4. Bridge in which the supports are replaced by balances "c" and "d."

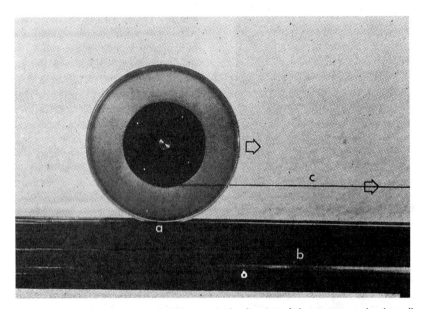

Figure 5. Spool resting on track "b" moves in the direction of the open arrow by the pull in string "c."

in the wheels to fit the ways of a wooden track "b." The track guides the spool as it rolls forward. By changing the angle which the string makes with the horizontal the spool can be made to roll forward, roll backward or slide. When the string "c" is pulled as indicated in the photograph, the spool will move in the direction of the arrow. When the string is pulled vertically upward, the spool will roll backward. When the string is pulled in such a direction as to line up with the contact point "a" between the spool and the track, the spool will not turn but simply slide forward.

The contact point "a" is at rest even though the spool is turning. This concept seems to be a new one to many students. This point can be verified by the student by taking a flash picture of one of the tires of a moving automobile as the car goes by at high speed. If the picture is properly taken, the photograph will clearly show the markings on the lower part of the tire while the upper part will be a blur or streak.

Now it should be easy to see why the spool behaves as it does. In the first case, where the pull is horizontal, the moment of force due to the string will be clockwise about point "a" and the spool will roll forward. In the case where the string is pulled upward, the moment of force about "a" will be counter clockwise and the spool will roll backward. Finally when the string lines up with point "a" there will be no turning effect about this point and the spool will slide.

The bicycle demonstration is, in principle, the same as the spool just described. This can be seen from the following explanation. The back wheel makes more turns than the foot pedal in the ratio of the number of teeth on sprocket "b" to those on sprocket "e." Thus in effect, the radius of the back wheel is increased by a factor equal to this ratio. The behavior of the bicycle then becomes equivalent to that of the spool described previously in which the radius of the central part of the spool corresponds to the pedal arm of the bicycle and the large part of the spool to the calculated size of the bicycle wheel with no chain drive. Thus when the string "d" is pulled in the direction indicated by the photograph, the bicycle moves toward the left.

A very interesting variation to this experiment is to attach an extension arm to the pedal arm at "c" and then tie the string to various points along this extension. When the distance from the center of the pedal arm to the point where the string is attached is equal to the enlarged radius of the back wheel as calculated above, the back wheel will slide as the string is pulled. If the string is moved to a point farther down on the extension arm and then pulled horizontally, the bicycle will roll away rather than toward the person performing the experiment. The reason the string is pulled backward rather than forward is to eliminate the problem associated with the coaster brake on most bicycles. The added features to the bicycle are supports "f" to keep it from falling and clamp "a" to prevent the front wheel from turning to the left or right.

The final photograph, Figure 7, illustrates most vividly that identical transverse waves passing through each other in opposite directions at the same speed result in so-called stationary or standing waves. The two long boards, "c" and "d," in the upper part of the photograph slip easily in the guides "a" and "b." Initially the boards "c" and "d" are adjusted until the zeros on both boards line up with each other and point "e" on the graph board below. In this position one adds up the vertical displacements of the two curves and marks the result point by point on the graph board. After this is done, the two boards "c" and "d" are moved so that point 1 on each board lines up with "e," and the plotting is repeated. In succession, points 2, 3, etc. are likewise aligned with "e" and the plotting procedure repeated for each setting. The loops drawn on board "e" are the results of such plotting and dramatically show the familiar characteristics of a standing wave.

Figure 6. Bicycle, held up by supports "f" on both sides of the rear wheel, moves in the direction of the open arrow by pull in string "d." See TPT, Vol. 1, pages 92 and 193.

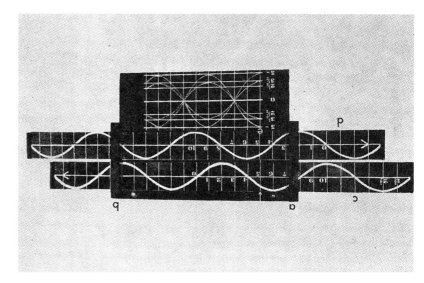

Figure 7. Two sine curves painted on moveable boards "c" and "d" move in opposite directions in successive and equal steps to give the series of resultant graphs shown on board "e."

Backwards clock

Thomas W. Norton
Linton High School, Schenectady, New York 12308

Most electric clocks are self-starting. Many of these have a small opening in the face which exposes a red warning flag when the current has been interrupted and shows that the clock no longer indicates the correct time. After one has reset the clock and inverted it momentarily, the warning signal springs back out of sight.

Another type, the impulse-start clock, remains stopped when the current is interrupted. In order to restart this type, a small knob on the back must be rotated in the indicated direction. Such clocks may be started in the reverse direction and they will continue to run backwards with the same accuracy as they have when running forward. Figure 1 shows the front of such a clock for which a new face and enlarged hands (glued to existing hands) were made for mounting on the classroom wall.

It is interesting to note the various reactions of students (and teachers) to the clock the first time they see it. Some don't notice any difference. They will read the time by reference to the position of the hands relative to the normal or expected position and ignore the numerals on the face. Thus, if the correct time is 3 o'clock, they would read the time from the backwards clock as 9 o'clock. Similarly, when the hands are in the normal 1:25 position, the correct time is 10:35. This typical misreading of data brings to the students' attention an important lesson. The observer has bias which must be overcome to ensure the successful collection of correct data. Without vigilance, the more common measurental tasks become subject to so casual an attitude that blunders can occur with relative ease.

An interesting discussion will usually develop regarding what is meant by clockwise and counterclockwise. One readily sees how problems in communication can arise as a result of different viewpoints. Without previous agreement, discrepancies can and do occur which interfere with understanding.

It takes some students in a class less than a week to become familiar enough with the clock to be able to read it correctly without taking an excessive amount of time. Some avoid its use altogether.

Fig. 1. *What time is it?*

Photo by L. Huppert

Drinking Duck Shutter

Dr. James Boyden, a senior scientist with Electro-Optical Systems, Inc., solved a problem encountered in a laser project by use of the Drinking Duck [see *The Physics Teacher* **4**, 121 (1966)]. To study propagation effects, the infrared radiation from a CO_2 laser is transmitted from the laboratory to a mirror situated high on a mountain and back to the laboratory. It was necessary to discriminate between this reflected beam and scattered sunlight, so a shutter was inserted in front of the mirror to interrupt the laser light, and, thus, identify it by producing a timed, lighthouse-like signal. A battery operated shutter could be used, but that would necessitate a climb up the mountain every week to replace the batteries. An answer was the Drinking Duck with a metal plate attached to the back (see Figure below). When the bird stands erect, the laser beam is interrupted; when it leans down to drink, the light strikes the reflecting mirror behind the bird. An unexpected problem arose when the Drinking Duck acquired a real life companion. A wire mesh screen over the door of the bird house was the solution.

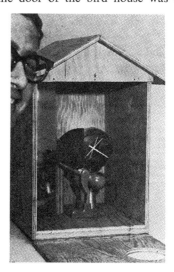

An Introvert Rocket or Mechanical Jumping Bean

A simple mechanical rocket which "lifts itself by its own bootstraps" two or three times its own height is described. Energy and momentum considerations are discussed and its uses as a demonstration device or as a laboratory instrument are detailed.

Richard M. Sutton
California Institute of Technology
Pasadena, California

While the world is still looking for a "clean atomic bomb," why not also for a "clean rocket" that doesn't spew out hot gases or eject any mass? The mechanical jumping bean described herein is a partial answer to that need. It is designed to show in paradoxical fashion how the principles of conservation of energy and conservation of momentum are intimately related and numerically applied. Chesterton said of Paradox, "It is Truth standing on her head to call attention to herself." This is just what any good demonstration should do. In this case, Truth (as embodied in this rocket) lifts herself by her own high heels.

The rocket as described may be used for *quantitative* measurements in the laboratory with rather good effect, inasmuch as the height to which it will lift itself is predictable to a reasonable degree of accuracy by simple measurements made prior to its firing.

Construction

A sturdy but light mailing tube about 40 cm long and 5 cm diameter is equipped with cork stoppers at top and bottom (Fig. 1). Through the top stopper are drilled 8 holes at 45° intervals parallel to the axis of the tube to match 8 holes in a cylindrical steel block M_1. Let M_2 represent the mass of the tube and stoppers. A screw-eye S in the bottom of the inner weight allows one to latch on a metal hook H introduced through the bottom stopper so as to draw down M_1 to the bottom of the tube against the strong pull of 8 rubber bands. A small window W cut through the side of the tube allows one to see how to engage the hook in S. A simple hook latch L pivoted at P is provided to retain the heavy cylinder in its lowest position until it is to be released, say, by the burning of a loop of string.

When the rocket is set in its vertical position, resting on the table with M_1 held at the bottom, it is ready for its remarkable jump. The string holding the latch is burned, there is a bit of noise followed by "pop" and the tube jumps suddenly to two or three times its own height.

Discussion

How can this object pick itself up so abruptly and dramatically? Evidently, its source of energy is one's muscles: work is done by the operator to store potential energy in the stretched rubber bands. At the moment of release, the bands are free to accelerate the inner cylinder M_1 upwards while they pull downward on the top of the tube. The momentum imparted to table and earth is not evident, although such transfer does occur; and the momentum imparted to the inner object on its upward trek is not evident either because this mass

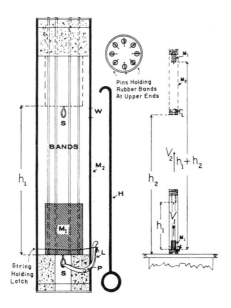

Fig. 1 Rubber-band driven rocket and its performance.

Richard M. Sutton was born in Denver, Colorado in 1900. He obtained the B.S. degree (1922) at Haverford College and the Ph.D. at California Institute of Technolgy in 1929. Three years, 1922-25, were spent teaching at Miami University, Ohio. Later teaching assignments include California Institute of Technology 1925-31, Haverford College 1931-56, with one year 1941-2 at the University of Minnesota. After two years, 1956-58, at Case Institute of Technology, he returned to Caltech as Professor of Physics and Director of Relations with Secondary Schools.

He is a past president of A.A.P.T. (1940) and was the Oersted Medalist in 1953. He was editor of the well-known book *Demonstration Experiments in Physics* and co-author of a college physics text (Mendenhall, Eve, Keys and Sutton). Numerous articles in the *American Journal of Physics* bear his name as author.

He is a member of Phi Beta Kappa, Fellow of the American Physical Society, the Board of Directors of N.S.T.A. and a member of the Advisory Committee on Science, State of California, Department of Education.

The Rockies of Colorado still hold a great attraction for him, although he now enjoys downhill climbing more than in the other direction!

is hidden from view. Only at the instant when the upward moving mass strikes the top stopper does the presence of this momentum make itself known in the upward jump of the whole tube and its load. So, without ejecting any mass in a direction opposite to the motion produced (as is usual in rockets), this rocket gains a net momentum in the upward direction in a somewhat mystifying manner: the equal and opposite momentum given during the pull of the bands is forgotten in the insignificant and unobserved motion of table and earth and observers.

One can, of course, devise ways to make this large reaction force evident, a force which is initially about 65 newtons acting on a mass M_1 of only 0.4 kg., thus giving it an initial upward acceleration of the order of 150 m/sec^2 relative to the table. The bands also pull the tube downward against the table with this large force. Two simple ways suggest themselves for showing this. First, let the rocket stand on one pan of a sturdy two-pan balance, the other pan being loaded with five or six kilogram weights. When the rocket is fired, these weights should be given a momentary but observable disturbance and the rocket should not rise quite so high from its movable launching pad. Second, lay the rocket on its side on a light board resting on two lab rods to serve as rollers. Now, when it is fired, it will make a sudden horizontal jump in the direction of the hook-latch end. In this case, there is no net motion of the center of mass of the system: the tube jumps lengthwise a large fraction of its length but stops as suddenly as it starts.

The rocket is *quantitatively* predictable in behavior. Let E_B be the energy needed to stretch the bands. This may be measured by pulling the bands down with increasing loads and plotting force versus extension to find the net amount of work done in stretching the bands (Fig. 2). The work done in stretching the rubber bands, that is the product of the force times the displacement, may be found by determining the area under the curve between two values of the extension. Since the rubber bands do not follow Hooke's law, the curve of load vs. extension is not a straight line and one cannot merely use one-half final force times extension as a measure of the work done. Data for the curve may be obtained in a separate experimental arrangement in which the hook H is used to pull down the rubber bands by loads placed on the lower ring and the corresponding extension noted on a suitably placed vertical rule. The tube M_2 will need to be supported by a ring stand or a similar arrangement may be devised. In these measurements, remember the mass M_1 and the hook H are part of the load on the rubber bands.

Let M_1gh_1 be the increase of potential energy of M_1 as it rises through the available height h_1. Let V_1 be the velocity acquired by the inner cylinder at the instant it strikes the top of the tube, and V_2 the velocity of take-off of the rocket from its launching pad. Three simple equations now pertain to the motion.

First, the kinetic energy acquired by M_1 is found by equating its kinetic energy to the difference of two potential energies, E_B of stretch, and M_1gh_1 of gravitational position:

$$\tfrac{1}{2} M_1 V_1^2 = E_B - M_1gh_1 \quad (1)$$

(There may, of course, be some error introduced by friction of M_1 against the walls of the tube or of the rubber bands in their channels. Hence V_1 may be somewhat less than computed

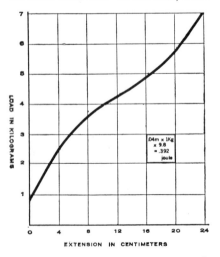

Fig. 2 Force vs. Extension Plot for Rubber-band driven rocket. Each square represents 0.392 joules of stored energy. Energy E_b is found from area under curve.

"GEORGE" By G. Freier

from Eq. 1.)

Second, at the moment of impact, there is conservation of momentum, but *not* conservation of kinetic energy, as shown by

$$M_1 V_1 = (M_1 + M_2) V_2. \quad (2)$$

Kinetic energy is lost in this inelastic collision process, but momentum is not; that is to say, there is fractional loss of energy in heat, to the extent of $M_2/(M_1 + M_2)$ which can be calculated and compared with experiment. Finally, the height to which an object starting with initial upward velocity V_2 rises is given by the well-known equation,

$$V_2^2 = 2gh_2. \quad (3)$$

Thus, one has all the ingredients with which to compute h_2 in advance, to predict how high the rocket will rise when M_1 is released, and one can proceed to compare this with experience by setting a marker at the predicted height and comparing actual performance with prediction. Solving (1), (2) and (3) for h_2, we have:

$$h_2 = \frac{M_1 (E_B - M_1 g h_1)}{g(M_1 + M_2)^2}. \quad (4)$$

As for energy lost in impact of M_1 with the top of the tube, this can also be computed. Let E_3 represent the final gravitational potential energy of tube and payload above table, namely $M_1 g h_1 + (M_1 + M_2) g h_2$. The "efficiency" of the rocket can then be called E_3/E_B, and this can be related to the computed fraction of the kinetic energy remaining after an inelastic collision between a moving object of mass M_1 with a stationary object of mass M_2, namely $M_1/(M_1 + M_2)$.

The design of the equipment is expressly such as to assure an inelastic collision with no unused space and to have M_1 and M_2 move with the same velocity after collision. This was the reason for taking the trouble to bore holes clear through both the top stopper and the inner metal cylinder M_1, so that the rubber bands would be under slight tension even when M_1 was held against the stopper. This removed the need for a dart-like tip on M_1 to hold it in the cork stopper, as was first tried.

Because rubber bands break or rot, it has been found best to have them individually retained by pins above the stopper and below the cylinder M_1. This is far more convenient than having them held by split rings at top and bottom, as was first tried.

Some sample data on the operation of the rocket follow:

Overall height	0.36 m
M_1	0.41 kg
$M_2 + M_1$	0.51 kg
h_1, free space through which M_1 is drawn down	0.23 m
E_B, energy stored in stretching of 8 bands through distance h_1 (by area under curve, Fig. 2)	8.5 joules
h_2, *predicted* height of jump	1.20 m
Observed height h_2 (average of 10 tries) or 2.34 times the length of tube	0.84 m
Computed efficiency from $M_1/(M_1 + M_2)$	0.80
Observed efficiency from E_3/E_b	0.61

Thus, one sees that the observed performance falls short of the ideal. One might be intrigued by the idea of a two-stage rocket of this kind, but the author forewarns of disappointment inasmuch as the second stage would not have the whole earth to react against and any gain would scarcely be worth the effort and additional complications.

A Possible Solution to the Energy Crisis?

I - The Archimedes Wheel

Albert A. Bartlett
Joseph Dreitlein
University of Colorado Boulder, Colorado 80302

The "Archimedes Wheel" consists of a large smooth cylinder C which is free to rotate about its axis, A in Fig. 1. It is in a tank in which the left side and right side are separated by a vertical partition P which touches the surface of the cylinder so that the partition and the cylinder divide the tank into two compartments. Water is shown in the left one and mercury in the right one. Thus the right half of the cylinder displaces mercury while the left half displaces water. According to Archimedes' Principle the buoyant force on a submerged object is equal to the weight of the displaced fluid. Since the density of mercury is 13.7 times the density of water, the buoyant force of the right half of the cylinder will be 13.7 times the buoyant force on the left half. This should produce a counterclockwise torque which would set the cylinder in rotational motion. The rotating cylinder could then be used to drive electric generators to run our industries.

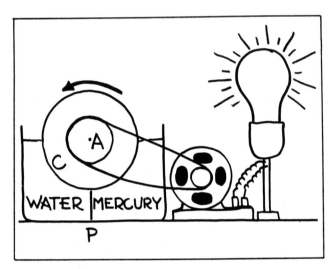

Fig. 1. The Archimedes Wheel

II — The Capillary Pump

The "Capillary Pump" is a capillary tube of inside radius r in which water from vessel A of Fig. 1 will rise by capillary action to a height h given by

$$2\pi r T \cos \alpha = mg = \pi r^2 \rho g h$$

where T is the surface tension of the water-air interface in ergs/cm^2, α is the angle of adhesive contact between the water and the glass wall of the capillary, ρ is the density of the water and g is the acceleration of gravity. For pure water and clean glass, the angle of adhesive contact is so small that one can set $\cos \alpha \cong 1$. In this case

$$h = \frac{2T}{\rho r g}$$

When the capillary column has reached its height h we open a side arm on the capillary tube below the top of the water column so that the water can pour out into the vessel B. The process is repeated with another "capillary pump" to raise the water from vessel B to vessel C. It is clear that if one of these capillary pumps will work as pictured here, then any number of them can be arranged in series to raise water to any desired height and any number of the pumps could be placed in parallel to generate any needed volume of flow. One objection that might be raised is that when the water first rises in the capillary tube it may not climb past the side arm. The "pump" can be primed by temporarily closing the side arm and pouring water in the top of the tube to overcome this difficulty.

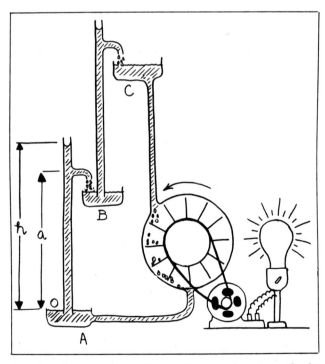

Fig. 1. The Capillary Pump. The explanation of the Capillary Pump will be found on page 179.

Do-at-home energy exercise

Varghese D. Pynadath
Fulton Montgomery Community College, Johnstown, New York 12095

In these times, when energy and environmental issues are of major world concern and energy conservation is the order of the day, educators have a moral responsibility to bring out an awareness of these issues and instill in students a positive attitude toward energy conservation. Because energy is a basic concept of physics, physics courses can play a key role in this regard. Presented below is a do-at-home laboratory exercise which the author has found to be of help in partially meeting this responsibility.

Each student is asked to do the following:

1. Measure the amount of electric energy used at home during one week by reading the electric meter at the beginning of the week and again at the end of the week.
2. Record the wattage ratings of all lights and appliances at home. Roughly estimate how long each one of them is on during the week and calculate the total energy used by each of the items during the week.
3. The amount of energy available per pound of coal is 12 000 BTU. The efficiency (output/input) of the electric power plant is 0.35 and the efficiency of electric energy transmission is 0.85. Using the above information, calculate the amount of coal that was burned to produce the amount of electric energy that was used during the week. Also, calculate the amount of energy lost at each stage of the energy conversion process at the power plant, starting with coal, using the following as the efficiencies of the different stages: boiler 0.88; turbine 0.40; generator 0.99. Discuss the environmental implications of the energy conversion processes at the power plant.
4. Carefully investigate the areas in which your family and you can cut down energy consumption.
   A. Take steps to minimize consumption of energy in all forms. Estimate how much reduction in energy consumption per week results from each of the steps taken. If your family and you had (before the assignment of the laboratory exercise) already taken steps to conserve energy, they should also be stated specifically.
   B. State the additional steps your family and you are planning to take in the near future to reduce energy consumption.
   C. State any additional steps one could possibly take (even though your family and you are unwilling to undertake them yourself for some reason or other) to reduce energy consumption.

Through this exercise students learn to read electric meters. It is astonishing to find out the number of high school graduates who have never before bothered even to look at the meter and who do not know how to read it. Students should be provided with the definitions of kilowatt-hour and BTU, and how to convert one to the other. It will be helpful if the students understand the three methods of heat energy transfer and also the basic processes involved in a heat engine.

The author has assigned this exercise to science and nonscience majors with equal success. The exercise is instructional and enjoyable. Most often it becomes a family project. The process of having the whole family involved makes them better educated and energy conscious. Most of all, the exercise leads, without doubt, to significant energy conservation measures at homes. As one student stated: "Through this experiment I gained an awareness of how much waste is prevalent in a household. If others could take the time to do a similar experiment, they too could gain the same perspective." The author could not agree more.

Electric Stop Clock

To use an ordinary electric clock as a precise stop clock, connect the simple adapter, shown in Fig. 1, between the line cord and the ac source. A silicon rectifier applies direct current to the clock motor when the switch is operated in the *stop* position. Eddy currents produce a braking effect, which combines with frictional losses to slow the rotor until it is about one-half cycle out of step with the pulsating dc magnetic field in the motor. At this time the dc magnetic field opposes rather than encourages rotation, and finally locks the rotor in a stationary position. This happens rapidly enough to cause nearly instant stopping, with negligible coasting. If automatic timing of a moving object is desired, substitute a low voltage relay and battery for the switch. For measuring short intervals of time, a 300-rpm synchronous motor may be substituted for the clock. Attach a strong but lightweight pointer to the motor shaft and calibrate a dial so a complete revolution is indicated in 0.2 sec.

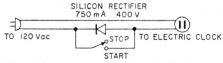

Fig. 1. Circuit for converting an electric clock to a stop clock.

A stirring experiment

Bruce Jones
Kentucky Country Day School, Louisville, Kentucky 40222

Since we do not have available the store-bought apparatus for determining the mechanical equivalent of heat we have improvised a rather simple and straightforward alternative. Using a blender borrowed from the physics teacher's kitchen we set up the experiment. Meters to measure the voltage across, and the current through the appliance are provided. It is also necessary to obtain a Celsius thermometer calibrated to 0.1 C° and a stopwatch.

The water is heated by the stirring effect of the blender and the following quantities are determined. (Sample data are given.)

I_o = current required to run blender without water = 1.10 A
I = current required to run blender with water = 1.37 A
V = voltage across appliance = 119 V
M = mass of water used = 0.300 kg
ΔT = change in temperature resulting from stirring = 2.2 C°
t = time period for which blender runs = 90.0 s
c = specific heat of water = 1.00 kcal/kg C°

The electrical energy used in stirring, $(I - I_o)Vt$, is then compared to the quantity of heat resulting, $cM\Delta T$. If the former is plotted against the latter the graph is linear and the slope gives the relationship between joules and calories. Our students' results were all within 15% of the accepted 4.18×10^3 joules/kcal. To attain this degree of accuracy it is critical that the net heat exchanges through the sides of the plastic container be kept to a minimum. This is achieved by allowing the water to come to room temperature before doing the experiment.

A suggestion would be to follow this experiment with the standard one in which the mechanical process is omitted, and the water is heated directly by a resistance coil. If a similar joule/kcal ratio is obtained then one can argue that apparently there is no appreciable energy loss in the mechanical process in the "blender" experiment.

SCOOPED

One of my students now knows what it's like to be scooped. We had been working for a month or so on calculating the mechanical equivalent of heat by using a blender. You can imagine his surprise when he and I read "A Stirring Experiment," [Phys. Teach. 18, 671 (1980)]. Students often hear in class that two researchers make identical discoveries at exactly the same time but they really don't appreciate how this can happen. This experience was definitely a convincing one for my class.

Rodney LaBrecque, *Westminster School, Simsbury, Connecticut 06070*

Demonstration: A nail driven into wood

Dennis P. Zicko
Marshfield High School, Marshfield, Massachusetts 02050

If you told your class that you were going to drive a nail through a piece of ¾-in. pine board using only your bare hands, they would most probably think two things: One, you're crazy; and two, it's impossible. With this little demonstration you can prove them wrong on both counts and demonstrate some interesting physics.

The basic question to answer is how much energy is required to drive a nail through a piece of wood? The Project Physics course filmloop "A Method of Measuring Energy: Nails Driven into Wood,"[1] can be used to show that the depth of penetration is approximately proportional to the energy supplied. Using this as a basis, some quantitative data can now be gathered.

I do the same experiment as is depicted in the filmloop, being certain to use the same pine board and nail that I intend to use later. I drop a 1-kg mass from a height of 20 cm onto the head of the nail and measure the nail's penetration. To keep the mass from rotating, I guide its path by placing it inside a glass tube. After a series of trials, the energy per depth of penetration is calculated. A typical value for a pine board and a six-penny nail is 2.2 J of energy for every 1-cm of nail penetration. In order to drive the nail completely through the board, the penetration must be about 2.5 to 3.0 cm. The energy required, therefore, is roughly 6.6 J. (Actually it is less, for once the nail completely penetrates the board, the energy per centimeter decreases.)

I now make the assumption that my hands, which hold the nail, act in the same fashion as a pile-driver. Therefore it is the moving mass of the hands that supply the kinetic energy to the nail. Estimating the mass of two clasped hands as 1 kg (a good approximation), and using the kinetic energy relationship, the velocity that the hands must develop in order to generate 6.6 J of energy is 3.6 m per sec. This is not an abnormally high velocity for hands accelerating downward.

In doing the demonstration, the ends of a 30 cm long, 2 cm thick (¾ in.) pine board are set on two blocks approximately 10 cm high. A six-penny common nail is grasped firmly between the index finger and the middle finger of one hand. The head of the nail is abutted to a piece of cloth held firmly in the palm of the hand. A fist is made such that the nail is pointed downward, perpendicular to the board. The free hand is clasped firmly around

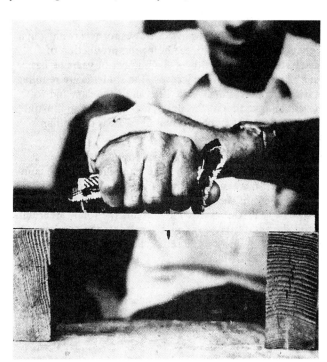

the wrist of the nail-holding hand. The two hands are raised about 60 cm above and over the board and then thrust sharply downward — similar to a karate chop.[2] If all goes well, the nail will penetrate the board (Fig. 1).

There are two difficulties to overcome before this can be performed in the classroom. The first is a natural tendency to restrain the velocity with which the hands and nail move downward. The second is a tendency to have the nail strike the board at an angle other than 90°. With a little practice, these two obstacles can be overcome and you will have a fine and dramatic demonstration.

References

1. "A Method of Measuring Energy: Nails Driven into Wood," Ealing Filmloops No. 80-3791, Cambridge, Mass.
2. Michael S. Feld, Ronald E. McNair, and Stephen R. Wilk, "The Physics of Karate," *Scientific American*, April, 1979.

Graphing Henry Aaron's home-run output

Herbert Ringel
Manhattan Community College, New York, New York 10019

One example used in my physics classes for reviewing graphing techniques is to graph the home-run production of Henry Aaron as a function of the number of years he has been playing in the major leagues. The students are required to draw the best straight line that can fit the data and to present an equation of this line. This example also illustrates the application of graphical analysis to areas other than physical science and mathematics.

The data table shows Aaron's home-run production on a yearly basis and cumulatively. The graph indicates the cumulative output as a function of the year. An equation which represents the data in years 2 through 19 fairly accurately is $Y = 37X - 37$ when Y represents the total career output at the end of the year and X represents the number of years played. In more advanced sections, a least-squares analysis can be utilized.

Years in Majors	Calendar year	Home runs in that year	Total career home runs
1954	1	13	13
1955	2	27	40
1956	3	26	66
1957	4	44	110
1958	5	30	140
1959	6	39	179
1960	7	40	219
1961	8	34	253
1962	9	45	298
1963	10	44	342
1964	11	24	366
1965	12	32	398
1966	13	44	442
1967	14	39	481
1968	15	29	510
1969	16	44	554
1970	17	38	592
1971	18	47	639
1972	19	34	673
1973	20	40	713

Data on Henry Aaron

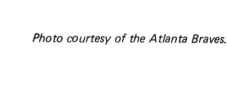

Photo courtesy of the Atlanta Braves.

HOME-RUN OUTPUT OF HENRY AARON

CAN THIS BE EXTRAPOLATED?

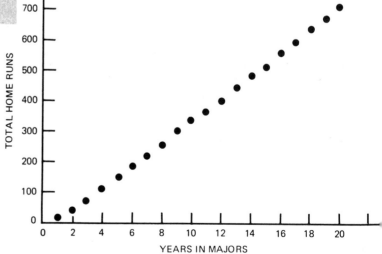

A POTPOURRI OF PHYSICS TEACHING IDEAS—ODDS AND ENDS

REMOVING THE BUOYANT FORCE

In discussions of Archimedes' principle, students learn that the buoyant force on an object is due to the upward hydrostatic force on the bottom of the object. However, alert students will sometimes ask, "What if a piece of wood is sitting on the bottom of a container? There should be no hydrostatic force upward then, and the wood should not float." They do not always find credible the answer that water seeps under the object and therefore manages to raise the wood.

Unfortunately, neither wooden blocks nor glass containers generally have surfaces that are flat and smooth. We got around this problem by gluing a cork to a coin-shaped piece of aluminum (approximately 1 in. diam) which had been machined flat to 0.0005 in. The surface was then hand-polished using wet 400-grain emory cloth until a surface finish of approximately 4 microinches was obtained.

A second, larger piece of aluminum was similarly prepared. The experiment is performed by placing the larger metal piece in a beaker and setting the cork-aluminum piece on it, metal-to-metal. Water is added slowly, and, *Voila!* If good contact is initially obtained between the metal surfaces, the cork stays submerged for several seconds (perhaps even a minute or two) before bobbing to the surface. Adding the water will sometimes cause a disturbance large enough to permit the water to seep between the surfaces. This can be avoided by first filling the container, putting the block in place, and then pushing the cork-aluminum firmly down on top.

We have found that this demonstration generates a great deal of interest, particularly if the students are drawn into a discussion to predict the result before the experiment is run.

Incidentally, aluminum machined to a flatness of 0.0005 in. prevents entry of water but does not result in significant intermolecular forces between the aluminum discs. However, this might be a problem if one were to use steel at a flatness of 10^{-6} in.

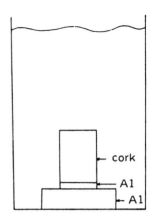

Fig. 4. As long as water cannot seep between aluminum discs, the cork experiences no buoyant forces.

GORDON E. JONES
W. PAUL GORDON
Department of Physics
Mississippi State University
Mississippi State, Mississippi 39762

L'EGGS* DEMONSTRATIONS

One popular brand of ladies' panty hose (L'eggs) is marketed in hollow egg-shaped containers. Several interesting demonstrations can be made with these containers.

A demonstration illustrating the effects of viscosity and internal friction is often performed with a raw and a hard-boiled egg. When spun rapidly, then stopped suddenly followed by a quick release, a raw egg will begin to spin again while a hard boiled egg remains stationary. To avoid the cost and problems associated with fresh eggs I use these egg-shaped containers. The two halves of a L'eggs panty hose container can be permanently sealed with epoxy. With a 1/2-mm hole drilled in one end this hollow "egg" can then be filled almost full of water using a hypodermic syringe. The hole is sealed with epoxy and we have a raw egg.

The hard-boiled L'egg is obtained by pouring hot paraffin into the two halves of another L'eggs container. When the wax has cooled and is slightly contracted, a thin layer of paraffin is poured into the depression in each half and allowed to cool to a semi-soft state. The two halves are then pressed together to form a hard-boiled L'egg.

A hollow L'egg container can be spun on its side. It will quickly reorient itself so the spin axis is that of its principal axis of smaller inertia. This behavior is similar to that of a spinning football but the L'egg, which is about 10 cm long, is easier to grasp for demonstrators with weak or small hands than a football about 30 cm long.

The explanation of this behavior lies in the principle of a system seeking its state of lowest total energy. Hence, as the center of mass rises, thereby increasing the potential energy, the kinetic energy must decrease by *more than* that amount. It does this by assuming a new axis of rotation about which the moment of inertia is less. For a solid ellipsoid of semi-axes of length a, b, and c, and assuming $b = c$, the principal moments of inertia are $I = m/5\ (a^2 + b^2)$ and $I = m/5\ (b^2 + c^2)$. L'eggs have a mass when filled with paraffin of 222 g and $a = 5$ cm with $b = c = 3.5$ cm. This yields $I = 1654$ g cm^2 and 1088 g cm^2 for the principal axes.

Empty L'eggs are approximately hollow ellipsoids and $I = m/3\ (a^2 + b^2)$ and $I = m/3\ (b^2 + c^2)$ yielding values of 291 g cm^2 and 191 g cm^2 for I.

Because of the differences in the moment of inertia of the solid vs the hollow L'eggs compared to the required changes of potential energy, the paraffin filled L'egg does not change axes as readily as the hollow L'egg unless it is spun at a higher initial angular velocity.

A hollow L'egg can also be used in a vertical airstream instead of the usual ping-pong ball or beach ball. Because of its egg shape it bounces, twists, and tosses more than a spherical object thereby lending additional interest to the students.

D. RAE CARPENTER, Jr.
Virginia Military Institute
Lexington, Virginia 24450

---

* L'eggs — A brand of panty hose distributed nationally by L'eggs Products, Inc., P.O. Box 2495, Winston-Salem, N.C. 27102

"Reciprocating" Engine

Kemp Bennett Kolb
The Haverford School
Haverford, Pa.

The Drinking Duck, often available from novelty stores or easily made by a glass blower, may be the operating unit of a reciprocating engine. The Duck also illustrates a number of principles of physics.

The lower bulb (Fig. 1) contains a liquid that has a relatively high vapor pressure at room temperature. Sometimes ethyl ether is used. Anyone building or working with a Duck should remember that ethyl ether is highly flammable and should not be brought near flames or sparks.

The Duck is supported by a horizontal shaft or pin at (A) and may swing about this pin as an axis. When it is not operating the system will come to rest with the center of mass (B) directly below (A). The construction of the system is such that the center of mass (B) is below the center line of the tube representing the body of the bird. The center of mass may be outside the body of the bird as indicated in Fig. 1. The tube is sealed to the lower bulb at (C) and extends into the bulb (D) containing the liquid. Thus a region (E) is

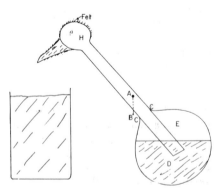

Figure 1

sealed off from the upper end of the system when the bird is erect and the lower end of the tube is below the surface of the liquid.

If now the head with the felt covering is moistened with water at room temperature and at a place where the relative humidity is not 100% so that evaporation can take place, the head is cooled. The vapor pressure of the liquid inside the tube and head is then reduced, and because of the higher vapor pressure in the lower space (E), the liquid is pushed up into the tube as shown in Fig. 2. This results in the center of mass of the system moving higher to (B^1) and the system tipping about the axis (A) so that (B^1) is

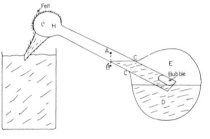

Figure 2

vertically below (A). The bird dunks his head. Some arrangement is usually made such that the bird cannot tip beyond the horizontal position—the liquid does not enter his head and the center of mass does not move above (A).

The amount of the liquid in (D) and the length of the central tube extending into (D) are such that just before the approximately horizontal position is attained, the lower end of the tube rises to the surface of the liquid in (D) and a vapor bubble (Fig. 2) moves up the tube. The pressure difference between upper and lower sections is reduced, in fact the pressures are equalized and the liquid runs out of the tube and back into the bulb (D). The center of mass of the system returns to (B), the initial position, the bird raises his head and the cycle is repeated.

A simple, very economical, low power reciprocating engine (Fig. 3) can be made with the Drinking Duck. It puts out an estimated 10^{-8} horse-

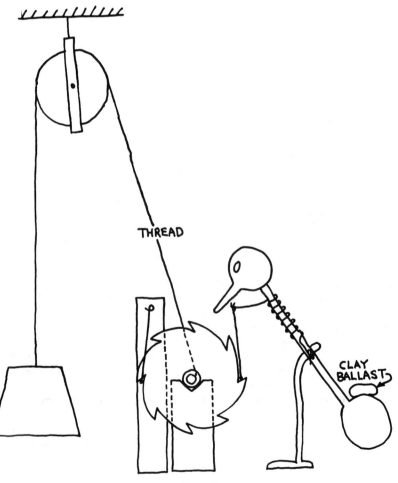

Figure 3

power when harnessed to a pulley system and performs roughly 10^{-3} calories of work per gram of water disappearing from the surface of the glass. The efficiency is even greater for the water actually evaporating from the head of the duck.

The wheel is made of cardboard and it rotates on two miniature ball bearings taped to notched wood blocks. The ratchets are made of stiff wire and paper clips. All components should be rigidly fastened to the base. A little adjustment is necessary but the operation is quite stable. In a display case the shaft rotated at about 15 rph, which gives some indication of how long it takes to lift the aluminized cork.

Here we have an engine doing work. Does the system violate any of the statements of thermodynamics which say that we cannot have an engine which merely abstracts heat from a source and yields mechanical energy? We usually say that for a heat engine to operate we must have a source of heat at one temperature and a "sink" at a lower temperature. We cannot simply connect an engine to the waters of the Pacific Ocean, which has an enormous store of energy and have work done without other changes than absorption of heat from the ocean taking place. The source for the Duck is the atmosphere. It and the entire system start operation at the same temperature. Does it violate the laws of thermodynamics?

This device operates because we do not have a complete system in equilibrium. In such a system we may sometimes obtain work without an initial difference of temperature between any of its parts. For example, a tank filled with compressed air is such a system. Work may be obtained by expansion of the air down to atmospheric pressure without having heat passing from a hot to a cold body. Indeed the temperature of the air may be lowered while it is doing work.

The statements of thermodynamics refer to a system in equilibrium. We do not have a system in equilibrium in the case of the Drinking Duck, since equilibrium does not exist between the water on the Duck's head and the water vapor in the space around his head. When the vapor pressure and density rise so that as many molecules enter the liquid state of water as leave it—that is when we have equilibrium between the water and its vapor—the engine will not operate even though the atmosphere has a large amount of energy associated with it.

If alcohol is substituted for the water in the tumbler or even merely added to the water, the Duck operates much faster. He is a toper!

Photo of the engine.

A Device for Demonstrating Half-Thickness of Shielding Materials

Richard O. Thomas
Amarillo College
Amarillo, Texas

A SIMPLE device has been developed for demonstrating two exponential curves: 1) that found when radiation received at a point is plotted *versus* the thickness of material used for shielding the point from the source, and 2) the curve of radioactive intensity *versus* time.

White light is used as the radiation source, tinted glass is the shielding material, and an exposure meter serves as a detector of radiation.

Figure 1 shows the essential parts of the apparatus and illustrates the placement of the increments of glass in the light path. The dimensions of the box are not critical, with ours measuring about six inches by eighteen inches by four inches. For our apparatus, it was found that a full-scale deflection of the exposure meter was obtained when a frosted 60-watt bulb was used as the light source. If more intensity had been required, a larger bulb in series with a rheostat could have been used.

To use the apparatus, the light is turned on, the top placed on the box, and the initial reading of the light meter recorded. The meter we used reads directly in foot-candles, but other meters should serve as well. After obtaining the initial reading, panes of tinted glass are placed in the light path and successive meter readings recorded for each pane used. It has been found that the light intensity is cut by one-half for each increment of glass, and we have been able to obtain readings from 70 foot-candles down to around 2 foot-candles with a fair degree of accuracy.

The most critical portion of the device is the glass used as the absorbing medium. Our local glass shop furnished the scraps of glass used at no cost to us. The glass is plate, carrying the trade name of Solar Gray. It is manufactured by the Pittsburgh Plate Glass Company, and is used in commercial installations where shielding from the sun is desired.

There are several advantages of this approach to illustrating half-thickness. The materials used are inexpensive and easily obtained, the construction details are few, the apparatus is large enough for classroom demonstration and small enough for individual use, there are no concealed "gimmicks" or black boxes, there is no danger from exposure to radioactivity, and the accuracy is enough that a fair exponential curve is obtained.

By referring to the glass plates as periods of time rather than thicknesses of material, an analogy may be made with the half-life of radioactive substances, with the same exponential curve being plotted.

Figure 1

USING THE OVERHEAD PROJECTOR IN SIMULATION OF THE RUTHERFORD SCATTERING EXPERIMENT

Rudolph J. Eichenberger

Eureka College, Eureka, Illinois 61530

The following demonstration is designed for use on the overhead projector to simulate the Rutherford scattering experiment.

Place a 19 cm × 30 cm Pyrex glass baking dish on the stage of the overhead projector. Then mold a "volcano shaped potential hill from plastic modeling clay (see Fig. 1) and place it near the center of the glass dish. The modeling clay will stick to the bottom of the glass dish and hold its position well during the demonstration. Cut a piece of cardboard slightly wider than the clay potential hill and long enough so that it will span the narrower dimension of the glass dish (about 4.5 cm × 20 cm). The cardboard will simulate the gold foil of the Rutherford experiment, and the clay potential hill will simulate a nucleus in the gold foil. Therefore, the cardboard must be placed directly above the clay hill. Use marbles or ball bearings to simulate the alpha particles from the Rutherford experiment. The marbles can be rolled at random from one end of the dish to the other and often will be undeflected. Because the marbles pass under the cardboard, to the viewer, they appear to be going through the simulated gold foil, see Fig. 2. Occassionally the marble will be scattered through a small or large angle depending upon how it strikes the clay hill, simulating the nucleus hidden under the cardboard, see Fig. 3 for more complete explanation. A very thin layer of clay (2 mm thick) can be used to catch the bombarding marbles (simulated detector).

This demonstration could also be used as an experiment similar to Exp. 13-1, pp. 299–306 from *Physical Science For Nonscience Students* to determine indirectly the radius of the nucleus (clay hill) given the radius of the bombarding alpha particle (marble).[1] One would have to use a smaller number of nuclei (clay hills), such as two or three, because of the narrower width of the glass dish, about 20 cm. Of course, rolling 200 trials could consume considerable time, but one could obtain student involvement during the laboratory period. One must be sure the overhead projector stage is level to avoid trouble when doing the demonstration or experiment.

Fig. 1. Modeling clay potential hill.

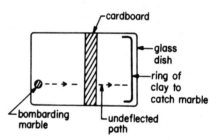

Fig. 2. View as seen on overhead projector.

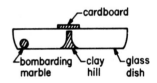

Fig. 3. Side cross sectional view of apparatus.

References

1. PSNS Project Staff, *Physical Science For Nonscience Students* (Wiley, New York, 1970), pp. 299–306.
2. J. Richards and M. Wehr, *Physics of the Atom* (Addison-Wesley, Reading, Massachusetts, 1967), 2nd ed. pp. 89–94.

A game for facilitating the learning of units in physics

James E. McGahan
Northwest High School, Grand Island, Nebraska 68801

It is often a difficult task to motivate students to learn the numerous units required for successful unit analysis of physics problems. In an attempt to remove the drudgery of learning units, the author devised the following simple game. Interest in learning appropriate physical units has skyrocketed since the introduction of this units game.

Purpose: To demonstrate that most metric units used in beginning physics are composed of six fundamental units; to apply a game strategy for the learning or review of physical units.

Materials:

A. Two sets of *Player* cards, each set of 17 cards consisting of the following:

4 - meter (m)	2 - degree celsius (°C)
2 - kilogram (kg)	2 - Coulomb (C)
3 - second (sec)	2 - lumen (lm)
2 - calorie (cal)	

B. One set of 22 *Dealer* cards consisting of the following:

length	impulse	temperature
mass	work	specific heat
time	power	heat capacity
velocity	density	current
acceleration	area	potential energy
force	volume	pressure
momentum	heat	illumination
linear expansion		

C. One *Dealer* answer board showing the correct units for each of the dealer cards.

D. Two 24 in. x 24 in. cardboard playing boards with a dark line across the center of the cardboard to divide the cardboard into a numerator section and a denominator section.

Rules:
1. The game is designed for three participants. Two are players and one is a dealer.
2. Each player lays his cards *face up* on the edge of his playing board.
3. The dealer stands his answer board in a position so the players cannot see the answers and then shuffles the dealer cards.
4. Play begins as the dealer turns up the first dealer card and the player on the dealer's left attempts to place the proper units on his playing board to designate the quantity on the dealer card.
5. If this player is successful, he is awarded points as described in the scoring section below; if he cannot assemble the proper unit within 20 seconds, the dealer allows the other player an equal time to attempt the task. If neither player can complete the task, no points are awarded.
6. The dealer turns up the second dealer card, and the player on the dealer's right now attempts to assemble the units. As before, if he can't complete the task in 20 seconds, the other player receives equal time to assemble the units.
7. Play continues until all 22 dealer cards have been turned up and attempted.
8. At this point, the players rotate in a clockwise direction so the person on the dealer's right becomes the dealer, and the dealer becomes the left player. The cards are reshuffled, and a new game is played as before.
9. This process is completed one more time so that each person has played two hands and dealt one hand. The game is now complete, and the high total score wins.
10. To score the game, the dealer gives one point to a player for each card used in a successful answer plus an extra point if the player can state a name given to that certain unit.

Example:
Dealer card shows the word WORK. Player plays cards (Kg), (m), and (m) on the numerator of the playing board, and he plays two (sec) cards on the denominator of the playing board. The player then calls it a joule. This is correct; so the dealer awards the player 5 points for the cards and one point for the proper name for a total of 6 points.

One can vary the length and difficulty of the game by adding or deleting the appropriate player cards and dealer cards. Blank cards, which are sold by many bookstores and educational materials suppliers as flash cards, make excellent cards for this game. The appropriate information can be written on each card with a felt-tipped pen.

Radiation safety in the lab

G. Stroink

Department of Physics, Dalhousie University, Halifax, Nova Scotia, Canada B3H 3J5

One of the lab experiments in our course "Physics for the Life Sciences" involves the use of a simple Geiger counter and a small radioactive source, commercially available from almost any supplier of lab equipment. We have used this experiment for about three years without any difficulty, but recently we encountered a few problems which made us more aware of the possible dangers in handling the sources.

The five μCi sources used in this lab experiment are counted before and after each lab session, and only handed out in exchange for a student I.D. card. In one incident, two students who had obtained permission to do the experiment outside the scheduled hours decided to keep the source in a pocket during a lunch break, after which they continued with the experiment. Their action was prompted by the absence of the instructor, who had the students' I.D. cards, at the moment they went for lunch. The students decided to take the source with them to make sure that they got their I.D. cards back — attaching more value to the I.D. s than to the radioactive source.

I found this incident disturbing, not only because of the obviously undesirable situation when sources are handled without supervision outside the lab environment, but also because of my obvious failure to communicate to students a sense of responsibility in handling this radioactive material, although the concept of rem's and safety doses had been discussed at length.

Since then we have had a discussion in class during which I told the students what had happened and asked them to visualize possible dangerous situations when a source is taken outside the lab, and to calculate what would be the maximum possible whole-body dose someone could receive under the worst circumstances. Possible dangerous situations include cutting the source open to see what is inside, a child accidentally swallowing the bright orange disk, or carrying a source in one's pocket for months or years.

The maximum possible whole-body dose can be calculated as follows. The source available in the lab is a 5 μCi Cesium 137 source, emitting γ-rays with an energy of 0.66 MeV = 1.1×10^{-13} J. Five μCi is equivalent to 1.85×10^5 decays per second, or 6.6×10^8 decays per hour. Each decay produces one γ-ray. If the source is swallowed, each γ-ray can interact with the body. The half value thickness of tissue for 0.66 MeV γ-rays is about 7.2 cm, so one can estimate that about half of the 6.6×10^8 γ-rays will be absorbed by the body. The total energy absorbed per hour is, then, $3.3 \times 10^8 \times 1.06 \times 10^{-13}$ J = 3.5×10^{-5} J. This amount will be about the same if someone carries the source in his pocket. For a 50-kg person, the whole-body dose received in one hour is then

$$\frac{3.5 \times 10^{-5} \text{ J}}{50 \text{ kg}} = 7 \times 10^{-7} \text{ J/kg or } 0.07 \text{ mrem}$$

For one day he will receive a whole-body dose of 1.68 mrem and during one year about 613 mrem. The recommended maximum dose for students is 100 mrem in any one year.[1] It should be pointed out that under normal circumstances the student receives a dose of 0.2 mrem, at the most, during a three-hour lab period. This is about two hundred times less than what would be received by an average x-ray chest examination.[1]

After calculating the dose that could be received by someone exposed to the source for long periods of time and its possible consequences, the class agreed to assist in implementing the following rules aimed at making it almost impossible to remove a source from the lab:

a). Sources should be counted before and after each lab.
b). Sources should be numbered and signed out so that each source can be traced.
c). Sources should be locked up between experiments.
d). Sources not used during the lab period should be locked up.
e). A staff member should always be present.

This note is intended to increase the awareness of teachers and students to the possible dangers of the educational radioactive sources used in the lab. I am not advocating the abolition of such labs; the concepts that can be taught are important and the topic is relevant to the present discussion on nuclear safety. However, everything possible should be done to keep the sources in the lab and to store them safely.

Reference:

1. S. C. Bushong, "Radiation exposure in our daily lives," Phys. Teach. 15, 135 (1977).

THOSE BETAS CAN MAKE A DIFFERENCE

I found the article by G. Stroink, "Radiation Safety in the Lab" [Phys. Teach. 18, 207 (1980)] very interesting and informative. The following comments should prove helpful to the general reader to make the message in the above paper more complete.

A cesium-137 "source" gives off both betas and gammas. Cesium-137 decays to barium-137 emitting a negative beta with a maximum energy of about 0.5 MeV. The half life is 27 years. The resulting barium-137 is usually left in an excited state which then decays to the ground state by emitting a 0.662 gamma. The half life is 2.6 minutes. If the source is damaged so that the cesium is not shielded with plastic, and then swallowed, perhaps about one-half of the gamma rays will interact with the body as stated; but it should be noted that so will ALL of the betas. The beta range is several centimeters in air and much less in tissue. The calculated dose would increase by approximately a factor of three times that for gammas alone. When discussing this point, the subject of decay schemes can be introduced to actually show the decay process.

When a beta or alpha source is ingested, the source, or part of it, could remain localized in a vital organ or bone marrow and create, due to the short beta range, extensive local radiation damage to body cells. The damage would be much more, in fact, than if the source were a pure gamma-ray emitter where the region of interaction would not be so localized due to the infinite interaction distance associated with gamma rays.

Ronald A. Kobiske, *Department of Physics, Milwaukee School of Engineering, Milwaukee, Wisconsin 53201*

Absorption of Radiation — A Lecture Sized Demonstration

Dean Zollman
Kansas State University
Manhattan, Kansas 66506

The energy crisis has renewed interest in how solar energy is absorbed by various materials. In the laboratory[1] or at home[2] this physical concept can be investigated by comparing the temperature rise in a white object with the rise in a black object. Radiation of energy from the objects can be compared by watching the temperature fall when the energy source is removed.

While these experiments are very instructive, they are not readily adapted to the lecture situation. A very similar demonstration can be made by using two 18 inch dial thermometers.[3] These thermometers are painted white on the back. I removed the white paint from one of them and painted it black. The two thermometers were then mounted on a piece of plywood with large holes cut in it so that the backs of the thermometers were exposed. To distinguish them when viewed from the front, the wood around the black thermometer was painted black while the wood around the other one was painted white. In lecture the students are shown the backs of the thermometers and then watch the front while radiation is incident on the backs. (One hundred watt light bulbs work fine.) Within a matter of minutes the "black" thermometer is at a considerably higher temperature than the "white" one. Since the thermometers are large, the result is easily visible to the entire class.

During a recent cold but sunny day, I placed the two thermometers outside. Students and faculty were easily hooked into discussions when they saw two seemingly identical thermometers — one registering nearly 60°F and the other about 20°F.

The thermometers have a list retail price of $13.50 but have been available in local discount stores for $8.95. Thus, for less than thirty dollars, including wood, I have an easily seen demonstration which seems to interest a large number of students.

References

1. H.R. Crane, *The Heat Balance* (unpublished).
2. F. Brunschwig, AAPT Announcer 2, 25 (1972).
3. Manufactured by Ohio Thermometer Sign Co., 33 Walnut St., Springfield, Ohio 45501.

Simulating radioactive decay with dice

Ludwik Kowalski
Department of Physics-Geoscience, Montclair State College, Upper Montclair, New Jersey 07043

There is an old[1] game which demonstrates how the random nature of a decay process leads to an exponential decrease in the activity and in the number of radioactive atoms. Take a large number of identical dice and let one face of each die be recognizable as "d-face," where d stands for the decay or death. Throw all of them on a flat surface and count those which fall with the d-face up. Those N_d dice stand for atoms which decayed in the time interval represented by the throw. The remaining N_s dice represent the "survivors"; that is, atoms which did not decay.

Remove the decayed dice and throw all the survivors on a flat surface. Again there will be N_d decays and N_s survivors. By repeating the procedure over and over again, one gets a sequence of decreasing numbers N_d and a sequence of decreasing numbers N_s. As long as these numbers remain large, the decrease in each sequence is exponential.

The above game can easily be simulated on a computer with the fall of each die being controlled by a random number generator. For example, a student who simulates the game may obtain the following sequence of N_s: 1000, 841, 696, 582, 474, 403, 339, 282, 238, 202, 163, etc. Plotted versus the number of throws, these values of N_s follow a nearly perfect exponential curve. But there is a puzzle, the values do *not* agree with the formula $N_s = N_o \cdot e^{-\lambda t}$; where t is the number of time intervals, $\lambda = 1/6$ is the decay probability of each die per throw, and $N_o = 1000$ is the initial number of dice. The simulated curve drops to any predetermined value of N_s about 8% faster than would be expected from the above formula. Why?

This challenging question can be presented to a student. A hint can be given by noticing that dividends in a bank grow faster when compounded daily than when compounded every quarter, or every year. The key to the answer is in calculating the time interval, Δt, represented by each throw. When dice are cubical, 5/6 of them are expected to survive in every throw. On the other hand, the value of t in the exponential formula corresponding to $N_s/N_o = 5/6$ is 1.08 so that the theoretical sequence of N_s decreases 8% slower than the simulated one. The discrepancy between the two curves depends on the decay probability per unit time, λ, which in turn is determined by the shape of each die. For example, the discrepancy is smaller for the 14-sided "dice" ($\lambda = 1/14$),[2] and larger for the 2-sided "dice" (coins, where $\lambda = 1/2$). In other words, the simulated sequence approaches the theoretical one when only a small fraction of dice decay during each throw. It is in the limit of $\Delta t \to 0$ that a step-wise decrease can be identified with the exponential function.[3]

All this can be tested numerically with a properly written program, and the entire game can be a useful exer-

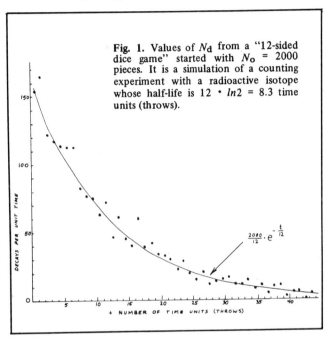

Fig. 1. Values of N_d from a "12-sided dice game" started with $N_0 = 2000$ pieces. It is a simulation of a counting experiment with a radioactive isotope whose half-life is $12 \cdot ln2 = 8.3$ time units (throws).

cise for students who know how to program a computer. Naturally, the same model can be used to simulate exponential growth instead of decay. In this case the number of dice equal to N_d is *added* to the population after each throw. One can also create a situation in which random oscillations are superimposed on an upward or downward trend of N_s. To simulate this we sometimes add N_d to N_s and sometimes subtract it randomly, according to a certain probability. This offers a chance for exploring various "stock market strategies" in both stable and unstable situations, and opens the door for numerous scientific applications of the so-called "Monte Carlo" method[4] in which any desired trend can be imposed on a randomly behaving system.

References

1. C. S. Swartz, *Used Math* (Prentice Hall, Englewood Cliffs, 1973) p. 60.
2. A computer-simulated 14-sided die is just a function, but in the world of real objects it can be visualized as a pencil-like roller with 14 faces.
3. L. Kowalski, "Radioactivity and Nuclear Clocks," Phys. Teach. **14**, 409 (October 1976).
4. R. Ehrlich, "Physics and Computers," Houghton Mifflin Co., Boston, 1973.

Two new experiments

Bob Kimball
Crestwood High School, Dearborn Heights, Michigan 48128

1. The Speed of Sound:

Two pool balls connected to strings of equal length, with the other ends attached to a ring, can be used as a timing device. Each ball is drilled along a diameter, and a string is passed through and knotted. Lengths of about 6 in. seem to work well. When the ring is moved up and down the balls "clack" together. Of course, energy must be supplied at the frequency of the "clacker." This frequency is reasonably reliable in the middle range of amplitudes.

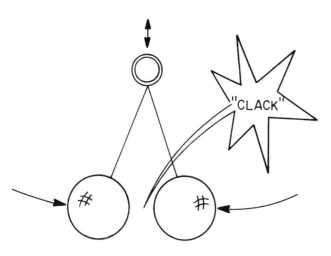

Fig. 1. *A pool-ball "clacker."*

The period is easily found by measuring the time for one hundred clacks and dividing. With this in mind, a student can stand in front of a brick wall and listen for the echo. At one point, the returning clack is simultaneous with the production of the next clack. Measurement of the distance to the wall allows the student to calculate the speed of sound. (See Table I.)

Table I. *The velocity of sound determined by a "clacker."*

Period	Distance	Velocity	Error
0.30 sec	54 m	$\frac{108}{0.3}$ = 360 m/sec	5%

2. An Analog Experiment:

The Millikan experiment is a classic, but often the central idea is clouded by the difficulty of setting up the equipment and maintaining it from one class period to the next. Last year, I did the experiment by analogy and observed the movement of the latex spheres qualitatively. The students entered the room to find a large beaker containing some 50 penny wrappers. Each wrapper contained a random number of pennies from one to ten. The students were asked to weigh several of these packets and write the results on the board. Slowly, a pattern emerged. Distinct columns were formed with the difference between columns equal to the weight of one penny. The final discussion revolved around the actual Millikan experiment and the corresponding parts and ideas.

A POTPOURRI OF PHYSICS TEACHING IDEAS—ODDS AND ENDS

Throwing dice in the classroom

John L. Roeder
The Calhoun School, 433 West End Avenue, New York, New York 10024

In spite of Einstein's well-known disbelief that events in the universe are determined by the throw of dice, the fact remains that this is the way most of us interpret happenings at the submicroscopic level. If that is so, we should be using those dice in our classrooms to make the probabilistic nature of submicroscopic processes a more vivid experience for our students.

Dice are, of course, associated with games. What I have done is to devise games to illustrate two of the better known submicroscopic processes: the flux of neutrons in a reactor, and radioactivity. These games are so easy to fabricate and play that you can be using them in your classroom tomorrow. Directions follow.

Nuclear Reactor game

Equipment: Two dice, plus about 50 squares (about 1.5 cm on a side) with the letter "n" written thereon; and about 15 squares (about 2.5 cm on a side) bearing the symbol "^{239}Pu". This should suffice for about six players, although any number can play. (The squares are most conveniently cut from a piece of oak tag and labeled with a narrow black felt-tipped pen.)

Object of the game: The object of the Nuclear Reactor game is to learn the different things that can happen to neutrons in a nuclear reactor. Neutrons resulting from the fission of a nucleus of uranium-235 (^{235}U) or plutonium-239 (^{239}Pu) are accompanied by the release of energy. They can also cause other nuclei to fission and thereby release even more energy. However, other things can also happen to neutrons; they can escape or be absorbed. Accordingly, a player who loses all his neutrons is out of the game; the last player to "survive" is the winner.

Beginning the game: Each player is given five "n" squares representing neutrons. The remaining n squares and all the ^{239}Pu squares (representing plutonium-239 nuclei) are placed at the center of the table.

Rules of play: Each player takes a turn, in order, as follows: First he "antes" one n square into the center of the table. This represents the neutron whose fate is to be decided by the throw of the two dice, according to Table I. It may escape or be absorbed, or it can cause fission, resulting in the release of additional "new" neutrons. After a player has thrown the dice and carried out the corresponding instructions, the next player takes a turn. The game is over when all the players but one (the winner) have lost all their neutrons.

Note: The formation of ^{239}Pu from ^{239}U after ^{238}U absorbs a neutron is the process of "breeding" nuclear

Table I

Number thrown on dice	Result
2	Neutron escapes: lose neutron*
3	^{239}Pu nucleus fissions: exchange one ^{239}Pu for three neutrons (drawn from the center of the table)**
4	Neutron absorbed by moderator: lose neutron*
5	^{238}U nucleus absorbs neutron: draw one ^{239}Pu nucleus from the center of the table
6	^{239}Pu nucleus fissions: exchange one ^{239}Pu for two neutrons (drawn from the center of the table)**
7	^{235}U nucleus fissions: draw two neutrons from the center of the table
8	Neutron absorbed by cadmium rod: lose neutron*
9	^{238}U nucleus absorbs neutron: draw one ^{239}Pu nucleus from the center of the table
10	Neutron absorbed by coolant: lose neutron*
11	^{235}U nucleus fissions: draw three neutrons from the center of the table
12	^{239}Pu nucleus fissions: exchange one ^{239}Pu for two neutrons (drawn from the center of the table)**

* The "lost" neutron is the neutron which has *already* been anted into the center of the table.

**A player may not collect neutrons when he throws a 3, 6, or 12 unless he has a ^{239}Pu square which he has already collected by throwing a 5 or 9.

fuel in a reactor. While this produces added fuel, it must be noted that the fission of ^{239}Pu made in this way costs twice as many neutrons as the fission of a ^{235}U nucleus: one neutron to make a ^{239}Pu nucleus—and a second to fission it.

Radioactivity game

Equipment: One die, one marking token bearing the name of each player, plus a game board consisting of the portion of the chart of the nuclides connecting ^{240}Pu, ^{237}Np, ^{241}Am, ^{238}U, ^{242}Cm, ^{243}Cm, and ^{239}U to their ultimate decay products: ^{208}Pb, ^{209}Bi, ^{206}Pb, and ^{207}Pb. This can also be laid out on a sheet of oak tag, and a good idea is to draw arrows to illustrate each of the four fundamental decay chains with a different color.

Object of the game: The object of the radioactivity game is to learn about the four fundamental radioactive decay chains. Heavy nuclei decay through a series of alpha decays (loss of two protons and two neutrons) and beta decays (neutron converted to proton or vice versa) until ^{206}Pb, ^{207}Pb, ^{208}Pb, or ^{209}Bi is attained. The player whose nucleus decays to the end of the chain first wins.

Beginning the game: Each player places his marking token on the square representing one of the starting nuclei: ^{240}Pu, ^{237}Np, ^{241}Am, ^{238}U, ^{242}Cm, ^{243}Cm, or ^{239}U. There is no limit to the number of tokens on each starting square and no limit to the total number of players.

Rules of play: Each player takes a turn, in order, as follows: First he throws the die. If a 1 or 2 turns up, there is no radioactive decay, and the turn is over. If a 3, 4, or 5 turns up, beta decay will occur for nuclei known to undergo it, and the marking token is moved to the daughter nucleus in the event that beta decay does occur. The same considerations apply to alpha decay if a 6 turns up. A player's turn continues as long as radioactive decay continues to occur.

I have used these games in teaching students at both junior high and high school levels. The response has always been enthusiastic, with students eagerly seeking to increase their supply of neutrons or exhorting their radioactive nuclei to decay.

Throwing dice in the classroom II

John L. Roeder
The Calhoun School, New York, New York 10024

A few years ago I advocated using dice in the classroom to teach students about submicroscopic processes that are governed by probabilities.[1] I also presented directions for two games employing this principle. These two earlier games concerned heavy elements: radioactive decay chains and fission reactors. The present note presents directions for two additional games, both concerning light elements: stellar evolution and fusion reactors. These games can be easily fabricated and played in your classroom tomorrow. Directions follow.

Stellar evolution game

Equipment: Two dice, a set of squares marked Q (25 per player), plus a set of ten markers for each player (these may be pieces of dried corn, peas, beans, or pieces of colored paper, but must be different for each player) and a game board consisting of the portion of the chart of the nuclides through $^{15}_{7}$N (only the squares for $^{3}_{2}$He, $^{4}_{2}$He, $^{12}_{6}$C, $^{13}_{6}$C, $^{14}_{7}$N, and $^{15}_{7}$N are used, but the intervening squares serve to illustrate the relationships among the nuclides involved.)

Object of the game: The object of the game is to learn the thermonuclear processes which occur in the evolution of a star through the formation of $^{15}_{7}$N (listed in Table I). This includes the formation of $^{4}_{2}$He from protons (believed to be the original nuclear material from which "first generation" stars are made), otherwise known as the "proton-proton" cycle; the formation of $^{12}_{6}$C from $^{4}_{2}$He; and the formation of $^{4}_{2}$He via the "carbon-nitrogen" cycle in "second generation" stars which contain $^{12}_{6}$C at the outset. The evolution of each player's "star" is determined by the throw of two dice, in accordance with Table I. If a player can move one of his markers in accordance with Table I, he does so — and his "star" evolves accordingly.

A POTPOURRI OF PHYSICS TEACHING IDEAS—ODDS AND ENDS

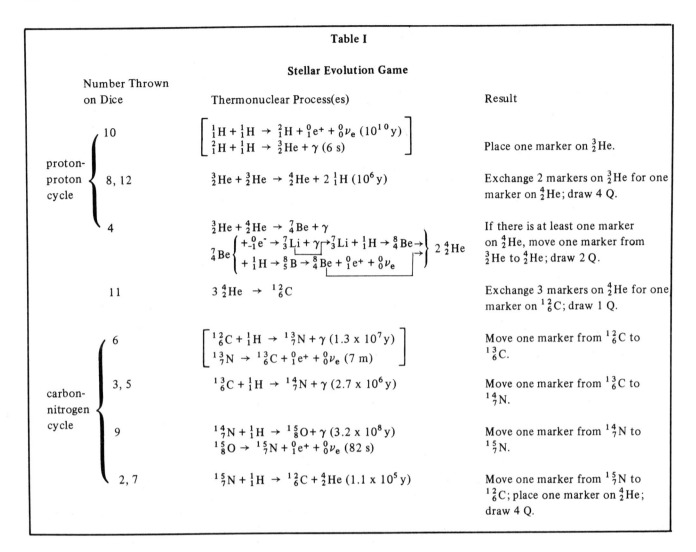

Table I

Stellar Evolution Game

Although the probabilities for the thermonuclear processes in Table I are not proportional to the probabilities for the respective dice throws, they are coordinated in that the more probable processes (as indicated by shorter time scales in Table I) correspond to more probable dice throws. Players will quickly note that they must throw a 10 to begin the stellar evolution process (the first step of the proton-proton cycle). Although this may frustrate players with many actionless turns, the important lesson here is that the probability for this step is relatively low. In fact, the only lower probability in this game is for the formation of $^{12}_{6}C$ from 3 $^{4}_{2}He$ (made low by the fact that the intermediate state of $^{8}_{4}Be$ is unstable relative to 2 $^{4}_{2}He$).

The formation of each $^{4}_{2}He$ and $^{12}_{6}C$ releases energy (represented by the drawing of Q squares). The number of Q squares drawn is proportional to the amount of energy released, one Q square representing about 7 MeV. The player with the largest number of Q squares at the end of the period wins.

Rules of play: Each player takes a turn, in order, as follows: He throws two dice to determine what process may occur in his "star" and moves one of his markers in accordance with Table I if he can. If he cannot move a marker (i.e.,

the process corresponding to the number thrown cannot occur), the turn is over. If he can move a marker, he continues to throw the dice and move his markers accordingly until he throws a number which does not allow him to move a marker. Play continues until the end of the period (or for how long the teacher or players decide). The player with the largest number of Q squares is the winner.

Fusion reactor game

Equipment: Two dice, plus squares marked n (10 per player), Q (20 per player), $^{3}_{1}H$ (5 per player), and Li (2 per player). The n squares may be the same as those used in the nuclear reactor game described in Ref. 1 and the Q squares may be the same as those in the stellar evolution game described above.

Object of the game: The object of the game is to learn the thermonuclear processes which are expected to occur in a fusion reactor (listed in Table II). This includes the fusion of deuterium nuclei ($^{2}_{1}H$) with both other deuterium nuclei and tritium nuclei ($^{3}_{1}H$). Because the probability for the latter process is about ten times that for the former, the probabilities for the dice throws are related in the same way. However, while deuterium is very abundant on earth,

Table II

Fusion Reactor Game

Number Thrown on Dice	Thermonuclear Process	Result
3	$^2_1H + ^2_1H \rightarrow ^3_2He + ^1_0n$	Draw 1 n and 1 Q.
11	$^2_1H + ^2_1H \rightarrow ^3_1H + ^1_1H$	Draw 1 ^3_1H and 1 Q.
6, 7, 8, 9	$^2_1H + ^3_1H \rightarrow ^4_2He + ^1_0n$	Exchange 1 ^3_1H for 1 n and 5 Q.
4, 5, 10, 12	$^7_3Li + ^1_0n \rightarrow ^4_2He + ^3_1H + ^1_0n$	Exchange 1 Li for 1 ^3_1H.
2	$^6_3Li + ^1_0n \rightarrow ^4_2He + ^3_1H$	Exchange 1 Li and 1 n for one ^3_1H. Draw 1Q.

tritium must be made, usually in a reaction with lithium. The probabilities for the dice throws corresponding to the reactions with lithium to produce tritium are also in the same ratio as that of isotopic abundances: $^6_3Li/^7_3Li = 1/11$. However, lithium is also in limited supply. Thus the squares marked Li for lithium and ^3_1H for tritium are needed to keep track of nuclides of limited abundance. In fact, this is the major lesson to be learned from the game: that when supplies of lithium and tritium run out, fusion power must rely on the lower probabilities of deuterium-deuterium reactions. This will be realized as players lose their Li and ^3_1H squares. The other important lesson, which players will also readily recognize, is that the reactions in a fusion reactor yield a surplus of neutrons. These have been seen by some as a means to breed plutonium from uranium for use in fission reactors.

The object in the fusion reactor game is the same as the object of the fusion reactor: the release of energy, represented by the drawing of Q squares. As in the stellar evolution game, the number of Q squares drawn is related to the amount of energy released. A deuterium-tritium fusion reaction releases a little more than five times the energy released in a deuterium-deuterium fusion reaction. The player with the largest number of Q squares at the end of the period wins.

Beginning the game: Each player is given two Li squares (representing lithium nuclei) and two ^3_1H squares (representing tritium nuclei). The remaining squares (Li, ^3_1H, n, and Q) are placed at the center of the table.

Rules of play: Each player takes a turn, in order, as follows: He throws two dice to determine which process may occur in his "fusion reactor" according to Table II and carries out the instructions in Table II if he can. There is only one dice throw per turn. Play continues until the end of the period (or for how long the teacher or players decide). The player with the largest number of Q squares is the winner.

Reference

1. John L. Roeder, Phys. Teach. 15, 428 (1977).

"Mission Improbable" problems

Doug Jenkins
Warren Central High School, Bowling Green, Kentucky 42101

For the past four years I have used humorous extra-credit problems, intended to provide a bit of relaxation after students have taken a test, and, at the same time to sharpen their skills at working word problems. Usually the actual problem is not very difficult, once the student has managed to separate it from the story's details and bad puns. These problems have been well received by students.

All of the problems revolve around this cast of characters:

Dr. J — A physics teacher and part-time C.I.A. agent. As the central character, Dr. J struggles through trial and tribulation to show that truth, justice, and physics will always prevail.

Tripod — Dr. J's trusted three-legged dog and faithful companion. When not helping Dr. J in his crusade for physics, he stands on the lawn in front of the house as a constant reminder that there are three feet in a yard.

J, Jr. — Dr. J's only son and chief helper in the cause of physics. A college graduate himself with degrees in Physics and Puns, he is a valuable asset to his father.

S.A.P. — Students Against Physics, a rival group with a habit of constantly appearing to oppose Dr. J and his helpers. The opposition usually results in some interesting situations and unusual solutions.

Timex — Dr. J's trusted lab assistant. Refusing a salaried position, he prefers to work for an hourly rate. When not conducting physics experiments he helps with the janitorial duties — a sort of second-hand sweep.

Here are three sample problems:

The Pet Rock

In this thrill-packed episode, Dr. J takes a trip to Boulder, Colorado, to pick up his new pet rock (think about it). He bought it from a mail-order catalog, *"Hard Things to Buy and Sell."* Little did he know what he was in for. The catalog price was $1.98 with a delivery charge of $200.00. Therefore, Dr. J decided to travel to Boulder to pick it up himself in order to save the delivery charge. When he arrived, he was shocked to learn that his "pet" weighed 1000 newtons! (I guess he took it for *granite* that it would fit in his pocket.) After renting a truck, he decided to use his knowledge of physics in order to load it. Using a small rock as a fulcrum and a long pipe, he fashioned a lever with which to pry the rock. Try as he might, however, he couldn't apply enough torque to budge the boulder. Fortunately for Dr. J, a local elephant named Nate came along. With great ease Nate lifted the rock 1.2 m into the bed of the truck. (The moral of this story is of course: "Better Nate than lever.") If Nate charges $50 per kilowatt-hour, how much did Dr. J have to pay for the work Nate did?

The Cycle Shop

In this shocking episode, Dr. J and his crew invest in a new part-time business. With the high price and scarce availability of gasoline, many of the rental car companies have started renting motorcycles. Of course the leader in this field is Hertz (after all, a cycle per second is a fantastic rate of business). On the other hand, Avis and some of the other companies have shown a *resistance* to this *current* trend because of their concern over the *potential* of business. Dr. J has decided to go one step further and rent electric motorcycles. It seems like a natural; J, Jr. will handle the renting; Timex will serve as the mechanic (his specialty is timing engines); and even Tripod can show first-time customers how to keep their balance. Dr. J designed these electric motorcycles himself. The only drawback to them is the speed. Whereas a gas-guzzling Harley can reach 70 MPH and over, Dr. J's model can only reach 7 MPH (he named it a "Hardly"). A problem arose while assembling the first of these cycles. There were three 10-Ω resistors left, but the circuit diagram showing the proper way to wire them was blurred. After thinking a bit, Timex said, "How many ways can you do this, anyway?" Dr. J said that there are seven different combinations (Not every combination need use all three resistors.) Find the total resistance for each of these seven combinations.

Hackensack

In this episode Dr. J tries his hand at milking a new breed of cow. The one he has chosen is named Hackensack. (Yes, he's milking Hackensack, the new Jersey.) However, Hackensack failed to realize that Dr. J has as much pull as he does and as a result developed an odd reaction to the whole process. Every time Dr. J milked her, she would have trouble with her moo. It sounded like "M-m-m-moo." (A sort of udder-stutter you might say.) Finally, she couldn't take it any longer. Starting from rest, she accelerated at 2 m/s for 4 s out the barn door. She then ran at a constant velocity for 5 min until Dr. J succeeded in throwing a rope around her neck after which she slowed down at a rate of 1 m/sec^2. After the ordeal, Dr. J admitted that it was a *moo*ving experience, indeed.

A. Make a distance-time graph of Hackensack's trip.
B. Make a velocity-time graph of Hackensack's trip.
C. What was the total distance traveled by Hackensack?

The author will send a selection of similar problems on receipt of a stamped addressed envelope.

The history of physics in a high school physics course

Michael Grote
Princeton High School, Cincinnati, Ohio 45246

Under a grant from the American Association of Physics Teachers sponsored by the Carnegie Foundation, I produced a set of slides and transparencies on the history of physics.[1] The purpose of the project was to generate more interest in the people behind major discoveries and the processes they followed to make these discoveries. The materials produced were used as the basis for a multimedia presentation for selected topics.

The slides were made by photographing pictures from books and magazines using a Kodak Instamatic with a special lens.[2] Pictures of physicists and the equipment they used as well as the places they worked were sought.

The transparencies listed major accomplishments of the physicist as well as important events in his life. After investigating various methods, I made a transparency master using commercial lettering sheets such as Format or Letraset (see Fig. 1). Using this master I then produced a negative transparency with Escotherm negative transparency film.[3] The negative transparency (lettering is clear, background is opaque) allows one to project phrases using the

Fig. 1. Isaac Newton display

ISAAC NEWTON	
1642	Born Dec. 25
1654	Attends King's School
1658	Returns Home
1661	Enters Trinity College
1665	Great Plague
1669	Chair in Mathematics
1672	Elected to Royal Society
1676	Hooke-Newton Controversy
1679	Nervous Breakdown
1684	Halley Visits
1687	PRINCIPIA Published
1696	Warden of Mint
1703	President of Royal Society OPTICS Published
1705	Knighted
1727	Dies

Fig. 2A. Slide alone

Fig. 2B. Slide and transparency projected simultaneously.

Fig. 2C. New slide with more of transparency revealed.

overhead projector on top of a slide image without washing it out (see Fig. 2). This is really a very dramatic effect and there is quite a reaction the first time students see the words appear on the slide.

Using such materials I present a 15 to 20 minute introduction to the accomplishments of physicists such as Galileo, Kepler, Newton, Bohr, and Einstein.

Before using this method for presenting the history portion of the course, I generally met with considerable apathy about it. This is no longer the case (see graphs in Fig. 3). The bar graphs indicate student opinion of the presentations as well as the change in attitude which has taken place in seeing the history of physics as a necessary prelude to understanding our present physical concepts. The 1975-76 school year was the first year I used these materials. Although there are several reasons for the 50% increase in enrollment we have for 1976-77 (increasing from 98 to 144), these presentations are certainly a contributing factor.

References

1. Thanks to Frances Lestingi of SUNY at Buffalo who suggested this idea.
2. Ektagraphic Visual Maker Kit-Kodak, includes camera, lenses and stands. About $120.
3. From Seneca Blueprint Co., Cheektowaga, N.Y. 14225. No. 101 black negative film (Escotherm). $51/box of 100.

Fig. 3. Four graphs showing student opinion.

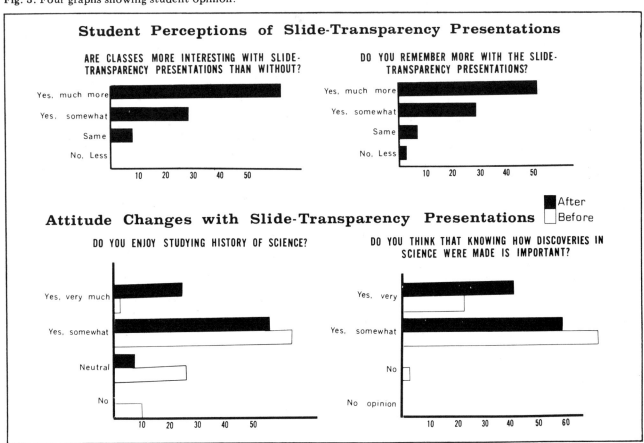

Legibility in the lecture hall

Albert A. Bartlett
Michael A. Thomason
Department of Physics, University of Colorado, Boulder, Colorado 80309

The larger the physics lecture hall, the greater the attention the lecturer must pay to the legibility of material that is written or drawn on the chalkboards. Obviously, the size and scale of the writing must be appropriate for the distance to the students who are seated farthest from the chalkboard.

There are some technical details that will improve the legibility. These have to do with increasing the *contrast* between the lines drawn with chalk and the background.

1. The chalkboards should be black. The pastel colored chalkboards that are popular with designers are bad for large lecture halls because of the reduced contrast.

2. Chalkboards should be wet-washed prior to every lecture. The dry erasers leave a gray surface with greatly reduced contrast. We find that a large cotton bath towel dripping wet works very well.

3. The chalk should be large and soft. The normal pieces of "dustless chalk" (9-mm diam, 80 mm long) seem to be "hard" in the sense that they typically leave a narrow light line which, because of its small width and its low intensity, has low contrast. We wish to call attention to a large diameter soft chalk that leaves a wide high-intensity line. Material written or drawn on a chalkboard with this chalk has vastly improved legibility compared to material written with the normal dustless chalk. This chalk is called *Railroad Crayon* #888 ("for railroads, mines, mills and foundries")[1] and it is available from distributors of industrial equipment and supplies.[2] The pieces are 25-mm diam and 105 mm long. They are available in white, red, yellow, green, and blue and they cost about $0.16 apiece. The volume of a piece of this Railroad Crayon is approximately ten times the volume of a piece of standard chalk but the cost is only about four times the cost of a piece of the standard chalk so the Railroad Crayon is inherently inexpensive. Because of its size the Railroad Crayon rarely breaks and it can be used without difficulty down to the point where the remaining piece is roughly spherical and $\sim$ 20 mm diam. Figure 1 shows the contrast in size between regular chalk on the left and Railroad Crayon on the right. Figure 2 shows a section of our chalkboard. The material on the left was written with Railroad Crayon (the colors are not apparent) while that on the right was written with regular "dustless" chalk. Because of the improved legibility we use Railroad Crayon almost exclusively in our lecture halls.[3]

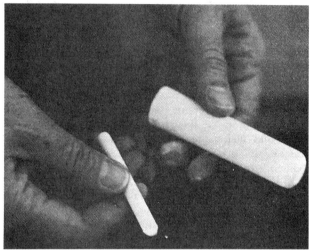

Fig. 1. Contrast the size of Railroad Crayon on the right and regular chalk on the left.

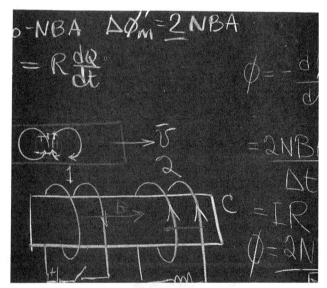

Fig. 2. The left part of the board has material written with Railroad Crayon while the material on the right was written with regular chalk.

References

1. Joseph Dixon Crucible Co., 167-T Wayne St., Jersey City, NJ 07303, 201-333-3000.
2. We purchase from: M. L. Foss, Inc., 1900 Lawrence St., Denver, CO 80201, 303-296-2200.
3. A. A. Bartlett, Am. J. Phys. **41**, 1233, 1973.

A POTPOURRI OF PHYSICS TEACHING IDEAS—ODDS AND ENDS

Blitz quiz

Herbert A. Perlman, Sr.
Bethel High School, Bethel, Connecticut 06801

Here is a "Blitz Quiz" that I give my General Physics class every year just before Christmas. You will notice that if the student answers all the questions correctly it will spell out, MERRY CHRISTMAS AND A HAPPY NEW YEAR. It is fun to watch them descend the list of questions and eventually realize the answers to this trick quiz. This is my way of wishing my students a Merry Christmas and a Happy New Year.

1. The acceleration due to gravity is ......
2. The ..... is a unit of heat energy.
3. Density is defined as mass per unit ......
4. Boyle's law states that the product of pressure and ..... for an ideal gas is constant.
5. Force equals mass times ......
6. A ..... indicates both magnitude and direction.
7. The man who measured the mechanical equivalent of heat was ......
8. If other things are kept constant, increasing the temperature of a gas also increases the ......
9. ..... improved the reciprocating steam engine.
10. The heat of fusion of water is ......
11. The heat of vaporization of water is ......
12. Weight in newtons is the product of mass in kilograms and ......
13. Objects are pulled toward the earth by ......
14. To melt ice requires ......
15. Newton first stated the law of ......
16. 32°F = ... °C.
17. The expression for kinetic energy is ......
18. The force holding a satellite in orbit is caused by ......
19. One newton-meter is a ......
20. On the moon the force of ..... is one-sixth of that on the earth.
21. ..... demonstrated that atmospheric pressure decreases with increasing altitude.
22. A law of hydraulics is known by the name of ......
23. ..... can be expressed in meters/second².
24. 273°K = ... °C.
25. One ..... will increase the temperature of one kilogram of water one degree Celsius.
26. ..... is a good science course.
27. Change in velocity per unit time is ......
28. The energy obtained from food is expressed with the unit ......
29. The force that keeps a pendulum swinging is supplied by ......
30. The product of length, width and height is ......

A	–	Gravity	F	–	Mgh	K	–	Energy
B	–	Galileo	G	–	Momentum	L	–	Elvis
C	–	Vector	H	–	Joule	M	–	9.8 m/sec²
D	–	½mv²	I	–	Watt	N	–	Zero
E	–	Calorie	J	–	6.02	O	–	Dyne
						P	–	Pascal
Q	–	Friction	V	–	Pizza
R	–	Volume	W	–	Physics
S	–	80 cal/g	X	–	Yard
T	–	540 cal/g	Y	–	Acceleration
U	–	Optics	Z	–	Wavelength

PROBLEMS

In the past THE PHYSICS TEACHER has published problems of various degrees of difficulty to challenge and stimulate our readers. The editors think that this is a fine way to learn physics, and invite authors to send us problems and solutions which they feel are particularly instructive. To expedite the transfer of this material into the classroom, solutions will appear in the same issue, but on a different page.

Galileo Revisited

Terrence P. Toepker
  Department of Physics, Xavier University, Cincinnati, Ohio.

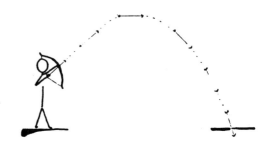

1. Why does an arrow, which is pointed upward from the horizontal when it is launched, always manage to rotate itself about an axis perpendicular to its length so that the point strikes the ground first? (See Fig. 1.)

2. Consider two pieces of dowel rod (2 ft by ¾ in.). Rod A has nothing special done to it; one end of rod B has been drilled out with a ½-in. bit to a depth of 4–5 in., molten lead is poured in, and a small wooden plug is placed over the lead. Externally both rods are now identical. Would it be possible for an observer to distinguish rod A from rod B by merely watching an assistant drop first one rod and then the other? (See Fig. 2.) (Prior to dropping, each rod is held horizontal by exerting pressure on the ends of the rod with the palms of the hands. The rod is released by pulling the hands away.)

3. Recalling Galileo's experiment, a student drops a bowling ball and a much lighter plastic ball of the same radius simultaneously from the same height *in the atmosphere*. Which one will strike the ground first if the height is sufficiently high?

4. The bowling ball and plastic ball in Problem (3) are now connected together with a very light rod, as shown in Fig. 3. Discuss the motion of the dumbbell if it is released with the rod horizontal (parallel to the ground). Will there be any rotation again assuming a sufficient height?

5. Discuss Problems (1), (3), and (4) if they were performed on the moon where there is no atmosphere.

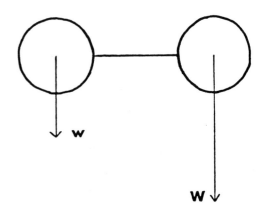

I have found these questions to be both interesting and instructive to students who are encountering for the first time the concepts of free fall, limiting velocity, buoyant force, center of mass, and torque.

Problems Solutions

1. The forces acting on an arrow while it is in flight are its (a) weight, (b) viscous forces, and (c) buoyant force. Since the buoyant force is very small it can be neglected. The weight acts through the center of gravity and thus cannot cause rotation. Therefore the rotation must come from the viscous forces. If one thinks of the guinea and feather experiment that is found in many textbooks, one sees that the viscous force retards the motion of the feather much more than that of the guinea. Thus for a given height in an atmosphere the guinea will strike first. If the guinea is started from a greater height than the feather, sufficient time (height) must be provided so that the guinea may catch up. By drawing a line from the feather to the guinea, one has a first approximation to an arrow. The important thing to note about an arrow is its shape (*not* the fact that one end is heavier than the other). One could make an arrow out of a piece of aluminum rod by merely pinching out some aluminum fins at one end. The center of gravity would still be at the same place, yet the fins would shift the distribution of the viscous forces from the center of gravity of the rod toward the tail end, and the arrow would then tend to rotate when shot in an atmosphere.

2. When the weighted rod is released the same three forces as mentioned in Problem (1) act on it: (a) a viscous friction force, (b) a buoyant force which acts at the center of gravity of the displaced fluid (in this case, at the geometric center of the rod), and (c) the weight which acts at the center of gravity. No torque about the center of gravity can result from the weight. However, the buoyant force does not act through the center of gravity because the center of gravity has been shifted toward the end of the rod with the lead weight, and the viscous forces are not uniformly distributed about this new center of gravity. Thus a torque about the center of gravity exists and rotation should result. However, since both the buoyant force and the viscous forces are small, the amount of rotation will be small for a short height. A more obvious example of the action of the viscous forces can be seen by attaching a few paper clips to one end of an envelope. If dropped with the plane of the envelope parallel to the floor, the viscous forces will cause the envelope to rotate so that the end with the paper clips will strike first. This would not happen in a vacuum.

3. Since both the bowling ball and the plastic ball have the same radius, the buoyant forces on each are equal. The weights are obviously not equal. Thus the initial acceleration of the bowling ball will be greater than that of the plastic ball since:

$$Mg - BF = Ma_1 \qquad mg - BF = ma_2$$
$$g - BF/M = a_1 \qquad g - BF/m = a_2$$

where $BF/M < BF/m$. Thus $a_1 > a_2$.

Assuming a viscous force R given by Stokes' law ($R = -6\pi\eta rv$) or more simply, $R = -kv$, where $k = 6\pi\eta r$, then in general (dropping the subscripts for a moment) the force equation becomes:

$$W - BF - kv = Ma. \qquad (1)$$

When the velocity is large enough so that $R = (W - BF)$, the acceleration is zero and a terminal speed is reached:

$$v_t = (W - BF)/k. \qquad (2)$$

Obviously the terminal speed of the bowling ball is greater than that of the plastic ball:

$$(W - BF)k > (w - BF)/k,$$

since BF and k are constant and $W > w$. Thus:

$$v_{t_M} > v_{t_m}. \qquad (3)$$

Rewriting Eq. (1) after dividing by k gives:

$$(Mg - BF)/k - v = (M/k)a. \qquad (1a)$$

Substituting Eq. (2) gives:

$$v_t - v = (M/k)a. \qquad (4)$$

Separating variables and integrating then gives

$$-\int_0^t (k/M)dt = \int_0^v (dv)/(v - v_t),$$

$$v = v_t(1 - e^{-(k/M)t}). \qquad (5)$$

Now $v_{t_M} > v_{t_m}$ and $e^{-(k/m)t} > e^{-(k/M)t}$ for any t. Therefore $v_M > v_m$ for all t, and the heavier ball will strike the ground first.

4. Since it was shown in Problem (3) that the heavier ball will fall faster than the lighter one, there will be a rotation of the dumbbell as it falls. The situation is obviously similar to Problem (2).

5. Referring to Problems (1), (3), and (4), the removal of the atmosphere will remove the viscous and buoyant forces. Thus the only remaining force that acts is gravity. This force by definition acts at the center of gravity. Therefore there will be no rotation about the center of gravity. Thus an arrow shot on the moon at some angle above the horizontal will hit the ground tail feathers first.

"PHYSICS IS PHUN" QUIZ

Match the numbered item on the left with the description on the right.

1. 4 ft	a fish	36. fluid	vicious pig belonging to a boy
2. Δ	in Rome directory	37. area	small Italian child
3. Gauss	quadriped	38. arc	proclaiming oneself a fool
4. quadrilateral	Japanese song	39. πc	destined to become mine
5. Hertz	(Dick Cavet) (Dick Cavet)	40. Nobel	inflamed swelling on skin
6. physics, physiology	end of a river	41. isotope	delayed April fool's
7. $\lambda\nu$	singular for smoked salmon	42. Fizeau	device that measures taxes
8. Galileo	part of an opera	43. Neutrally	used by veterinarian to put bull to sleep
9. Niels Bohr	Halloween dress		
10. Fermi	poetically old fashioned	44. Avogadro's #	subconscience of cold virus
11. γ	results in need for digel	45. ammeter	Italian word for sport
12. Yukawa	noise made by donkey	46. MC^2	sheep in the shade
13. $\alpha\alpha$	draft	47. oscilloscope	funny viewer
14. $\cos(2m)$	condition making it necessary to knock	48. λ	pea as main course
		49. Brahe	what the sun gives
15. proteins	four football passes	50. millimeter	paper on the offense of God's law
16. $\beta\rho$	command to a dog		
17. Boyle	Genie	51. acceleration	command to bow
18. $\tan(2u)$	professional adolescents	52. inverse square	truck full ot statistics
19. Laplace	pains	53. frequency	radicalization
20. Locke	give her more food	54. Coulomb	Japanese boy
21. Ptolemy	inform me	55. red shift	intoxication indicator
22. steam engine	study of effervescence	56. Mars	soft drink
23. synthesis	beat of a wood block	57. Van de Graff	baseball regulation
24. Neptune	king's heart burn	58. barometer	French for particular position
25. logarithm	release our 6th president	59. little chromosome	cow's food
26. Tycho	maryda hadda little	60. entropy	wagon train defense
27. rad	knowledgeable sheep		
28. induction	if you're sinking to dent		
29. solute	renovated streetcar		
30. oxidation	done in the military		
31. slide rule	Leo's girlfriend		
32. Leyden	two radii minus di		
33. matrix	combination of gas and liquid		
34. ring stand	Cockney greeting		
35. noble gas			

Hint: 6 3 1 9 5 2 7 4 8

Answers may be obtained by sending a self-addressed envelope to

JAMES H. NELSON
*Harriton High School
600 North Ithan Avenue
Rosemont, Penna. 19010*

SOLUTION TO "PHYSICS IS PHUN" QUIZ, USING HINT

The hint is based on the number of letters needed to spell the numbers 1 to 60 in order starting with one. Consider a similar matching puzzle with nine numbers. We would start with nine blanks:

___ ___ ___ ___ ___ ___ ___ ___ ___

Start at the first blank and begin by spelling the number 1 (i.e. O-N-E). Move one space to the right for each letter. When you finish spelling the word, put the number in the last space counted.

```
___  ___  _1_  ___  ___  _2_  ___  ___  ___
 o    n    e    t    w    o    t    h    r
```

When you reach the last blank go back to the beginning and continue counting blanks as you spell. DO NOT count a blank after it is filled with a number.

```
___  _3_  _1_  ___  ___  _2_  ___  _4_  ___
 o    n    e    t    w    o    t    h    r
 e    e    -    f    o    -    u    r
```

If you continue in the manner, you will generate the number given as a hint:

6 3 1 9 5 2 7 4 8

Since the "PHYSICS IS PHUN" quiz has 60 matching items, the answer sheet has 60 blanks. You generate the answer key by counting the letters in the words one to sixty as in the example above. NOTE: Do not count the hyphen in numbers like twenty-two.

Fizzicks Quizz

Fun for Your Party
(Author Unknown)

This is a matching test. Each item in the left column is to be matched with one in the right column. Items in the right column are utilized only once. Some aid is given by publishing two answers. A complete list of answers will be furnished if you request them and send a stamped addressed envelope.

(1) Atom	(4)	1) To act
(2) Battery	( )	2) An officer of the county
(3) Millimeter	( )	3) Judge at conjecture
(4) Dyne	( )	4) The first man
(5) Gas	( )	5) Listerine
(6) Molar solution	(5)	6) Used in driving
(7) Dew	( )	7) Tom Mix
(8) Light rays	( )	8) Please repeat
(9) Ion	( )	9) Scotchman's pocket
(10) Lens	( )	10) Funny paper
(11) Liter	( )	11) A dead parrot
(12) Shooting star	( )	12) A lie one likes to hear
(13) Rain	( )	13) Our parents
(14) Tangent	( )	14) More legs than a centipede
(15) Ellipse	( )	15) Larva of a fly
(16) Focus	( )	16) One who conducts
(17) Conic section	( )	17) Agile
(18) Magnet	( )	18) Pitcher and catcher
(19) Polygon	( )	19) A heavy metal in raisins
(20) Prism	( )	20) Small salary increase
(21) Corona	( )	21) To take sustenance
(22) Nicol prism	( )	22) To loan
(23) Nimbus	( )	23) Ethiopian
(24) Watt	( )	24) His speech impediment
(25) Complement	( )	25) Jail
(26) Directrix	( )	26) Deep scratches
		27) A baby's trousers
(27) Density	( )	28) Devices for cleaning fish
(28) Harmonic oscillator	( )	29) The Pope's section of Rome
(29) Phase	( )	30) A female ghost
(30) Pole	( )	31) Duty of the undertaker
(31) Mass	( )	32) Pertaining to a part of a flower
(32) Desiccate	( )	33) Where one lives
(33) Ferrous	( )	34) Wife
(34) Normal solution	( )	35) A female comedian
(35) Diopter	( )	36) False belief
(36) Wind	( )	37) To divide again
(37) Vacuum	( )	38) Where street cars are kept
(38) Choke coil	( )	39) The wrong answer
(39) Scalar	( )	40) Dumbness
(40) Spectra	( )	41) Place of business
(41) Period	( )	42) The front side of one's head
(42) Ohm	( )	43) One who kisses harmoniously
(43) Orifice	( )	44) A deep dish to hold soup
(44) Corollary	( )	45) To profane
(45) Node	( )	46) Savage
(46) Refraction	( )	47) The olfactory organ
(47) Dilution	( )	48) Conversation
(48) Torque	( )	49) Bob-tailed comma
(49) Barium	( )	50) Moving air
(50) Carbon	( )	

POSSIBLE SOLUTIONS TO FIZZICKS QUIZZ PUZZLE

(1) Atom
(2) Battery
(3) Millimeter
(4) Dyne
(5) Gas
(6) Molar solution
(7) Dew
(8) Light rays
(9) Ion
(10) Lens
(11) Liter
(12) Shooting star
(13) Rain
(14) Tangent
(15) Ellipse
(16) Focus
(17) Conic section
(18) Magnet
(19) Polygon
(20) Prism
(21) Corona
(22) Nicol prism
(23) Nimbus
(24) Watt
(25) Complement
(26) Directrix
(27) Density
(28) Harmonic oscillator
(29) Phase
(30) Pole
(31) Mass
(32) Dessicate
(33) Ferrous
(34) Normal solution
(35) Diopter
(36) Wind
(37) Vacuum
(38) Choke coil
(39) Scalar
(40) Spectra
(41) Period
(42) Ohm
(43) Orifice
(44) Corollary
(45) Node
(46) Refraction
(47) Dilution
(48) Torque
(49) Barium
(50) Carbon

(4) The first man
(18) Pitcher and catcher
(14) More legs than a centipede
(21) To take sustenance
(6) Used in driving or (26) Deep scratches
(5) Listerine
(1) To act
(20) small salary increase
(19) A heavy metal in raisins
(22) To loan
(16) One who conducts
(7) Tom Mix
(6) Used in driving/(3)judge at conjecture
(23) Ethiopian
(24) His speech impediment
(13) Our parents
(10) Funny paper
(15) Larva of a fly
(11) A dead parrot
(25) Jail
(2) An officer of the county
(9) Scotchman's pocket
(17) Agile
(8) Please repeat
(12) A lie one likes to hear
(34) Wife
(40) Dumbness
(43) One who kisses harmoniously
(42) The front side of one's head
(44) A deep dish to hold soup
(26) Deep scratches/(3)judge at conjecture
(45) To profane
(46) Savage
(39) The wrong answer
(27) A baby's trousers
(50) Moving air
(29) The Pope's section of Rome
(35) A female comedian
(28) Devices for cleaning fish
(30) A female ghost
(49) Bob-tailed comma
(33) Where one lives
(41) Place of business
(32) Pertaining to a part of a flower
(47) The olfactory organ
(37) To divide again
(36) False belief
(48) Conversation
(31) Duty of the undertaker
(38) Where street cars are kept

Tricks of the Trade

If a picture is worth a thousand words, then an observed event is surely worth many pictures. Physics teachers have learned to trust this idea, with the result that an introductory physics course without demonstrations is the exception rather than the rule. We also rely on certain controlled events that we call experiments to give reality to our words. Whether physics teachers become gadgeteers because of this or chose to teach physics because they are gadgeteers to begin with is a "chicken or egg" question.

At any rate, the response to our invitation to submit short notes on tricks in using apparatus for teaching physics was enthusiastic and we present the results below. As the reader will find, classification of these ideas was nearly impossible, and as contributors will find, some of the submissions have been sternly edited. No claim of originality is intended, and with this in mind we have in some cases deleted the author's gracious acknowledgement of "where-he-saw-it-first". —Ed.

HOW TO STOW IT

Hang Those Meter Sticks

The storage of meter sticks can be a nuisance in the physics laboratory. Instead of laying them horizontally — hang them vertically! All one needs is a metal rod several millimeters in diameter and about two feet in length and as

many "spring clips" as there are meter sticks. Mount one end of the rod to the laboratory wall and slide the clip with attached meter stick over the other end of the rod.

B.M. GUY
Cleveland State Community College
Cleveland, Tennessee 37311

Assemblies of Single Concept Demonstrations

When a single concept can be demonstrated by the use of several small devices and miscellaneous supplies, it is convenient to store these in a plastic or plywood tray with labeled compartments. Thus, a surface tension box might contain a flat-bottom projection dish, needles, razor blades, clay pipe, soap solution, wetting agent, capillary tubes, wire frames, etc. The missing or misplaced components are readily spotted. The single concept sets are particularly useful when several instructors teach multiple sections of a course.

HAYM KRUGLAK
Western Michigan University
Kalamazoo, Michigan 49001

Storage and Display Box for a Diffusion Cloud Chamber

A plywood box (25 x 25 x 20cm) is convenient for storing a diffusion cloud chamber (Atomic Laboratories Raymaster) with all its accessories. The box bottom is

recessed about 5cm: this serves as a tray for the dry ice and its insulation when the empty box is inverted. Two wooden stepwedges may be used under the box sides to keep the chamber level.

HAYM KRUGLAK
Western Michigan University
Kalamazoo, Michigan 49001

Protecting Electronic Tubes

One easy method of protecting electronic tubes (such as canal ray tubes, luminescence tubes, etc.) is to wrap a turn of shock-absorbing tape around the glass at each end of the tube. The tape is self-adhesive. It is available in many hardware stores. Should the tube be dropped, the tape will absorb much of the shock.

GLEN SPIELBAUER, student
East Texas State University
Commerce, Texas 75428

Small Part Storage

Plastic shoe boxes are good for storage of all kinds of small tools and apparatus. I like being able to see what is inside. Watch for sales of these boxes in drug stores and notion departments.

SISTER MARTHA RYDER
Clarke College
Dubuque, Iowa 52001

A POTPOURRI OF PHYSICS TEACHING IDEAS — TRICKS OF THE TRADE

A Rack for Test Tubes

For economy of materials and space-saving storage of test tubes simple racks are suggested (See figure). Each hole is

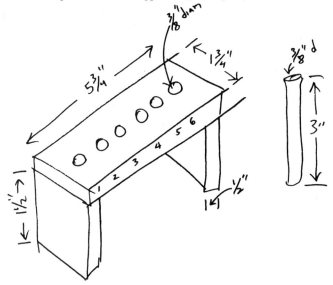

numbered on both edges of the top so a student can identify a tube from either side. The tubes rest on the table, and will not fall out (because of their lips) when the rack is lifted.

R. KENNEDY CARPENTER
*Butler High School
Butler, New Jersey 07405*

Storage of Support Cable from High Ceilings

We store our multi-stranded support cable from our lecture hall ceiling by screwing it into a bracket that is mounted on the wall, using the same lock-nut that is used to secure the cable to the bowling ball. This provides convenient storage and eliminates the need of a ladder. A surprising number of people, including faculty, are amazed to learn that the cable had been hanging in the room all the time.

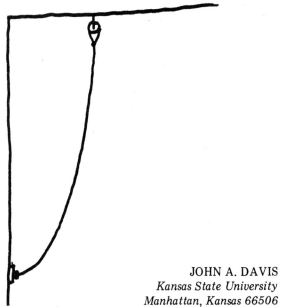

JOHN A. DAVIS
*Kansas State University
Manhattan, Kansas 66506*

Tool Board Outlines

Instead of painting the outlines of tools on pegboards, cut a silhouette from "Contact" paper, available in many different colors in dime stores and hardware departments.

SISTER MARTHA RYDER
*Clarke College
Dubuque, Iowa 52001*

HOW TO DO IT

Emptying Ripple Tanks

If you dislike the messy job of emptying ripple tanks try using a wet-dry shop vacuum cleaner. Five liters of water can be removed in about 10 seconds.

SPENCER LIVINGSTON
*Natrona County High School
Casper, Wyoming 82601*

Heat Shrinkable Tubing

Tubing that shrinks when heated ranges in size from a fraction of an inch to several inches in diameter and can be ordered in a variety of colors from most electronic suppliers. It can be used, for example in place of tape to cover a splice between wires. A small one-inch length of this tubing is placed over the repair and heated. If a high temperature air heat gun is not available, a lighted match can be passed under the tubing while rotating the wires. The tubing will shrink to form an excellent covering over the repair. The tubing can also be used to cover metal rods or small parts in physics apparatus to insulate, or protect them. In home made apparatus, the workmanship given a professional appearance.

RONALD A. KOBISKE
*Milwaukee School of Engineering
Milwaukee, Wisconsin 53201*

Color Anodizing

After machining and constructing physics apparatus from aluminum, one can protect the metal from oxidation, provide some electrical insulation, and give the surface a professional appearance with almost any color one could wish for. The process is called "Color Anodizing." The yellow pages directory in any medium to large size city will provide information on companies that provide this service. To get some idea of what the result looks like, stop at a key shop and view the large variety of color-anodized aluminum key blanks that are available.

RONALD A. KOBISKE
*Milwaukee School of Engineering
Milwaukee, Wisconsin 53201*

A Perfect Circle Without a Compass

Face the blackboard with a piece of chalk in your right hand. Slightly cock the elbow and hold it and the wrist rigid. Touch the chalk to the blackboard at 7 o'clock and rotate the arm clockwise from the shoulder *only*. Follow through quickly to the starting point not allowing the chalk to leave the blackboard. For a person who is left handed this procedure should be reversed. With a little practice you will be able to make perfect circles almost every time. The size of the circle is controlled with a little additional practice.

EDWIN PAUL HEIDEMAN
Pleasantville High School
Pleasantville, New York 10570

Reuseable Chalk Board Graphs

On the surface of a *wet* chalk board construct rectangular coordinates with a meter stick and white dustless chalk. Allow the board and chalk markings time to dry. Data may be plotted with any color chalk and erased many times before the original lines begin to fade. To remove the original pattern simply sponge off the board. This procedure is also excellent for superimposing data upon partially-completed maps, diagrams, drawings and equations.

EDWIN PAUL HEIDEMAN
Pleasantville High School
Pleasantville, New York 10570

"Glue" for Plexiglass

To make plexiglass glue, break plexiglass into small pieces, add an equal volume of the organic liquid 1,2-dichloroethane and cover tightly to prevent evaporation. After an overnight waiting period, you will be able to judge if you should add a little more plexiglass or a little more of the solvent. When the glue is used, the solvent evaporates — allow several hours — and a joint of pure plexiglass is left. I keep a supply of the glue in a stubby test tube, closed with a "cork" cork. The 1,2-dichloroethane can be purchased from Chemical Supply Houses.

SISTER MARTHA RYDER
Clarke College
Dubuque, Iowa 52001

Fused Ammeters

If "expensive" ammeters are mounted on a block so that a visible "cheap" fuse can be connected in series with the ammeter one can save a number of burned out ammeters in beginning labs. The fuse is best placed in a very accessible position so that it is easily replaced and a student witnesses his error.

S. WINSTON CRAM
Kansas State Teachers College
Emporia, Kansas 66801

Removing Scratches from Clear Plastics

Any apparatus which is constructed wholly or partially of clear lucite or other plastics eventually becomes scratched and unattractive. Most of these scratches can be removed without a trace by buffing with a cloth and a mild abrasive, such as ordinary toothpast. Toothpaste has been particularly effective on the scratched plastic lenses of stopwatches and transparent models for overhead projection.

HERBERT H. GOTTLIEB
Martin Van Buren High School
Queens Village, New York 11427

A Red Hot Demonstration

When demonstrating the effect of vapor pressure on the boiling point of water, add some very visible red dye to the water in a heavy-walled round-bottom flask.[1] Bring the water to a boil. After a few minutes when the air has been replaced by water vapor, remove the heat source and quickly stopper the flask securely. Invert it in a supported ring and wipe it with ice cubes. The colder you make it, the more violently the water boils. Introduce your explanation with the remark, "I take some *red hot* water in ... or say "Of course it's boiling, it's red hot." '

[1] CAUTION — I had a flat-bottom Florence flask implode.

R. KENNEDY CARPENTER
Butler High School
Butler, New Jersey 07405

The Speed of a Pendulum Bob

To determine how fast a pendulum is going at the bottom of its swing a student suggested burning the pendulum string at the bottom of the arc. Use a five or six inch piece of nichrome wire carrying four or five amps of current to make it red hot. The wire is mounted perpendicular to the pendulum swing and adjusted so it cuts the thin support thread just above the bob.

The horizontal distance of travel and the vertical fall are measured to determine the velocity of the bob when the string is cut.

RICHARD F. KOTHEIMER
St. Francis de Sales High School
Chicago, Illinois 60617

Photocurrent Detection with an Oscilloscope

The commercial photoelectric Planck's constant apparatus normally requires the use of a dc current amplifier and galvanometer for detection of the weak photocurrent. We have found that an oscilloscope with one megohm input impedance works very well in place of these two devices and provides very stable monitoring of the photocurrent. The light intensity of the mercury light source used with this apparatus is conveniently modulated at 120 Hz by the

A POTPOURRI OF PHYSICS TEACHING IDEAS — TRICKS OF THE TRADE

60-Hz line voltage, automatically providing the time varying signal required for oscilloscope detection. It was found that a coaxial cable connection between the photocell and oscilloscope was useful for reducing noise.

MONTE GILES
Missouri Western State College
St. Joseph, Missouri 64507

Latex Sphere Clusters in Millikan's Experiment

The commercially-prepared latex spheres for use in the Millikan oil-drop experiment have a known diameter. Results obtained using these spheres often fail to show charge quantization clearly. We have measured the distribution of zero-field terminal velocities of these droplets, and found that these velocities tend to group themselves around several discrete values, suggesting a possible sphere clustering. We tentatively identify those droplets in the slowest group of the distribution as single, unclustered spheres. An analysis of results obtained using only these slow droplets showed charge quantization convincingly and yielded satisfactory results for the charge on the electron.

ERNEST JOHNSTON AND MONTE GILES
Missouri Western State College
St. Joseph, Missouri 64507

Electrostatic Charging of Fur and Silk

One can obtain a greater electrostatic charge by grasping the fur or silk at the upper edge and then forcefully whipping the rod against it. The downward stroke of the rod will cause the fur or silk to wrap around it, thereby greatly increasing the contact. Several vigorous strokes should result in a strongly charged rod. Care should be taken not to rap one's knuckles!

JOHN A. DAVIS
Kansas State University
Manhattan, Kansas 66506

HOW TO MAKE IT

Ripple Tank Wave Absorbers or Shoals

Rubberized hair used by upholsterers is an inexpensive, readily available, and effective material for wave dampers or shoals in projection ripple tanks. The material is also known under the trade name of PARATEX. The wave dampers may be easily cut to size with shears and attached to the bottom and sides of the tank with rubber cement.

HAYM KRUGLAK
Western Michigan University
Kalamazoo, Michigan 49001

Projectile Launchers for Electric Trains

Here are two methods of launching projectiles vertically from moving carts of model trains.

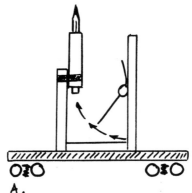
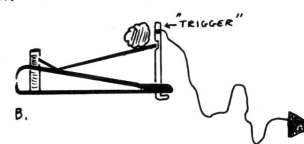

(1) Mount a mousetrap in a position such that when sprung, the moving portion hits the bottom of a pencil supported in a vertical tube.

(2) Build a trigger out of coathanger wire that will hold a pair of nail clippers closed. Place the projectile on the press plate and trip the trigger.

Both launchers can be tripped by a small stopper tied to the trigger with a slack string.

ROBERT Q. KIMBALL
Crestwood High School
Dearborn Hts., Michigan 48127

Plaster of Paris Mold Making

When several objects of the same size and shape but different mass are needed, a plaster of paris mold can be used.

I have used materials from lead to styrofoam, producing a range of masses. When making the mold, almost any material can be used for the "master" but it must be well oiled to prevent the plaster from sticking. Most molds can be made in two steps. First pour the bottom half and, when this is set and oiled, pour the top half.

Caution should be exercised when using molten lead.

ROY COLEMAN
Morgan Park High School
Chicago, Illinois 60643

Scope Shade

A three pound coffee can makes a very neat and acceptable shade for an oscilloscope if you (1) open the bottom end smoothly with a kitchen can opener, (2) put a hole in the removable plastic top which is slightly smaller than the flange on the graticule holder (bezel), and (3) push the plastic top onto the can and onto the oscilloscope.

For photography it is helpful also to paint the inside of this can black.

J. GERARD ANDERSON
Univ. of Wisconsin — Eau Claire
Eau Claire, Wisconsin 54701

A Quickly Made Galvanometer Shunt

Dual banana plugs like the Pomona #MDP Double Banana Plug or the G.C. Electronic #33-010 are stackable. In addition they have two terminals which connect the usual pair of leads to the plug. If instead, a resistor is connected across these terminals, the dual banana plug then becomes a shunt for any galvanometer which happens to have a dual banana socket input. With a little luck the shunt resistance required (by the internal resistance of the meter and the desired full scale reading) will be just a little lower than some standard resistor. Then a second relatively high resistance placed in parallel will usually bring the shunt within a couple of percent of its desired value.

JAMES REYNOLDS
University of Nebraska
Lincoln, Nebraska 68508

Center of Mass

Irregular figures for center-of-mass determinations can be cut from cardboard. Punch three holes close to the edge of each with maximum distance between each pair. For orientation mark a large letter close to one hole. Support the object from a paper clip hook tied near the top of a weighted string. Students observe the relative location of the plumb line at the bottom edge and draw a corresponding straight line from the suspension hole to the opposite edge on a replica of the object on a worksheet. This allows the unmarked objects to be reused. Turn the object to each of the remaining holes and repeat the process. The three lines cross at the center of mass. A triangle's size at the intersection indicates the experimental error.

R. KENNEDY CARPENTER
Butler High School
Butler, New Jersey 07405

Fast Square Wave Generator

Several popular college physics text books show circuits with switches that can be used to determine RL or RC time constants. To permit steady-state viewing on an oscilloscope, one can substitute a high speed mercury-wetted relay for the switch. The relay can be driven from the 60-Hz line through a 6.3-volt transformer. D.C. supplied to the input is chopped by the relay into square-wave pulses. The mercury vapor in the switch prevents contact bounce and aids in producing clean pulses with typical rise times of one to ten nsec. The relay that we employ was salvaged from a surplus computer board. It is a C.P. Clare relay, HGSM5133, and is available from most electronic suppliers. Actually, many of the HGSM series relays will work well for this application. The relay and transformer can be mounted in a small mini-box provided with input terminals for the dc. When mounting the relay, one must observe the upright position. Failure to do this could result in relay contact freeze. If all parts are purchased new, the total cost should be under $12.00.

RONALD A. KOBISKE
Milwaukee School of Engineering
Milwaukee, Wisconsin 53201

Simple Accelerometers

I. Use a piece of thread and some tape to attach a cork or a piece of styrofoam to the inside of a jar cover. Fill the jar with water and carefully screw on the cap with the attached cork. Invert the jar and you have an accelerometer.

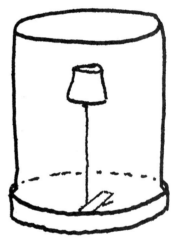

II. *Almost* fill a test tube with water and a drop of liquid detergent. Add a cork and Presto! You have a bubble accelerometer.

ROBERT GARDNER
The Salisbury School
Salisbury, Connecticut 06068

Series and Parallel Electric Circuits

For experiments in series and parallel electric circuits, attach Fahnstock clips to a board so that 10 watt resistors can be connected in various series and parallel arrangements to investigate voltage and current relations.

BRO. JAMES MAHONEY, C.F.X.
Malden Catholic High School
Malden, Massachussetts 02148

A Resistor Board

Equipment for storing and using resistors for a Wheatstone

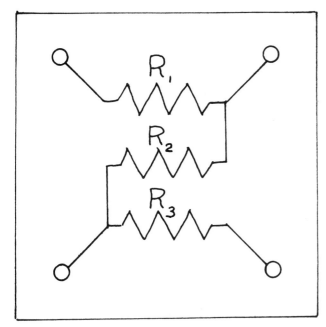

Bridge experiment is made by mounting three resistors on a board with single and double Fahnestock clips and supplying the combination with four terminals for connections (Fig.). This equipment provides the student with three resistance values and the possibility of five different combinations of series and parallel circuits achieved by using one or two additional leads. The resistor board also provides the student an excellent opportunity to translate a schematic diagram to hardware.

WALTER C. CONNOLLY
Appalachian State University
Boone, N.C. 28608

Sound-Proofed Box for Air Track Blower

Have you tried to out-shout an air track blower lately? Try housing the blower in a wooden box lined with styrofoam, fiberglass insulation or ceiling tile. Our box is fitted with an internal electrical receptable and an external switch and power cord and is highly mobile on its casters. Appropriate hangers and hooks allow for storage of hose and power cord when not in use. The top provides convenient storage and transportation of air track gliders. Try it, it cuts the noise level to about one-third!

JOHN A. DAVIS
Kansas State University
Manhattan, Kansas 66506

(*Ed. note: Be sure not to restrict the flow of air too much; this might result in a damaged blower motor.*)

Simple Lab Power Distribution System

The problem of supplying multiple stations in a physics laboratory with varied types of electrical power and/or signals is expensive to solve by any fixed system.

Our tables have standard pedestals equipped with an ac outlet and two pairs of five-way terminals. We wired corresponding pairs of terminals of the different tables in parallel, providing two separate circuits *connected to no source*. In use an appropriate supply is plugged into any terminal pair and all stations have access.

When voltage drop or station interaction is a problem, e.g. under low-voltage, high-current conditions, we supplement the system with individual battery eliminators.

WILLIAM H. PORTER
Gilman School
Baltimore, Maryland 21210

Inexpensive Immersion Heater

Resistors commonly available in radio parts stores can be used to construct inexpensive immersion heaters suitable for a student experiment on the electrical equivalent of heat. The figure shows an immersion heater made with a 25

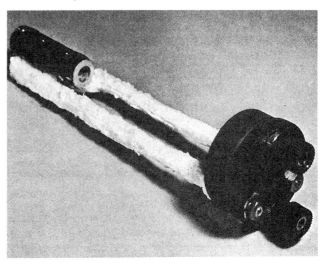

ohm 25 watt, fixed wire-wound resistor on a ceramic core with vitreous enamel coating. Rigid supporting leads position the resistor in the calorimeter and binding posts facilitate electrical connections. Electrical and thermal insulation is provided by a silicone rubber product usually used for home bathroom caulking.

FRANK G. KARIORIS
Marquette University
Milwaukee, Wisconsin 53233

Simple Resistance Thermometer

A relatively accurate, rugged, inexpensive resistance thermometer suitable for beginning student use can be made from a thermistor. This is a solid state resistor having markedly decreasing resistance with increasing temperature. The units are available in a wide range of resistances, power ratings, and physical forms. The ones used with some success by my students are small, 1-mm beads. Insulated wires are soldered to the leads and the bead epoxied into a small brass tube. Resistance is typically measured with a Wheatstone bridge, but an ohmmeter suffices for approximate measurements.

The thermistor can be used to investigate the resistance-temperature characteristic of the semiconducting material itself, or it may be used as an accurate remote temperature sensor once calibrated. Other applications are limited only by the ingenuity of the user.

L.C. CORRADO
University of Wisconsin
Manitowoc, Wisconsin 54220

HOW TO SHOW IT

Brownian Movement

Obtain from the Biology Department half a teaspoon of the red powder known as "Carmine (Alum Lake)". This dye, which can also be ordered from chemical supply companies, is used for staining slides in histology and cytology. With a dry toothpick or wooden splint, add a very small amount of the powder to a drop of water on a microscope slide. Brownian movement of the pretty red particles can be easily observed with an ordinary microscope with a 10x ocular and a 43x objective.

SISTER MARTHA RYDER
Clarke College
Dubuque, Iowa 52001

Projection of Diffraction Plates With A Microscope

Large images of diffraction plates such as slits, wires and optical crystals can be produced on a screen by placing the diffraction plate in a low power laser beam followed by a measuring type microscope. The microscope, which is used for projection purposes *only* focuses an erect image on a screen which reveals the details of the diffraction plate. When out of focus the microscope projects an enlarged Fresnel, near field, diffraction pattern. When the microscope is removed from the system the Fraunhofer, fair field diffraction pattern is displayed on the screen.

RONALD R. BERGSTEN
University of Wisconsin-Whitewater
Whitewater, Wisconsin 53190

Bichromatic Light Source For Diffraction Demonstrations

An intense bichromatic point light source is effective in demonstrating the dependence of holographic images on wavelength as well as showing the wavelength dependence of diffraction patterns produced by slits, gratings and optical crystals. Light from a point source such as a Chicago miniature 1630 bulb is filtered by a Gaertner L541E three-element filter with the blue element removed, resulting in a yellow light which disperses into its red and green components.

RONALD R. BERGSTEN
University of Wisconsin-Whitewater
Whitewater, Wisconsin 53190

Vectors With a Chalk Line

Vector composition of forces may be demonstrated on the chalk board with good results. Two spring scales are hung approximately a meter apart on hooks above the chalk board. A piece of cotton string 1-1/2 to 2 meters long coated with chalk is attached to the scales and an unknown weight is suspended near the center of the string. Record the force on each scale and snap the string against the chalk board. Remove the apparatus and measure the angle between the chalk lines. From the data a graphic and mathematical solution can be developed.

EDWIN PAUL HEIDEMAN
Pleasantville High School
Pleasantville, New York 10570

$F = q v \times B$; A Simple Demonstration

Suspend a rubber covered, flexible braided copper wire between two ring stands so that it sags down near the table. At its lowest point the wire passes near or between the poles of a strong permanent magnet resting on the table. When a 1-1/2 volt cell is connected very briefly between the ends of the wire, the wire will "jump" as the current pulse passes through it. Reversing the polarity of the battery, of course, reverses the direction of the force and changing the orientation of the magnetic field to vertical makes the wire swing sideways. By proper timing of the current pulses the amplitude of the "swings" can be made very large as the wire oscillates as a pendulum.

ROBERT M. DONNER
University of Wisconsin
Baraboo, Wisconsin 53913

Simple AC Demonstrator

Here's an "oldie" that can bear repeating. Take a piece of regular line cord at least two meters long and install a plug on one end of it. Now to the other end attach an NE-2 neon lamp in series with a 150-200 KΩ 1/4 watt resistor. For insualtion use silicon glue which is available from hardware stores. Plug the device in and the neon bulb should glow. Spin the bulb in a large circle and its appearance leaves no doubt that ac does indeed turn on and off. Of course, this demonstration is most dramatic in a darkened room.

ROBERT F. NEFF
Suffern High School
Suffern, New York 10901

Convex Lens as a Diverging Lens

To show that a convex lens can act as a diverging lens tape together the outer edges of two watch glasses with masking tape, immerse the watch-glass lens in water, and shine a light beam through the lens. The beam will diverge.

BRO. JAMES MAHONEY, C.F.X.
Malden Catholic High School
Malden, Massachusetts 02148

A POTPOURRI OF PHYSICS TEACHING IDEAS — TRICKS OF THE TRADE

Three Dimensional Magnetic Field

A magnetic field can be easily investigated by students using a piece of paper-clip wire suspended by a thread. Glue

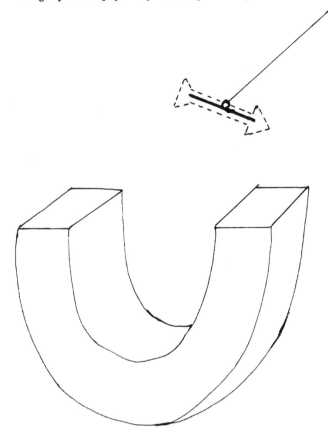

a paper arrow to it after making a loop in the wire and attaching the thread. A strong magnetic field can be studied qualitatively by moving the arrow from pole to pole at various distances from the magnet.

WALTER C. CONNOLLY
Appalachian State University
Boone, North Carolina 28608

The Vertical View

For some lecture demonstrations, such as Chladni figures, it is desirable to have a vertical, as well as a horizontal, view of the apparatus. This can be easily accomplished by attaching a large mirror to either a moveable lecture-table section or framework. Useability is increased by allowing for differing angles of tilt for the mirror.

JOHN A. DAVIS
Kansas State University
Manhattan, Kansas 66506

Translucent Screens for Optical Projections

For some demonstrations, such as showing fringes from interferometers and the shadow image of an electroscope it is very convenient if one uses a translucent screen — thereby allowing the students to observe on one side what the instructor sees on the other. The screen is made by stretching drafting vellum across a frame.

JOHN A. DAVIS
Kansas State University
Manhattan, Kansas 66506

HOW TO ADAPT IT

Regelation Made Easier

To demonstrate regelation I use a block of ice made by removing the dividers from a regular ice cube tray and support the ice with strong nylon straps. The straps come from a discarded lawn chair. They usually come in two or three lengths and some even have a grommet in each end. They not only hold the ice is a sling arrangement but are good thermal insulators as well.

ROBERT F. NEFF
Suffern High School
Suffern, New York 10901

A Simple Timing Device

Many animated window displays use an electric pendulum operated by a 1-1/2 volt dry cell. Often, such cast away displays are yours for the asking. I find them more useful than a simple pendulum because by counting the audible clicks a student is able to time, observe, and gather experimental data unassisted.

EDWIN PAUL HEIDEMAN
Pleasantville High School
Pleasantville, New York 10570

Inexpensive Tuning Forks

Turnbuckles that are readily available in hardware stores ($.30 and up, depending on size) can be cut to make tuning forks. One eye from the turnbuckle is cut short to become the handle, threaded back in and held tightly with a nut

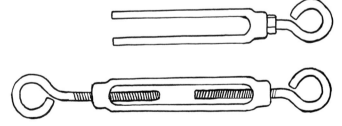

The body of the turnbuckle that becomes the vibrating prongs is a hard aluminum alloy with a nice finish, easy to cut and file, and attractive when completed.

These forks are as loud or louder than the regular steel tuning forks, but do not sound for as long after being struck. With so little money invested in them, one feels free to change their frequencies, shortening the prongs to raise, filing at the base to lower the frequency.

ROBERT H. JOHNS
The Academy of the New Church
Bryn Athyn, Pennsylvania 19009

Using Polaroid Negatives

Large-scale use at Polaroid photography in student labs may produce a film bill on the order of several thousand dollars per year. To reduce the costs one of the lab partners of each pair can use the negative as his record of the data taken. To preserve the negative and enhance the contrast ratio we initially tried dipping them in sodium sulphite.[1] The sodium sulphite, however, does not increase the contrast ratio sufficiently and in addition, creates quite a mess. Our solution was to use Kodak Rapid Fixer, at film strength, in used Dippit containers.[2] The student removes the chemical pods at the ends of the negative and trims the edges with a pair of scissors. The negative is inserted in the Dippit container with about 0.5 cm sticking out. The container is shaken and left on its side for three to five minutes. The negative is then pulled out through the rubber seal which squeegees off most of the Rapid Fixer, then rinsed in running water and rubbed gently until the soapy feeling disappears, after which it is patted with paper towels and allowed to dry.

[1] Harold A. Daw, Amer. J. Phys. 36, 1022 (1968).
[2] Polaroid Dippit #646, see your local photo supply house.

B. G. EATON
University of Minnesota
Minneapolis, Minnesota 55455

Better Spreader For Polaroid Colorpack II Cameras

The Polaroid Colorpack II camera is no longer available and has been replaced by a more expensive model, the Colorpack 100 which features a developer spreader. With the CP 100 spreader the film is passed out of the camera between rollers. The resulting reduction in friction means smaller forces exerted on the camera and a lower rate of development malfunctions. One can order the roller spreader separately from Polaroid[1] and replace the non-roller spreader in the CP II camera.

[1] Polaroid Corp., Part #995995, 19-1 Spreader, $4.23 ea., Camera Products Serv. Dv., 895 Second Ave., Waltham, Mass. 02154.

B. G. EATON
University of Minnesota
Minneapolis, Minnesota 55455

Cartesian Diver

Instead of a commercial cartesian diver, a 3-in. dropping pipette or eye dropper can be used. It is filled with as much water as it will draw up and tested to see if it will float in a large beaker or can. At first it usually sinks. It is made lighter by letting out one drop of water at a time until it will just float. Then it can be used in a tall cylinder or bottle. Just the pressure from the palm of your hand over the cylinder will make it dive. This makes it easier for students to see how it works.

RICHARD F. KOTHEIMER
St. Francis de Sales High School
Chicago, Illinois 60617

Poor Man's Mercury Source

Use an old tube from a "sun lamp" as a mercury light source. These often "burn out" in a non-final way. In many such tubes there is a small inner quartz tube which contains the mercury and in series with this is a ballast resistor in a helium atmosphere — all enclosed by the outer envelope. Failure often occurs through breakage or oxidation when air leaks into the outer bulb. It is the lead of the resistor that breaks and hence the small quartz tube is intact and can be set up with an appropriate power supply and ballast — giving the mercury spectrum. Because ultraviolet is present the eyes must be protected and skin exposure strictly limited.

K.E. DAVIS
Reed College
Portland, Oregon 97202

Miniature Diffraction Gratings

Make diffraction gratings for the student lab by purchasing a six foot roll of diffraction grating material and cutting it into half inch squares. Tape each square over a hole punched through an index card as shown. Since

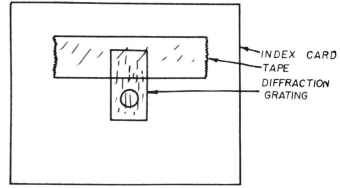

the cost of each grating is less than 1/2 cent, students can be allowed to keep their gratings after lab for further experiments at home. A 6 ft by 8-1/2 in. roll of diffraction material (13 400 grooves per in.) may be purchased from Edmond Scientific Company, Barrington, N.J. 08007. Stock No. 50,180, $5.95 postpaid.

HERBERT H. GOTTLIEB
Martin Van Buren High School
Queens Village, New York 11427

A Precise Time Base for the Physics Lab

Enthusiasm and confidence grow if beginning physics students can make measurements that compare favorably with known values. A good measurement of g is possible with a modified PSSC vibrating recording timer. Remove the contacts, and remove about one-third of the vibrating mass to bring the natural period to 60 per second. Add a silicon rectifier in series with the coil, and operate from 1-4 volts 60-cycle ac. The coil gets 60 half-wave pulses each second, to a precision better than most other equipment in the laboratory. The rest is up to the student.

JOHN G. JOHNSON
McLean High School
Arlington, Virginia 22205

A POTPOURRI OF PHYSICS TEACHING IDEAS—TRICKS OF THE TRADE

Constructing an Hyperbola

Marvin Ohriner
Elmont Memorial High School, New York

The construction of an ellipse using a piece of string is probably well known to most mathematics and physics teachers. The construction of a parabola using a string and a tee-square is described in some texts. It is possible to extend this use of a string (and a meterstick) to construct a hyperbola by the method described below.

Three half-inserted staples are placed in a meterstick at A, B, and C (see Fig. 1). A single strand of string is arranged so that one end is attached to a movable weight. This string is then run through the half-inserted staple C, continued through staple A, and then (past the movable point X) through staple B. Finally this same string is run back through staple C, and its other end is attached to the movable weight.

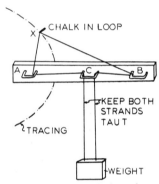

Fig. 1. Construction of a hyperbola using a weight and a length of string.

Make a slipknot in the string at point X and loop it around a chalkstick. Move the chalkstick in such a manner that the two ends of the string attached to the weight lift the weight with *no slack in either strand*. The path traced by the chalk will be part of an hyperbola. (Removing the slipknot and retying it at another point will produce another of a whole family of hyperbolas, all of whose foci are located at A and B.) A pair of rubber suction cups attached to each end of the meterstick will hold it firmly in place while tracing the hyperbola.

The hyperbola is produced because there is always a constant difference between points XA and XB. This demonstration should be most useful when used as an introduction to interference patterns produced by wave motion.

A slightly different method of constructing a hyperbola and other curves was published four years ago.*

The Handy Freon Can

Charles L. Roberts,
State University of New York, Morrisville, New York 13408.

A recent addition to our stockroom shelves is the can of Freon gas. This small can of compressed gas shows great possibilities, and it is hoped that the two uses discussed here will sow the seeds for more innovations.

Freon gas is readily obtainable and safe to use. A refrigeration service man or an appliance dealer will sell a 15-oz can for about $1.60. The gas used for these demonstrations was sold by Virginia Chemicals* under the label Can-O-Gas Freon 12.

Freon is an organic compound known to chemists as dichlorodifluoromethane (Cl_2CF_2). It is inexpensive, noncorrosive, nonflammable, nonirritating, has a faint odor, and very little toxic action, unless it is decomposed by hot metal or flame. Chemically it is a safe compound, but the liquid causes rapid freezing, and it would be unwise to bring it in contact with the eyes.

The physical properties of Freon make it helpful to the science teacher. Since the molecular weight is 120.91, it is useful in demonstrations that require a dense gas. The liquid boils at $-29°C$ under normal pressure which makes it especially useful as a refrigerant and for demonstrations dealing with heat.

The speed of sound depends on the medium being used. In the New York State *Physics Handbook*, this fact is illustrated by the following demonstration**:

3.11. Speed of Sound Depends on Medium

A whistle is operated from the fuel gas supply through a long rubber tube. While there is still air in the tube, the pitch of the sound is normal, but when the gas reaches the whistle the pitch rises sharply. The tube is removed from the gas jet, and the demonstrator blows through it. The pitch drops suddenly when his breath reaches the whistle.

After a class has learned of the factors determining the frequency of a tube, this demonstration may be shown without explanation, challenging the students to apply their knowledge to the new situation.

If this demonstration is carried out using Freon, the effect is even more pronounced since Cl_2CF_2 has more than twice the molecular weight of bottled gas (propane). It is also safer because it is nonflammable, and it does not have a bad odor.

Another use for the Can-O-Gas is to invert the can so that some liquid Freon is expelled into a dish of water. The Freon boils as the water freezes.

The Can-O-Gas shows possibilities for development as a "frictionless puck" and should lend itself well to other heat experiments.

---

* Virginia Chemicals, Inc., West Norfolk, Va.

** *Physics Handbook* (The New York State Education Department, Albany, 1959), p. 101.

MINIMIZING GLASS BREAKAGE

When students are using the resonance apparatus to determine the speed of sound, it is not at all uncommon for them to get the tuning fork too close to the top of the glass tube and chip or break the glass. I managed to stop this breakage by cutting an inch-wide band out of a small size bicycle inner tube and stretching it over the end of the glass tube.

Theoretically, the antinode at the top of a glass tube should be at a distance of 0.6 r above the end of a tube, with radius r. The rubber does change this value but it does not affect the overall accuracy of the experiment, that is, determining the speed of sound.

I also stretch a piece of rubber over the end of the Kundt's tube apparatus to prevent breakage of the glass tube. Used bicycle inner tubes can usually be picked up free at almost any bicycle shop.

John G. McCaslin
Montana College of Mineral Science and Technology
Butte, Montana 59701

A CURE FOR THE FICKLE RIPPLE-TANK MOTOR

Like many others, I've had my problems with ripple-tank motors. There are always a few in the class set that randomly change speeds, causing endless frustration to student and instructor. I've tried changing power supplies, dissembling the motors, and lubricating the motors, but it always seems to come down to admonishing the students to do what they can with what is available.

Last fall, I took a suspect motor to the back room and attached an ammeter and a voltmeter to it to check its power consumption. It became apparent that, when the motor is loaded, i.e., with light finger pressure, the current goes up dramatically as the speed is reduced by the load. By serendipity, I noticed that when the motor was loosely held, the current increased. Apparently, if the motor can vibrate, its load is significantly increased and its speed changes. Hold the motor tightly, and the current drops to a stable value. In checking the motor brackets on the ripple tanks, I found several that had slightly loose screws. When the screws were tightened, the clamp held the motor more tightly and reduced the load on the motor, thus producing a more stable frequency.

If you have had trouble with random speed fluctuations in your ripple tanks, try tightening the screw that holds the motor clamp.

Thomas J. Senior, *Radnor High School, Radnor, Pennsylvania 19087*

Storing Slinky
Frederick H. Giles, Jr.
University of South Carolina
Columbia, S. C.

Since the introduction of the P.S.S.C. course, the "slinky" has become a prominent and prevalent part of the laboratory apparatus. This neat little coil spring has, however, an almost irascible tendency to get tangled and twisted. Furthermore, no supplier has yet come up with a simple way of storage which will help frustrate this pernicious habit.

A suggestion: store the slinkies upright, by threading each one of them over a number 6 dry cell battery (see diagram). These batteries are common in the laboratory, are sufficiently long to prevent "spillage" of the coils over the top, and they are sufficiently massive so as not to be easily toppled while on the shelf.

RIPPLE-TANK PERIOD MEASUREMENTS

Measurement of the period of water waves in ripple-tank experiments is straightforward if your department has a calibrated stroboscope available. But one may also utilize an oscilloscope and an inexpensive photocell as an alternative.

By illuminating the ripple-tank from above with a high intensity lamp, shadows from the water waves appear on a screen placed underneath. A photocell (connected to the y-input of an oscilloscope) placed on the screen can monitor the troughs of water as peaks in the oscilloscope trace. The period may then be determined by measuring the separation of the oscilloscope trace peaks and multiplying by the time-base factor.

LES D. BURTON and IAN G. ELLIS
Jefferson Community College
Louisville, Kentucky 40201

REJUVENATING BAR MAGNETS

To quickly and safely magnetize or demagnetize ferrous materials, I use the solenoid from a magnetic field balance (Welch — Cat. No. 2339) and an ac-dc power pack (Ranges ac 0 - 24 V, dc 0 - 13 V).

To demagnetize simply connect the terminals of the solenoid to the ac output of the power supply. Insert the object to be demagnetized, such as a wrist watch or bar magnet, in the solenoid, turn the voltage to full and then slowly to the off position. Whatever magnetic material that has been placed in the solenoid will now be demagnetized.

To magnetize or rejuvenate bar magnets, connect the solenoid to the dc terminals of the power pack.

VINCENT BUCKWASH
Unionville High School
Unionville, Pennsylvania 19375

TIPS FOR LASER HOLOGRAPHY

While setting up a holography club in my high school, I discovered a few tricks which may be of interest to other high school teachers. If you have a small budget and a limited amount of time, do get this book before starting to setup a holography laboratory: *A Study Guide on Holography*, by Tung H. Jeong, Lake Forest College, Lake Forest, Illinois. The book can be ordered from Integraf, 745 North Waukegan Road, Lake Forest, Illinois 60045, for about $3.00.

A child's wagon can be used as a sand table or as a base for a sand box in which to set up for holograms. The wagon can be easily moved for storage and the rubber tires help to absorb unwanted vibrations.

MADELINE V. FORREST
Baltimore Polytechnic High School
Baltimore, Maryland 21209

HEAVY WEIGHTS AT ZERO COST

Replacement of "noisy" fluorescent ballasts at this institution is running at the rate of about 100 each year. These ballasts weigh about 4 lb each and, due to their size and shape and the conveniently situated mounting holes, are readily adaptable to all purposes where large weights are needed. Suitable hangers and hooks are easily fashioned from discarded wire coathangers as shown in Fig. 6.

Each one should be marked with its weight. In experiments they have the incidental advantage of impressing upon the student the point that the independent variable does not have to be a regular series of integers.

BERNARD SCOTT
Pennsylvania State University
Erie, Pennsylvania 16510

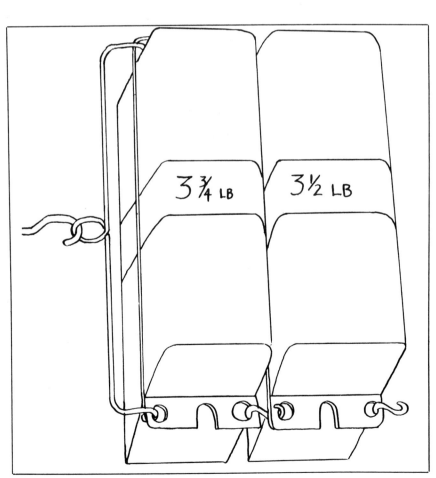

Fig. 6. Four discarded ballasts are temporarily hooked together to provide a total weight of 14¾ lb (65.6 N).

GLASS TUBING MICROSCOPE

I have used glass-tubing microscopes to arouse interest very successfully since 1965, when Professor Mitchell of New Mexico State University demonstrated it to the N.S.F. participants at a summer institute in physics.

Here are the steps to follow:

Cut tubing
Cut off a piece of glass tubing (I use 6-mm o.d. tubing) 10-cm long.

Stretch tubing
Heat tubing at the center with a Bunsen burner flame. Turn the tubing constantly between your fingers and when heated properly, pull the tubing apart.

Form lens
Place the tubing on an asbestos pad and allow to cool. Take one of the pieces of tubing and hold it with a test-tube tongs directly over the flame of the Bunsen burner. Form a bead of glass at the end of the tubing. This will serve as your convex lens. It is a good idea to wrap the handle of the test-tube tongs with a wet paper towel or it will get too hot to hold.

Fit lens into end of tubing
After the tubing has cooled, use a triangular file and cut off the lens at the end of your tubing.

Place the lens inside of the tubing and mark the outside of the tubing with a wax pencil where it needs to be cut. See Fig. 2 to know where to mark the tubing.

Take out the lens and cut the tubing at your mark. Now place the lens back inside of the cut-off tubing, making sure that the tip on the end of the lens is pointing down. (Fig. 3)

Seal lens in end of tubing
Hold the tubing with the test tube tongs directly over the Bunsen burner flame. Heat the end carefully until the lens is fused into the end of the tubing and you have a smooth convex surface with all of the tip melted.

After the tubing is cool, use a tri-

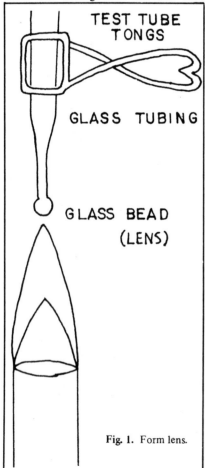

Fig. 1. Form lens.

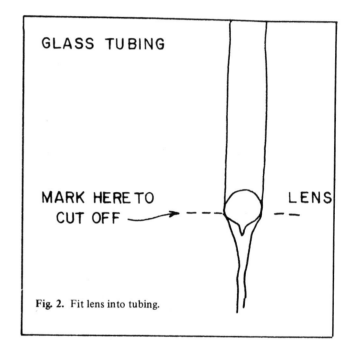

Fig. 2. Fit lens into tubing.

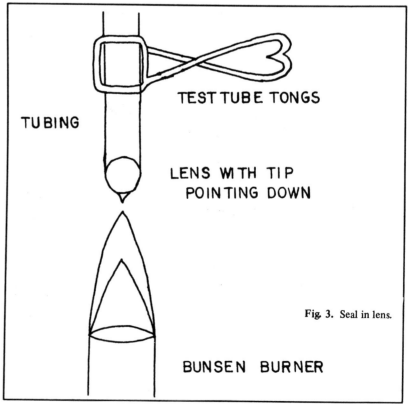

Fig. 3. Seal in lens.

Take out the lens and cut the tubing at your mark. Now place the lens back inside of the cut-off tubing, making sure that the tip on the end of the lens is pointing down. (Fig. 3)

Seal lens in end of tubing

Hold the tubing with the test tube tongs directly over the Bunsen burner flame. Heat the end carefully until the lens is fused into the end of the tubing and you have a smooth convex surface with all of the tip melted.

After the tubing is cool, use a triangular file and cut off the tubing so that, measuring from the lens, it is only 2.5 cm long.

Use microscope

To use your glass-tubing microscope, place a small drop of pond water or culture of protozoans (see the biology teacher) in the tube. Hold the lens very close to your eye while you look at an exposed light bulb.

With any luck, you will see Anton Van Leeuwenhoek's "wee cavorting beasties."

WAYNE O. WILLIAMS
Corona del Sol High School
Tempe, Arizona 85284

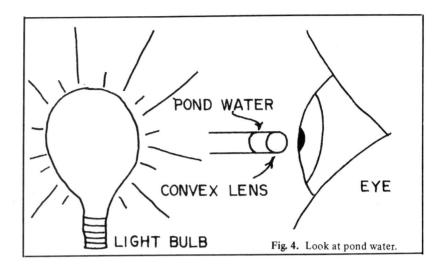

Fig. 4. Look at pond water.

USING DUCO CEMENT AS REPLACEMENT CROSSHAIRS

In the April 1976 issue of *The Physics Teacher* you asked for suggestions on how to replace the crosshairs on an eyepiece. I have been using the following scheme for a number of years and it is much easier than trying to use spider webs.

Place a bead of Duco cement on the end of a toothpick. Touch it to the rim of the reticle and then draw the cement across the opening to the other side. Let this first "crosshair" harden and then repeat the process with a second at right angles. You can make the crosshair as fine as you wish by using the least amount of cement necessary. An excessive amount of cement will produce a nonuniform crosshair.

When you get good at it, you can dispense with the toothpick and draw the cement directly from the dispenser.

KATASHI NOSE
University of Hawaii
Honolulu, Hawaii 96822

REPLACING CROSSHAIRS WITH DENTAL FLOSS FIBERS

In response to the question in the April 1976 edition of *The Physics Teacher*, here is a quick way of replacing the crosshairs in a prism spectroscope.

Remove the crosshair holder from the spectroscope. For the material of the crosshair, use a fiber from dental floss; the single fiber is quite fine. Glue the fiber across a slot cut in an index card with a dot of glue at each end of the fiber (Fig. 6). With the crosshair holder lying on a flat surface, adjust the fiber in the correct position and add a spot of glue to attach the fiber to the rim of the holder. Snip off the extending ends of the fiber and

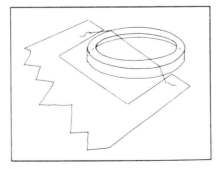

Fig. 6. Dental floss fiber cemented across notch in an index card is positioned on crosshair holder for taut and precise mounting.

repeat the process for the second crosshair.

I do not know the originator of this process, but I learned of it through notes left here by emeritus professor Dr. Harry Hill.

WILLIAM FOLAND
Washington and Jefferson College
Washington, Pennsylvania 15301

WORK-ENERGY APPARATUS

A quick demonstration of work-energy relationships can be shown using three standard cylindrical masses and a half-meter stick. Place a 200-g mass on its side and lay the half-meter stick over it so part of the meter stick extends into the air while the end of the longer part rests on the demonstration table. Rest a 100-g mass on the lower end of the stick and drop a 500-g mass, from a height of approximately 70 cm, so that it strikes the short end of the stick. The impact will lift the 100-g mass to a height greater than the original height of the 500-g mass. With a little practice, the flying 100-g mass can be caught.

The resulting discussion could bring out the work-energy and energy to energy conversions involved.

NOAH M. ROSENHOUSE
(Retired)
Martin Van Buren High School
Queens Village, N.Y. 11427

LENS MOUNT FOR LASER

I found that the top of an Air Wick solid air freshener will fit perfectly into the front of a Metrologic Laser (Fig. 3). Since this top is plastic, a hole can be drilled in its center, and a small diverging lens can be flued in the hole. Thus the laser beam is diverged and the laser is not only useful for holography but it becomes a much safer instrument.

MADELINE V. FORREST
Baltimore Polytechnic Institute
Baltimore, Maryland 21209

Fig. 3. When a hole is drilled through the top of an Airwick solid air freshener container, it can be used as a lens holder that just fits into the mounting aperture of a Metrologic HeNe Laser.

REPAIRING GOLD LEAF ELECTROSCOPES

To answer your question in the April 1976 issue of *The Physics Teacher*: "How does one replace the fragile leaves of a gold leaf electroscope," I offer my technique with no idea of its novelty or value; it works for me. I have a pack of gold sheets from Cenco with sheets separated by a nonstick paper. I fold double a piece of this nonstick paper and suspending the pack by its spine, trap a piece of free-hanging foil as if with a tweezer, with a straight edge of the foil against the fold. (If there are no straight edges the accompanying description can be generalized to creating one.) Once the foil is trapped I cut the paper foil sandwich to the appropriate size, then opening the paper to form a trough, jiggle it so that an end of the cut leaf slides slightly out one end. I reclamp and, applying a piece of Scotch tape to the exposed end, lift the foil free by it. I tape the foil to the fixed electroscope blade, wrapping the tape slightly around the fixed blade to insure continuing electrical contact, and that's it.

BURTON BRODY
Bard College
Annandale-on-Hudson
New York 12504

ELECTRON BEAM DEFLECTION BY A ROTATING MAGNET

Demonstrating the deflection of an electron beam in a static magnetic field involves holding a magnet over a cathode-ray tube (Sargent-Welch: Cat. No. 2145) while manipulating the magnet by hand. A simple extension of this demonstration to illustrate the effects of an alternating magnetic field is to attach a standard bar magnet to a pivoted support (Sargent-Welch: Cat. No. 1840) and spin the magnet near the cathode-ray tube. The correspondence of the angular speed of the rotating magnet and the frequency of the alternating deflection can be observed. A similar effect can be observed if the magnet is spun near the side of a conventional cathode-ray oscilloscope. In the horizontal sweep mode, the alternating vertical deflection produces the usual sinusoidal pattern on the screen.

ABEL ROSALES
ROBERT PRIGO
University of California
Santa Barbara, California 93106

RUST PROOFING

Do you have steel apparatus (tuning forks, steel bars, etc.) that tend to rust if not oiled regularly? Try using Vaseline Petroleum Jelly on them. A thin film of Vaseline will repel moisture and will not stain your clothes.

REINFORCING SPADE LUGS

Spade lugs for banana plugs (H. H. Smith No. 218) are not tough enough to withstand flexing. After flexing two or three times, the spade breaks off. One way to extend their usage is to put some solder at the place where the fork of the spade begins. This adds rigidity to the spade.

SANTOS A. RAMIREZ
Rice University
Houston, Texas 77001

A POTPOURRI OF PHYSICS TEACHING IDEAS—INDEX

MECHANICS

Walk-a-graph, run-a-graph
Robert Gardner
Nov 1975, 498 1

A quantitative demonstration of relative velocities
Dean Zollman
Jan 1981, 44 2

Kinematics of a student
James H. Nelson
Sept 1983, 386–387 3

A demonstration of Newtonian and Archimedian forces
Harley J. Haden
April 1964, 176–177 5

A nonlinear spring
R. M. Prior
Nov 1980, 19601 7

Cookie sheet friction
Patton H. McGinley
Sept. 1973, 362 8

Functional dependence of elongation of a coil spring on the load applied
Francis W. Sears
Feb 1966, 1984 8

Motivation for the force vector
Michael Bernstein
March 1971, 148 9

Further motivation for the force vector
Alfred M. Eich
Feb 1972, 95 9

Saw-horse on teeter-totter for one person
Terry Lee Templeton
March 1967, 138 10

A "Universal" mechanics demonstration
George W. Ficken, Jr. and Angelo A. Gousios
March 1975, 169–170 11

Loops from cove molding
Michael L. Berry
Sept 1977, 368 12

Which way is up?
Lester Evans and J. Truman Stevens
Nov 1978, 561–563 13

Balancing
Robert N. Jones
Nov 1972, 469 15

Hot wheels physics
Stanley J. Briggs
May 1970, 257–259 16

Balanced dynamic carts
Herbert H. Gottlieb
April 1967, 173 18

A demonstration of constant acceleration
Ernest Hammond
Feb 1966, 1977 19

Inelastic collisions using velcro
G. Scott Hubbard
Nov 1972, 478 19

Velocity measurements outdoors with a tape recorder
Donald J. Pruden
Jan 1972, 49 19

Kinematics and the driver
Barton Palatnick
April 1975, 229–230 20

Drag strip timer, an extended use
Donald J. Pruden
Feb 1972, 100 21

Friction in a moving car
Fred M. Goldberg
April 1975, 234–236 22

Driving safety and straight-line kinematics
Van E. Neie
April 1975, 236–237 24

To stop or not to stop—kinematics and the yellow light
J. Fred Watts
Feb 1981, 114–115 26

Automobile stopping distances
L.J. Logue
May 1979, 318–320 28

On the strength of butterfly wings
George W. Ficken, Jr.
Feb 1976, 114 30

The answer is obvious. Isn't it?
Robert Beck Clark
Jan 1986, 38–39; Letters—Oct 1986, 392–393. 31

Another Newton's sail-boat
Robert J. Brown, Jr.
Dec 1972, 488 35

Sailboat demonstration
Paul G. Hewitt
Feb 1968, 1979–80 35

The simple pendulum experiment
Robert Kern Curtis
Jan 1981, 36 36

A whipped-cream pendulum
Edward H. Leonard
Feb 1966, 1984 37

Superball problem
G. Stroink
Oct 1983, 466 38

A bouncing superball—the poor man's projectile
Paul Latimer
Nov 1977, 485 39

Spectacular rocket experiment
Evan Jones and P. Peter Urone
Feb 1976, 112–113 39

Rocket engine analog
Robert F. Coutts
Oct 1969, 407 41

Snowball fighting: A study in projectile motion
Peter N. Henriksen
Jan 1975, 43 42

Predicting the range of a spring cannon
William E. Cooper
Feb 1972, 94–95 43

Range of a dart gun
Edwin Paul Heideman
Sept 1973, 362–363 45

The hunter and the monkey
James Moore
May 1971, 282 46

Monkey and hunter in slow motion
William P. Brown
Sept 1977, 368–369 47

Capturing the projectile in the monkey and gun experiment
Kenneth Wright
May 1972, 263 47

The water drop parabola
Billy Tolar
May 1980, 371–372; Letter—Oct 1980, 488 ... 48

The spin on baseballs or golfballs
R. D. Edge
April 1980, 308–309 50

Catapulting the interest and success of physics students
Charles Hartman
Dec 1986, 556–557 52

The cylinderical wing, why does it fly?
Gary Ronald Login
Dec 1978, 19662; Letter and Answer—May 1979, 286, 334 54

The tippy-top
George D. Freier
Jan 1967, 36–38 56

Spinning footballs and class rings
Matthew Treptow, Mike Flynn, and Ralph Baierlein
Sept 1986, 361–363 59

The Coriolis effect and other spin-off demonstrations
Lester Evans
Feb 1982, 102–103 62

Whirlpool in a swimming pool
George W. Ficken, Jr.
Sept 1975, 348 63

An experience in observation
Merle Fisher
March 1969, 166–167 64

Large scale use of a liquid-surface accelerometer
Lawrence E. L'Hote
Feb 1975, 100–101 66

Teacher's Pets II—Circling carts
Robert Gardner
Dec 1975, 552–553 67

A personal application of physics
Dorothy Russell
April 1974, 234–235 69

Spin art: A rotational effects demonstration
Dean Zollman
Feb 1975, 106–107 71

Centripetal force using a hand rotator
David Chesnut
Sept 1980, 467 72

A penny for your thoughts
John Dixon
Jan 1966, 38–39 73

The great American vector game
Thomas Ritter
May 1982, 310–311 74

The wicked king and the beautiful princess
Samuel Derman
Oct 1971, 387–388; Letter—Jan 1972, 196 76

A tensile strength lab-contest
Arnold Gorneau
Jan 1973, 41–42 78

Using blocks to demonstrate inertia, center of gravity, and friction
James R. Keady
Sept 1967, 292 79

Dropping a string of marbles
R.D. Edge
April 1978, 233 80

The falling meter stick
J. Thomas Dickinson
Sept 1971, 336–337 81

Catch a dollar bill
William Schnippert
March 1976, 177 82

Improved suspension for acceleration of g apparatus
Sam Hirschman
Nov 1967, 387 83

Free fall using audio recording tape
Charles Rudisill
April 1968, 179 83

Weightlessness and other ideas
R.D. Edge
March 1981, 190 84

Weightlessness and free bodies
D. Easton
Nov 1983, 521 85

The measurement of "g" in an elevator
Haym Kruglak
Nov 1972, 466–468 86

Apparent weight changes in an elevator
Larry Jensen
Oct 1976, 436–439 88

Using a laser to investigate free fall
Edward V. Lee
March 1974, 168–169 91

Physical effects of apparent "weightlessness"
Haym Kruglak
April 1963, 34–35 92

A center of gravity demonstration
Terrence P. Toepker
April 1977, 241 94

Ways to demonstrate center of gravity
Vincent Buckwash
Jan 1976, 39–40 95

With a grain of salt
Richard Morandi
March 1979, 152 96

Center of mass revisited
Ernie McFarland
Jan 1983, 42–44 97

Center of mass of a rotating object
Marvin Ohriner
March 1980, 230–231 100

More center of gravity
Terrence P. Toepker
Nov 1976, 499–501 100

Center of mass
R.D. Edge
Nov 1984, 535–537 103

Center of gravity of a student
G. Stroink
April 1979, 254–255 106

FLUIDS AND HEAT

The water can paradox
Lester G. Paldy
Sept 1963, 126 107

The water can explored again
Roy H. Biser
Oct 1966, 304–305 108

Floaters and sinkers
Terrence P. Toepker
March 1986, 164 110

How to cool a book by its cover
Robert Everett Vermillion
March 1975, 165; Letter—Sept 1975, 324 111

"How to cool a book by its cover" revisited
Edsel M. Langdon and Gerald M. Straks
Oct 1976, 443–444 112

Newton's law of cooling or is ten minutes enough time
for a coffee break?
Chandler M. Dennis, Jr.
Oct 1980, 532–533 113

Tensiometer
Julius H. Taylor
Nov 1972, 478 115

Brownian motion
Wallace A. Hilton
Dec 1972, 534 116

A question of air pressure
Walter Thumm
April 1973, 242 117

Boyle's law using a vacuum gauge
William Carlson
Nov 1967, 387–388 118

Archimedes' principle meets Charles' law
Clyde J. Smith
March 1979, 187 119

Elasticity shown with mirror blank and marble
Harry H. Kemp
Oct 1977, 420 119

Bernoulli demonstration
R. E. Worley
Oct 1965, 320 20

Repairing thermometers with split mercury columns
Charles L. Roberts
Nov 1968, 427–428 120

The cartesian diver
Robert N. Jones
Sept 1973, 345; Letters—Feb 1975, 1968–69; Jan 1974, 51 121

Cartesian diver with pressure head
Edward V. Lee
Sept 1981, 416 122

Elasticity demonstration
Bernard Scott
Sept 1976, 373 122

Elasticity of glass
Salvatore J. Rodano and James D'Amario
Dec 1979, 595; Letter—March 1980, 176 123

Teachers' pet III: How thick is a soap bubble?
Robert Gardner
Jan 1976, 41 124

Barroom physics, part I
James T. Schreiber
Sept 1975, 361, 378 125

Barroom physics, part II
James T. Schreiber
Oct 1975, 418, 428 126

Divergent barroom physics
George B. Barnes
Jan 1976, 41–42 127

An application of Archimedes' principle: Eureka! I'm 28% fat
Roland A. Hultsch
Sept 1981, 408–409 128

ELECTRICITY AND MAGNETISM

A large-scale electroscope
Sheldon Wortzman
Oct 1981, 481 120

Negative charges from an electrophorous
John E. Girard
Sept 1978, 402–403 130

Electroscope shadowgraph
William J. Muha
April 1968, 179 131

Pith ball substitute
Alfred Romer
Nov 1972, 477–478 131

Electrostatic pong
Robert J. Krohl
May 1982, 330 132

Electroscope discharge rate
Bro. James Mahoney, C.F.X.
Dec 1972, 534–535 132

Electrostatic lobby display
Menno Fast
Feb 1972, 100–101 133

Kelvin water dropper revisited
Lester Evans and J. Truman Stevens
Dec 1977, 548–549 134

Is a swimmer safe in a lightning storm?
Lester Dwyer and Captain Bobby N. Turman
May 1980, 388–389 136

Dissectible Leyden jar
G. Bradley Huff
May 1986, 292 ; Letter—Nov 1986, 460 137

Coulomb's law on the overhead projector
John B. Johnston
Jan 1979, 1960–61 39

The Oersted effect on the overhead
Sam J. Cipolla
Nov 1975, 475................................ 140

Lenz's law
Harry H. Kemp
Dec 1977, 543................................ 141

Electrostatic charges and copying machines
Robert P. Bauman
Dec 1977, 543................................ 141

Electrical figures
Colin Pounder
Nov 1972, 468–469........................... 142

Demonstration of Gauss' law for a metal surface
T.W. Haywood and R.C. Nelson
Dec 1979, 596................................ 143

A motor is a generator and vice versa
John A. Johnson and Franklin Miller, Jr.
Jan 1976, 36–37.............................. 144

Force between parallel currents on the overhead projector
R.C. Nicklin
Sept 1978, 402 144

Sealed batteries
Fritz G. Will
Nov 1979, 539................................ 145

The smoke detector
J.R. Young
Oct 1979, 467 146

Ohm's law mnemonic
Carl H. Hayn
Sept 1977, 364 146

The Omega competition
Robert P. Lanni
Oct 1978, 483 147

The volt competition
Donna A. Berry
Nov 1982, 549................................ 148

Why is the ac power line grounded?
Courtney Lantz and Charles H. Anderson
April 1980, 314 149

Electric field using an overhead projector
Ian Mennie and Cyril Snook
Dec 1975, 558–559 150

Turn-by-turn transformer demonstration
Joe L. Ferguson
Jan 1979, 59 151

Static electricity demonstration
Philip E. Highsmith
Nov 1968, 427................................ 152

Lenz's law demonstration
Thomas D. Miner
Nov 1968, 427................................ 152

Voltage surge protection
Robert P. Barrett
March 1979, 205............................. 152

A circuit demonstration
Renato Lichtenstein
Jan 1978, 35 153

Neon lamps and static electricity
John W. Layman and Delbert J. Rutledge
Jan 1972, 49–50.............................. 153

A simple transistor demonstration
David L. Mott
Sept 1980, 460 154

ac made visible
Lloyd Harrich
Oct 1984, 448 154

Mysterious lights in series and parallel
Clifton Keller
Sept 1980, 464 155

Shape of an electric field
Mike Weiss
Sept 1967, 286 156

Force on current carrying aluminum foil
Herbert H. Gottlieb
Nov 1967, 387................................ 156

Standing waves by a current carrying conductor
Marvin Ohriner
Sept. 1967, 287 156

Touch-panels in elevators, and idiosyncrasies of gas tubes
H. Richard Crane
Sept 1983, 402, 404 157

A magnetic tripole
Michael Davis
Jan 1976, 34 158

A magnetic tripole—what caused it?
Gladys F. Luhman, Arthur R. Quinton, Dean Hartman, Rod Fisher
May 1976, 261 159

A three-pole bar magnet?
Jerry D. Wilson
Sept 1976, 365–366 160

There's still a lot we don't know (magnetic tripole)
Dwight L. Barr, Sr.
Dec 1976, 534................................ 161

A POTPOURRI OF PHYSICS TEACHING IDEAS — INDEX

Recycling a magnet or The little magnet that could
Feb 1973, 114 162

Construction of a simple compass
Liberio Mar and Carlos Hernandez
April 1976, 247 162

Ceramic magnets
Herbert H. Gottlieb
March 1976, 181 163

Three dimensional views of magnetic effects
Phillip E. Miller
Oct 1965, 320 163

Why are so few substances ferromagnetic?
Jean Frazier, Henry H. Kolm
March 1982, 183, 185 164

A question of magnets and keepers
Walter Thumm
Nov 1972, 483 165

Experiments with nickles and magnets
R.D. Edge
Feb 1981, 124–126 166

The field strength of a permanent magnet
Brother James Mahoney
Nov 1975, 507 168

Parallel circuit
Paul Hewitt
March 1986, 179 169

Power lines
Paul Hewitt
Feb 1986, 112 170

OPTICS AND WAVES

A choice observation
Ronald A. Brown
March 1977, 173–174 171

Blue sky and red sunsets
Marla H. Moore
Oct 1973, 436–437 172

A simplified sunset demonstration
Haym Kruglak
Dec 1973, 559 173

Dispersion and inversion
Samuel Hirschman
Feb 1969, 116 173

Atmospheric refraction
John B. Johnston
May 1977, 308–309 174

Colored lights and shadows
Robert Gardner
Oct 1978, 477 175

Making rainbows in the classroom
Fred B. Royalty
Nov 1984, 523 176

Solar spectrum projection
Samuel Derman
Jan 1978, 58 177

Making rainbows with a garden hose
Helene F. Perry
April 1975, 197 177

A pinch of coffee-mate
Ben M. Doughty
March 1974, 132 177

Using a video projector for color-mixing demonstrations
Richard A. Bartels
April 1982, 247–248 178

A simple reflection experiment
T.T. Crow
May 1973, 309 179

Recombination of spectral colors
Fred T. Pregger
Sept 1982, 403 180

Color mixing for a large audience
William A. Butler, A. Douglas Davis, and Charles E. Miller, Jr.
Jan 1979, 43 181

A different way to use Newton's color wheel
Gordon R. Gore
Feb 1982, 101 181

Light box, inexpensive but versatile
Robert Gardner
Jan 1975, 50–51 182

Why is the string colored?
Sue Gray AlSalam, R.D. Edge
Oct 1980, 518 183

Reflection on the study of flat mirrors: two demonstrations
Jim Nelson
Sept 1984, 388–389 184

A favorite experiment
R. Kennedy Carpenter
Oct 1973, 428–430 186

An introduction to pinhole optics
Wallace A. Hilton
Feb 1973, 112 188

That can't be: A virtual comment
John W. Layman
April 1979, 253–254 189

Why is your image in a plane mirror inverted left-to-right but not top-to-bottom?
Kenneth W. Ford
April 1975, 228–229 190

Physics at home and in the backyard (Part III)
George Ficken
Dec 1977, 550 192

"His specs — Use them for burning glasses"
C. Bob Clark
Dec 1973, 522 193

Pinhole glasses?
Russell Patera
Sept 1978, 383 193

To make a camera obscura
Reuben Alley
Dec 1980, 19638 194

Inversion of an image on the retina
George Barnes
Oct 1981, 499 195

Inversion of shadows on the retina
Karl C. Mamola
May 1983, 332–333 195

Behind the eye
George W. Ficken, Jr.
Feb 1982, 1972......................... 196

Image from a pinhole
E. Scott Barr
March 1981, 154......................... 196

Optics of the rear-view mirror: a laboratory experiment
R.D. Edge
April 1986, 221–223 197

Mirrors in air and water
Robert Gardner
Feb 1976, 114–115 199

An optical puzzle that will make your head spin
Samuel Derman
Sept 1981, 395 200

Markers for Young's Experiment (PSSC version)
William H. Porter
March 1974, 178.......................... 201

Cylinderical mirrors
Paul E. Wack
Dec 1981, 581............................ 201

Recording timer tape for interference demonstrations
Fr. Earl R. Meyer
Sept 1972, 334 202

Drawing wave diagrams for the interference of light
Frank A. Anderson
March 1973, 175.......................... 203

Long lasting soap-gelatin films
Paul A. Smith
Feb 1967, 1987 205

Soap film interference projection
John A. Davis
March 1974, 177–178...................... 206

Water lens
John B. Johnston
Feb 1973, 114 206

Standing waves on the overhead projector
William Warren
Dec 1980, 19674–675 207

Sine wave analog
Donald H. Trollope
Feb 1979, 130–131 208

Images from a piece of a lens
Samuel Hirschman
April 1969, 246 208

The psychedelic student-getter
William R. Franklin
April 1969, 227 209

Optics in a fish tank
Richard Breslow
April 1976, 234–235 209

Optical effects in a neutral buoyancy simulator
Van E. Neie
Jan 1981, 53 211

A problem in image formation
Hyman A. Cohen
Oct 1965, 324 212

Ray models of concave mirrors and convex lenses
Robert Gardner
Nov 1980, 19608 213

Binocular vision—A simple demonstration
Richard Didsbury
Nov 1980, 595............................ 214

Physics for automobile passengers
Elizabeth A. Wood
April 1973, 239 214

The Cheshire cat
Walter C. Connolly
April 1983, 263 214

The disappearing dropper
Walter C. Connolly and Thomas L. Rokoske
Sept 1980, 467 215

Ripple tank projection with improved contrast
Robert W. Smith
Dec 1972, 533 215

How the world looks underwater—A demonstration for nonswimmers
Samuel Derman
Oct 1982, 474–475 216

The physics of visual acuity
Michael J. Ruiz
Sept 1980, 457, Letter—Nov 1980, 560....... 217

Pyrex "vanishing solution"
F.J. Wunderlich, D.E. Shaw, M.J. Hones
Feb 1977, 118 219

A safe pyrex "vanishing solution"
Donald K. Day
Oct 1977, 438 220

Optical astigmatism model
April 1972, 201 221

Wave motion demonstrator
Marc D. Levenson
Jan 1974, 47–48........................... 222

Standing wave analogy using pocket combs
Robert Prigo and Richard Wormsbecher
March 1977, 187–188...................... 224

Beat production analogy using pocket combs
Peter Melzer
Feb 1976, 120 225

Hot standing waves
Harold C. Jensen
May 1968, 254 225

Versatile mount for slinky wave demonstrator
John F. Spivey
Jan 1982, 52 226

Cigar box spectroscope
Brother Shamus, C.F.X.
April 1967, 173 226

Finding the principal focus with laser and thread
Kenneth Fox
Oct 1981, 480–481 227

Paper waves
Richard A. Lohsen
Nov 1983, 532 228

Traveling waves on a rope
Keith Fillmore
May 1983, 314 229

Laser gimmic
William H. Porter
April 1974, 239 230

Laser beam splitting for the student lab
Robert J. Julian
Feb 1972, 101 230

Laser safety goggles: An unnecessary expense
Herbert H. Gottlieb
Jan 1974, 51 230

Long-playing diffraction grating
Scott R. Welty
March 1981, 187 231

Calibrating an inexpensive strobe light
Alfred M. Eich
March 1972, 153 232

Mercury vapor street lamps
Zygmunt Przeniczny
March 1972, 153–154 232

Diffraction of light apparatus
Samuel Hirschman
March 1976, 180–181 233

Classical demonstration of polarization
Robert P. Bauman and Dennis R. Moore
March 1980, 214–215 234

Demonstrating interference
H.L. Armstrong
April 1979, 255 235

Can a spot of light move faster than c?
Arthur Eisenkraft and Peter G. Bergmann
Feb 1981, 127 236

An op art wave demonstration
R.D. Edge
Dec 1970, 521–524 236

SOUND

A sound-level meter: A sound investment
A. David Chandler
April 1979, 251–253 240

Demonstrating resonance by shattering glass with sound
Willard C. Walker
May 1977, 294–296 242

Seeing the Science of Sound
Brother Shamus Mahoney CFX
April 1969, 246–247 244

Transmitting sound by light
J. B. van der Koii
Jan 1979, 43–44 245

Velocity of sound using an oscilloscope
Herbert H. Gottlieb
April 1967, 173–174 245

Resonance tube for measurement of speed of sound
Franklin R. Muirhead
Feb 1979, 131 246

Sound on a light beam
William M. Zeitz and Terrence P. Toepker
May 1969, 301 247

Colorful Chladni
R.C. Nicklin
May 1973, 312 247

Binaural hearing
Morris G. Hults
Oct 1980, 509 248

Sound waves
Morris G. Hults
Dec 1980, 19671 248

Resonance and foghorns
Carl J. Naegele
Dec 1977, 559 249

Sound in a vacuum demonstration
Alfred M. Eich, Jr.
Nov 1968, 428 249

Demonstrations of reflection and diffraction of sound
Lester Dwyer
May 1968, 252–253 250

Physics on a dinner plate
Jeffrey May
Dec 1980, 19668 251

How to hear a mouse roar
James Burgard
Nov 1971, 457 252

A classroom apparatus for demonstrating the Doppler effect
Ernest Hammond
Sept 1967, 286 252

Vibrations in pipes
R.D. Edge
May 1980, 383–384 253

Demonstration experiment to show the effect of pressure variation on the velocity of sound in a gas
Kenneth Geo. Keith
Jan 1965, 30 255

An acoustics demonstration for students interested in music
J.S. Faughn and S.W. Slade
March 1973, 171–172 256

Sound intensity and good health
Hugh Haggerty
Oct 1974, 421–423 257

Magnetically driven sonometer
Ronald B. Standler
Dec 1972, 533–534 259

Oscilloscope measurement of the velocity of sound
Robert H. Johns
Dec 1972, 533 260

The harmonica, an audio-frequency generator
Robert H. Johns
Dec 1975, 557 261

A "Sound" crossword
Veralee Falkenberg
Jan 1972, 44-45 261

TOYS

Children's toys
D.P. Jax Mulder
Feb 1980, 134-135 263

What makes it turn?
Susan S. Welch
May 1973, 303 265

A mechanical toy: The gee-haw whammy-diddle
Gordon J. Aubrecht, II
Dec 1982, 19614-615 266

A "perpetual motion" toy?
Donna A. Berry
May 1982, 319 267

More toy store physics
David P. Koch
Nov 1976, 506 268

The physics of toys
Henry Levinstein
Sept 1982, 358-365 269

ODDS AND ENDS

Ideas for the amateur scientist
Jearl Walker
Nov 1978, 544-549 277

Humor in the physics classroom
Ivars Peterson
Dec 1980, 19646-650; Letter—Feb 1981, 1982
283

Proclamation
Michael Scott
Nov 1975, 464, 507 287

Pocketbook science
James Watson, Jr. and Nancy T. Watson
April 1982, 235, 238 288

The physics "flub stub"
Wayne Williams
Sept 1976, 369 290

Hot dog conduction, an edible experiment
Robert F. Neff
Jan 1975, 50 291

Hot dog physics
George B. Barnes
Dec 1976, 578 291

Lab-cover art
Craig Fletcher
May 1986, 286-287 292

Scaling and paper airplanes
Robert H. Johns
Dec 1971, 541 294

Enchanting things to think about
Julius Sumner Miller
May 1977, 296-297 294

Further enchanting things to think about
Julius Sumner Miller
Sept 1979, 383-384 295

A game of gravity
Gus A. Sayer
Sept 1976, 359 297

An experiment on population growth and pollution
M. Jeffries
Dec 1971, 536-537 298

The trigonometry "laboratory"
Chandler M. Dennis, Jr.
April 1981, 246-247 299

Lecture demonstrations for the high school science teacher
Thomas J. Parmley, J. Irwin Swigart, and Ray L. Doran
Jan 1966, 36-38 301

Backwards clock
Thomas W. Norton
Oct 1972, 401 304

Drinking duck shutter
James Boyden
Oct 1967, 342 304

An introvert rocket or mechanical jumping bean
Richard M. Sutton
Sept 1963, 108-110 305

A possible solution to the energy crisis? I—The Archimedes wheel
Albert A. Bartlett and Joseph Dreitlein
Feb 1975, 108 308

A possible solution to the energy crisis? II—The capillary pump
Albert A. Bartlett and Joseph Dreitlein
March 1975, 175 308

Do-at-home energy exercise
Varghese D. Pynadath
Jan 1977, 45-46 309

Electric Stop clock
Herbert H. Gottlieb
April 1967, 173 309

A stirring experiment
Bruce Jones
Dec 1980, 19671; Letter—March 1981, 154... 310

Demonstration: A nail driven into wood
Dennis P. Zicko
Jan 1980, 50 311

Graphing Henry Aaron's home-run output
Herbert Ringel
Jan 1974, 43 312

Removing the Buoyant Force
Gordon E. Jones and W. Paul Gordon
Jan 1979, 59-60 313

L'Eggs Demonstrations
D. Rae Carpenter, Jr.
March 1977, 188 313

"Reciprocating" Engine
Kemp Bennett Kolb
March 1966, 121-122 314

A device for demonstrating half-thickness of shielding materials
Richard O. Thomas
Feb 1966, 1977 316

Using the overhead projector in simulation of the Rutherford scattering experiment
Rudolph J. Eichenberger
March 1972, 147 317

A game for facilitating the learning of units in physics
James E. McGahan
April 1977, 233-234 318

Radiation safety in the lab
G. Stroink
March 1980, 207; Letter—Sept 1980, 406 319

Absorption of Radiation—A lecture sized demonstration
Dean Zollman
Dec 1974, 563 320

Simulating radioactive decay with dice
Ludwik Kowalski
Feb 1981, 113 321

Two new experiments (the speed of sound, an analog experiment)
Bob Kimball
Sept 1973, 353 322

Throwing dice in the classroom
John L. Roeder
Oct 1977, 428-429 323

Throwing dice in the classroom II
John L. Roeder
April 1980, 302-304 324

"Mission Improbable" problems
Doug Jenkins
Oct 1981, 482 327

The History of physics in a high school physics course
Michael Grote
Feb 1977, 102-103 328

Legibility in the lecture hall
Albert A. Bartlett and Michael A. Thomason
Nov 1983, 531 330

Blitz quiz
Herbert A. Perlman, Sr.
Dec 1981, 19617 331

Problems: Galileo Revisited
Terrence P. Toepker
Feb 1968, 1976, 1988 332

"Physics is Phun" Quiz
James H. Nelson
Jan 1972, 44 334

Fizzicks Quizz
March 1966, 143 336

TRICKS OF THE TRADE

HOW TO STOW IT:

Hang those meter sticks
B.M. Guy
Dec 1974, 568 338

Assemblies of single concept demonstrations
Haym Kruglak
Dec 1974, 568 338

Storage and display box for a diffusion cloud chamber
Haym Kruglak
Dec 1974, 568 338

Protecting electronic tubes
Glen Spielbauer
Dec 1974, 568 338

Small part storage
Sister Martha Ryder
Dec 1974, 568 338

A rack for test tubes
R. Kennedy Carpenter
Dec 1974, 569 339

Storage of support cable from high ceilings
John A. Davis
Dec 1974, 569 339

Tool board outlines
Sister Martha Ryder
Dec 1974, 569 339

HOW TO DO IT:

Emptying Ripple tanks
Spencer Livingston
Dec 1974, 569 339

Heat Shrinkable tubing
Ronald A. Kobiske
Dec 1974, 569 339

Color anodizing
Ronald A. Kobiske
Dec 1974, 569 339

A perfect circle without a compass
Edwin Paul Heideman
Dec 1974, 570 340

Reuseable chalk board graphs
Edwin Paul Heideman
Dec 1974, 570 340

"Glue" for plexiglass
Sister Martha Ryder
Dec 1974, 570 340

Fused ammeters
S. Winston Cram
Dec 1974, 570 340

Removing scratches from clear plastic
Herbert H. Gottlieb
Dec 1974, 570 340

A red hot demonstration
R. Kennedy Carpenter
Dec 1974, 570 340

The speed of a pendulum bob
Richard F. Kotheimer
Dec 1974, 570 340

Photocurrent detection with an oscilloscope
Monte Giles
Dec 1974, 570–571 340

Latex sphere clusters in Millikan's experiment
Ernest Johnston and Monte Giles
Dec 1974, 571 341

Electrostatic charging of fur and silk
John A. Davis
Dec 1974, 571 341

HOW TO MAKE IT:

Ripple tank wave absorbers or shoals
Haym Kruglak
Dec 1974, 571 341

Projectile launchers for electric trains
Robert Q. Kimball
Dec 1974, 571 341

Plaster of Paris mold making
Roy Coleman
Dec 1974, 571 341

Scope Shade
J. Gerald Anderson
Dec 1974, 572 342

A quickly made galvanometer shunt
James Reynolds
Dec 1974, 572 342

Center of mass
R. Kennedy Carpenter
Dec 1974, 572 342

Fast square wave generator
Ronald A. Kobiske
Dec 1974, 572 342

Simple accelerometers
Robert Gardner
Dec 1974, 572 342

Series and parallel electric circuits
Bro. James Mahoney, C.F.X.
Dec 1974, 572 342

A resistor board
Walter C. Connolly
Dec 1974, 572–573 342

Sound-proofed box for air track blower
John A. Davis
Dec 1974, 573 343

Simple lab power distribution system
William H. Porter
Dec 1974, 573 343

Inexpensive immersion heater
Frank G. Karioris
Dec 1974, 573 343

Simple resistance thermometer
L.C. Corrado
Dec 1974, 573–574 343

HOW TO SHOW IT:

Brownian Movement
Sister Martha Ryder
Dec 1974, 574 344

Projection of diffraction plates with a microscope
Ronald R. Bergsten
Dec 1974, 574 344

Bichromatic light source for diffraction demonstrations
Ronald R. Bergsten
Dec 1974, 574 344

Vectors with a chalk line
Edwin Paul Heideman
Dec 1974, 574 344

$F = q v \times B$; A simple demonstration
Robert M. Donner
Dec 1974, 574 344

Simple AC demonstrator
Robert F. Neff
Dec 1974, 574 344

Convex lens as a diverging lens
Bro. James Mahoney, C.F.X.
Dec 1974, 574 344

Three dimensional magnetic field
Walter C. Connolly
Dec 1974, 575 345

The vertical view
John A. Davis
Dec 1974, 575 345

Translucent screens for optical projections
John A. Davis
Dec 1974, 575 345

HOW TO ADAPT IT:

Regelation made easier
Robert F. Neff
Dec 1974, 575 345

A simple timing device
Edwin Paul Heideman
Dec 1974, 575 345

Inexpensive tuning forks
Robert H. Johns
Dec 1974, 575 345

Using polaroid negatives
B.G. Eaton
Dec 1974, 576 346

Better spreader for Polaroid Colorpack II cameras
B.G. Eaton
Dec 1974, 576 346

Cartesian diver
Richard F. Kotheimer
Dec 1974, 576 346

Poor man's mercury source
K.E. Davis
Dec 1974, 576 346

Minature diffraction gratings
Herbert H. Gottlieb
Dec 1974, 576 346

A precise time base for the physics lab
John G. Johnson
Dec 1974, 576 346

Constructing a hyperbola
Marvin Ohriner
Sept 1967, 286–287 347

The handy freon can
Charles L. Roberts
April 1968, 178–179 347

Minimizing glass breakage
John G. McCaslin
March 1980, 231 348

A cure for the fickle ripple-tank motor
Thomas J. Senior
Jan 1982, 52 348

Storing slinkies
Frederick H. Giles, Jr.
Oct 1965, 320 348

Ripple tank period measurements
Les D. Burton and Ian G. Ellis
Dec 1975, 559 349

Heavy weights at zero cost
Bernard Scott
Dec 1975, 559 349

Rejuvenating bar magnets
Vincent Buckwash
Dec 1975, 559 349

Tips for laser holography
Madeline V. Forrest
March 1977, 190 349

Glass tubing microscope
Wayne O. Williams
March 1979, 204–205 350

Using duco cement as replacement crosshairs
Katashi Nose
March 1977, 190 351

Replacing crosshairs with dental floss fibers
William Foland
March 1977, 190 351

Work-energy apparatus
Noah M. Rosenhouse
Dec 1976, 581 351

Lens mount for laser
Madeline V. Forrest
Dec 1976, 581 352

Electron beam deflection by a rotating magnet
Abel Rosales amd Robert Prigo
Dec 1976, 581 352

Repairing gold leaf electroscopes
Burton Brody
Sept 1976, 374 352

Rust proofing
Santos A. Ramirez
Sept 1976, 374 352

Reinforcing spade lugs
Santos A. Ramirez
Sept 1976, 374 352

NOTES

NOTES

NOTES

NOTES

NOTES

NOTES

NOTES

NOTES